ENVIRONMENTAL _____
ECOLOGY _____

ENVIRONMENTAL ECOLOGY

The Impacts of Pollution and Other Stresses on Ecosystem Structure and Function

Bill Freedman

Department of Biology and
School for Resource and Environmental Studies
Dalhousie University
Halifax, Nova Scotia, Canada

ACADEMIC PRESS, INC.

Harcourt Brace Jovanovich, Publishers

San Diego   New York   Berkeley   Boston
London   Sydney   Tokyo   Toronto

Front cover—*top*, jaguar (*Panthera onca*). (Photo courtesy of World Wildlife Fund.) *Center,* decline of red spruce (*Picea rubens*) on Camels Hump Mountain in Vermont. (Photo courtesy of H. W. Vogelmann.) *Bottom,* rosy periwinkle (*Catharantus roseus*). (Photo courtesy of Steele, World Wildlife Fund.) Back cover—*top*, giant panda (*Ailuropoda melanoleuca*). Photo courtesy of MacKinnon, World Wildlife Fund.) *Center,* a sparsely vegetated serpentine barrens in central California. (Photo courtesy of A. R. Kruckeberg.) *Bottom,* American ginseng (*Panax quinquifolium*). (Photo courtesy of Steele, World Wildlife Fund.)

Academic Press, Inc.
San Diego, California 92101

United Kingdom Edition published by
Academic Press Limited
24–28 Oval Road, London NW1 7DX

Library of Congress Cataloging in Publication Data

Freedman, Bill
    Environmental ecology : the impacts of pollution and other stresses
on ecosystem structure and function / Bill Freedman.
        p.   cm.
    Bibliography: p.
    Includes index.
    ISBN   0-12-266540-6   (alk. paper)
    1. Pollution--Environmental aspects. 2. Man--Influence on
nature. 3. Ecology.   I. Title.
QH545.A1F74   1989
574.5'222--dc19   88-7629
                                                       CIP

Printed in the United States of America
89 90 91 92   9 8 7 6 5 4 3 2 1

"Man, false man, smiling, destructive man."

Nathaniel Lee 1653(?)–1692

CONTENTS

PREFACE

The intended audience for this book is students taking an upper undergraduate or graduate course in environmental ecology, in which the ecological impacts of pollution and other stresses are the primary focus. Since 1979 I have been teaching a course in this discipline at Dalhousie University. I have not found a textbook that in my idiosyncratic opinion properly deals with this subject matter, and I have therefore had to rely on collections of reprints and other primary source material to supplement and expand upon my lectures. Hence, my perception of a need for this book.

In addition to its primary student target, I expect that this book will be useful as a supplemental source of information for more general university-level courses in environmental studies, and to ecologists and other professionals whose specific interests fall within the multi- and interdisciplinary field of environmental science. For some readers who have not had a prior course in ecology or other relevant experience, it will occasionally be necessary to consult the book's glossary in order to deal with specialized terminology.

A number of acknowledgments are in order. First, I gratefully thank my "main mentor," T. C. Hutchinson of the University of Toronto. Tom has been most responsible for opening my eyes to the anthropogenic ravaging of ecosystems that has been and is now taking place. He also made available an opportunity to build a career in which my professional activities focused on the effect of humans on the natural environment. Of course, my ecological thinking has also been strongly influenced and molded by other friends and colleagues. Most nota-

ble have been P. M. Catling of the Biosystematics Research Institute in Ottawa and J. Svoboda of the University of Toronto.

Many colleagues selflessly provided me with an invaluable service by reviewing one or more chapters of this book and by making important reference material available to me. As is inevitable, I was not always able to incorporate all of their criticisms and suggestions because they did not always jive with my own way of thinking. The great majority of suggestions made by these people resulted in changes in my preliminary manuscripts, however, and they immeasurably improved the quality, and sometimes the accuracy, of my writing. I am truly grateful to these scientists for their insightful assistance with this book. Listed in alphabetical order, these people are: S. Beauchamp, D. Busby, A. R. O. Chapman, R. P. Cote, R. Cox, C. Cronan, R. W. Doyle, D. Eidt, E. Greene, M. Havas, R. Hughes, T. C. Hutchinson, J. Kerekes, D. MacLean, I. A. McLaren, D. Patriquin, M. Patterson, J. Svoboda, J. Vandermeulen, R. Wassersug, P. Wells, A. Westing, and H. Whitehead. Eight anonymous reviewers of chapter manuscripts also provided invaluable criticisms. I do not know who they are, but I am grateful for their help. Speaking of which, if any readers have comments or criticisms of the material or approach used in this book, I'd be most grateful if you would communicate your ideas to me.

A number of the professional staff at Academic Press have also been of great help to me. The editorial assistance of Charles Arthur has made the book more readable and better able to communicate key ideas. The production work of Julia Hagen has given the book a pleasing and interesting appearance.

Lastly, I thank my two young children, Jonathan and Rachael. They were of no direct help with this book, but they inspired and motivated me in mysterious ways, and they did allow me to work on the manuscript when they were in school or asleep.

Bill Freedman

1

THE ECOLOGICAL EFFECTS OF POLLUTION AND OTHER STRESSES

1.1
THE SCOPE OF THIS BOOK

Like ecology, environmental science is a multi-disciplinary and interdisciplinary field of scientific endeavor. The three major subdisciplines of environmental science can be appreciated from the title of the influential book, "Ecoscience: Population, Resources, Environment" (Ehrlich *et al.*, 1977).

1. *Population.* The remarkable increase in the size of the human population that has recently taken place, and that is continuing apace, is arguably the key cause of the environmental crisis (see below). The cumulative anthropogenic impact on the biosphere is a function of (a) the size of the human population and (b) the environmental impact per person, which varies greatly among and within countries, largely depending on the nature and degree of industrialization.

2. *Resources.* The resources that are available to sustain human endeavors include (a) nonrenewable natural resources, which are present in a finite supply that is diminished by mining and use (although the availability of some resources, especially metals, can be extended by the recycling of discarded material), and (b) potentially renewable natural resources, which if husbanded suitably can provide a flow of material or energy that is sustainable on a long-term basis. However, the inappropriate management or overexploitation of a potentially renewable resource can degrade it irretrievably. In such a situation, the resource is effectively "mined"—that is, it is managed as if it were a nonrenewable resource.

3. *Environment.* This subdiscipline of environmental studies deals with the ways in which anthropogenic or natural influences affect the quality of the inorganic and biotic com-

1

ponents of the biosphere, and of the purely human environment. Included within this topic are the effects of pollution and other stresses on natural and managed ecosystems; this is the primary focus of this book. Also falling within the purview of "environment" are the direct effects of pollution and other disturbances on human health, and on the built environment of humans (e.g., the deterioration of buildings and statues caused by acidifying air pollution, or the destruction of cities by warfare). Strictly speaking, the latter are not ecological issues because they do not deal with ecosystems (unless we consider the "human environment" of buildings and settlements to be components of a contrived and artificial sort of ecosystem). However, it would have been remiss to totally ignore these important human topics in this book, and therefore several key aspects receive a limited discussion, notably with respect to air pollution, toxic elements, pesticides, and warfare.

Therefore, of the three major subdisciplines within environmental science, this book focuses on the last—the impacts of stress, with particular reference to the ecological impacts of anthropogenic stresses.

1.2
GROWTH OF THE HUMAN POPULATION

For perspective, it is important to consider the remarkable increase that has taken place in the size of the human population, as well as the anticipated future changes. As mentioned previously, this population growth is critically important because it essentially drives the aggregate environmental impact of human activities. However, our consideration of this topic is fairly superficial. For more detailed information on the past and projected changes in the human population, see Ehrlich *et al.* (1977), Barney (1980), Peterson (1984), Shuman (1984), Ehrlich (1985), or World Resources Institute (WRI) (1986, 1987).

At the present time, by far the most numerous large animal (i.e., weighing more than about 40 kg) on Earth is *Homo sapiens*, with a total population of more than 5 billion individuals. It is remarkable that, within historical times, no other large animal has achieved such a great abundance. Furthermore, it is quite possible that for such a large animal, the tremendous population of humans on Earth is unprecedented. It has been estimated that prior to its overhunting the American bison (*Bison bison*) may have numbered as many as 60 million individuals (McClung, 1969; Wood, 1972). At present, the most populous large animal in the wild is probably the crabeater seal (*Lobodon carcinophagus*) of the Antarctic, with an estimated 15–30 million individuals (Bengston and Laws, 1985), equivalent to less than 1% of the present human abundance. Other large wild animals whose populations number in the millions include [Walker *et al.*, 1968; Roff and Bowen, 1983; Northridge, 1984; Organization for Economic Cooperation and Development (OECD), 1985; Dobson and Hudson, 1986; WRI, 1987; Bergerud, 1988] (1) gray and red kangaroos (*Macropus gigantea* and *M. rufus*) of Australia, with an aggregate abundance of about 19 million individuals; (2) the ringed seal (*Phoca hispida*) of the Arctic, with 6–7 million; (3) the caribou or reindeer (*Rangifer tarandus*) of the Arctic, with about 3 million; (4) the harp seal (*Phoca groenlandica*) of the North Atlantic, with 2–3 million; (5) the striped and spotted dolphins (*Stenella* spp. and *Lagenorhynchus* spp.) of the east Pacific, with > 2 million each; (6) the northern fur seal (*Callorhinus ursinus*) of the North Pacific, with 2 million; (7) the wildebeest (*Connochaetes taurinus*) of Africa, with 1.4 million; and (8) the sperm whale (*Physeter macrocephalus*) of all oceans, with 1 million. Some large fish species may also have populations that number in the millions, but hard data are not available. Clearly, for a large animal, humans are uniquely populous on Earth.

Prior to the development of agriculture around 10,000 years ago, the global abundance of humans was about 5 million individuals (after Ehrlich *et al.*, 1977). The development of an agricultural technology allowed a slow but steady growth of population, since farming could support more people than could the hunting and gathering of wild foods. The rate of population increase was also influenced by further technological innovations that enhanced the ability of humans to control and exploit their environment. For example, the discovery of the properties of metals and their alloys allowed the development of superior tools, the discovery of the wheel made it possible to easily move large quantities of materials, and the progressive domestication and genetic "improvement" of plants and animals increased agricultural yields.

Each of these innovations increased the effective carrying capacity of the environment for humans, and permitted additional population growth. As a result, there were about 300 million people in 0 A.D., and 500 million in 1650. At about this time the rate of population growth quickened markedly, and this trend has continued to the present. The most important reasons for this recent, unparalled growth of the human population have been (1) the development of an increasing sophistication of medical treatment and sanitation, which decreased death rates; (2) technological developments, which further increased the carrying capacity of the environment; and (3) the "discovery" by Europeans of exploitable lands rich in natural resources, which were quickly colonized and used to sustain and absorb additional population growth. As a result of these developments, the global population of humans increased to 1 billion by 1850, then to 2 billion by 1930, to 4 billion in 1975, and 5 billion in 1987 (Ehrlich *et al.*, 1977; WRI, 1986, 1987).

The human population continues to grow rapidly. In the mid 1980s, the rate of global increase was just less than 2% per year (at 2%/year the population would double in only 35 years). Predictions of future population growth must take into account the likely rates of change in fecundity, death, and other uncertain demographic variables. Hence it is

impossible to fortell accurately the future size of the human population. Nevertheless, the most sophisticated population models have projected that the global abundance of humans could reach about 6.1 billion by the year 2000, 9.5 billion by 2050, and 10.2 billion by 2100, and that it could stabilize at about that level (Peterson, 1984; WRI, 1986, 1987). Therefore, barring some intervening catastrophe such as a nuclear holocaust, an unprecedented pandemic, or an environmental collapse, it appears that the global population of humans will at least double from the late 1980s level, before it stabilizes.

Clearly, the human population has been growing exponentially in recent centuries, although it is unlikely to continue to do so in the forseeable future. This population increase has caused a concomitant environmental degradation by deforestation, by the overharvesting of wild animals and plants, by pollution, by the rapid mining of nonbiological resources, and by other detrimental influences. It is intuitive that the cumulative anthropogenic impact on the environment must increase with an increase in the abundance of our species. However, it must also be stressed that during the most recent century or so of particularly vigorous technological development and innovation, there has been a great intensification of the per capita environmental impact of humans. This phenomenon is illustrated in Table 1.1, which clearly shows that the global per capita economic output and fuel consumption have both increased much more rapidly than has the human population. Of course, this trend has been especially marked in relatively industrialized countries. Together, human population growth and the resulting more-than-linear environmental deterioration have changed the biosphere on a scale and intensity that is comparable to the effects of such massive geological events as glaciation.

The remarkable growth of the human population should be kept in mind whenever environmental problems are considered. The environmental impacts of people can be modified by such technological strategies as pollution control and the conservation of natural resources. However, the size of the

Table 1.1 Global population, economic output, and consumption of fossil fuels in 1900, 1950, and 1986[a]

Year	Population ($\times 10^9$)	Economic Activity		Fossil Fuel Consumption (10^9 tonnes coal equiv.)	
		Gross World Product (GWP) (1980 dollars $\times 10^{12}$)	GWP per Capita (dollars $\times 10^3$)	Total	Per Capita
1900	1.6	0.6	0.4	1	0.6
1950	2.5	2.9	1.2	3	1.2
1986	5.0	13.1	2.6	12	2.4

[a]Modified from Brown and Postel (1987).

human population remains a root cause of the degradation of our environment.

1.3
THE ECOLOGICAL EFFECTS OF STRESS

In this book, stress is simplistically defined as "any environmental influence that causes a measurable ecological change." The stressing agent can be quite foreign to the ecosystem (e.g., pesticides, toxic metals, etc.), or it could represent an accentuation of some preexisting environmental factor beyond the point where ecological change is caused (e.g., wind, thermal or nutrient loading). Often implicit within the concept of stress is an anthropocentric judgement about the quality of the ecological change, that is, whether the resulting effect is "good" or, more usually, "bad" from the human perspective.

There are several discrete classes of environmental stress, which are definable according to the causative agent. The most prominent of these are as follows.

1. *Physical stress* is an episodic effect (or disturbance) that is caused when an intense loading of kinetic energy disturbs an ecosystem, often for only a brief period of time. Examples include the effects of blast from a volcanic eruption or from anthropogenic explosions, windstorms such as hur-

ricanes and tornadoes, oceanic tidal waves or tsunamis, earthquakes, and (over geological time) glacial advance.

2. *Wildfire* is another episodic stress, caused by the rapid combustion of much of the biomass of an ecosystem. Once ignited by lightning or by an anthropogenic source, the biomass burns in an uncontrolled fashion until the fire either runs out of fuel or it is quenched.

3. *Pollution* occurs when some chemical agent is present in the environment at a concentration that is sufficient to have a physiological effect on organisms, and thereby cause an ecological change. Chemicals that are commonly involved in toxic pollution include the gases sulfur dioxide and ozone, elements such as mercury and arsenic, and pesticides. Pollution by nutrients can enhance ecological processes such as productivity, as in the case of eutrophication. Note that environmental contamination by a potentially toxic agent does not necessarily connote pollution; because of the great sophistication of modern analytical chemistry, contamination by vanishingly small concentrations of potentially toxic agents can often be measured at levels that are much lower than the thresholds at which physiological detriments can be demonstrated.

4. *Thermal stress* takes place when heat energy is released into an ecosystem, causing

ecological change. Thermal pollution is usually associated with the dissipation of waste heat from power plants and other industrial facilities, but it is also naturally present in the vicinity of hot springs and hot submarine vents.

5. *Radiative stress* occurs when there is an excessive load of ionizing energy. Examples include the effects of radiation flux from nuclear waste or explosions, and the experimental exposure of ecosystems to ionizing gamma radiation.

6. *Exploitative stress* is the selective removal of particular species or size classes of organisms from an ecosystem. Exploitation can be anthropogenic, as in the harvesting of wild animals or trees, or it can be natural, as in outbreaks of defoliating insects or disease-causing pathogens. The primary effect of exploitative stress is the removal of individuals, but there may also be secondary consequences if, for example, the removal of nutrients entrained in the biomass is sufficiently large to cause a reduction in site nutrient capital, with a consequent decrease in fertility, or if the harvested organisms were somehow critically important in the structure and function of their ecosystem.

7. *Climatic stress* is caused by an insufficient or excessive regime of temperature, moisture, and/or solar radiation.

When an ecosystem is disturbed by an event of physical stress or wildfire, it quickly suffers a great deal of mortality of its component species, structural disruption, and other ecological damage. Primary or secondary succession then begins, and this process of recovery often restores an ecosystem that is similar to the one that was present prior to the disturbance.

Stress by pollution, thermal energy, or ionizing radiation causes ecological change by exerting physiological effects on organisms. Depending on the intensity of the stress, organisms may suffer chronic or acute toxicity, and individuals may die.

If there is genetic variation for the susceptibility of individuals to the stress, then evolution will result in an increased tolerance at the population level. At the community level, relatively susceptible species may be eliminated from stressed sites. Their ecological role may then be preempted by other more tolerant members of the community, or by invading species that can exploit a stressful but weakly competitive habitat.

If an episode of pollution, thermal energy, or ionizing radiation disturbs an ecosystem, but the resulting stress is then alleviated, succession will subsequently take place. However, if the stress is exerted continuously, then a long-term ecological change will result. Consider a situation where chronic, severe atmospheric or soil pollution is caused by the construction and operation of a smelter in a remote, forested landscape. If the ensuing stress is sufficiently severe and/or cumulative, it will cause a progressive replacement of the original mature forest by an open woodland, followed by shrub-sized vegetation, and a herbaceous ecosystem. Ultimately, a complete ecological collapse can result in a totally devegetated landscape. Involved in these effects are changes in species composition, in the spatial distribution of biomass, and in such ecosystem functions as productivity, litter decomposition, and nutrient cycling (Smith, 1981, 1984; Bormann, 1982; Rapport *et al.*, 1985; Schindler, 1988). Because pollution stress decreases exponentially with increasing distance from a point source, these ecosystem changes become manifest in a continuous gradient of community change along a transect that originates at the source of pollution.

Several ecologists have described the attributes of ecosystems that have been stressed for a short or a long period of time. In general, the advent of severe stress to an ecosystem causes mortality, a reduction of species richness, a relatively uncontrolled export of nutrients and biomass, a rate of community respiration that exceeds net production, and other unstable conditions (Auerbach, 1981; Odum, 1981, 1985; Bormann, 1982; Smith, 1981, 1984; Rapport *et al.*, 1985; Schindler, 1988; see also Table 12.1 in Chapter 12). In comparison, chronically stressed ecosystems are more stable, they are relatively simple in terms of their species richness and structural complexity, they are comprised of long-lived species with a small standing crop of biomass, and they have low rates of productivity, decomposition, and nutrient cycling (Grime, 1979).

Stress can be caused by natural agencies, with great ecological effects. Because many cases of natural pollution and other stresses are ancient, consideration of their ecological impacts can allow a measure of insight into the potential longer-term effects of modern anthropogenic stresses. For this reason, natural cases are dealt with frequently in this book, and several are described as detailed case studies. Some notable examples of natural sources of stress include the following.

1. An erupting volcano can emit a large quantity of sulfur dioxide, particulates, and other pollutants to the atmosphere. The global volcanic emission of SO_2–S averages 2–5 million metric tons/year (10^6 MT/year), and individual eruptions can emit more than 1×10^6 MT (Cullis and Hirschler, 1980; Moller, 1984). In 1883, the cataclysmic eruption of the Indonesian volcano Krakatau injected an estimated 18–21 km^3 of particulates as high as 50–80 km into the atmosphere, causing spectacular sunsets and other visual phenomena throughout the world (Thornton, 1984). The even larger eruption of Tambora in Indonesia in 1815 resulted in the injection of an estimated 300–1000 km^3 of ash into the atmosphere. Some of the finer particulates made their way into the stratosphere, where they formed a reflective veil that affected the Earth's albedo and caused global cooling. The year 1816 was known as the "year without a summer" in Europe and North America because of unusually cool and wet weather, including frost and snowfall events during the summer (Stothers, 1984). A volcanic event can also initiate a devastating oceanic wave or tsunami—the 30-m-high wave caused by the eruption of Krakatau travelled as quickly as 25 m/sec,

and killed an estimated 36,000 people (Thornton, 1984). In addition, a large eruption can physically damage a great expanse of forest. For example, the 1980 explosion of Mount St. Helens in Washington blew down 21,100 ha of conifer forest, killed another 9700 ha of forest by heat injury, and otherwise damaged another 30,300 ha. There was also devastation caused by great mudslides, and a very large area was covered by particulate tephra that settled from the atmosphere to a depth of up to 50 cm (Rosenfeld, 1980; Del Moral, 1983; Means and Winjum, 1983).

2. A wildfire can kill mature trees over millions of hectares, after which a secondary succession ensues. Fire is especially frequent in boreal forests and in other seasonally droughty ecosystems such as prairie and savannah. For example, an average of about 3 million ha of forest burns each year in Canada, mostly as a result of natural ignition by lightning (Wein and MacLean, 1983; Honer and Bickerstaff, 1985). Even moist tropical rainforest will sometimes burn, as occurred over 3.5 million ha of Borneo during relatively dry conditions in 1982–1983 (Malingreau *et al.*, 1985). In addition to the direct effects on ecosystems, forest fires emit a large quantity of pollutants to the atmosphere (Smith, 1984). In Canada, forest fires emit an estimated 730 $\times 10^6$ kg/year of particulates to the atmosphere (Anonymous, 1981). Wildfires also oxidize most of the organic nitrogen of the combusted biomass, and this is emitted to the atmosphere as gaseous oxides of nitrogen (Chapter 2).

3. An outbreak of pathogens or pests can devastate susceptible ecosystems. For example, in 1975 spruce budworm (*Choristoneura fumiferana*) had severely defoliated more than 55 million ha of conifer forest in eastern North America, causing much tree mortality and other ecological change (Kettela, 1983; Chapter 8). An outbreak of herbivorous invertebrates can damage mature

marine ecosystems in ways that parallel the effects of forest defoliators. For example, off the rocky coast of Nova Scotia, an outbreak of a sea urchin (*Strongylocentrotus droebachiensis*) overgrazed a mature kelp "forest" dominated by *Laminaria* spp. and *Agarum* spp., and changed the ecosystem to a "barren ground" with a much smaller standing crop and productivity (Breen and Mann, 1976; Mann, 1977; Chapman, 1981). After a collapse of the urchin population was caused by a disease epizootic induced by unusually warm water, the kelp forest rapidly reestablished (Scheibling, 1984; Scheibling and Stephenson, 1984). Another example is the introduced chestnut blight fungus (*Endothia parasitica*) that virtually eliminated the American chestnut (*Castanea dentata*) as a canopy species in the deciduous forest of eastern North America, resulting in its replacement by other shade-tolerant species (Hepting, 1971; Spurr and Barnes, 1980). A similar pandemic of the introduced Dutch elm disease fungus (*Ceratocystis ulmi*) is removing native elms (*Ulmus* spp., especially *Ulmus americana*) from deciduous forests of North America and elsewhere (Fowells, 1965; Hepting, 1971). A final example concerns outbreaks of toxin-releasing marine phytoplankton, which occasionally bloom and cause ecological damage by the release of biochemicals that are toxic to a broad spectrum of organisms. For example, a bloom of the flagellated phytoplankton *Chrysochromulina polylepis* in the Baltic Sea in 1988 caused a mass mortality of various species of macroalgae, invertebrates, and fish (Rosenberg *et al.*, 1988).

4. Other natural agents that cause a severe physical disturbance of ecosystems are hurricanes and tornadoes, flooding, and over a geological period of time, glaciation. Other natural sources of toxic stress include surface metal mineralizations (Chapter 3), the bioaccumulation of toxic elements such as mercury by large fish, and selenium by certain plants (Chapter 3), the spontaneous combustion of bituminous material with a resulting release of gaseous SO_2 (Chapter 2), and the natural acidification of land and water caused by biological processes or by the oxidation of sulfide minerals in soil (Chapter 4).

However, by far the major emphasis of this book is on ecological damages caused by anthropogenic sources of stress. As is described below, the various topics are aggregated by the type of pollutant or other stress. This seemed to be a preferable approach to the separate consideration of topics by ecosystem or landscape type (e.g., as aquatic versus terrestrial pollution, etc.). Each chapter is organized with a particular, suitable structure, an approach that allows the individual chapters to stand alone. However, where necessary there is integration across chapter topics. For example, it will later become clear that the topic of acidification (Chapter 4) cannot be considered without some knowledge of the biogeochemistry and ecological effects of gaseous air pollutants (Chapter 2) and toxic elements (Chapter 3). Furthermore, all three of these topics may be relevant to an understanding of the decline of forests that is taking place in many parts of the industrialized world (Chapter 5).

The first class of stress that is considered in this book is gaseous air pollutants (Chapter 2). For each of the most important pollutant gases (i.e., sulfur dioxide, oxides of nitrogen, ozone, and some others), the sources of emission, chemical transformations, and toxicity are described. This is followed by a brief consideration of the effects of air pollution on human health. Three detailed case studies of the ecological effects of toxic gases are then described. The first concerns a situation in the Arctic where a natural emission of SO_2 is damaging the tundra, the second case describes ecological damage in the vicinity of SO_2-emitting smelters near Sudbury, and the third examines regional forest damage caused by ozone in southern California. The last topic in this chapter deals with the phenomenon of an increasing concentration of carbon dioxide in the atmosphere, and the ecological effects that may be taking place through global climatic change and CO_2 fertilization of plants.

Toxic elements (Chapter 3) are initially described in terms of their biogeochemistry and toxicity (especially to plants). Several examples are considered of the ecological effects of natural pollution by toxic elements. This is followed by a description of some of the most important types of anthropogenic contamination, with particular emphasis on pollution by agriculture and by the metal mining and processing industry.

Consideration of acidification (Chapter 4) begins with a description of the chemistry of precipitation, and the dry deposition of acidifying substances from the atmosphere. We then examine the chemical changes that take place as precipitation percolates through the forest canopy, the soil, and then to surface water. This is followed by an examination of the effects of acidification on terrestrial and aquatic biota, and of management practices such as liming that can be used to counter acidification.

The next topic examines the phenomenon of forest decline (Chapter 5). This syndrome has recently afflicted forests in many parts of the industrialized world, where its causal agents may include gaseous air pollutants, toxic elements, and acidic deposition. However, the precise etiology of forest decline is uncertain, and in some cases the syndrome appears to be natural in origin.

Oil pollution (Chapter 6) is dealt with by consideration of the nature of hydrocarbons, the sources and weathering of oil in the environment, and the toxicity of crude oil and its common refined products. The ecological effects of oil spills are then described using a case study approach, with particular reference to marine spills from wrecked supertankers and drilling platforms, and the effects of oil spilled in arctic environments.

The next topic is the excessive enrichment of ecosystems with nutrients, with particular reference to the eutrophication of fresh water (Chapter 7). The causes of eutrophication are considered by the examination of whole-lake fertilization experiments, the anthropogenic enrichment of Lake Erie and some other lakes, and the recovery of eutrophied waterbodies after the diversion of waste nutrients.

The next chapter describes some of the ecological effects of pollution by pesticides (Chapter 8). Pesticides are first classified and described by their chemical characteristics and uses. Their ecological impacts are then examined using a case study approach, focusing first on the insecticde DDT, then on the spraying of spruce budworm-infested forest with insecticides, and finally on the use of herbicides in forestry.

The ecological impacts of the harvesting of forests are discussed in Chapter 9. The major topics considered here are the effects of nutrient removals on inherent site fertility, the effects of disturbance of the watershed on leaching of nutrients, erosion, and hydrology, and the impacts of harvesting on wildlife.

Changes in global species richness caused by natural and anthropogenic extinction are examined in Chapter 10. A case study approach is used to illustrate species that have been made extinct or endangered by overharvesting or by the destruction of their habitat. The chapter closes with examples of endangered species that have been brought back from the brink of extinction by vigorous conservation efforts.

The environmental impacts of warfare are examined in Chapter 11. This topic is divided into three sections by separate consideration of the effects of conventional, chemical, and nuclear warfare.

The final chapter of the book synthesizes convergent patterns among the ecological impacts of various types of stress. Major topics include the nature of damage caused by stress on both the short and the long term, the spatial pattern of ecological damage that is observed around point sources of stress, and ecological recovery after the alleviation of stress.

2

AIR
POLLUTION

2.1

INTRODUCTION

There are many natural sources of gaseous air pollution. These include volcanoes, forest fires, and out-gassings from anaerobic sediment and soil. In some cases the magnitude of natural sources can rival or exceed the emissions by human activity. In other cases anthropogenic emissions are more important, and are increasing in magnitude because of the growth of the human population, and because of technological developments that result in the emission of pollutants. Of particular note in the latter respect is the emission of waste gases from fossil-fueled power plants and automobiles; neither of these sources existed prior to the twentieth century.

Even in ancient times, the activities of humans caused air pollution. Smoky wood fires used for cooking and space heating were an early source of air pollution that must have caused poor air quality inside caves, sod houses, and other badly ventilated dwellings.

Later on, when industrialization became an increasingly dominant aspect of human culture, air pollution became much more extensive. In particular, the burning of coal for energy caused severe air pollution by sulfur dioxide and soot in the cities of Europe. This problem became severe in the nineteenth century, when it was first noted as a cause of damage to human health and ecosystems. After these effects were recognized, mitigative actions were progressively taken, including (1) the construction of tall smokestacks that spread emissions over a wider area, so that ground-level fumigations were less frequent and less intense (this is the "dilution solution to pollution"); (2) the switching to "cleaner" hydrocarbon fuels such as methane and oil; (3) a reduction of total emissions by the centralization of energy production (e.g., the construction of power plants, to supplant some of the relatively dirty burning of coal in fireplaces to heat homes); and (4) the removal of pollutants from waste gases before they are vented to the atmosphere.

A contamination of the atmosphere by sulfur dioxide and soot was characteristic of the initial phase of urban air pollution during and following the industial revolution of the 1800s. This type of air pollution is often called reducing, or London-type, smog (the word "smog" is an amalgam of "smoke" plus "fog"). Smog occurred during episodes of stability of the lower atmosphere, which prevented the mixing of the polluted ground-level air mass with cleaner air from a higher altitude. As recently as the 1950s, reducing smog caused hundreds of human deaths and a high frequency of acute respiratory stress during pollution episodes in London, Glasgow, and other industrial centers of Europe, and around Pittsburgh in the United States.

To some degree, reducing smog has now been supplanted in importance by oxidizing, or Los Angeles-type, smog. Oxidizing smog occurs in sunny locations where there are large emissions of hydrocarbons and nitric oxide from automobiles and industrial sources, and where atmospheric temperature inversions are frequent. The primary emitted pollutants are transformed by complex photochemical reactions into secondary pollutants, most notably ozone and peroxyacetyl nitrate. It is these secondary gases that are harmful to people and vegetation.

Secondary pollutants are also formed from emitted sulfur dioxide and oxides of nitrogen. Ultimately, these gases are respectively oxidized to the ions sulfate and nitrate, and in this form they can be delivered to terrestrial and aquatic ecosystems as acidic deposition (described in Chapter 4).

In this chapter, the chemical characteristics, transformations, toxicity, sources, and sinks of gaseous air pollutants will be examined. This is followed by the description of several case studies of ecological damage caused by environmental pollution with toxic gases. The chapter ends with a consideration of the increasing atmospheric concentration of carbon dioxide and other radiatively active gases. Because these gases can interfere with the processes by which the Earth dissipates absorbed solar radiation, they may be capable of causing global climatic change. Such an effect would have a great impact on both natural and anthropogenic ecosystems.

2.2

EMISSION, TRANSFORMATION, AND TOXICITY OF AIR POLLUTANTS

The most important gaseous pollutants are sulfur dioxide (SO_2), hydrogen sulfide (H_2S), oxides of nitrogen (NO_x), ammonia (NH_3), carbon monoxide

(CO), carbon dioxide (CO_2), methane (CH_4), ozone (O_3), and peroxyacetyl nitrate (PAN). In addition, there are pollutant vapors of hydrocarbons and elemental mercury, and particulates with small diameter (<1 μm) that behave aerodynamically like gases and remain suspended in the atmosphere for a long time. These particulates include inert siliceous or other minerals; dusts containing toxic elements such as arsenic, lead, copper, nickel, etc.; organic aerosols emitted as smoke from combustions; and high-molecular-weight condensed hydrocarbons such as polycyclic aromatics.

Most of these air pollutants have both natural and anthropogenic sources of emission. However, O_3 and PAN are not emitted directly, but are produced secondarily in the atmosphere by complex photochemical reactions.

Sulfur Gases

Characteristics, Emissions, and Transformations

Gaseous sulfur is largely emitted as SO_2 and H_2S. SO_2 is a colorless but pungent gas that can be tasted at 0.3–1 ppm. H_2S is a gas with the foul smell of rotten eggs, which can be detected by smell at <1 ppb (Urone, 1976). In the atmosphere H_2S has a residence time of <1 day, as it is rapidly oxidized to SO_2 (Table 2.1). Atmospheric SO_2 is transformed ultimately to the anion sulfate (SO_4^{2-}). The rate of oxidation of SO_2 ranges from $<1\%$ to 5% per hour during the day, and it is influenced by the intensity of sunlight, humidity, and by the presence of nitrogen oxides, hydrocarbons, strong oxidants, and catalytic metal-containing particulates (Meszaros, 1981; Newman, 1981; Wilson, 1981; Anlauf *et al.*, 1982; Liebsch and de Pena, 1982; Fox, 1986). Because of its moderately long residence time (about 4 days), most SO_2 is transported a long distance from its point of emission before it is oxidized or deposited to the surface of the landscape.

The atmospheric SO_4^{2-} that is produced by the oxidation of SO_2 is balanced electrochemically by various cations. In eastern North America, most particulate sulfate occurs as ammonium sulfate [$(NH_4)_2SO_4$], a major component of pollution haze

Table 2.1 Global emission and other characteristics of important air pollutants[a]

Pollutant	Anthropogenic Emissions (10⁶ MT/year)	Natural Emissions (10⁶ MT/year)		Background Concentration (ppb)	Atmospheric Residence Time	Typical Concentration (ppm)	
						Clean Air	Polluted Air
SO_2	146–187	5		0.2	4 days	0.0002	0.2
H_2S	3	100		0.2	<1 day	0.0002	—
CO	304	33		100	<3 years	0.1	40–70
NO/NO_2	53 (as NO_2)	NO:	430	0.2–2	5 days	<0.002	0.2
		NO_2:	658	0.5–4		<0.004	(as NO_2)
NH_3	4	1160		6–20	7 days	0.01	0.02
N_2O	0	590		250	4 years	0.25	—
Hydrocarbons	88	200		<1	?	<0.001	—
CH_4		1600		1.5×10^3	4 years	1.5	2.5
CO_2	14,000	1,000,000		340×10^3	2–4 years	340	400
Particulates	3900	3700		—	—	—	—
O_3	—	—		—	—	0.02	0.5

[a]Modified from Kellogg *et al.* (1972), Robinson and Robbins (1972), Urone (1976), Whelpdale and Munn (1976), Cullis and Hirschler (1980), and Moller (1984).

(Ferman *et al.*, 1981; Hosker and Lindberg, 1982). If there are not sufficient cation equivalents other than H^+ to balance the SO_4^{-2}, the latter will be present as a strongly acidic and hygroscopic sulfuric acid aerosol (H_2SO_4), which contributes to the acidity of precipitation (see Chapter 4).

The largest natural sources of SO_2 are volcanic emission and forest fires. The average volcanic emission of sulfur is estimated as 2–5 million metric tons per year (2–5×10^6 MT/year), but a single large eruption has been estimated to emit more than 1 million metric tons (Cullis and Hirschler, 1980; Moller, 1984). About 90% of the global volcanic emission of sulfur is SO_2 and 10% is H_2S. The annual global emission of SO_2 from the oxidation of organic sulfur during forest fires has not been well quantified.

The anthropogenic emission of SO_2 to the atmosphere is much larger than the natural emission, and has been estimated as 146–187×10^6 MT/year (Table 2.1). The largest source of SO_2 is the burning of fossil fuels, accounting for 54% of the total anthropogenic emission (Moller, 1984). These fuels contain sulfur in both mineral and organic forms, and during combustion more than 90% of the sulfur is oxidized to gaseous sulfur dioxide. Typical sulfur concentrations of fossil fuels are: hard coals from eastern North America, 1–12%; softer coals from western North America, 0.3–1.5%; lignite, 0.7–0.9%; crude oil, 0.8–1.0%; residual fuel oil, 0.3–0.4%; kerosene, 0.4%; and motor fuels, 0.04–0.05%. Manufacturing processes (23% of anthropogenic emissions) and the smelting of sulfide ores (7%) are other large sources of SO_2 emission (Dvorak and Lewis, 1978; Moller, 1984).

The anthropogenic emission of SO_2 has increased greatly in the past century (Table 2.2), from approximately 5 million tonnes in 1860 to more than 180 million tonnes/year in the mid 1980s. In the near future, the increasing demand for electric power will be at least partly met by the construction of additional fossil-fueled power plants. This will result in an even larger emission of SO_2, unless there is an increased effort toward emission reductions by flue-gas SO_2 removal, fuel desulfurization, fuel switching, or energy conservation (Moller,

Table 2.2   Changes in the anthropogenic emission of SO_2 over time[a]

Year	Emissions of SO_2-S (10^6 MT/year)			
	Coal	Oil	Others	Total
1860	2.4	0.0	0.1	2.5
1880	5.6	0.0	0.5	6.1
1900	12.6	0.2	1.3	14.1
1920	21.2	0.7	3.4	25.0
1940	24.2	2.3	6.2	32.7
1960	30.4	8.3	10.7	48.6
1970	32.4	17.6	12.0	62.0
1977	37.2	24.0	13.7	74.9
1985	48	25	17	90
2000	55	23	22	100

[a]Modified from Moller (1984).

1984). Changes in the nature and scale of industrialization and urbanization over the last century have also increased the area of terrain that is impacted by SO_2 pollution. This is partly due to the use of increasingly taller smokestacks as a means of SO_2 dispersal, a practice that causes a more regionalized pollution as a consequence of the long-distance transport of emissions.

The emission of SO_2 and other pollutants varies greatly between and within countries, because of differences in the population density, degree and type of industrialization, quantity and type of fuel used, etc. This difference is particularly large in any comparison between relatively developed and underdeveloped nations. However, neighboring industrialized countries can also differ. For example, the total U.S. emission of SO_x (almost all of which is SO_2) is 5.1 times larger than that of Canada (Table 2.3). However, since the U.S. population is about 10 times greater (237.7 million versus 25.6 million, in 1984), the Canadian per capita emission is 2 times larger. Of the total U.S. emissions, 77% is from stationary sources such as fossil-fueled power plants, and 20% is from large industrial sources. In Canada, only 23% of the total SO_2 emission is from power plants, while 75% is from industrial sources, especially sulfide metal smelters. The most important reasons for these differences are (1) a relatively large proportion of Canadian electricity

Table 2.3 Comparison of United States and Canadian emissions of air pollutants by source (1974 data, 10^6 MT/year)[a]

Source	SO_x (as SO_2)	%	NO_x (as NO_2)	%	Hydro-carbons	%	CO	%	Partic-ulates	%
United States										
Transportation	0.8	2.6	10.7	47.6	12.8	42.1	73.5	77.7	1.3	6.7
Stationary fuel combustion	24.3	77.4	11.0	48.9	1.7	5.6	0.9	1.0	5.9	30.2
Industrial processes	6.2	19.7	0.6	2.7	3.1	10.2	12.7	13.4	11.0	56.4
Solid waste disposal	<0.1	<0.1	0.1	0.4	0.6	2.0	2.4	2.5	0.5	2.6
Miscellaneous[b]	0.1	0.3	0.1	0.4	12.2	40.1	5.1	5.4	0.8	4.1
Total	31.4		22.5		30.4		94.6		19.5	
Canada										
Transportation	0.1	1.6	1.3	61.9	1.3	52.0	10.8	70.1	0.1	4.2
Stationary fuel combustion	1.4	23.0	0.5	23.8	0.1	4.0	0.2	1.3	0.3	12.5
Industrial processes	4.6	75.4	0.2	9.5	0.2	8.0	1.3	8.5	1.5	62.5
Solid waste disposal	<0.1	<0.1	<0.1	<0.1	<0.1	<0.1	0.4	2.6	<0.1	<0.1
Miscellaneous[b]	<0.1	<0.1	0.1	4.8	0.9	36.0	2.7	17.5	0.5	20.8
Total	6.1		2.1		2.5		15.4		2.4	

[a]Modified from Council on Environmental Quality (CEQ) (1975) and Anonymous (1976a); see Benkovitz (1982) for a more detailed comparison. % = percent of total by country.

[b]Includes oil and gas production.

Table 2.4   Global emission of sulfur in 1976[a]

Source	Emission of S (10^6 MT/year)		
	Northern Hemisphere	**Southern Hemisphere**	**Total**
Natural			
Volcanoes	3	2	5
Sea spray	19	25	44
Biogenic (land)	32	16	48
Biogenic (oceans)	22	28	50
Total	76	71	147
Anthropogenic			
Coal	59	2	61
Petroleum	24	1	25
Nonferrous ores	8	3	11
Others	7	<1	7
Total	98	6	104
Total emission	174	77	251

[a]Modified from Cullis and Hirschler (1980).

generation is by nuclear and hydro technology, which do not emit SO_2, and (2) the metal smelting and refining industry is a relatively important industrial sector in Canada.

In contrast to the pattern for SO_2, the global emission of gaseous sulfides is predominantly from natural sources, especially emissions of H_2S from anaerobic sediments of shallow coastal and inland waters, and the emission of dimethyl sulfide by marine phytoplankton (Moller, 1984; Charlson *et al.*, 1987). The total natural emission of H_2S has been estimated as 100×10^6 MT/year (Table 2.1). Anthropogenic emissions of H_2S are from the chemical industries, sewage treatment facilities, and animal manure. These total about 3×10^6 MT/year (Table 2.1).

By combining information for SO_2, H_2S, and other sulfur gases, the global emission of sulfur to the atmosphere can be calculated. Of the total sulfur emission of 251×10^6 MT/year in 1976, 59% was from natural sources (Table 2.4). The natural emissions were approximately equally divided between the northern and southern hemispheres, and the biogenic emission was evenly split between oceanic and terrestrial sources. Ninety-four percent of the

anthropogenic emissions was from the northern hemisphere, reflecting the global distribution and industrialization of the human population. Interestingly, about 80% of the average total sulfur content of the atmosphere is estimated to be in the form of the relatively unreactive gas carbonyl sulfide, even though it comprises < 3% of the global emission of sulfur gases (Moller, 1984).

The typical concentrations of SO_2 and H_2S in clean, unpolluted air are each about 0.2 ppb (Table 2.1). The concentration of SO_2 in polluted air is extremely variable, but it averages about 0.2 ppm in urban air (Table 2.1).

Toxicity

H_2S is seldom present in a sufficiently large concentration to damage plants. The phytotoxicity of SO_2 is well known, and there are many field and laboratory observations of injury and reduced yield of cultivated and native plants. The toxic effects include (1) acute injury, in which there is tissue damage such as necrosis, usually in response to a short-term exposure to a large SO_2 concentration; (2) chronic injury, in which a loss of yield is associated with less severe injury such as chlorosis or premature abscission of foliage, usually in response to a longer-term exposure to a smaller SO_2 concentration; and (3) "hidden injury," in which a loss of yield takes place in the absence of visible symptoms of injury (Jacobson and Hill, 1970; Heck and Brandt, 1978; Roberts, 1984).

Roberts (1984) reviewed the effects of SO_2 on plants. He concluded that it is difficult to generalize about the phytotoxic threshold of SO_2 in air, because of large differences in susceptibility (1) between species, (2) among varieties within species, and (3) under varying conditions of stress caused by drought, nutrient supply, and other pollutants. Studies of the effects of short-term exposures to SO_2 on yield are most relevant to conditions near a point source, where fumigations are relatively sporadic but usually intense. Roberts concluded that where fumigations occurred during about 10% of the growth period, the average concentration of SO_2 must be >185 ppb to affect most crop species, and >170 ppb to affect most tree species (for particularly sensitive species or varieties the thresholds are lower).

Roberts also reviewed studies involving the continuous exposure of plants to an ambient concentration of SO_2. These studies are most relevant to field conditions where the concentration of SO_2 is regionally elevated, for example in a large city. In studies of the pasture grass *Lolium perenne* exposed to unfiltered or charcoal-filtered (which removes SO_2) air in open-top chambers at various localities in Britain, a long-term average SO_2 concentration of >38 ppb was required to measurably reduce yield.

Roberts also reviewed laboratory and field studies of the effects of controlled SO_2 fumigation on the yield of a wide range of species, and concluded that the exposure to (1) 76–150 ppb SO_2 for 1–3 months produced a statistically significant decrease in yield in most species; (2) 38–76 ppb for several months produced a small decrease in yield for some but not all species; and (3) <38 ppb variously produced beneficial, no, or minor effects on yield. Notably, the concentrations described in (1) and (2) are

considerably smaller than the U.S. air quality standard for SO_2 in the atmosphere, which is 500 ppb for 3 hr. This standard is based on the threshold concentration that causes acute foliar injury to a variety of plants (Roberts, 1984).

Therefore, SO_2 can be phytotoxic to some plants, even in a moderately polluted atmosphere. In very polluted environments SO_2 has caused great ecological damage, as will be described later in several case studies. Other consequences of SO_2 exposure are (1) there has been selection for SO_2-tolerant populations of plants in chronically polluted environments (Roose *et al.*, 1982), and (2) in some agricultural environments with high NPK fertilization, sulfur deficiency may limit crop productivity, and sulfur uptake from the atmosphere may partially compensate for this as long as the sulfur does not occur at a phytotoxic concentration (Coleman, 1966; Beaton *et al.*, 1971, 1974; Terman, 1978). In North America, situations of sulfur deficiency are most frequent in western grain agriculture and on calcareous soil. Sulfur deficiency in agriculture is rare in the eastern United States and southeastern Canada, where sulfur dioxide pollution and sulfate deposition are important environmental problems (see also Chapter 4).

Humans are generally more tolerant of exposure to SO_2 than are most plants. The American Conference of Governmental Industrial Hygienists has recommended a concentration no higher than 2 ppm SO_2 as an average occupational exposure, and 5 ppm for a short-term exposure [American Conference of Governmental and Industrial Hygienists (ACGIH), 1982; Henderson-Sellers, 1984]. However, exposure to an SO_2 concentration of <1 ppm can cause respiratory distress in sensitive individuals, especially people with a history of asthma or other respiratory diseases (Goldsmith, 1986). It should be noted that there is controversy over the effects of chronic exposure to the smaller concentrations of SO_2 and sulfate particulate aerosol that are commonly present in the urban environment. Some epidemiological studies have shown a statistically significant effect on human health at ambient urban concentrations of these pollutants, while other studies have not (Amdur, 1980; Goldsmith, 1986). More research is required to sort out this potential environmental problem.

Nitrogen Gases

Characteristics, Emissions, and Transformations

From a pollution perspective, the most important nitrogen gases are ammonia (NH_3), nitric oxide (NO), nitrogen dioxide (NO_2), and nitrous oxide (N_2O). Together, NO and NO_2 are often abbreviated as NO_x.

Ammonia is a colorless gas. Its major source is natural emissions from wetlands, where NH_3 is produced during biological decay. The total natural emission of NH_3 is estimated as $>10^9$ MT/year (Table 2.1). The anthropogenic emission of NH_3 is much smaller, and sources include coal combustion (3×10^6 MT/year), oil and gas combustion (1×10^6 MT/year), and cattle feedlots (0.2×10^6 MT/year) (Whelpdale and Munn, 1976). Ammonia is oxidized to NO_x in the atmosphere, where it has a residence time of 7 days (Table 2.1).

N_2O is a colorless and nontoxic gas. It is used in medicine as a mild anesthetic, and it is sometimes called "laughing gas" because it produces a mild euphoria. The background concentration of N_2O in the atmosphere is 0.25 ppm, and because it is relatively unreactive it has a long residence time of 4 years (Table 2.1). There are no large industrial emissions of N_2O. Natural emissions result from microbial denitrification of nitrate in soil and water (Box, 1982; Aneja et al., 1984). Fertilized agricultural soil can have a particularly large rate of N_2O emission, and it has been estimated that modern agriculture has increased the global N_2O emission by 50% (Aneja et al., 1984). The global emission of N_2O is about 590×10^6 MT/year (Table 2.1).

NO is a colorless, odorless, and tasteless gas, while NO_2 is reddish-brown, pungent, and irritating to respiratory membranes. The background atmospheric concentration of NO is 0.2–2 ppb, while that of NO_2 is 0.5–4 ppb. In a polluted atmosphere

these gases are present at $\sim$0.2 ppm (expressed as NO_2). In the atmosphere, NO is oxidized relatively rapidly to NO_2 by reactions that are described later when we deal with photochemical air pollution. NO_2 is eventually oxidized photochemically and catalytically to nitrate, an anion that contributes to the acidity of precipitation (Chapter 4).

Both NO and NO_2 have large natural sources, which in aggregate exceed the total anthropogenic emission. However, there is considerable uncertainty over the magnitude of the natural flux of NO_x to the atmosphere, largely because of difficulties in the estimation of the emission resulting from denitrification in soil. The global natural emission of NO has been estimated as 430×10^6 MT/year, and natural NO_2 as 658×10^6 MT/year (Table 2.1; see also Guicherit and van den Hout, 1982; Logan, 1983; Hov, 1984). The largest natural sources of NO_x are (1) emission from soil and water as a result of bacterial denitrification of nitrate, (2) dinitrogen fixation by lightning (i.e., the oxidation of N_2 to NO_x at high temperature), and (3) combustion of biomass (i.e., the oxidation of organic N to NO at high temperature, plus the oxidation of N_2 during combustion) (Urone, 1976; Box, 1982; Logan, 1983; Hov, 1984).

There is less uncertainty about the magnitude of the anthropogenic emission of NO_x, with recent estimates ranging from 36 to 60×10^6 MT/year of NO_x expressed as NO_2 (Urone, 1976; Whelpdale and Munn, 1976; Logan, 1983; Hov, 1984). The combustion of fossil fuels is the largest anthropogenic source. This produces NO_x by the oxidation of organic N during combustion, and by the oxidation of atmospheric N_2 to NO_x. The latter process is especially important if the combustion occurs at a high temperature and pressure, as it does in the internal combustion engine of automobiles. NO is the primary NO_x gas that is produced by these oxidations; NO_2 is produced secondarily in the atmosphere by the oxidation of emitted NO.

Toxicity

NH_3 and the NO_x gases are capable of injuring plants. However, the concentrations that are required to cause this effect are considerably larger than are usually present in the field, except in an exceptionally polluted environment near a point source (Amundson and MacLean, 1982). The threshold for acute injury to plants by NO_2 is ≥ 0.15 ppm (Heck and Brandt, 1978). As is discussed later, the importance of the NO_x gases with respect to plant damage lies in their role in the photochemical production of ozone.

Similarly, the ambient concentration is rarely large enough to affect humans. The guideline for a long-term occupational exposure to NO is 25 ppm and for NO_2 it is 5 ppm, while those for short-term exposure are 35 ppm and 5 ppm, respectively (Henderson-Sellers, 1984).

Hydrocarbons

Characteristics, Emissions, and Transformations

Hydrocarbons are a chemically diverse group of air pollutants, ranging from gaseous methane (CH_4), through various vapor-phase hydrocarbons, to complex, high-molecular-weight types such as polycyclic aromatic hydrocarbons (see Chapter 6).

A typical background concentration of CH_4 in the atmosphere is 1.5 ppm, while all other hydrocarbons together comprise <1 ppb (Table 2.1). The emission of CH_4 is largely natural. The most important source is microbial fermentation in anaerobic wetlands, but smaller sources include outgassing from natural gas and coal deposits, and methane produced by the incomplete oxidation of organic matter during forest fires (Urone, 1976; Whelpdale and Munn, 1976; Altshuller, 1983). Estimates of the global emission of CH_4 range from about 300×10^6 MT/year (Whelpdale and Munn, 1976; Altshuller, 1983) to 1600×10^6 MT/year (Table 2.1; from Urone, 1976). The wide variation reflects the typical uncertainty involved in the estimation of the global rates of cycling of air pollutants.

Estimates of the natural emission of nonmethane hydrocarbons range from 200×10^6 MT/yr (Table 2.1; after Urone, 1976) to 830×10^6 MT/year (Zimmerman, 1979), while the anthropogenic

emission has been estimated as 65×10^6 MT/year (Duce, 1978). The largest natural emissions of nonmethane hydrocarbons are from living vegetation, and outgassing from natural gas and coal deposits of light hydrocarbons such as ethane, propane, butane, and pentane (Dimitriades, 1981; Altshuller, 1983; Salop et al., 1983; Yokouchi et al., 1983).

Certain hydrocarbons such as isoprene and various monoterpenes are emitted from forest vegetation. The rate of emission in the temperate zones is much larger during the growing season than in cooler months, especially on hot, sunny days (Altshuller, 1983; Salop et al., 1983; Isidorov et al., 1985). Emissions vary greatly among plant species and forest types. Salop et al. (1983) studied the emission of hydrocarbons from forested terrain in Virginia, and found much larger rates for angiosperm stands or species, compared with pines (Table 2.5). The specific hydrocarbons that are emitted differ greatly among plant species (Altshuller, 1983; Isidorov et al., 1985).

The most important anthropogenic emission of nonmethane hydrocarbons is from automobiles, aircraft, and the hydrocarbon extracting and refining industry (Table 2.3). The rate and type of hydrocarbon emission from vehicles depends on the type of engine (e.g., diesel versus conventional internal combustion engine), vehicle weight, and specific engine conditions such as spark timing, exhaust gas

Table 2.5 Emission factors for nonmethane hydrocarbons from forest types and tree species of a 6.3×10^5 ha area of Virginia[a]

Forest/Tree	Emission Rate	
Deciduous forest	58	g/ha-hr
Mixed-wood forest	59	g/ha-hr
Conifer forest	17	g/ha-hr
Sweetgum (Liquidambar styraciflua)	61	µg/g-hr
Oaks (Quercus laurifolia, Q. michauxii, Q. nigra, Q. alba)	25	µg/g-hr
Black gum (Nyssa sylvatica)	10	µg/g-hr
Red maple (Acer rubrum)	6.5	µg/g-hr
Loblolly pine (Pinus taeda)	5.5	µg/g-hr

[a]Modified from Salop et al. (1983).

recirculation, air: fuel ratio, and engine temperature. These all affect the efficiency of combustion, and therefore the rate of hydrocarbon emission. Nelson and Quigley (1984) found that the most prominent hydrocarbons emitted by automobiles were ethylene, 11.2% of total emissions; toluene, 10.2%; acetylene, 8.7%; *m,p*-xylenes, 6.5%; benzene, 5.0%; propylene, 5.0%; and *i*-pentane, 4.8%, with a variety of other types comprising the remainder.

Another anthropogenic source of hydrocarbon emissions is fossil fuel-fired power plants. Although they are a relatively small source [<2% of the total anthropogenic emission; National Research Council (NRC), 1976], power plants emit some toxic hydrocarbons. These include polycyclic aromatic hydrocarbons such as benzo[a]pyrene, which are mainly vented to the atmosphere as small particulates <3 μm in diameter (Freedman, 1981).

Toxicity

The natural and anthropogenic emissions of hydrocarbons are large. However, except in the exceptionally polluted vicinity of large point sources, there is no evidence of acute plant or animal damage caused by exposure to these chemicals in the atmosphere. In terms of ecological impacts, atmospheric hydrocarbons are most important for the role they play in photochemical reactions involving NO_x and O_3 (described below).

Photochemical Air Pollutants

Characteristics and Transformations

By far the most damaging photochemical air pollutant is ozone (O_3). Peroxy acetyl nitrate (PAN), hydrogen peroxide (H_2O_2), and aldehydes and other oxidants play relatively minor roles. All of these are secondary pollutants, that is, they are not emitted, but are formed in the atmosphere by photochemical reactions involving emitted gases, especially NO_x and hydrocarbons.

Ozone is a bluish gas, 1.6 times as heavy as air, and very reactive as an oxidant. Ozone is present in a relatively large concentration in the stratosphere, in an atmospheric layer higher than 8–17 km altitude. The stratospheric O_3 concentration averages ~0.2–0.3 ppm, compared with ≤0.02–0.04 ppm in background situations closer to ground level (Urone, 1976; Singh *et al.*, 1980; Grennfelt and Schjoldager, 1984; Skarby and Sellden, 1984). Stratospheric O_3 is formed naturally by ultraviolet photochemical reactions (after Haagen-Smit and Wayne, 1976):

$$O_2 + h\nu \ (200 \ nm) \rightarrow O + O \qquad (1)$$

$$O + O + M \rightarrow O_2 + M \qquad (2)$$

$$O + O_2 + M \rightarrow O_3 + M \qquad (3)$$

$$O_3 + h\nu \ (290\text{--}200 \ nm) \rightarrow O_2 + O \qquad (4)$$

where M is an energy-accepting third body. To summarize: O_2 interacts with ultraviolet radiation to form O atoms (reaction 1), which can either recombine to O_2 (reaction 2) or combine with O_2 to form O_3 (reaction 3). Ozone can be consumed by a variety of reactions, including an ultraviolet photodissociation (reaction 4), and reaction with trace gases such as NO_x, N_2O, and ions or simple molecules of chlorine and fluorine.

Because of a large increase in the anthropogenic emission of some of these O_3-consuming substances or their precursors, concern has been raised over a potential upset of the dynamic equilibria among stratospheric O_3 reactions, with a consequent decrease in O_3 concentration. Because stratospheric O_3 absorbs much of the incoming solar ultraviolet (UV) radiation [it is an especially effective absorber at 210–293 nm (Rowland, 1988)], it serves as a UV shield. As such it helps to protect organisms on the earth's surface from some of the deleterious effects of this high-energy electromagnetic radiation [the UV-wavelength range of 280–320 nm is especially damaging (Rowland, 1988)]. If not intercepted, the ultraviolet radiation could disrupt genetic material, since DNA is an effective absorber of UV wavelengths <320 nm. Damage to DNA could cause increased rates of skin cancers and heritable mutations [National Academy of Sciences (NAS), 1979a; Cicerone, 1987]. One atmospheric process model estimated a 16.5% reduction in the stratospheric concentration of O_3 in response to the 1977 rate of emission of chlorofluorocarbons (CFCs) (NAS, 1976, 1979a). In the mid-1980s, a springtime decrease in the concentration of stratospheric O_3 (called ozone "holes") has been observed at high latitudes. This phenomenon has been most notable over Antarctica, where the holes develop between September and November, and where the average decrease in stratospheric ozone between the late 1970s and the late 1980s was 30–40%. In October, 1987 the average decrease of stratospheric ozone over Antarctica was 50%, and it was 95% in the hardest-hit altitudinal zone of the lower stratosphere at 15–20 km (Kerr 1988b). Although the cause of this phenomenon has not yet been determined, it is strongly suspected that chlorine atoms or simple chlorine compounds such as ClO may play a key role, and that these O_3-consuming chemicals may have an indirect, anthropogenic origin through the emission of CFMs (Farman *et al.*, 1985; Crutzen and Arnold, 1986; Farmer *et al.*, 1987; Hoffman *et al.*, 1987; Solomon, 1987; Solomon *et al.*, 1987; Zafra *et al.*, 1987; Bowman, 1988; Rowland, 1988).

During an event of great turbulence in the upper atmosphere, stratospheric O_3 can enter the troposphere. Usually this affects the upper troposphere, although there have been rare observations of an incursion reaching ground level for a short (<2 hr) period of time (Singh *et al.*, 1980; Chung and Dann, 1985).

However, most tropospheric O_3 is formed and consumed by endogenous photochemical reactions, and these are the source of oxidants in Los Angeles-type smog. The most important of these reactions are summarized below (following Grennfelt and Schjoldager, 1984):

As described previously, the O_3-forming reaction is

$$O + O_2 + M \rightarrow O_3 + M \qquad (5)$$

Most atomic O is formed by the photodissociation of NO_2:

$$NO_2 + h\nu \ (<440 \ nm) \rightarrow NO + O \qquad (6)$$

NO reacts with O_3, regenerating NO_2:

$$NO + O_3 \rightarrow NO_2 + O_2 \qquad (7)$$

If other reactions convert NO to NO_2 (as in reaction 8, below) the O_3 can accumulate, since this operates in competition with reaction 7 for NO:

$$NO + RO_2 \rightarrow NO_2 + RO \qquad (8)$$

The reaction species RO_2 includes various peroxy radicals. These are formed by the degradation of organic molecules (RH) by reaction with hydroxy radicals (OH) (reaction 9a), followed by the addition of molecular O_2 (reaction 9a):

$$RH + OH \rightarrow R + H_2O \qquad (9a)$$

$$R + O_2 \rightarrow RO_2 \qquad (9b)$$

The concentration of OH is maintained by photodissociation of O_3 to produce atomic O (reaction 10a), followed by reaction with H_2O to form OH (reaction 10b):

$$O_3 + h\nu \ (<320 \ nm) \rightarrow O + O_2 \qquad (10a)$$

$$O + H_2O \rightarrow 2OH \qquad (10b)$$

Other, more complex photochemical reactions have been described, including those by which PAN and aldehydes are formed, but these will not be described here (see Haagen-Smit and Wayne, 1976; Guicherit and van den Hout, 1982; Grennfelt and Schjoldager, 1984; Fox, 1986).

Therefore, the formation of oxidizing smog involves a complex group of photochemical interactions between anthropogenically emitted pollutants (NO and hydrocarbons) and secondarily produced chemicals (O_3, NO_2, PAN, aldehydes). The concentrations of these chemicals exhibit a pronounced diurnal pattern, depending on their rate of emission, and on the intensity of solar radiation and the atmospheric stability at different times of the day. This pattern is illustrated in Fig. 2.1 for important air pollutant gases in Los Angeles. NO is the emitted NO_x, and it has a morning peak of concentration at 06:00–07:00, largely due to emissions from vehicles during the morning rush of traffic. Hydrocarbons are emitted from vehicles and refineries, and

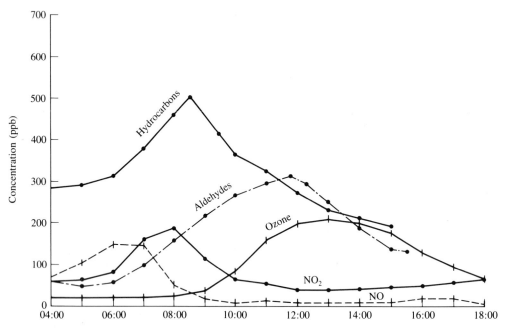

Fig. 2.1 The average concentration of various air pollutants in the atmosphere of Los Angeles during days of eye irritation. Modified from Haagen-Smit and Wayne (1976).

they have a similar temporal pattern to that of NO except that their concentration peaks slightly later. In bright sunlight the NO is photochemically oxidized to NO_2 (reactions 7 and 8), resulting in a decrease in NO concentration and a peak of NO_2 at 07:00–09:00. Photochemical reactions involving NO_2 produce O atoms (reaction 6), which react with O_2 to form O_3 (reaction 5). These result in a net decrease in NO_2 concentration and an increase in O_3 concentration, peaking broadly at 12:00–15:00. Aldehydes are also formed photochemically, but they peak earlier than O_3. As the day proceeds, the various gases decrease in concentration as they are diluted by fresh air masses, or are consumed by photochemical reactions (e.g., reaction 10 and others), including the formation of nitrated organics, peroxides, aerosols, and other terminal products. This cycle is repeated daily, and it is typical of an area that experiences photochemical smog (Urone, 1976).

Hov (1984) calculated a global tropospheric O_3 budget (Table 2.6). The considerable range of the estimates reflects uncertainty in the calculation of the O_3 fluxes. On average, stratospheric incursions account for about 18% of the total O_3 influx to the troposphere, while endogenous photochemical production accounts for the remaining 82%. About 31% of the tropospheric O_3 is consumed by oxidative reactions at vegetative and inorganic surfaces at ground level, while the other 69% is consumed by photochemical reactions in the atmosphere.

Various countries have set standards for allowable O_3 concentration, below which little acute damage to vegetation is expected (Table 2.7). Some standards have been set rather low in comparison to the ambient concentration in certain areas, and this means that the limits may frequently be exceeded. Until 1979 the U.S. standard for the maximum 1-hr average O_3 concentration was 80 ppb. However, the standard was increased in that year to 120 ppb, because 80 ppb was so frequently exceeded in many jurisdictions (Skarby and Sellden, 1984). Because the original ozone standard was not enforceable in a practical sense, it was raised.

Table 2.6   A tropospheric ozone budget[a]

	Northern Hemisphere (kg/ha-year)	Southern Hemisphere (kg/ha-year)
Transport from stratosphere	13–20	8–10
Photochemical production	48–78	28–73
Destruction at ground	18–35	10–20
Photochemical destruction	48–55	28–30

[a]Modified from Hov (1984).

In the Los Angeles basin the 1-hr peak concentration of O_3 can reach 580 ppb, and it typically exceeds 100 ppb for more than 15 days during the growing season (Roberts, 1984). Elsewhere the O_3 concentration is usually smaller. In other North American cities the maximum annual 1-hr peak concentration is typically 150–250 ppb, and in London, England, it is 90–180 ppb (Roberts, 1984). However, during the dry and sunny summer of 1976, O_3 concentrations >200 ppb occurred in southern England (Skarby and Sellden, 1984). In West Germany, an hourly mean concentration of 50–125 ppb is frequent over a large area during the growing season, but during the dry and sunny summer of 1976 a maximum 1-hr concentration of >150 ppb occurred (Skarby and Sellden, 1984).

Toxicity

Ozone is the most toxic constituent of photochemical smog, and it has caused considerable damage to agricultural and native plants in many locations. Ozone causes a distinctive acute injury to plants (Jacobson and Hill, 1970), resulting in a loss of photosynthetic area. It is often assumed that the resulting yield decrement is proportional to the percentage loss of leaf area to acute injury. However, in some cases this relationship is moderated by a compensatory increase of photosynthesis in undamaged foliar tissue (Jacobson, 1982a; Roberts, 1984).

A general threshold for yield decrease caused by the exposure of crops to O_3 is about 100 ppb (Roberts, 1984). However, in sensitive species both

Table 2.7   Ozone standards that have been set or recommended by different countries in order to prevent significant damage to vegetation[a]

Country	Maximum 1-hr Average Concentration (ppb)	Frequency of Exceeding	Status
United States	120	Once a year	Legal standard
Japan	60	Not to be exceeded	Long-term goal
Canada	50	Max. desirable	Guideline
	80	Max. acceptable	Guideline
	150	Max. tolerable	Guideline
Sweden	60	Once a month	Proposed guideline
Norway	50–100		Recommended
World Health Organization	60		Recommended

[a]Modified from Grennfelt and Schjoldager (1984).

acute and hidden injury can be caused at a smaller concentration (Jacobson, 1982a; Roberts, 1984; Skarby and Sellden, 1984). For example, in a laboratory fumigation experiment tobacco (*Nicotiana tabacum*) was acutely injured by 50–60 ppb during a 2- to 3-hr exposure, and spinach (*Spinacea oleracea*) was injured by 60–80 ppb for 1–2 hr (Skarby and Sellden, 1984).

A number of field studies have examined the growth of agronomic plants in chambers receiving either ambient air, or charcoal-filtered air without O_3. These studies have frequently observed a substantial decrease in yield caused by the exposure to O_3 in unfiltered, ambient air. Thompson and Taylor (1969) found that the yield of fruit of lemon and orange trees (*Citrus limonia* and *C. sinesnis*) was ~50% smaller in unfiltered air in southern California. Studies in New York reported a 30% decrease in the yield of tomato (*Lycopersicum esculentum*) and snap beans (*Vicia faba*) (MacLean and Schneider, 1976). Studies of soybean (*Glycine max*) in Maryland found a 20% reduction in yield (Howell *et al.*, 1979). Other studies in Maryland between 1972 and 1979 found an average reduction in yield of 12% (range 4–20%) among potatoes (*Solanum tuberosum*), soybeans, corn (*Zea mays*), snap beans, and tomatoes (Heggestad, 1980).

In a detailed series of field experiments at five sites in the United States, several important crop plants were exposed in open-top chambers to either ambient air, or 25 ppb O_3, a typical background concentration (Heck *et al.*, 1982). The average annual concentration of O_3 in ambient air at the study sites was, for the northeast site, 40 ppb; southeast A, 56 ppb; southeast B, 31 ppb; central, 42 ppb; and southwest, 106 ppb. Symptoms of acute O_3 injury were observed in crop species at each of these sites. Across all sites, the yield decrease due to ozone was 10% in soybean, 14–17% in peanut (*Arachnis hypogaea*), 7% in turnip (*Brassica napa*), 53–56% in lettuce (*Lactuca sativa*), and 2% in kidney bean (*Phaseolus vulgaris*). It was estimated that at the current U.S. standard for O_3 in air (i.e., 120 ppb for a 1-hr exposure), the crop loss would be about 2–4% of the total potential yield, with a direct economic value in 1982 dollars of as much as $5 billion per year (Heck *et al.*, 1982, 1986).

Because the threshold concentration for yield reduction caused by O_3 is so close to or even less than the ambient concentration in many parts of North America, it is the economically most damaging air pollutant in agriculture (Heck *et al.*, 1982, 1983, 1986; Jacobson, 1982b; Howitt *et al.*, 1984; Skarby and Sellden, 1984; Reich and Amundson, 1985; Rowe and Chestnut, 1985). Effects of ozone on forests are described later as a case study.

Humans are very sensitive to ozone. The gas causes irritation and damage to membranes of the respiratory system and eyes. The guideline for a long-term occupational exposure to ozone is 0.1 ppm, while for a short-term exposure it is 0.3 ppm (ACGIH, 1982). However, relatively sensitive people can suffer respiratory distress at smaller concentrations than these, including those that commonly occur during events of oxidizing smog.

Fluorides

Characteristics and Emissions

Atmospheric fluorides exist in a variety of gaseous and particulate forms. The gases that are emitted anthropogenically in largest quantity are hydrogen fluoride (HF) and silicon tetrafluoride (SiF_4). Fluorides are also vented naturally from volcanoes and fumaroles; identified gases include ammonium fluoride (NH_4F), silicon tetrafluoride, ammonium fluorosilicate [$(NH_4)_2SiF_6$], sodium fluorosilicate (Na_2SiF_6), potassium fluoroborate (KBF_4), and potassium fluorosilicate (K_2SiF_6) (NAS, 1971; Urone, 1976). A secondary aerosol is the strong acid fluorosilicic acid (H_2SiF_6), which is formed in the atmosphere by the hydrolysis of SiF_4. Important particulates include the minerals fluorospar (CaF_2), cryolite (Na_3AlF_6), and fluorapatite [$Ca_{10}F_2(PO_4)_6$], which can be released as dusts from industrial sources (NAS, 1971).

Volcanoes are the only quantified natural source of fluoride, emitting about 7.3×10^6 MT/year (Urone, 1976). The anthropogenic emission of F is much smaller, amounting to about 0.4×10^6 MT/year. The total anthropogenic emission in the United States was 1.2×10^5 MT/year in 1971, with the largest sources being steel manufacturing (33%

of the total emission), phosphate fertilizer processing (16%), brick and tile manufacturing (15%), aluminum reduction (13%), and coal combustion (13%) (NAS, 1971).

The atmospheric concentration of F in rural areas is generally very small (<0.3 ppb; Urone, 1976). Urban air has a somewhat larger concentration, generally <3 ppb (NAS, 1971; Urone, 1976). Close to an industrial point source, the concentration in air can be much larger; up to 250 ppb was measured in an area in Florida that was contaminated with dust from the mining and processing of phosphate rock that contained ~3–4% F (NAS, 1971).

Toxicity

The threshold concentration for injury and reduction of yield of vegetation depends on the susceptibility of plant species or varieties, and on environmental conditions. For a 1-day exposure, the threshold concentration of HF in air that is required to cause acute injury to sensitive plants is 4–5 ppb, and >13 ppb for more tolerant species, while for a 1-month exposure the concentrations are 0.6 ppb and 1–4 ppb, respectively (NAS, 1971). Sensitive plants exhibit symptoms of acute injury at a foliar concentration of 20–150 ppm F, while resistant accumulator species can tolerate >4000 ppm. Gladiolus (*Gladiolus communis*) is very sensitive to F, and it is injured by only >20 ppm in foliage, whereas tea (*Camellia sinensis*) is fairly tolerant, accumulating as much as 200 ppm without suffering acute injury. Cotton (*Gossypium hirsutum*) is very tolerant, and it can accumulate as much as 4000 ppm in its foliage without suffering acute injury (NAS, 1971).

2.3
AIR POLLUTION AND HUMAN HEALTH

On rare occassions, the natural emission of toxic gases has caused human deaths. One remarkable incident occurred in 1986, when a large volume of CO_2 was unexpectedly released from Lake Nyos in Cameroon (Freeth and Kay, 1987; Kling *et al.*, 1987). This release was apparently triggered by a seismic event, which caused an overturn of the stratified, 200-m-deep volcanic lake. This brought a large volume of CO_2-supercharged deep water to the surface, where it quickly degassed. The resulting dense atmospheric mass was laden with water vapor and CO_2, and it flowed into a surrounding lowland and killed 1700 people, 3000 cattle, and uncounted wildlife by CO_2 asphyxiation. Most of the humans were killed as they slept. Essentially no plant damage was caused by this event of CO_2 pollution.

There have also been incidents in which anthropogenic air pollution has caused a marked increase in human mortality, particularly within high-risk groups of people with chronic respiratory or heart disease. These toxic pollution events took place during periods of prolonged atmospheric stability, which prevented the dispersion of emissions. This resulted in a buildup of a large concentration of SO_2 and particulates, often accompanied by fog. The term "smog" was originally coined to label these air pollution events.

Coal smoke has long been recognized as a pollution problem in England, from at least 1500 (Wise, 1968; Chambers, 1976). In London, dirty pollutant-laden fogs occurred quite frequently, and they were known as "pea-soupers." The first conclusive linkage of an air pollution event with a large increase in human mortality was in Glasgow in 1909, when 1063 deaths were attributed to a noxious accumulation of pollution during a period of atmospheric stagnation (Chambers, 1976). Another episode occurred in the Meuse Valley of Belgium in 1930 and resulted in 60 deaths. In that toxic episode, a very stable atmospheric inversion developed. The continued emission of toxic gases and particulates from steel works, a sulfuric acid plant, glass factories, and zinc works caused the accumulation of a large concentration of pollutants, especially SO_2 (Perkins, 1974).

Another notable episode occurred in October 1948 in Donora, a town located in a valley some 45 km from Pittsburgh, Pennsylvania. In this case, a fog was coincident with a temperature inversion that persisted for 4 days. Continued emissions from a steel mill and zinc and sulfuric acid plants caused severe air pollution, with SO_2 estimated at 0.5–2 ppm, and a large but unmeasured concentration of particulates. A very high rate of mortality was associated with the smog (20 deaths out of a population of 14,100). There was also great morbidity, with about 43% of the population made ill, 10% of them severely so. Typical symptoms were irritation of the eyes, nose, throat, and lower respiratory tract, along with coughing, headache, and vomiting (Perkins, 1974; Davidson, 1979).

Probably the best known air pollution event is the so-called "killer smog" that affected London, England, in December of 1952. In this case, an extensive temperature inversion was accompanied by a "white fog" throughout southern England. In the vicinity of London, this transformed into a noxious "black fog" with virtually zero visibility, as the concentrations of SO_2 and particulates built up as a result of emissions from coal combustion for home heating because of the cold temperature at the time, and for the generation of electricity and other industrial purposes. Visibility was so poor that many people lost their way while driving cars, others walked off wharves and fell into rivers, and theaters were closed because the projection screen was not visible to most of the audience. Of course, there were also serious disruptions of activity in workplaces and in commerce. During the 4-day episode, the maximum daily concentration of SO_2 was 1.3 ppm, and total suspended particulates were 4.5 mg/m^3. In total, there were 18 days of greater-than-usual human mortality. Overall, 3900 "excess" deaths were attributed to the episode, mainly among the elderly, the very young, and persons with preexisting respiratory or coronary disease (Wise, 1968; Perkins, 1974; Amdur, 1980).

Episodes such as this were frequent in the industrialized cities of Britain, Europe, and the northeastern United States, and they were in large part due to the burning of coal. Apart from effects on human health, there were many instances of death of livestock, and of damage to vegetation. It was widely known that only plants that were very tolerant of pollution could be grown in most large cities. Such species included privet (*Ligustrum vulgare*), plane tree (*Platanus acerifolia*), Norway maple

(*Acer platanoides*), and tree-of-heaven (*Ailanthus altissima*). In Britain, the passing (and enforcement) of the Clean Air Act in 1956 resulted in greatly reduced emissions of air pollutants. This caused an improvement of air quality in cities such as London, so that severe reducing smogs may now be historical occurrences.

It should be stressed that the above cases represent very severe air pollution episodes. The effects on human health of more typical, nonepisodic urban air quality have also been investigated. However, the conclusions with respect to enhanced morbidity or mortality are much more tentative and controversial (Coffin and Stokinger, 1978; Goldsmith and Friberg, 1977; Imai *et al.*, 1985; Goldsmith, 1986).

2.4
CASE STUDIES OF THE ECOLOGICAL EFFECTS OF TOXIC GASES

In the following, selected case studies of ecological damage caused by gaseous air pollution will be examined. The first case describes the effects of naturally occurring SO_2 pollution at a remote location in the Arctic. We then examine damage caused by the anthropogenic emission of SO_2 from point sources, focusing on the large smelters at Sudbury, Canada. Next we examine a case where regional air pollution by ozone has caused extensive damage to coniferous forests in California. Finally, we discuss the potential effects of climatic change that may be caused by the increasing concentration of CO_2 in the atmosphere.

Effects of Natural SO_2 Pollution at the Smoking Hills

The Setting

The Smoking Hills are at 70°N on the mainland seacoast of the Canadian Arctic. The area is remote, pristine, and virtually uninfluenced by humans. At a number of locations along 30 km of seacoast, strata of bituminous shale in a 100-m-high seacliff

View of a plume contaminated with sulfur dioxide at the Smoking Hills in the Canadian Arctic. Bituminous shales have ignited in a discrete slump of the 100-m-high coastal seacliff. An older burned-out slump is located to the left of the burn. The plume is about 100 m wide where it reaches the top of the cliff. Note that the plume fumigates the ground surface as it moves inland. (Photo courtesy of B. Freedman.)

A sparsely vegetated site at the Smoking Hills. All of the plant clumps are the herb *Artemisia tilesii*, which is very tolerant of sulfur dioxide, acidity, and toxic metals such as aluminum. (Photo courtesy of B. Freedman.)

A grassland community at the Smoking Hills, in a zone of moderate air and soil pollution. The dominant plant is the pollution-tolerant grass *Arctagrostis latifolia*. (Photo courtesy of B. Freedman.)

have spontaneously ignited. As a result, the adjacent tundra is fumigated with SO_2 and other pollutants (Hutchinson *et al.*, 1978; Havas and Hutchinson, 1983a).

The area has a bedrock of shale, covered with 10 m or less of calcareous glacial drift and fluvial deposit. Interbedded in the shale are dark-colored bituminous strata. When this material is exposed to the atmosphere by slumping of the seacliff, pyritic sulfur undergoes an exothermic oxidation to sulfate. In some cases the heat can accumulate sufficiently to ignite the bituminous shale, which proceeds to burn into the cliff until the oxygen supply becomes insufficient to support further combustion. This process is similar to that proposed for the spontaneous ignition of bituminous material in coal waste piles and mines, where pyrites are also exposed to oxygen (Sussman and Mulhern, 1964; Mathews and Bustin, 1984).

The earliest documented sighting of the Smoking Hills was by British explorers in 1826, but the burns are undoubtedly more ancient than this. The area was not covered by ice during the most recent glaciation about 10,000 years ago, and the tremendous piles of ash and oxidized shale at the bottom of the seacliff suggest great antiquity.

Atmospheric Pollution

During the growing season, the prevailing wind direction at the Smoking Hills is onshore. As a result, the sulfurous plumes are carried inland where they impact the tundra ecosystem. Gizyn (1980) found that the largest concentration of SO_2 was present within 20 m of the edge of the sea cliff, just where the plume begins to roll inland (10-min average values were as high as 1.7 ppm SO_2). Along transects running inland from the sea, the longer-term average concentration of SO_2 decreased rapidly (Table 2.8).

Atmospheric sulfur was also sampled by exposing *Sphagnum* moss in a nylon bag to the atmosphere. With this technique, gaseous and particulate pollutants are adsorbed, filtered, and otherwise

An acidic pond at the base of the seacliffs at the Smoking Hills. This pond is frequently fumigated by plumes having a large concentration of sulfur dioxide, and it also receives acidic drainage from its small watershed. Consequently, the pond has a very acidic pH of 1.8, and very high concentrations of sulfate and of soluble, toxic forms of aluminum, manganese, zinc, and other metals. The observer is standing on a slope of roasted shales. At the far upper left are more whitish, roasted shales and an active burn. Beyond the observer is sea ice. (Photo courtesy of B. Freedman.)

accumulated. Moss placed for 14 days near an active burn increased in sulfur concentration from an initial 0.05% S, to 8–21%. The sulfur was so acidic that the nylon bag and *Sphagnum* were partially digested! Along a transect from the cliff edge in a major plume, the sulfur concentration in the moss decreased from 2.5% at the cliff edge, to 1.4% at 80 m, 1.1% at 160 m, 0.39% at 640 m, and 0.08% at 4.8 km (Hutchinson *et al.*, 1978). Sulfate deposition was measured with bucket-type bulk collectors, and was 0.39 mg m^{-2} per 30 days near the cliff edge, compared with 0.041 mg m^{-2} per 30 days at a remote site (Gizyn, 1980). The bulk-collected rainwater was very acidic near the cliff edge (pH 2.6–2.7). The acidity was due to sulfuric acid, since the concentration of equivalents of SO_4^{-2} was virtually equal to that of H^+ (Gizyn, 1980).

Suspended particulates in the atmosphere were also sampled. Sites within a major plume had 120–160 ng S/m^3, compared with <0.3 ng S/m^3 at an unfumigated site. Other elements with an elevated concentration were selenium (130–270 ng/m^3 versus <30), arsenic (13–36 ng/m^3 versus <1), manganese (53–109 ng/m^3 versus 2), and bromine (55–73 ng/m^3 versus 5) (Hutchinson *et al.*, 1978).

Soil and Water Pollution

The most important chemical effects of air pollution at the Smoking Hills have been the acidification of soil and fresh water, and the subsequent solubilization of toxic metals. Surface (0–2 cm) soil in a large fumigated area has a pH as acidic as 2.7–3.2, compared with pH 7.2 at an unfumigated site (Gizyn, 1980). Sulfur concentration is also high in soil at fumigated sites (1.0–1.5%, compared with 0.44%

Table 2.8 Concentration of SO_2 along a transect running inland from the edge of the seacliff at one of the burn sites at the Smoking Hills[a]

Distance from Edge of Cliff (m)	SO₂ Concentration (ppm)	
	1975	1977
20	0.27	0.61
40	0.20	0.53
80	0.17	0.43
160	0.10	0.25
320	0.06	0.16
640	0.02	0.11
1280	0.02	0.06
1920	0.02	0.05
2560	0.01	0.04

[a] SO_2 concentration was calculated from the rate of sulfation, a simple field measurement in which SO_2 reacts with PbO_2 to form $PbSO_4$ at a rate proportional to the concentration of SO_2 [Huey, 1968; American National Standards Institute (ANSI), 1976]. The data are mean values for an 8-day sampling period in 1975, and 14 days in 1977. Modified from Gizyn (1980).

for reference soil). The very acidic condition has caused the leaching of basic cations from surface soil. This effect is especially obvious for calcium, which decreased from 1.8% at a reference site, to 0.5% in fumigated surface soil (Gizyn, 1980).

Some ponds at the Smoking Hills are very acidic, with pH values as low as 1.8 occurring in ponds found at the base of the seacliff (Table 2.9). These ponds receive acidic water that percolates through deposits of shale and ash, where pyrite is actively being oxidized (this is analagous to acid mine drainage, described in Chapter 4). The only reports of similarly acidic pH values in natural water are for volcanic crater lakes in Japan in which pHs as low as 1.0 to 1.4 have been measured (Yoshimura, 1933; Ueno, 1958; Takano, 1987), and surface waters with pH < 2 caused by drainage from coal mines and coal waste disposal areas (Riley, 1960; King *et al.*, 1974).

The tundra ponds at the top of the seacliff at the Smoking Hills do not receive the very acidic drainage that has percolated through pyrite-containing shale and ash. Instead, these water bodies have been acidified by (1) the direct deposition from the atmosphere of sulfuric acid mist, rain, and snow; (2) the dissolution into water of atmospheric SO_2, which is later oxidized to sulfuric acid (Terraglio and Manganelli, 1967; Beilke and Gravenhorst, 1978; Meszaros, 1981); and (3) acidic surface runoff from their small watershed. The lowest pond-water pH at the top of the seacliff is 2.4. The tundra pond waters are either acidic with pH < 4.5, or nonacidic with pH > 6.5, with very few ponds in the weakly buffered pH range of 4.5–6.5. A bimodality of pH distribution in surface waters has also been observed in areas affected by acid mine drainage (King *et al.*, 1974), or by acidic deposition from the atmosphere (Wright and Gjessing, 1976).

Table 2.9 Chemistry of shallow tundra ponds in the vicinity of the Smoking Hills[a]

pH Range	n	Al	Fe	Mn	Zn	Ni	Cd	As	Ca	SO₄
1.8–2.5	4	270	500	61	14	6.3	0.52	0.13	301	8200
1.8–3.5	14	5.5	18	15	0.45	0.21	0.022	0.005	157	890
1.8–4.5	9	1.1	1.2	3.6	0.12	0.04	0.011	0.005	44	156
1.8–5.5	1	<0.6	0.5	2.3	0.03	0.04	0.001	0.004	182	813
1.8–6.5	1	<0.2	0.2	1.8	0.05	0.06	0.012	0.006	249	713
1.8–7.5	4	<0.8	<0.04	0.7	0.08	0.02	0.001	0.004	90	360
1.8–8.5	8	<0.7	0.1	<0.5	0.04	0.004	0.003	0.005	49	106
1.8–9.5	3	<0.2	<0.04	<0.2	<0.03	0.007	<0.001	0.006	31	34
1.8–9.7	2	<0.2	0.2	1.2	5.3	0.01	<0.001	0.005	16	29

[a] The sample size refers to the number of ponds within each pH class. The data are geometric means (mg/l). Modified from Hutchinson *et al.* (1978) and Havas and Hutchinson (1983a).

Other chemical constituents are present in a high concentration in the acidic pond water (Table 2.9). The high concentration of sulfate results from atmospheric deposition and in some cases from acidic drainage from the terrestrial part of the watershed. The high concentrations of other elements such as aluminum result from the acidic dissolution of minerals in pond sediment and watershed soils—this process causes a large concentration of soluble metals in all acidic environments (Chapter 4).

Ecological Effects

Intensely fumigated sites at the Smoking Hills are toxic to most species. The most important factors that affect the terrestrial vegetation are SO_2, acidic soil, and large concentrations of soluble aluminum, manganese, and other metals. Within 40 m of the sea cliff in the major fumigation zone there is no vegetation whatsoever—ecological degradation is complete (Hutchinson et al., 1978; Gizyn, 1980; Freedman et al., 1988b). From about 80 to 320 m there is a very depauperate plant community, with less than 3% cover and few species present. At 80 m the only plant species is the perennial dicot herb Artemisia tilesii; at 320 m this species is accompanied by the grass Arctagrostis latifolia and the lichen Cladonia bellidiflora, with the moss Pohlia nutans present at low frequency. All of these plants are widespread in this region of the Arctic, but they are minor components of the typical, undisturbed vegetation. Therefore, species that are tolerant of pollution stress have replaced plants that are characteristic of unpolluted habitat, such as the mountain avens (Dryas integrifolia), arctic willow (Salix arctica), and at least 70 other plant species.

A pollution-tolerant biota has also developed in the acidic ponds. Even the most acidic pond (pH 1.8) has six species of algae. The most prominent alga is Euglena mutabilis, which can be present as a green benthic edging just below the water surface (Sheath et al., 1982; Havas and Hutchinson, 1983a). Laboratory experiments with E. mutabilis from the Smoking Hills showed that it could survive at pH 1.0, although the optimum pH for its growth was 4.0 (Hutchinson et al., 1981). Other algae in the pH 1.8 pond are Chlamydomonas acidophila,

Euonotia glacialis, Nitzschia communis, N. palea, and Cryptomonas sp. (Sheath et al., 1982). Some of these are part of a widely distributed, acid-tolerant algal flora. Euglena mutabilis has been reported at pH 1.8 in coal mine drainage in England and the United States (Van Dach, 1943; Hargreaves et al., 1975; Hargreaves and Whitten, 1976). Similarly, Chlamydomonas acidophila occurs in pH 1.7 volcanic lakes in Japan (Ueno, 1958), and in pH 1.8 mine drainage in Britain (Hargreaves and Whitten, 1976). In contrast to the acidic ponds at the Smoking Hills, the reference ponds are alkaline (pH 8.1–8.2) and have a rich algal flora comprising at least 90 taxa (Sheath et al., 1982).

Both the acidic and the alkaline pond floras are highly adapted to their respective environments. In a field experiment involving the in situ manipulation of the pH of pond water within Plexiglass enclosures, the live phytoplankton volume collapsed from 1.49 mm^3/l at the natural pH of 2.8, to 0.0 mm^3/l at pH 8 (Sheath et al., 1982). Similarly, the acidification of an alkaline pond water caused cell volume to decrease from 17.4 mm^3/l at pH 8.1, to 0.0 mm^3/l at pH 3.

A few invertebrates are found in acidic ponds above pH 2.8 (Havas and Hutchinson 1982, 1983b). The most prominent zooplankton species is the rotifer Brachionus urceolaris, also reported from a pH 3.0 volcanic pond in Japan (Ueno, 1958). The benthos is dominated by the chrironomid Chironomus riparius. Neither of these species are present in alkaline ponds, which have a much more diverse fauna dominated by acid-sensitive crustaceans such as the water fleas Daphnia middendorffiana and Diaptomus arcticus, the tadpole shrimp Lepidurus arcticus, and the fairy shrimp Branchinecta paludosa.

An important lesson to be learned from the Smoking Hills is that "natural" gaseous pollution can cause an intensity of ecological damage that is as severe as anything caused by anthropogenic pollution. (This does not, of course, in any way justify anthropogenic pollution and its ecological impacts.) Therefore, SO_2 can have a damaging effect on ecosystems irrespective of its source. Because pollution at the Smoking Hills has been present for a very long time, the rate of ecosystem change and adaptation by organisms has probably stabilized. Under a longer-term condition of toxic pollution stress, the tundra ecosystem is characterized by biological simplification, and by disrupted productivity and nutrient cycling. In addition, the pollution-tolerant biota is comprised of species that are not normally present or are rare in the nearby, unpolluted habitat.

Effects of Emissions from the Sudbury Smelters and Other Large Point Sources of SO₂

The Setting

The first discovery of a commercial ore body in the vicinity of Sudbury, Ontario, was in 1883 in a surface bedrock cut, by a worker employed in the construction of Canada's first transcontinental railroad.

The principal metals that are mined and processed at Sudbury are nickel and copper, although iron, cobalt, gold, and silver are also produced, as are the nonmetals sulfur and selenium. The Sudbury area now contains one of the world's largest mining and metal processing complexes, with many shaft mines and an open pit, two ore-crushing mills with associated tailings disposal areas, two smelters, several metal refineries, an iron ore recovery plant, sulfuric acid plants, and a host of secondary and supporting industries. These activities are the primary economic base for a population of more than 100,000 people.

Atmospheric Emissions and Environmental Pollution

From the beginning of the mining–smelting developments to the present, SO₂ and heavy metals have

A view of the O'Donnell roastbed in the vicinity of Sudbury, Ontario, circa 1925. The roastbed was prepared with a bottom layer of locally cut cordwood (foreground), much of which was salvage harvested from nearby fume-killed forests. Above this was placed a layer of sulfide nickel–copper ore (background), using the track-mounted gantry with its continuous-feed apparatus. (Photo courtesy of INCO Archives.)

An aerial view of the O'Donnell roastbed while ore is being oxidized, circa 1925. Once initiated, the roasting would proceed for several months. When the metal-concentrated substrate had cooled sufficiently, it was collected and sent on to a refinery. The ground-level, phytotoxic emissions that resulted from this primitive roasting procedure devastated the surrounding forests and acidified freshwaters. (Photo courtesy of INCO Archives.)

been emitted to the atmosphere in a very large quantity (Costescu, 1974; Freedman and Hutchinson, 1980c). Prior to 1928, smelting was done in open pits using a primitive roasting technique. Large heaps of sulfide ore were placed over locally cut timber. The wood was ignited in order to initiate the oxidation of the metal sulfides. These exothermic reactions soon generated enough heat to cause a self-sustaining oxidation of the ore. The roast beds typically burned for several months, after which the nickel and copper concentrate was collected and taken elsewhere for refining. This roasting process generated choking, ground-level plumes that contained SO_2, acidic mist, and metal particulates. The fumigations devastated the surrounding terrestrial vegetation. The devegetation caused severe erosion of soil from slopes and the exposure of granitic–gneissic bedrock, which was then blackened and pitted by the sulfurous fumes. During this early

phase of smelting by the use of roast beds, an estimated 2.7×10^5 MT/year of SO_2 plus a large but undocumented quantity of metal particulates were emitted at ground level, from as many as 30 roast beds (Holloway, 1917).

The open roast beds were eventually replaced by more efficient smelters, which vented waste gases into the atmosphere through tall smokestacks and thereby alleviated the ground-level pollution (Freedman and Hutchinson, 1980a,c). By 1928 the use of roast beds was forbidden by legislation, and all subsequent roasting was carried out at three smelters located near Sudbury at Copper Cliff, Coniston, and Falconbridge. Operation of the largest smelter at Copper Cliff began in 1929–1930, when most of its pollutants were vented through a 155-m smokestack. Two additional smokestacks were added in 1936, and in 1956 an iron ore recovery plant with a 191-m stack was built. In 1972 the three

A view of the "superstack" at the Copper Cliff smelter near Sudbury. This 381-m-tall smokestack vents pollutants very high into the atmosphere, and it has greatly alleviated the local air pollution since it was commissioned in 1972. The ecological damage that is visible in the foreground is due to presuperstack fumigations. (Photo courtesy of B. Freedman.)

smokestacks at the Copper Cliff smelter were replaced by a single 381-m "superstack," the world's tallest. At various times during the development of the Copper Cliff smelting complex, a degree of pollution abatement was achieved by the installation of flue-dust recovery units, wet scrubbers, electrostatic precipitators, and sulfuric acid plants. These various units remove some of the SO_2 and particulate pollution before the waste gases are vented to the atmosphere.

Two other smelters were constructed in the Sudbury area. The Coniston smelter operated from 1913 to 1972, and had smokestacks of 114 and 122

m. The third smelter has operated since 1928 at Falconbridge, and has smokestacks of 93 and 140 m.

The emission of SO_2 in the Sudbury area peaked around 1965–1970, when the total emission from all three smelters was about 2.7 million metric tons per year. This was equivalent to more than 4% of the global anthropogenic emission of SO_2 (c.f. Table 2.2), and ranked Sudbury as the world's largest point source of SO_2. Emissions have decreased substantially since then, to still-large quantities of $\sim 1.4 \times 10^6$ MT/year over 1973–1977, and 0.4–0.9×10^6 MT/yr over 1980–1984 (Freedman and Hutchinson, 1980a; Chan and Lusis, 1985; D. Yap, Ontario Ministry of the Environment, personal communication). This large decrease in emissions was caused by a downturn in nickel markets and hence decreased production, by interruptions of operations by strikes, by a shift to lower-sulfur ore in order to reduce SO_2 emissions per unit of metal produced, and by other pollution control measures such as the building of a sulfuric acid plant. However, the emissions of SO_2 and other pollutants are still very large (Table 2.10).

Several researchers have computed deposition budgets for the emitted SO_2 and metals. Chan *et al.* (1984a) estimated deposition within 40 km of the superstack between 1978 and 1980. Their study indicates that the tall stack disperses pollutants very effectively; locally, the "dilution solution to pollution" works quite well. This is especially true of gaseous sulfur. Only 1.3% of the total S emission was deposited within 40 km of the superstack, and only 0.2% if background deposition is accounted for in order to isolate the smelter influence (Table 2.11). Similar observations have been made by Muller and Kramer (1977) and Freedman and Hutchinson (1980a). Therefore, more than 99% of the sulfur emitted from this tall stack is exported beyond 40 km; these long-range transported air pollutants contribute to regional acidic precipitation (Chapter 4).

There are several reasons why so little of the emitted sulfur is deposited close to the superstack:

1. The tall stack emits pollutants sufficiently high into the atmosphere that local surface

A closer view of a devastated hillside located about 2 km from the Copper Cliff smelter. The worst of this damage was caused prior to 1929, during the period of roastbed use, and during the early smelter era before tall smokestacks were built (prior to about 1935). However, revegetation was also prevented by frequent fumigations from the smelters, even after relatively tall stacks had been constructed, and by severe soil toxicity caused by acidity and large plant-available concentrations of aluminum, nickel, copper, and other metals. Following the demise of the forest and the salvage clearcutting of dead trees to fuel local roastbeds, there was severe erosion of soil from slopes. This exposed the pinkish gneissic and granitic bedrock, which was then blackened and pitted by reaction with the acidic, sulfurous fumes. (Photo courtesy of B. Freedman.)

Table 2.10 Average annual emissions (MT/year) of pollutants in the Sudbury area between 1973 and 1981[a]

Source of Emission	SO$_2$	H$_2$SO$_4$	Total Particulates	Fe	Cu	Ni	Pb	As
INCO								
381-m "Superstack"	886,000	7270	11,400	990	245	228	184	114
194-m Stack of iron ore recovery plant	55,000	1664	2400	643	171	226	6	4
Low-level sources	12,000	88	600	70	242	31	1	—
Two 45-m stacks of pelletizer plant	—	—	4100	2354	—	—	—	—
Falconbridge—93-m stack	173,000	438	900	98	11	10	13	6
Total	1,126,000	9460	19,400	4155	669	495	204	124

[a]Modified from Chan and Lusis (1985). In addition to the pollutants summarized above, the total emissions of NO$_x$ from these sources were ~3 × 10^3 MT/year, and the total emissions of HCl were ~0.50 × 10^3 MT/year.

Table 2.11 Emission and wet and dry deposition of pollutants within a 40-km radius of the 381-m "superstack" at Copper Cliff, Ontario[a]

	Emission (1)	Total Deposition			Background Deposition			Total Deposition as Percent of Emissions (4/1)	Plume Deposition as Percent of Emissions (4 − 7)/1
		Wet (2)	Dry (3)	W + D (4)	Wet (5)	Dry (6)	W + D (7)		
H^+	449	2762	—	2762	2741	—	2741	615	4.7
SO_2^{-2}	2,334,000	—	16,100	16,100	—	12,700	12,700	0.69	0.15
SO_4^{-2}	49,100	21,300	2250	23,550	19,800	2200	22,000	48	3.2
S	1,183,000	7100	8794	15,894	6600	7080	13,680	1.3	0.19
Fe	1626	415	1344	1759	344	1132	1476	108	17
Ni	798	29.2	30.1	59.3	10.8	12.2	23.0	7.4	4.5
Cu	591	61.2	22.8	84.0	21.5	9.1	30.6	14	9.0
Al	406	313	533	846	292	502	794	208	13
Pb	480	86.7	37.2	123.9	76.5	35.0	111.5	26	2.6

[a]The background deposition does not include any smelter-derived pollutants. Emission and deposition data are in kg/day. Modified from Chan et al. (1984a).

An open community dominated by shrub-sized individuals of *Acer rubrum* and *Betula papyrifera*, located about 6 km from the Copper Cliff smelter. The sparse ground vegetation is dominated by the metal-tolerant moss *Pohlia nutans*. Note that the photo was taken in October, after the autumn leaf-fall. (Photo courtesy of B. Freedman.)

A forest dominated by *Betula papyrifera*, located about 15 km from the Copper Cliff smelter. Conifers are relatively sensitive to sulfur dioxide, and they are notably absent from this stand. Note that the photo was taken after the autumn leaf-fall. (Photo courtesy of B. Freedman.)

impactions are infrequent. Ground-level fumigations only tend to take place on warm, sunny days during May to September when large-scale convective turbulence causes a type of plume behavior termed "looping." Individual plume loops typically impact the surface at 3–10 km from the stack, and they cause a half-hour average SO_2 concentration of 100–300 ppb. At any particular site, these ground-level events occur several times per year (Chan and Lusis, 1985). During ground-level SO_2 episodes, dry deposition is relatively effective (Tang et al., 1987).

2. In the absence of a ground-level fumigation, SO_2 is not efficiently deposited to the surface by precipitation. The solubility of SO_2 in rainwater is rather low, particularly if the solution has a pH less than ~5.5. As a result, precipitation that has passed through the superstack plume is only slightly elevated in SO_2 concentration, and therefore wash-out is not an effective mechanism of deposition for this gas (Millan et al., 1982; Lusis et al., 1983).

3. The deposition of emitted sulfate is more efficient than that of SO_2, averaging 3% overall (Table 2.11), and about 48% during a precipitation event (Chan et al., 1984b). The plume sulfate that is washed out close to the superstack is mainly emitted sulfate. Sulfate produced secondarily in the plume by the oxidation of emitted SO_2 is of minor importance close to the source, since the SO_2 oxidation rate only averages 1% per hour or less (Millan et al., 1982; Chan et al., 1983).

These represent the modern, postsuperstack mechanisms of sulfur deposition. Ground-level fumigations by SO_2 and acidic mists were much more frequent and intense during the era of roast bed use, and to a lesser extent prior to 1972 when smelter emissions were from less tall stacks. It was this earlier emission of pollutants that caused the great ecological damage in the Sudbury area, along with other anthropogenic disturbances such as the harvesting of forests for lumber and to fuel the roast beds, and fires started by prospectors and by railroad engine sparks (Winterhalder, 1978). Although modern emissions are as large or greater than earlier emissions, they are dispersed into the atmosphere so effectively that acute injury to vegetation is now infrequent.

Unfortunately, there are no data from the roast bed era for SO_2 concentration in air. Without question, the ground-level roast bed plumes were extremely phytotoxic, and they devastated the surrounding vegetation. More recently, air-quality data collected prior to the commissioning of the superstack in 1972 showed that the local SO_2 pollution was considerably worse than at present. Dreisinger (1967) summarized 3×10^5 hr of air-quality monitoring between 1954 and 1963. He reported no detectable SO_2 for 87% of the time, >0.25 ppm SO_2 for 2% of the time (6000 hr), and >1 ppm for <0.1% of the time (257 hr). Before the superstack was commissioned, the half-hour SO_2 concentration during fumigation episodes reached 3.0 ppm at a station 7.7 km from Copper Cliff, and 3.6 ppm at a site 26 km away (McGovern and Balsillie, 1972). During 1970, a 5074-km² area surrounding Sudbury had a mean annual SO_2 concentration of <5 ppb, 3100 km² had from 5 to 10 ppb, 1157 km² had 10–20 ppb, and 266 km² had >30 ppb. These concentrations were sufficient to frequently cause acute damage to vegetation close to the SO_2 sources (Dreisinger and McGovern, 1971).

Ecological Damage

Several studies have demonstrated that the severity of the ecological damage decreases geometrically with increasing distance from the Sudbury point sources. The most important cause of the damage has been SO_2, but toxic effects of nickel, copper, aluminum, and acidification are also important (Hutchinson and Whitby, 1974; Whitby and Hutchinson, 1974; see also Chapter 3).

No studies were made of vegetation damage during the period of roast bed use. However, a palynological study of annually laminated sediment in a meromictic lake demonstrated that beginning in 1880–1885 there was a 51% decrease in the influx of *Pinus* pollen, a 14% decrease in *Picea* pollen, and a decrease in total pollen of 67% (Huhn, 1974). At the same time there was a 400% increase in smoke microspherules, indicating an increased frequency of wildfire. These observations indicate that a general decrease in forest cover was coincident with the onset of industrial activity in the Sudbury area. The forests were impacted by several factors, including toxic fumes, a lumber industry, clearcutting to fuel the roastbeds, and fires in fume-killed vegetation and logging debris (Winterhalder, 1978; Freedman and Hutchinson, 1980c). Huhn (1974) also demonstrated an increased influx of tree pollen (especially *Pinus*, *Picea*, and *Betula papyrifera*) beginning in 1930–1935. This coincides with a time of abatement of ground-level pollution, when the use of roast beds was stopped and most SO_2 was vented to the atmosphere through smokestacks.

Other studies of vegetation damage in the Sudbury area documented the cumulative impacts of the roast bed era, plus the effects of the smelters that replaced them. Watson and Richardson (1972) recorded 103 km^2 of "severely barren" land around the Sudbury smelters, and 363 km^2 of terrain with "impoverished" vegetation (notably lacking conifers). Linzon (1971) related the atmospheric SO_2 concentration to damage of white pine (*Pinus strobus*), a very sensitive conifer. He observed a reduced growth rate, high mortality, and severe foliar injury within an area of 1840 km^2, and foliar injury alone over 4100 km^2; almost all of the *P. strobus* observations were made beyond the impoverished zone of Watson and Richardson (1972). Therefore, in the presuperstack period, vegetation was injured over at least 6400 km^2 of terrain around the Sudbury smelters.

The first detailed study of plant communities was that of Gorham and Gordon (1960a), who examined a transect running northeast from the Falconbridge smelter. They observed a sharp decrease in species richness within 6.4 km of the smelter, with the most tolerant taxa being *Sambucus pubens*, *Polygonum cilinode*, *Acer rubrum*, and *Quercus rubra*. *Pinus strobus* and *Vaccinium myr-*

tilloides were only present at sites further than 18 km, indicating their particular sensitivity to pollutants. Needles of *P. strobus* were injured as far as 30 km from the smelter. At about the same time, a qualitatively similar pattern of damage was observed among epiphytic, arboreal lichens in the Sudbury area (LeBlanc and Rao, 1966; LeBlanc et al., 1972).

Freedman and Hutchinson (1980c) documented long-term effects on vegetation by examining 23 stands along a transect running south from the Copper Cliff smelter (Figs. 2.2 and 2.3). There were no forest remnants within 3 km of the smelter. Within this zone hilltops and slopes were denuded of vegetation and severely eroded, and the bedrock was pitted and blackened by reaction with the fumigations. The ground vegetation within this inner zone was depauperate, with <1% plant cover. The most prominent vascular species were the grasses *Deschampsia caespitosa*, *Agrostis hyemalis*, and *A. gigantea*, and the dicots *Polygonum cilinode* and *Acer rubrum*.

From 3 to 8 km, there were forest remnants where water was available and the topography gave some protection from fumigations. Either these sites were beside water bodies and dominated by *Populus tremuloides*, or they were on relatively protected, mesic lower slopes of hills where relatively pollution-tolerant woody angiosperms were dominant (e.g., *Acer rubrum*, *Quercus rubra*, *Populus tremuloides*, *P. grandidentata*, *Betula papyrifera*). The individuals of these "tree" species had a stunted growth form, usually with dead upper branches and much stump sprouting (see also James and Courtin, 1985). Denuded, blackened hilltops were common within this patchily vegetated zone.

Beyond 8 km the forest cover was almost continuous, but the tree species composition and basal area and the understory vegetation were relatively depauperate. Stands beyond about 20 km were little affected by pollution, and the forest was a mixed conifer–hardwood association, typical of the region. In this reference forest, conifers contributed >50–75% of the tree basal area—mostly of *Pinus strobus* and *Picea glauca* (see also Amiro and Courtin, 1981).

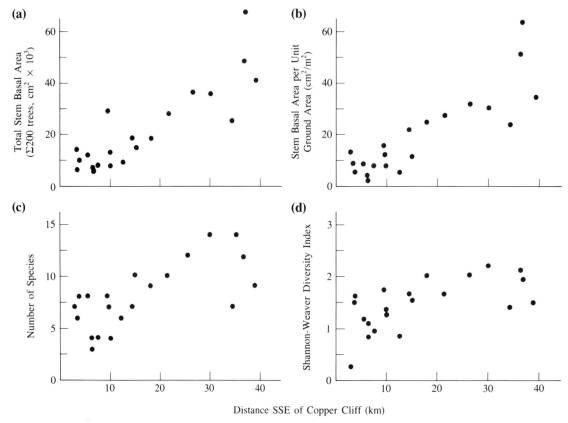

Fig. 2.2 Overstory vegetation characteristics as they vary with distance along a transect south of the Copper Cliff smelter. (a) Total stem basal area; (b) stem basal area per unit ground area; (c) species richness (total number species occuring among 200 randomly chosen trees); (d) Shannon-Weaver diversity ($-\Sigma\, p_i\, \ln p_i$, where p_i is approximated by relative frequency). [Modified from Freedman and Hutchinson (1980c); © 1980 National Research Council of Canada, with permission.]

The pattern of vegetation damage around the Copper Cliff smelter clearly paralleled the pattern of pollution intensity (Freedman and Hutchinson, 1980a). Pollutants that had accumulated in soil (i.e., nickel, copper, and to a lesser extent sulfur) exhibited a well-defined decrease in concentration with increasing distance from the smelter, as did the contemporary patterns of deposition of nickel, copper, iron, and sulfate in precipitation and dustfall, and SO_2 in air (see also Chapter 3).

Various studies have shown that lakes close to the Sudbury smelters have been acidified, and that they have large concentrations of sulfate, nickel, copper, and other elements (Gorham and Gordon, 1960b; Conroy *et al.*, 1975; Whitby *et al.*, 1976). In

parallel with the damage to terrestrial vegetation, the biota of impacted water bodies has responded by a general reduction of standing crop and productivity, a small species richness and diversity, and dominance by pollution-tolerant taxa. These general effects are particularly marked in the planktonic algae, zooplankton, and macrophytes; fish are not present in the most polluted lakes (Gorham and Gordon, 1963; Whitby *et al.*, 1976; Yan, 1979; Yan and Strus, 1980).

Ecological Recovery after the Abatement of Pollution

As mentioned earlier, there has been a great decrease in ground-level atmospheric pollution in the

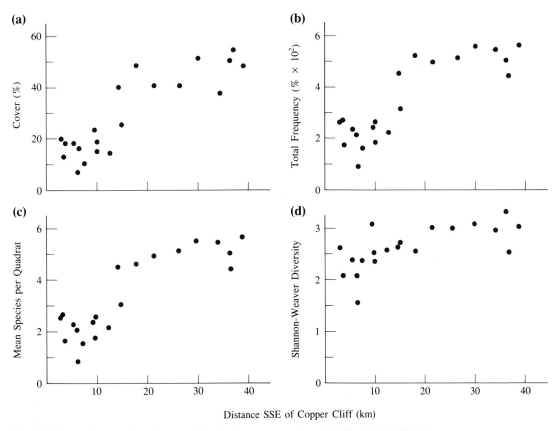

Fig. 2.3 Ground vegetation characteristics in the same stands as in Fig. 2.2. (a) Total cover per quadrat; (b) total frequency (0.5m × 0.5m quadrat); (c) species richness (mean number species per 0.5m × 0.5m quadrat); (d) Shannon-Weaver Diversity ($-\Sigma\ p_i \ln p_i$). [Modified from Freedman and Hutchinson (1980c); © 1980 National Research Council of Canada, with permission.]

Sudbury area since 1972. This has resulted in a dramatic recovery of terrestrial and aquatic ecosystems. In particular, there has been a vigorous spread of a natural grassland in mesic and wet sites where soil from eroded hilltops has accumulated, and along the border of water bodies. The most important grasses are *Deschampsia caespitosa* and *Agrostis gigantea*, both of which have been shown to be ecotypically tolerant of the residual soil pollution by nickel and copper (see Chapter 3). In addition, lakes in the vicinity of Coniston have recovered dramatically since the closing of that smelter. Over the period 1972–1984, the pH of one acidic lake increased from 4.1 to 5.8, while the concentrations of sulfate, copper, nickel, co-

balt, manganese, and zinc decreased by 60–90% (Hutchinson and Havas, 1985). Studies of the biota of the recovering lakes have not yet been made.

Comparison with Other Large Point Sources

The Sudbury example is one of the world's best-documented cases of extensive ecosystem damage caused primarily by the emission of toxic gases from smelters. Other metal smelters have also been studied. Although the ecological damage around them is smaller in scale than that near Sudbury, it is comparable in a qualitative sense. Examples of SO_2 damage include a lead–zinc smelter in Trail, British

Columbia (Katz, 1939), an iron-sintering plant at Wawa, Ontario (Gordon and Gorham, 1963; Scale, 1982), a copper smelter at Ducktown, Tennessee (Smith, 1981), and a copper smelter near Superior, Arizona (Wood and Nash, 1976). Less severe damage to terrestrial ecosystems has been described around sources of gaseous fluoride, particularly aluminum reduction smelters and phosphorus plants (Bunce, 1979, 1984; Thompson *et al.*, 1979; Murray, 1981; Taylor and Basake, 1984).

In all of these studies of point sources, a general observation is that ecological damage becomes progressively less severe as distance from the source increases. This pattern closely tracks the gradient of pollution. If a forested ecosystem is being impacted, then the tree stratum is generally affected first and is "stripped" away. As trees decline, shrubs are affected, and then the ground vegetation. This syndrome of sequential death of the vertical strata of the terrestrial vegetation was described as a "peeling" or "layered vegetation effect" by Gordon and Gorham (1963; see also Woodwell, 1970). This pattern of damage is probably caused by physical effects related to target surfaces, in that the canopy trees are impacted first, while the understory vegetation is initially protected by a boundary-layer effect. Of lesser importance are differences in physiological susceptibility to pollution by the component species of the vegetation strata.

Interestingly, the few studies that have been made in the vicinity of coal-fired power plants in North America and western Europe have only shown minor ecological damage, even though these facilities are large, point-source emitters of SO_2. For example, Jacobson and Showman (1984) surveyed vegetation in an area of the Ohio River valley with four coal-fired power plants. During a 12-year study period they only recorded one episode of acute injury to vegetation—this affected a few susceptible species in a 14×6 km oval around one of the power plants, and caused no longer-term damage. In addition, a survey of lichen distribution showed no effects that were attributable to the power plants. McClenahan (1978) studied seven forest stands along another 50-km stretch of the Ohio River with gradients of pollution by SO_2, Cl,

and F from a coal-fired power plant, three chemical plants, and an aluminum reduction plant. He only found a minor decrease in the species diversity of trees, shrubs, and ground vegetation in the vicinity of the sources. The tree *Acer saccharum* was less dominant in polluted sites, possibly indicating sensitivity to the gases, while tolerant woody plants included the tree *Aesculus octandra* and the shrub *Lindera benzoin*.

Rosenberg *et al.* (1979) studied a potentially more damaging situation in Pennsylvania, in which a coal-burning power plant located in a river valley had a 100-m stack that was barely higher than the surrounding ridgetops. They examined eight stands along an upwind–downwind transect, and found decreases in tree species richness, diversity, and basal area close to the source. However, these effects were minor in scale because they only occurred in stands within 2 km of the source. The maples *Acer saccharum* and *A. rubrum* were relatively prominent close to the power plant, while the conifers *Pinus strobus* and *Tsuga canadensis* were most prominent further away.

Because many lichen taxa are sensitive to toxic gases, these "species" (actually, a fungal–algal symbiosis) have been surveyed in many studies of regional and point-source air pollution. Only minor damage has been reported in studies done around coal-fired power plants. Showman (1975) surveyed 128 sites in the vicinity of a coal-fired power plant in Ohio. He found that the occurrence of terricolous and saxicolous species (growing on soil and rock substrates, respectively) was unaffected by proximity to the power plant. In contrast, distribution maps showed that two corticolous epiphytes (growing on tree bark) were sensitive—*Parmelia caperata* and *P. rudecta* had a void of occurrence of ~90 km² around the power plant. This area corresponded to a mean annual SO_2 concentration of >20 ppb. Other corticolous species were more tolerant. Showman (1981) resurveyed the study area following the abatement of pollution by the construction of taller stacks. He found a recolonization by *P. caperata* into its previous void, and the return to a near-normal distribution 8 years after the decrease in pollution. Interestingly, in the studies of

Jacobson and Showman (1984) of coal-fired power plants along the Ohio River, no effects were observed on the distributions of *Parmelia caperata* or *P. rudecta*, in contrast to their sensitivity in the Showman (1975) study.

In another study, Will-Wolf (1980a,b) examined lichen distribution and injury around a coal-fired power plant in Wisconsin. When she divided her 29 stands into high- and low-SO_2 exposure classes, she found a minor effect at the community level. Two species appeared to be relatively tolerant of SO_2: on average, *Bacidia chloroantha* apparently increased in frequency by 200% in the high-SO_2 sites and decreased by 31% in the low-SO_2 sites, while *B. chlorococca* differed by +180% and +24%, respectively. The most sensitive lichens were *Lepraria membranacea* (-82% at high-SO_2 sites but -7% at low-SO_2 sites) and *Parmelia caperata* (-17% and $+39\%$, respectively). Overall, these observations of lichens represent a measureable but ecologically minor effect on vegetation in the vicinity of coal-fired power plants. In contrast, relatively severe damage to lichens has been widely reported in urban areas with chronic air pollution (LeBlanc and De Sloover, 1970; Hawksworth and Rose, 1976; Johnsen and Sochting, 1980).

Oxidant Air Pollution and Forest Damage in California

The Setting

Ozone is a secondary pollutant that is of regional importance, rather than being concentrated around a point source of emission. A large area of natural and agricultural vegetation can be injured by ozone in a sunny location where there is a large concentration of NO_x and hydrocarbons, coupled with topographic and/or meteorologic factors that restrict air circulation.

A well-documented case of ozone-caused damage has affected conifer forests in southern California, especially along the western slope of the Sierra Nevada and San Bernardino Mountains. The ozone is produced photochemically from NO_x and hydrocarbons emitted in Los Angeles and other population centers to the west, and the polluted air masses are transported eastward to the mountains where vegetation is damaged (P. R. Miller *et al.*, 1972; Williams *et al.*, 1977; Williams, 1980, 1983).

Air Pollution

During the growing season, the diurnal concentration of ozone is largest between $\sim$12:00 and 22:00, with a much smaller concentration at night and in the early morning (Williams *et al.*, 1977). There is an especially large ozone concentration at the top of the atmospheric inversion layer that frequently develops on sunny days. As a result, sampling stations at this altitude measure a large concentration of oxidants (Walker, 1985), and forest damage is most severe at about 600–2600 m (P. R. Miller *et al.*, 1972; Williams *et al.*, 1977). Between 1973 and 1982 a sampling station at San Bernardino had an annual average O_3 concentration of 36 ppb, an average daily maximum of 158 ppb between May and October, 650 hr/year greater than 120 ppb, and 163 hours/year greater than 200 ppb (note that the EPA standard for the maximum 1-hr average concentration of ozone is 120 ppb). In comparison, the state of California overall has an annual average of 25 ppb O_3, an average daily maximum between May and October of 72 ppb, 85 hr/year $>$120 ppb, and 17 hr/year $>$200 ppb (Walker, 1985).

There has been an intensive effort in California to control the emission of oxidant precursors. For example, between 1973 and 1982 the emission standards for new vehicles were reduced by 88% for hydrocarbons and by 50% for NO_x. These efforts resulted in a 33% decrease in the atmospheric CO concentration between 1975 and 1981, a 29% reduction in the emission of transportation-derived hydrocarbons between 1975 and 1982, and a 7% reduction in ambient NO_x between 1976 and 1982. However, there was no measureable decrease in average O_3 concentration during that period. Moreover, only 10 of 65 air quality monitoring sites in California were able to meet the 120-ppb EPA standard between 1972 and 1982 (Walker, 1985).

Damage to Vegetation

Several conifer species are especially susceptible to ozone damage in the impacted forests. The most sensitive species are *Pinus ponderosa*, the dominant tree in most affected stands, and *P. jefferyi*. Within *P. ponderosa*, hypersensitive genotypes comprise 7–10% of the population (Williams, 1983). Less susceptible conifers in these forests include *Abies concolor*, *Pinus lambertiana*, and *Calocedrus decurrens* (Miller, 1973; Williams *et al.*, 1977; McBride *et al.*, 1985). The O_3–smog disease was first recognized in 1963, and it is characterized by an initial chlorotic mottling of foliage, followed by a necrosis that spreads from the leaf tip, premature leaf drop of older foliage, and, ultimately, death of the tree. Diseased trees are also relatively susceptible to secondary damage caused by bark beetle infestation (Williams *et al.*, 1977; Williams, 1980).

The degree of forest damage varies greatly among sites, depending on the relative exposure to ozone. In a survey of eight stands in the Sequoia and Los Padres National Forests, Williams (1980) found that 8–100% of the dominant trees were injured. At severely affected sites 15–25% of the current needles had acute injury symptoms, while 27–75% of 1- and 2- year-old needles were damaged and exhibited premature senescence and leaf drop. In relatively polluted sites there were no leaves older than 4 years, whereas in less contaminated sites foliage could be 6 years and older. Williams (1980) estimated that O_3 caused more than 3% excess mortality per year in these forests of southern California, and a reduction of up to 83% in the volume of wood production.

In spite of vigorous efforts to control the emission of air pollutants from automobiles and other sources in southern California, the ozone damage to *Pinus ponderosa* forests increased between 1975 and 1983. Williams and Williams (1986) resurveyed permanent forest plots in the Sierra Nevada mountains. They found that the incidence of ozone damage to the current foliage of *P. ponderosa* had increased from 15% of trees in 1975 to 24% in 1983, while on 1- year-old needles the incidence had increased from 44% to 61%. Only 21% of the

trees had 2 years of needle retention in 1983, and there was a 14% decrease in annual ring width between the two samplings. These observations indicate that a very substantial decrease of vigor is taking place in *P. ponderosa* populations in the $\frac{1}{2}$ million hectares of forest that are affected by oxidants in the Sierra Nevada.

Since tree species differ markedly in their susceptibility to O_3 damage, it is not surprising that there has been a large change in forest community composition. McBride *et al.* (1985) found that the dominance of *Pinus ponderosa* was decreasing at sites in the San Bernardino Mountains where the O_3 concentration is large, and that this pine was being replaced by relatively O_3-tolerant conifers such as *Abies concolor* and *Calocedrus decurrens*, to form an oxidant-caused disclimax.

Large changes have also taken place in the community of epiphytic lichens in these conifer forests. Sigal and Nash (1983) found that about 50% of the species recorded in a survey done in the early 1900s were now locally extinct, and that species richness was especially depauperate in relatively polluted sites. *Hypogymnia enteromorpha* was markedly injured at polluted sites. Identical symptoms developed when healthy specimens of this species were transplanted into high-oxidant sites.

Ozone Elsewhere

DeBauer *et al.* (1985) described a syndrome of damage and decline in high-elevation (3000–3500 m) pine forests in the mountains around Mexico City. The ozone injury symptoms and ecological effects were similar to those observed with *Pinus ponderosa* in southern California. *Pinus hartwegii* was the most sensitive species near Mexico City, while *Pinus montezumae* var. *lindleyi* was less susceptible.

Well-documented ozone injury to trees has also been described in the northeast and midwest United States. *Pinus strobus* is a sensitive species, but individuals within populations display a varying degree of damage depending on genotypic sensitivity (Berry, 1973; Houston and Stairs, 1973; Houston, 1974). Sensitive trees have a relatively small growth rate (Benoit *et al.*, 1982), but these effects have not been accompanied by a widespread mortality of white pine. During an episode of pollution, the O_3 concentration in this region of the United States is frequently 50–70 ppb, with episodal peaks ranging up to 100–200 ppb (Lioy and Samson, 1979; Benoit et al., 1982). Similarly, ozone events of up to 150 ppb occur in western Europe (Colbeck and Harrison, 1985). These oxidant events are as severe as but less frequent than those that occur in southern California.

2.5

CARBON DIOXIDE, CLIMATE CHANGE, AND PLANT GROWTH

The concentration of CO_2 in the atmosphere has been increasing progressively for at least the last century. The reason for this phenomenon is the release of gaseous CO_2 from two classes of anthropogenic activity: (1) the burning of fossil fuels, during which virtually all of the hydrocarbon carbon is oxidized to CO_2, and (2) the conversion of mature forest, characterized by a large standing crop of organic carbon, to an ecosystem with much smaller biomass, with the difference in carbon content being balanced by CO_2 emitted to the atmosphere by detritivore activity and fire. The change in the concentration of atmospheric CO_2 is very well documented, but there is much uncertainty over the ecological implications of this phenomenon. As is discussed below, the most likely effects are (1) global warming and other climatic changes caused by a "greenhouse effect," and (2) direct and probably beneficial infuences on plant physiology, especially a CO_2 fertilization effect, and a decrease in transpiration.

Anthropogenic Influences on Global Carbon Geochemistry

Emission of CO_2 by the Burning of Fossil Fuels

The rate of utilization of fossil fuels to produce energy for space heating, transportation, and indus-

trial purposes has increased remarkably during the last century (Fig. 2.4). In the decade of 1860–1869, the global emission of CO_2 from the oxidation of fossil fuels averaged 118×10^6 MT CO_2/year, almost entirely from the burning of coal and lignite. By 1973–1982 the emission had increased by a factor of 41, to 4887×10^6 MT/year. Of the total modern emission, 47% is from the burning of liquid hydrocarbons, 37% from solids, and 16% from gases (Gammon et al., 1985). There is much uncertainty in the prediction of the future emission of CO_2 by the combustion of fossil fuels, but the best-guess scenario is for a rate of about 8000–15,000 $\times$ 10^6 MT/year, or as much as three times larger than the current rate (Edmonds and Reilly, 1985).

Emission of CO_2 by the Clearing of Mature Forest

A mature forested ecosystem has a relatively large content of organic carbon, present in its living and dead biomass, and in the organic matter of the forest floor and mineral soil. This quantity of fixed carbon is considerably larger than that in either (1) a relatively young successional forest or (2) any other type of ecosystem, including anthropogenic ones. Therefore, when a unit of land covered by mature forest is cleared for any purpose (usually to provide fuel or fiber, or to create new agricultural land), it is replaced at least temporarily by an ecosystem that contains a much smaller quantity of fixed carbon. The difference in the carbon content of the ecosystems is balanced by a flux of CO_2 to the atmosphere, caused by the progressive oxidation of much of the mature forest carbon by fire and detritivore activity (Woodwell et al., 1978; Cooper, 1983; Houghton et al., 1983, 1985; Detwiler and Hall, 1988).

It is well established that human activities have caused a large reduction in the global coverage of mature forest, and that this effect has resulted in a large flux of CO_2 to the atmosphere. Between 1860 and 1980, the global reduction in area of mature forest plus other anthropogenic landscape changes have caused a net atmospheric CO_2 carbon emission of about 180×10^9 MT (Houghton et al., 1983). This quantity is approximately equal to the emission

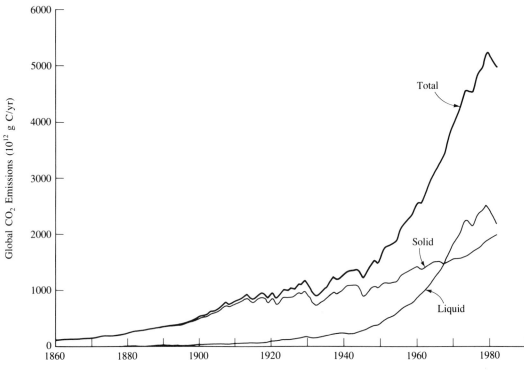

Fig. 2.4 The global emission of CO_2 as a result of the combustion of fossil fuels. Data are from Rotty and Masters (1985).

of CO_2 by the burning of fossil fuels over that same time period (about 160×10^9 MT CO_2 carbon). However, in recent years fossil fuels have been relatively more important. Between 1958 and 1980 the combustion of fossil fuels resulted in a CO_2 carbon flux to the atmosphere of 85.5×10^9 MT, while the net release from terrestrial ecosystems was 57.3×10^9 MT (Houghton *et al.*, 1983).

The global net flux of CO_2 carbon caused by land use practices between 1958 and 1980 is summarized in Table 2.12. The most important net source of carbon over that time period was the clearing of forest followed by use of the land for cultivation or grazing, since this practice causes a longer-term landscape conversion and a decrease in the amount of carbon present on the site. The net effect of forest harvest followed by regeneration is smaller, since the successional regrowth of a cut stand restores much of the carbon removed during the harvest. It is important to note that these global data conceal

large regional differences in the net CO_2 flux from terrestrial ecosystems (Table 2.13). The net flux of carbon from terrestrial ecosystems of North America and Europe was very large in 1920, but it is currently small or negative because the carbon content of forests within those regions is regenerating about as fast as it is being harvested, and because a large area of poor-quality agricultural land has been abandoned and is regenerating to forest. In contrast, the net flux of carbon in low-latitude, tropical regions of Latin America, Africa, and Asia has increased greatly because of deforestation since 1920 (see also Detwiler and Hall, 1988).

Effects on Global Carbon Geochemistry

The burning of fossil fuels and anthropogenic changes in vegetation cover have caused large and approximately equal net fluxes of CO_2 to the atmo-

Table 2.12   Estimated net flux of CO_2 carbon between 1958 and 1980 as a result of (1) changes in carbon storage in the soil and vegetation of terrestrial ecosystems caused by anthropogenic effects on ecosystem cover type and (2) the combustion of fossil fuels[a]

Category	1958–1980 Net Flux (10^9 MT C)	Total Reserves (10^9 MT C)
Terrestrial biota and soil		2500
Forest clearing and cultivation	−29.1	
Decay of wood from forests	−11.1	
Forest clearing and grazing	−5.9	
Harvest of forests	−58.3	
Regrowth of harvested forest	+45.8	
Net abandonment and afforestation	+1.3	
Terrestrial net flux	−57.3	
Burning of fossil fuels	−85.5	10,000

[a]Modified from Houghton *et al.* (1983).

sphere. These anthropogenic effects are summarized in Fig. 2.5, along with an estimate of the magnitude of other global CO_2 fluxes and the size of the largest compartments of carbon. The most important conclusions to be drawn from Fig. 2.5 are as follows.

1. Because the atmospheric CO_2 compartment is relatively small in magnitude, the anthropogenic emissions have caused a large change in its size. Atmospheric CO_2 has increased from about 580×10^9 MT prior to the development of industrial technology,

Table 2.13   Regional net flux of carbon in 1920 and 1980 resulting from forest harvest and other changes in land use[a]

Region	Net Flux (10^6 MT/year)	
	1920	1980
North America	290	10
Europe	100	−50
Latin America	140	800
Tropical Africa	100	700

[a]Modified from Houghton *et al.* (1983).

to 720×10^9 MT in 1982, a 25% increase. This change is reflected by an increase in the atmospheric concentration of CO_2 over that time period, from about 280 ppm to 341 ppm (see below for further details).

2. The net flux of carbon to the atmosphere is believed to have been close to zero prior to the era of an intense anthropogenic effect on the nature of the vegetation on the global landscape. Therefore, global gross primary production (GPP) approximately equalled global ecosystem respiration (ER), and there was no net accumulation of biologically fixed carbon. [Note that ER is the sum of the respiration of all autotrophs (AR) and heterotrophs (HR). The latter plus wildfire ultimately converted all global net primary production (NPP) to CO_2.] This dynamic balance has been upset by the anthropogenic conversion of forested landscapes (ALC in Fig. 2.5) to landscapes with a smaller carbon content. At present, the global flux of carbon from terrestrial ecosystems (ALC + ER) is about 2% larger than global NPP. Because of this effect, the carbon content of the global terrestrial landscape is smaller today than it was prior to

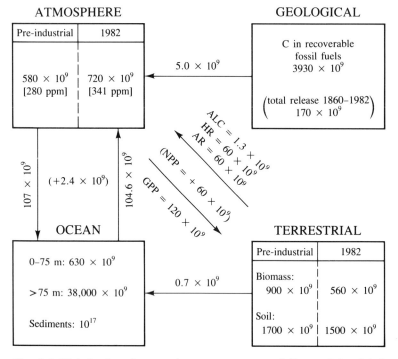

ATMOSPHERE

GEOLOGICAL

OCEAN

TERRESTRIAL

Fig. 2.5 Global values for some key compartments and fluxes of the global carbon cycle. The magnitude of compartments is in units of metric tons of carbon; fluxes are in units of MT C/year. Data are from Blasing (1985) and Solomon *et al.* (1985). See text for explanation of symbols.

the industrial era, by an estimated factor of 38% in the biota, and 12% in soil.

3. The most important sink for anthropogenically emitted CO_2 is the ocean, which currently has a positive net flux of carbon of 2.4×10^9 MT/year. This flux to the oceanic sink is large, but it is smaller than the total rate of anthropogenic emission, and hence the atmospheric CO_2 pool is increasing in magnitude. The ocean has a tremendous capacity to absorb atmospheric CO_2 by the formation of carbonic acid (after Baes *et al.*, 1985):

$$CO_2(g) + H_2O \leftrightarrows H_2CO_3(aq) \quad (11)$$

The carbonic acid dissociates to bicarbonate and carbonate, which respectively comprise about 90% and almost 10% of the total oceanic concentration of inorganic carbon:

$$H_2CO_3 \text{ (aq)} \leftrightarrows H^+ + HCO_3^- \quad (12)$$

$$HCO_3^- \leftrightarrows H^+ + CO_3^{-2} \quad (13)$$

The ionic carbon species can be taken up by biota and used for the construction of shells and other tissues containing calcium carbonate (or calcite), most of which eventually deposits to a sediment sink. The most frequent calcite-forming reaction is

$$Ca^{2+} + 2 HCO_3^- \leftrightarrows CaCO_3 + CO_2 + H_2O \quad (14)$$

The progressive increase in the concentration of atmospheric CO_2 is one of the best-documented longer-term trends in environmental science. The phenomenon is illustrated in Fig. 2.6, for an observatory at high altitude on the island of Hawaii. The annual periodicity of CO_2 concentration is caused by the large net uptake of CO_2 by vegetation during the growing season of temperate and higher lati-

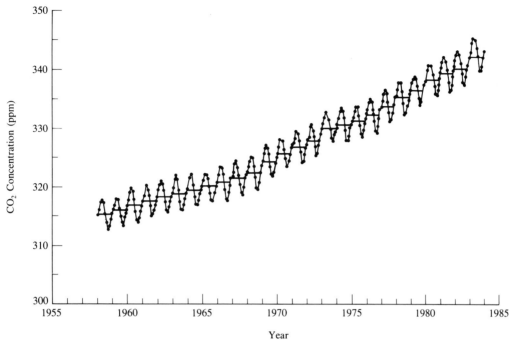

Fig. 2.6 The time series of atmospheric concentration of CO_2 measured at the Mauna Loa Observatory, Hawaii. The dots represent monthly averages, and the horizontal lines represent annual averages. From Solomon *et al.* (1985).

tudes of the northern hemisphere. This effect is illustrated in more detail in Fig. 2.7, which shows that the amplitude of the variation is largest at high latitude in the northern hemisphere (Barrow), and smallest in the southern hemisphere (Samoa and the South Pole). In spite of the large differences in annual variation of atmospheric CO_2 concentration among these four stations, between 1976 and 1982 they all exhibited an average increase in CO_2 of 9.0 ppm (Gammon *et al.*, 1985).

The Greenhouse Effect and Global Climate Change

A potentially important consequence of the increasing concentration of CO_2 in the earth's atmosphere is that of global warming due to changes in a physical process called the "greenhouse effect." The warming phenomenon is caused by interference with the process by which the earth dissipates ab-

sorbed solar radiation, as is briefly described below (after Gates, 1962, 1985; Odum, 1983; Luther and Ellingson, 1985; Solomon *et al.*, 1985).

Almost all of the extraterrestrial radiation incident to the earth has been radiated by its closest star, the sun. This influx is called the solar constant, and it has a magnitude of 2 cal cm^{-2} min^{-1}. About half of the incoming energy occurs as relatively short "visible" radiation within a wavelength band of about 0.4–0.7 μm, and half is comprised of "near-infrared" wavelengths between about 0.7 and 2.0 μm. There is a virtually perfect energetic balance between the quantity of electromagnetic energy that is incident to the earth and the amount that is ultimately dissipated back to outer space. Important components of the earth's energy budget are:

1. An average of about one-third of the incident radiant energy is reflected back to outer space by the atmosphere or by the earth's

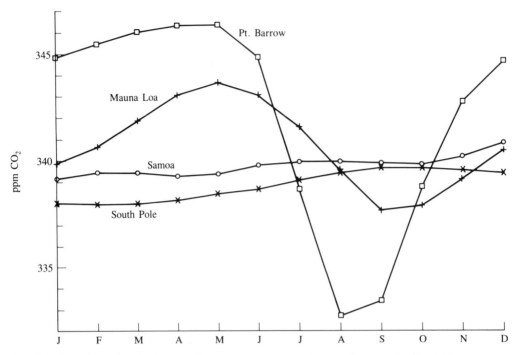

Fig. 2.7 Annual variation in the mean monthly concentration of CO_2 at four remote sites. Data are means over 1980–1983, and were taken from Gammon *et al.* (1985).

surface. This process is related to the earth's albedo, which is most strongly affected by cloud cover, by the quantity of small particulates suspended in the atmosphere, and by the character of the earth's surface, especially the type and amount of plant and water (including ice) cover.

2. About one-third of the incident energy is absorbed by atmospheric gases, converted to thermal kinetic energy (i.e., energy of molecular vibration), and then reradiated to space or to the earth's surface as longer-wavelength (7–14 μm) electromagnetic radiation (see also 3 below).

The remaining one-third of the incident electromagnetic energy from the sun is transformed or dissipated by:

3. Absorption at the earth's surface by various inorganic and living materials, and conversion to thermal energy, which increases the temperature of the absorbing surface. Over the medium term (i.e. days) and longer term (i.e., years) there is little net storage of heat, so that virtually all of the thermal energy is reradiated by the absorbing surface, as electromagnetic radiation of a longer wavelength than that of the incident radiation. The wavelength spectrum of typical reradiated electromagnetic energy from the earth's surface peaks at about 10 μm.

4. Some of the electromagnetic energy that penetrates to the earth's surface causes water to evaporate from plant and inorganic surfaces (evapotranspiration), or it causes ice and snow to melt.

5. A small amount (<1%) of the absorbed radiation drives mass transport processes that redistribute some of the unevenly distributed thermal energy of the earth's surface, that is, wind, water currents, and waves on the surface of water bodies.

6. A small but ecologically critical quantity of incoming energy, averaging about 1% or less, is absorbed by plant pigments and used to drive photosynthesis. As a result of this autotrophic fixation, some of the absorbed solar energy is "temporarily" stored in the interatomic bonds of biochemical molecules.

If the earth's atmosphere were transparent to the reradiated long-wave infrared energy from dissipation mechanisms 2 and 3, then that energy would travel unobstructed to outer space. However, certain so-called radiatively active gases in the atmosphere absorb within this infrared wavelength band, and they thereby slow the rate of radiative cooling of the earth. The most important of these atmospheric constituents are carbon dioxide and water, but the trace gases methane, nitrous oxide, ozone, and chlorofluorocarbons may also be important (Marland and Rotty, 1985; Ramanathan, 1988). For example, each molecule of methane is about 20 times as effective as carbon dioxide at absorbing longwave infrared radiation, while each molecule of the chlorofluorocarbon $CFCl_3$ is about 10,000 times as effective (Burnaby, 1988; Blake and Rowland, 1988). As was noted previously for CO_2, the atmospheric concentrations of various of these radiatively-active gases have increased significantly since pre-industrial times. Prior to 1850 the concentration of CO_2 in the atmosphere was about 280 ppm while in 1985 it was 345 ppm. Over the same time period CH_4 increased from 0.7 ppm to 1.7 ppm; N_2O from 0.285 ppm to 0.304 ppm; $CFCl_3$ from essentially zero to 0.22 ppb; CF_2Cl_2 from zero to 0.38 ppb; and CCl_4 from zero to 0.12 ppb (Ramanathan, 1988). For most of these gases, the rate of increase of their concentration in the atmosphere has been especially great since about 1950 (Blake and Rowland, 1988; Ramanathan, 1988; Rowland, 1988).

When these various substances absorb longwave infrared radiation, they develop a larger content of thermal energy, which is then dissipated by another reradiation (again, of a longer wavelength than that of the electromagnetic energy that was absorbed). Because some of the secondarily reradiated energy is directed back to the earth's surface, the net effect of the radiatively active gases is to slow the rate of cooling of the planet. This process is called the "greenhouse effect," because its physical mechanism is similar to the one by which a glass-enclosed space is heated, that is, the encasing glass and humid atmosphere of a greenhouse are transparent to incoming solar radiation, but they absorb much of the outgoing long-wave infrared radiation, and slow down the rate of radiative cooling of the interior.

The physical mechanism of the greenhouse effect is conceptually simple, and it is very well known that the concentration of CO_2 in the earth's atmosphere is increasing markedly. However, it has been difficult to conclusively demonstrate a warming trend of the earth's surface that can be linked to this increase of CO_2. It appears that since the initiation of detailed instrumental recordings of surface air temperatures around 1880, the four warmest years have all occurred in the 1980s (namely, 1980, 1981, 1983, and 1987), with 1987 averaging about 0.8°C warmer than the average for the decade of the 1880s (Hansen and Lebedeff, 1987, 1988). Overall, it appears that there has been a net increase in the earth's surface air temperature of about 0.5°C since about 1850 (Blasing, 1985; Jones *et al.*, 1986). However, the empirical temperature records on which these conclusions are based suffer from several problems: (1) many historical data are less accurate than contemporary records; (2) weather stations tend to be located in or near urban areas, so their data can be influenced by a "heat island" effect; (3) air temperature is extremely variable on both temporal and spatial scales (that is, there is a very low ratio of signal : noise), and this can make it difficult to interpret longer-term trend analyses of temperature data; and (4) global climate can change for many reasons other than a "greenhouse" response to elevated CO_2 concentration (Harrington, 1987; Kerr, 1988a).

Because of the "problems" associated with the measurement of climatic change using empirical data from the real world, sophisticated computer models have been used to predict the potential effects of changes in atmospheric CO_2. The most complex simulations are the so-called "three-dimensional general circulation models," which must be run on high-speed supercomputers. The GCM models attempt to simulate many of the complex mass-transport and other processes that are involved in atmospheric circulation, and the interaction of these with other variables that contribute to climate (NRC, 1982, 1983; Blasing, 1985; Mac-Cracken and Luther, 1985; Bolin et al., 1986). In order to perform simulation "experiments" with a GCM model, particular components are parameterized to take into account the likely physical effect of a particular increase of CO_2 concentration in the atmosphere. Many simulation experiments have been run, using a variety of GCM models, and the results vary according to the specifics of the experiment. However, the concensus of results of experiments involving a widely accepted (but not necessarily correct) CO_2 scenario (i.e., an eventual doubling of CO_2 from the present concentration of about 345 ppm) is for an increase in average global surface temperature of about 1.3–3.9°C. The climatic effect is predicted to be relatively large in polar regions, where the temperature increase could be about two to three times larger than in the tropics (NRC, 1983; Blasing, 1985).

Possible Ecological Consequences of an Anthropogenic Increase in Atmospheric CO_2

Effects of Climatic Change

The direct effects of climate change caused by an increased concentration of CO_2 would likely be restricted to plants. However, animals would be affected secondarily by any large changes in their habitat.

The anticipated increase in air temperature would probably be of little direct consequence to plants. However, if the temperature change were to cause a large change in the amount, spatial distribution, or seasonality of precipitation, then there would be a large ecological effect. There is much

uncertainty about the specifics of the potential changes in global rainfall, and therefore the effects on vegetation are uncertain. However, we can reasonably expect that any large changes in precipitation would cause a fundamental restructuring of the vegetation mosaic on the landscape (Blasing, 1985; Solomon and West, 1985; Bolin et al., 1986; Harrington, 1987).

For example, in an area where the climate becomes drier, there could be a decrease in the extent of closed canopy forest, and a concomitant expansion of the area of savannah or prairie. This sort of landscape change is believed to have taken place in the neotropics during the Pleistocene glaciations (Haffer, 1969, 1982; Prance, 1982; see also Chapter 10). At that time, the relatively dry climate may have caused a contraction of the presently continuous tropical rainforest into a number of relatively small and isolated refugia. It is believed that these forest remnants were interbedded in a landscape dominated by savannah and grassland. Obviously, such a great restructuring of the vegetational character of the tropical landscape would have a tremendous effect on the habitat of the multitude of rare and endangered species that occur in moist tropical forest (Peters and Darling, 1985; see also Chapter 10).

There would also be a profound change in the ability of the land to support certain crop plants. This is especially true of the great expanses of land that are tilled in regions of the globe that are marginal from a rainfall perspective, and that are susceptible to drought and to long-term degradation by desertification (Sheridan, 1981; Dekker et al., 1985; WRI, 1986; Table 2.14). For example, very important agricultural crops, particularly wheat (*Triticum aestivum*), are grown in areas of the interior of western North America that were formerly shortgrass prairie or semidesert. An estimated 40% of this 400×10^6 ha semiarid region has already been desertified as a result of anthropogenic activities (Table 2.14), and crop-threatening drought occurs sporadically over the entire area. This critical climatic limitation can be overcome by irrigation of the land, but there is a shortage of water for this purpose, and irrigation can cause secondary

Table 2.14 Extent of desertification in various geographic areas[a]

Region	Total Area of Productive Dry Land (10^6 ha)	Percent Desertified
Sudano–Sahelian Africa	473	88
Southern Africa	304	80
Mediterranean Africa	101	83
Western Asia	142	82
Southern Asia	359	70
U.S.S.R. in Asia	298	55
China and Mongolia	315	69
Australia	491	23
Mediterranean Europe	76	39
South America and Mexico	293	71
North America	405	40

[a]Modified from World Resources Institute (WRI) (1986).

environmental problems such as salinization (Sheridan, 1981).

Direct Effects of CO_2

In addition to the indirect climatic effects discussed above, it is likely that an increased concentration of atmospheric CO_2 would have direct effects on plants (Acock and Allen, 1985; Strain and Cure, 1985). The most important of these would probably be a stimulation of photosynthesis caused by CO_2 fertilization. This effect is pronounced under the optimized condition of laboratory studies, in which there is an adequate supply of other important nutrients that potentially limit plant growth (e.g., nitrogen, phosphorus, potassium), and where there is a minimal climatic limitation, particularly with respect to the availability of water. However, under relatively limiting conditions in the field, the CO_2 fertilization response is usually considerably smaller or absent. Experimental studies in the laboratory, and to a lesser extent in the field, have shown a

Table 2.15 Effect of CO_2 acclimation, leaf area index, and experimental CO_2 enrichment on CO_2 exchange, transpiration, and water-use efficiency of soybeans[a]

CO_2 Acclimation Treatment[b] (ppm)	Experimental CO_2 Exposure (ppm)	Total Daily CO_2 Exchange (mol CO_2/m²)	Total Daily Transpiration (mol H_2O/m²)	Water-Use Efficiency (10^{-3} mol CO_2/mol H_2O)
800a	800	1.64	346	4.73
	330	1.07	494	2.17
800b	800	1.79	380	4.71
	330	0.95	495	1.92
330a	800	1.84	294	6.26
	330	0.95	364	2.61
330b	800	1.72	282	6.10
	330	0.79	339	2.62

[a]Modified from Jones *et al.* (1985).
[b]Note that 800a,b had an average leaf area index (LAI) of 6.0, while 330a,b had an LAI of 3.3.

marked CO_2 fertilization effect on the productivity of many agricultural crop species, and of some native plants as well (Acock and Allen, 1985; Bazzaz *et al.*, 1985; Cure, 1985; Oechel and Strain, 1985). In fact, it is a frequent operational practice to enrich the atmosphere of commercial greenhouses with CO_2, in order to increase the productivity of such crops as cucumbers and tomatoes (*Cucumis sativus* and *Lycopersicum esculentum*, respectively).

Another important influence of increased CO_2 on plants is that of decreased transpiration (Acock and Allen, 1985; Strain and Cure, 1985). This effect takes place because stomatal aperture is partly under the control of the atmospheric concentration of CO_2, so that at a relatively large CO_2 concentration the stomata tend to close somewhat or entirely. A decreased loss of water from vegetation, coupled with increased productivity due to CO_2 fertilization, would lead to an increase in the so-called "water-use efficiency" of the ecosystem, that is, the amount of water evapotranspired per unit of biomass produced. In general, this effect would be looked upon favorably in agriculture, and it might also be beneficial in natural ecosystems.

The CO_2 fertilization and the decreased transpiration effects are illustrated in Table 2.15, for soybean (*Glycine max*) grown under laboratory conditions. The data show that the experimental enrichment of the atmosphere with CO_2 stimulated productivity, decreased transpiration, and increased water-use efficiency. However, this effect was somewhat smaller if the plants had previously been acclimated to 800 ppm of CO_2, and if they grew in a relatively competitive canopy with a large leaf area index.

It is impossible to accurately predict the ecological consequences of the well-documented increase of CO_2 and other radiatively active gases in the earth's atmosphere. However, there appears to be a general concensus among scientists that there would be a major readjustment of ecosystem boundaries in response to changes in rainfall and other climatic effects, and there could be an increase in the amount of desertification in currently semiarid regions. These ecological changes would probably be of much greater importance than the beneficial effects on plant physiology of CO_2 fertilization and decreased transpiration.

3

_____ **TOXIC**

_____ **ELEMENTS**

3.1
INTRODUCTION

Included among the toxic elements are heavy metals such as silver (Ag), cadmium (Cd), chromium (Cr), cobalt (Co), copper (Cu), iron (Fe), mercury (Hg), molybdenum (Mo), nickel (Ni), lead (Pb), tin (Sn), and zinc (Zn), as well as lighter elements such as aluminum (Al), arsenic (As), and selenium (Se). All of these substances are ubiquitous in at least a trace concentration in the environment. In addition, some of them are required by plants and/or animals as essential micronutrients, such as Cu, Fe, Mo, Zn, and possibly Al, Ni, and Se.

Under certain environmental conditions, these elements may accumulate to a toxic concentration, and cause ecological damage. Some instances of this elemental "pollution" are natural in origin. Often this involves a surface exposure of minerals with a high concentration of particular toxic elements, resulting in the contamination of soil, biota, and water. These natural situations can have an intensity of pollution by toxic elements that can match or exceed that caused by anthropogenic sources.

However, natural pollution is usually relatively local in extent. It was not until the modern era of industrialization that there was widespread environ-

mental contamination by toxic elements (and by other types of pollutants). This phenomenon can be illustrated by the increasing lead pollution of forests in a large area of the northeastern United States and elsewhere. The most likely source of the regional lead pollution is emissions from automobiles that use leaded gasoline. In addition, the anthropogenic emission of toxic elements from a point source can cause severe, but relatively localized, environmental contamination. For example, the contamination by toxic elements in the vicinity of the large smelters at Sudbury has damaged terrestrial and aquatic ecosystems.

There are instances where human health has been affected by toxic elements. Historians have speculated that the decline of the Roman Empire may have been partly caused by a decrease in the mental skills of the ruling class, as a result of lead intoxication. The Romans stored wine in pottery that was lined with lead, some of which was leached by the acidic beverage and subsequently ingested. In the nineteenth century, people working in the felt-hat industry in Britain frequently suffered neurological damage as a result of their exposure to mercury compounds used as a finish to top hats—hence Lewis Carroll's phrase, "Mad as a hatter." A more recent example of mercury poisoning is the death in

Table 3.1 Typical background concentrations (ppm = mg/kg, d.w.) of toxic elements in selected environmental compartments[a]

	Rocks					Soil
	Granite	Basalt	Shale	Limestone	Sandstone	
Ag	0.04	0.1	0.07	0.12	0.25	0.05
Al	77,000	87,600	88,000	9000	43,000	71,000
As	1.5	1.5	13	1	1	6
Cd	0.09	0.13	0.22	0.028	0.05	0.35
Co	1	35	19	0.1	0.3	8
Cr	4	90	90	11	35	70
Cu	13	90	39	5.5	30	30
F	1400	510	800	220	180	200
Fe	27,000	56,000	48,000	17,000	29,000	40,000
Hg	0.08	0.012	0.012	0.18	0.29	0.06
Mn	400	1500	850	620	460	1000
Mo	2	1	2.6	0.16	0.2	1.2
Ni	0.5	150	68	7	9	50
Pb	24	3	23	5.7	10	35
Se	0.05	0.05	0.5	0.03	0.01	0.4
Sn	3.5	1	6	0.5	0.5	4
U	4.4	0.43	3.7	2.2	0.45	2
V	72	250	130	45	20	90
Zn	52	100	120	20	30	90

(continued)

the 1960s of hundreds of people in Iraq, Iran, India, Pakistan, and elsewhere, caused by eating seed grain that had been treated with a mercury-based fungicide. The poisonous grain was intended for planting, not eating. Although the toxicity of the treated seed had been clearly labeled on the bags, the victims were illiterate, or they did not understand the implications of the message. Mercury intoxication has also caused deaths of humans in other situations. At Minamata, Japan, hundreds died and many more were acutely poisoned as a result of eating mercury-contaminated fish. In that situation, an acetaldehyde factory had dumped relatively nontoxic elemental mercury into Minamata Bay. However, microbes in anaerobic sediment converted the elemental mercury into methylmercury. This highly toxic and bioavailable compound of mercury entered the aquatic food chain, and caused a wide-scale poisoning of fish-eating birds, cats, and humans. Other toxic elements that have poisoned humans after being assimilated from food or from

some other environmental exposure include arsenic, cadmium, lead, nickel, and vanadium.

In this chapter several cases will be described of natural and anthropogenic contamination of the environment with toxic elements, with the resulting ecological consequences.

3.2
BACKGROUND CONCENTRATION IN THE ENVIRONMENT

The background concentration of toxic elements in selected compartments of the environment is described in Table 3.1. In all cases, the concentration in soil and rock is much larger than in water, and it is generally larger than in the biota. However, it is important to note that since the chemical form of toxic elements dissolved in water is generally relatively available to biota, even a seemingly small

Table 3.1   (*Continued*)

| Seawater | Fresh Water | Terrestrial Plants | Mammals | | Marine Fish |
			Muscle	Bone	
0.00004	0.0003	0.01–0.8	0.009–0.28	0.01–0.44	0.04–0.1
0.002	0.3	90–530	0.7–28	4–27	20
0.0037	0.0005	0.2–7	0.007–0.09	0.08–1.6	0.2–10
0.0001	0.0001	0.1–2.4	0.1–3.2	1.8	0.1–3
0.00002	0.0002	0.005–1	0.005–1	0.01–0.04	0.006–0.05
0.0003	0.001	0.03–10	<0.002–0.84	0.1–33	0.03–2
0.0003	0.003	5–15	10	1–26	0.7–15
1.3	0.1	0.02–24	0.05	2000–12,000	1400
0.002	0.5	70–700	180	3–380	9–98
0.0003	0.0001	0.005–0.02	0.02–0.7	0.45	0.4
0.0002	0.008	20–700	0.2–2.3	0.2–14	0.3–4.6
0.01	0.0005	0.06–3	0.02–0.07	<0.7	1
0.00058	0.0005	1–5	1.2	<0.7	0.1–4
0.00003	0.003	1–13	0.2–3.3	3.6–30	0.001–15
0.0002	0.0002	0.03	0.4–1.9	1–9	0.2
0.000004	0.000009	0.2–2	0.01–2	1.4	—
0.0032	0.0004	0.005–0.04	0.001–0.003	0.0002–0.07	0.04–0.08
0.0025	0.0005	0.001–0.5	0.002–0.02	0.003–0.03	0.3
0.005	0.015	20–400	240	75–170	9–80

[a]Modified from Bowen (1979).

aqueous concentration may exert a powerful toxic effect. In contrast, the relatively large concentration in soil or rock is largely insoluble. Hence, in these solid substrates the availability of the toxic element to plants and animals is much less than is implied from the "total" concentration, and there may not be a toxic effect.

Some elements are consistently present in a trace concentration in the environment, for example, cadmium, chromium, cobalt, copper, lead, mercury, molybdenum, nickel, selenium, silver, tin, and uranium (Table 3.1). However, all of these elements are potentially very toxic, and they can affect biota at a water-soluble concentration of less than 1 ppm. In contrast, other elements can be present in a very large concentration in some compartments of the biosphere—most notable are Al and Fe, both of which are important rock and soil constituents. Aluminum averages 8.2% of the mass of the earth's crust, in which it is the third most abundant element after oxygen (47%) and silicon (28%), while iron

averages 3–4% (Hausenbuiller, 1978; Bowen, 1979). However, as noted above, almost all of the Al, Fe, and other potentially toxic elements in soil and minerals is present in an insoluble form that is not readily taken up by the biota, and is considered to be generally "unavailable."

For example, aluminum is present in the soil in many chemical forms, most notably (1) tetra- and octahedral crystals of primary aluminosilicate minerals; (2) amorphous and crystalline clays and sesquioxides; (3) aluminum phosphates; and (4) ions bound to exchange sites on organic matter and clay surfaces, or free in the soil solution. In general, the amount of aluminum that is freely dissolved in the soil water is strongly influenced by acidity, with a much larger soluble concentration being present in an acidic solution. This characteristic of enhanced solubility in an acidic solution holds generally for most of the toxic elements, especially the metals. Moreover, the particular ionic species of Al is strongly dependent on solution pH. For example, ionic Al^{3+} is only present in very acidic solutions with pH < 4.5, while the predominant ionic species in mildly acidic solutions with pH 5.0–7.0 are $AlOH^{2+}$ and $Al(OH)_2^+$, and $Al(OH)_3$ and $Al(OH)_4^-$ are present in mildly to strongly alkaline conditions (Hausenbuiller, 1978). It is principally the ionic forms of Al that are toxic to biota, largely because the other forms are essentially insoluble in water. Therefore, consideration of the "total" quantity of Al in soil or in some other environmental compartment (usually determined after solubilization of the sample using a vigorous, heat-assisted, strong-acid digest) gives little direct information about biological availability or about potential toxicity [availability is usually measured in some sort of aqueous extract; see Black *et al.* (1965) or Allen *et al.* (1974) for a discussion of this topic]. In general, the "available" quantity of a toxic element in soil is a small part of the "total" quantity—usually $<10\%$ and frequently $<1\%$.

3.3
TOXICITY

As in most toxicological situations, the dose received by a target organism is not only a function of the concentration of the poison in the environment; it is also a function of the period of exposure. Therefore, in certain situations a long-term exposure to a small available concentration of some elements may cause a toxic effect. Often this takes place because of a progressive bioaccumulation of the element, until a toxic dose is reached.

The mechanism of toxicity is frequently damage to an enzyme system. This occurs when metal ions bind to the enzyme and cause a change in its three-dimensional configuration, with a resulting change or loss of its specific catalytic function. Another frequent toxic mechanism is damage to DNA via metal binding, resulting in genetic damage by a disruption of transcription, by an inability to produce specific proteins (especially enzymes), or by some other toxic effect. Symptoms of damage to biota can include an abnormal pattern or amount of growth and development, disease, and death.

Particular elements differ greatly in toxicity. Toxicity is also influenced by a number of other factors, including the following:

1. The chemical form of the toxic element. This is important because of its influence on aqueous solubility and hence on biological availability, and because of the inherently different toxicities of various ionic and molecular species. For example, in the ionic series previously described for aluminum, the most toxic species are Al^{3+} and $AlOH^{2+}$ (Havas, 1986).
2. The chemical environment, which can reduce or exacerbate toxicity. For example, the cooccurrence of calcium at a high concentration may decrease the toxicity of many toxic metals. In addition, (a) soil organic matter and clay can bind and thus partially immobilize ionic metals, (b) acidity frequently increases solubility and therefore the toxicity of metals, and (c) the cooccurrence of two or more toxic elements may cause a more-than-additive effect, that is, they are synergistic.
3. Differences in susceptibility among individuals and populations, which may be genetically based (see p. 69).

These various influences on elemental toxicity will not be described in detail. For an in-depth treatment see Goodman *et al.* (1973), Foy *et al.* (1978), Bowen (1979), Lepp (1981), or Freedman and Hutchinson (1986).

3.4
NATURALLY OCCURRING CONTAMINATION AND ITS ECOLOGICAL EFFECTS

Sources and Intensity of Natural Pollution

Natural emissions of toxic elements to the atmosphere can take place by volcanic outputs, and by vapor-phase outgassing of relatively volatile ele-

Table 3.2 Global emission of selected elements[a]

Element	Emissions (10^8 g/year)	
	Natural	**Anthropogenic**
Sb	9.8	380
As	28	780
Cd	2.9	55
Cr	580	940
Co	70	44
Cu	190	2600
Pb	59	20,000
Mn	6100	3200
Hg	0.40	110
Mo	11	510
Ni	280	980
Se	4.1	140
Ag	0.6	50
Sn	52	430
V	650	2100
Zn	360	8400

[a]Modified from Galloway *et al.* (1982).

ments such as As, Hg, and Se. The entrainment of contaminated soil dust may also be important. The natural rate of emission of various elements is summarized in Table 3.2, together with an estimate of their total anthropogenic emission. These two source categories are similar in magnitude (i.e., within a factor of 4) for Cr, Co, Mn, Ni, and V, while anthropogenic sources exceed natural ones by a factor of more than 10 for the other elements.

Surface and near-surface mineralizations containing toxic elements may give rise to a localized area of contamination. The occurrence of metal anomalies in soil, vegetation, and surface water and sediment overlying an ore body, sometimes associated with the presence of indicator plant taxa, has given rise to an exploration technique known as biogeochemical prospecting (Cannon, 1960; Allan, 1971; Wolfe, 1971; Fortescue, 1980).

These natural elemental contaminations can be comparable in concentration to the worst cases of anthropogenic pollution. For example, Boyle (1971) and Stone and Timmer (1975) reported a copper concentration as high as 10% in surface peat that was filtering Cu-rich spring water emerging

into a marsh in New Brunswick. Forgeron (1971) described surface soil with up to 3% Pb + Zn at a site on Baffin Island, Canada. Warren *et al.* (1966) reported a mercury concentration of 1–10 ppm in soil overlying a cinnabar (HgS) deposit in British Columbia, compared with <0.1 ppm in reference soil.

Predictably, the severe contamination of soil at some metalliferous sites has led to the occurrence of a very large metal concentration in vegetation. This phenomenon is especially pronounced in genetically adapted hyperaccumulator species, which are often endemic to metalliferous sites. For example, a nickel concentration as large as 10% has been found in *Alyssum bertolanii* and *A. murale* in the Soviet Union (Mishra and Kar, 1974, citing Malyuga, 1964). A nickel concentration of up to 25% occurs in the blue-colored latex of *Sebertia acuminata* from the Pacific island of New Caledonia (Jaffre *et al.*, 1976). In a chemical survey of 181 herbarium specimens of 70 *Rhinorea* species, Brooks *et al.* (1977) found specimens of *R. bengalensis* with up to 1.8% nickel, and *R. javanica* with up to 2200 ppm.

Similarly, copper indicators and copper endemics have been described. The labiate *Becium homblei* was important in the discovery of copper deposits in Zambia and Zimbabwe, where its presence is confined to soil with >1000 ppm Cu (Cannon, 1960). Reilly (1967) and Reilly and Reilly (1973) described *B. homblei* as a cuprophile, tolerant of >70,000 ppm copper in soil. Copper mosses were originally described from Scandanavia, and later from Alaska, the Soviet Union, and elsewhere (Persson, 1948; Shacklette, 1965). These bryophytes are specific to mineral substrates that have a large concentration of copper, and they have been used by prospectors as an indicator of surface mineralizations of this metal.

In another case, Reay (1972) described arsenic enrichment of aquatic macrophytes growing in New Zealand surface waters that were contaminated by geothermal springs. An arsenic concentration of up to 970 ppm was found in *Ceratophyllum demersum*, compared with 1.4 ppm in reference plants.

A chaparral community on serpentine-influenced soil in northern California. The codominant shrubs are *Quercus durata* and *Ceanothus jepsonii*, both of which are endemics tolerant of the stresses associated with serpentine minerals, whose distributions are restricted to such sites. On more normal soils without serpentine minerals, the local landscape is typically dominated by a conifer forest of pines (*Pinus ponderosa* and *P. lambertiana*) and fir (*Abies concolor*). (Photo courtesy of A. R. Kruckeberg.)

Serpentine Soil and Vegetation

Soil derived in part from serpentine minerals is a well-known example of natural metal contamination. This soil type is derived from ultramafic minerals, particularly olivine, pyroxenes, horneblende, and their secondary products such as serpentine minerals, fibrous amphiboles, and talc (Aumento, 1970; Douglas, 1970; Proctor and Woodell, 1975; Brooks, 1987). The serpentine minerals are a complex of layered silicates with the general formula $Mg_6Si_4O_{10}(OH)_8$. However, there is much substitution for magnesium by other elements, especially nickel, cobalt, and iron. Substantial amounts of nickeliferous pyrite and chromite may also be present in serpentine-derived soil. In many cases, serpentine-derived soil is toxic and inhibitory to the growth and survival of nonadapted plants. The toxicity is largely due to the large concentrations of available nickel, chromium, and cobalt, and the small concentrations of the nutrients calcium, potassium, phosphorus, nitrogen, and molybdenum.

A very sparsely vegetated serpentine barrens in the inner south Coast Range of central California. The widely spaced trees are digger pine (*Pinus sabiniana*). In the absence of soil influenced by serpentine minerals, the vegetation at this location would be an oak-dominated woodland. (Photo courtesy of A. R. Kruckeberg.)

Magnesium and iron are also present in large concentrations, but they are not thought to be directly toxic to serpentine vegetation (Whittaker, 1954; Proctor and Woodell, 1975).

Typical serpentine soils contain about 2000 ppm nickel, but as much as 25,000 ppm can be present at some sites (Proctor and Woodell, 1975; Brooks, 1987). Peterson (1975) described soil from a location in Zimbabwe with up to 125,000 ppm of chromium and 5500 ppm of nickel. As might be expected, a fraction of the toxic elements in serpentine soil is available for plant uptake, and a range of nickel concentration of 600–10,000 ppm in foliage ash has been observed, compared with 20–70 ppm in reference samples (Proctor and Woodell, 1975). Reeves *et al.* (1981) reported up to 16,400 ppm by

dry weight (d.w.) Ni in the tissue of the hyperaccumulator *Streptanthus polygaloides*, a cruciferous serpentine endemic from California. Jaffre *et al.* (1976) reported up to 11–25% Ni in the blue-colored latex of the hyperaccumulator *Sebertia acuminata* in the South Pacific island of New Caledonia. The poor crop growth on moderately serpentinized soil in Scotland, Zimbabwe and elsewhere has been ascribed to nickel and chromium toxicity (Hunter and Vergnano, 1952; Vergnano and Hunter, 1952; Soane and Saunder, 1959).

The natural vegetation that develops on serpentine soil is often strikingly low-growing and stunted (Brooks, 1987). In northern California serpentine the typical vegetation is an open chaparral, characterized by endemic sclerophyllous shrubs (how-

The perennial herb *Castilleja neglecta*, an example of a serpentine endemic. This species has a highly restricted distribution in the Bay Area of California, and it only grows on sites where the soil is strongly influenced by serpentine minerals. (Photo courtesy of A. R. Kruckeberg.)

ever, the range of vegetation types on Californian serpentine sites is from sparse barrens to open forest) (Kruckeberg, 1984). Serpentine tablelands in eastern Quebec and Newfoundland are tundra-like islands in an otherwise forested landscape (Scoggan, 1950).

Sites with soil having a strong serpentine influence may have a unique and biogeographically interesting plant community (Proctor and Woodell, 1975; Kruckeberg, 1984; Brooks, 1987). These sites often have a high percentage of endemics and ecotypes. On more productive and less toxic sites, the serpentine-adapted endemics and ecotypes are quickly eliminated through competition with species or populations that are better adapted to a more moderate habitat condition.

In eastern Canada, serpentine barrens are refugia for arctic species, but the vegetation also has dis-

junct temperate taxa and a few serpentine endemics (Scoggan, 1950). A notable example of the latter is the serpentine-indicating fern *Cheilanthes siliquosa*.

The vegetation of the eastern Canadian serpentine barrens is relatively young, since these sites were only deglaciated about 10,000 years ago. There is a considerably higher frequency of endemism in the much older serpentine vegetation of northern California. Kruckeberg (1984) estimated that 215 taxa of vascular plants are endemic to Californian serpentine formations, representing 14% of the total serpentine flora of 1544 taxa. Of the 1329 nonendemics, Kruckeberg classified 17% as local or regional indicators of serpentine soil, while the other 83% are indifferent. The most diverse genus of serpentine endemics in California is the crucifer *Streptanthus*, with 16 endemic taxa. Three of these have a particularly narrow distribution: *Streptanthus niger*, *S. batrachopus*, and *S. brachiatus* are all restricted to a few sites. *Streptanthus hesperidis* and *S. polygaloides* have a wider distribution, but they are nevertheless restricted to serpentine sites. *Streptanthus glandulosus* is particularly widespread and occurs on virtually all Californian serpentine sites, as does the endemic oak *Quercus durata*. In addition, the conifers *Pinus jeffreyi* and *Calocedrus decurrens* have altitudinally disjunct populations in some Californian serpentine sites, probably due to a relatively low intensity of competition in this type of stressed habitat. In the Sierra Nevada mountains the usual range of *P. jeffreyi* is ~1830–2745 m, whereas on serpentine it can occur as low as 915 m (Kruckeberg, 1984).

Not surprisingly, nickel-tolerant ecotypes of widespread species have been identified from serpentine sites in North America and Europe (Kruckeberg, 1954; Proctor, 1971a,b; Proctor and Woodell, 1975). Nickel tolerance is also, of course, a key feature of species that are endemic to serpentine sites.

Seleniferous Soil and Vegetation

In many semiarid areas, soil with a large concentration of selenium supports indicator plants that can hyperaccumulate this element. These plants are poisonous to livestock that graze upon them, causing a toxic syndrome known as "alkali disease" or "blind staggers."

In North America, the most important selenium-accumulating plants are legumes in the genus *Astragalus*. Of the 500 North American species of *Astragalus*, 25 are known to be selenium accumulators. These frequently contain thousands of parts per million of Se in foliage, to a maximum concentration of about 15,000 ppm (Trelease and Trelease, 1938, 1939; Rosenfeld and Beath, 1964; Davis, 1972; Stadtman, 1974). Frequently, accumulator and nonaccumulator species of *Astragalus* will grow together on a seleniferous site. In a location in Nebraska with 5 ppm of selenium in the soil, *A. bisulcatus* had 5560 ppm Se in its foliage, while *A. missouriensis* had only 25 ppm (Shrift, 1969).

The selenium-accumulating *Astragalus* spp. also cooccur with unrelated indicator species, including *Mentzelia decapetala*, *Oonopsis condensata*, *Stanleya* spp., and *Xylorhiza* spp., which can also contain selenium in a concentration of thousands of parts per million (Trelease and Martin, 1936). Selenium accumulators also occur outside of North America: the endemic Australian legume *Neptunia amplexicaulis* contains up to 5000 ppm Se and is known to poison cattle (Peterson and Butler, 1962, 1967). In addition, various widely cultivated varieties of the cabbage *Brassica rapa* can accumulate selenium to a concentration greater than 1000 ppm (Trelease and Martin, 1936).

Plants that hyperaccumulate a toxic element from the environment frequently turn out to have a physiological requirement for that element. This syndrome has been demonstrated for *Astragalus racemosus* and *A. pattersonii*, which grow much better in the presence of selenium in an experimental growth medium (Trelease and Trelease, 1938; Frost and Lish, 1975). Studies of some of the selenium-accumulating species of *Astragalus* have shown that they sequester much of their tissue Se in specialized amino acid analogues, especially selenomethionine (Trelease *et al.*, 1960; Virupaksha and Shrift, 1965). In addition, the accumulator *Astragalus* species are capable of emitting selenium-containing biochemicals to the atmosphere, including dimethyl selenide and dimethyl diselinide (Evans *et al.*, 1968). These specialized volatiles are responsible for the distinctive and unpleasant odor of these plants.

Mercury in the Aquatic Environment

In remote oceanic situations, mercury can accumulate naturally in marine fish, and in piscivorous birds and mammals. In North American offshore waters, fish species that are of particular importance with respect to mercury contamination include Atlantic swordfish (*Xiphias gladius*), Pacific blue marlin (*Makaira ampla*), bluefin (*Thunnus thynnus*), yellowfin (*Thunnus albacares*), and skipjack tuna (*Euthynnus pelamis*), Atlantic and Pacific halibut (*Hippoglossus hippoglossus* and *H. stenolepis*), and Atlantic and Pacific dogfish (*Squalus* spp.) and other sharks. All of these species accumulate mercury from a trace concentration in seawater (<0.1 ppb; Armstrong, 1979) to a level in edible flesh (Table 3.3) that exceeds the maximum acceptable concentration in fish for human consumption of 0.5 ppm fresh weight (f.w.) (Rivers *et al.*, 1972; Armstrong, 1979).

The contamination of oceanic fish by mercury is natural, and it is not a modern phenomenon. G. E. Miller *et al.* (1972) found no difference in the mercury contamination of contemporary tuna and seven museum specimens collected between 1878 and 1909. Within species, there is a strong tendency for larger, older fish to have a relatively high mercury concentration. For example, in a sample of 224 Atlantic swordfish, the average mercury concentration of animals weighing <23 kg was 0.55 ppm; for those between 23 and 45 kg the average was 0.86 ppm; and for those >45 kg it was 1.08 ppm (Armstrong, 1979). In some fish species only a small proportion of the total mercury is present as the particularly toxic compound, methylmercury. This probably indicates that demethylation reactions are an important adaptation that decreases the toxicity of mercury to the fish, since methylmercury is the form in which most mercury is actually absorbed from the aqueous environment (Rivers *et al.*, 1972; Bryan, 1976). In addition, the concentra-

tion of selenium tends to vary in direct proportion with mercury in these fish species. It has been suggested that selenium may somehow function to ameliorate the toxicity of mercury to fish, and also to fish-eating predators (Ganther *et al.*, 1972; Ganther and Sunds, 1974; Koeman *et al.*, 1973).

In some fresh-water ecosystems, a similar phenomenon of apparently natural mercury contamination of fish has been observed. For example, a mercury concentration exceeding 0.5 ppm f.w. is regularly measured in fish collected from remote lakes in many parts of Canada (McKay, 1985). In a lake in northern Manitoba, the average concentration in muscle was 2 ppm among a sample of 53 northern pike (*Esox lucius*), and one individual had a concentration of 5 ppm (McKay, 1985). Species of fresh-water fish that are top predators tend to have the highest concentrations of mercury, and within species older and larger individuals tend to be most contaminated (MacCrimmon *et al.*, 1983; McKay, 1985).

Large concentrations of mercury have also been found in piscivorous marine mammals at the top of the marine food web (Gaskin *et al.*, 1972; Buhler *et al.*, 1975; Jones *et al.*, 1975; Smith and Armstrong, 1978; Armstrong, 1979). The concentration of mercury in adult harp seals (*Phoca groenlandica*) in the Canadian Atlantic averaged 0.34 ppm in muscle but 5.1 ppm in the liver, while pups had 0.29 and 0.73 ppm, respectively (Armstrong, 1979). Since <10% of the total mercury in seal liver was present as the highly toxic methylmercury, it is likely that demethylation is a detoxification mechanism (Armstrong, 1979).

Analysis of feathers collected from fish-eating birds off coastal Peru revealed an average concentration of 2.0 ppm in Inca terns (*Larosterna inca*), 1.0 ppm in red-legged cormorants (*Phalacrocorax bougainvillii*), 0.81 ppm in Peruvian booby (*Sula variegata*), and 0.72 ppm in sooty shearwater (*Puffinus griseus*) (Gochfeld, 1980). However, these concentrations are smaller than the 5–10 ppm that was measured in three tern species (*Sterna* spp.) on Long Island, New York, where anthropogenic pollution is an important contributor to environmental mercury (Gochfeld, 1980).

Table 3.3 Average mercury concentration in the muscle tissue of some commercially important marine fish species[a]

Species	Mercury Concentration (ppm, f.w.)
Swordfish (>45 kg)	1.08
Bluefin tuna (>14 kg)	0.89
Yellowfin tuna (>32 kg)	0.62
Skipjack tuna (>4 kg)	0.21
Atlantic dogfish	0.41
Pacific dogfish	0.70
Pacific halibut (>45 kg)	0.42
Atlantic halibut (>45 kg)	0.80

[a]After Armstrong (1979).

3.5
ANTHROPOGENIC SOURCES OF TOXIC ELEMENTS

In this section, pollution caused by certain agricultural practices, by the metal mining and processing industries, and by the emission of lead from automobiles will be highlighted.

Agricultural Practices

The contamination of agricultural land by toxic elements has been caused by the longer-term use of inorganic pesticides, and by the use of contaminated sewage sludge as a soil conditioner.

Inorganic Pesticides

The use of inorganic pesticides has been particularly important in fruit orchards, where chemicals such as lead arsenate, calcium arsenate, and copper sulfate were used to control fungal pathogens and arthropods for more than a century (their use has now been largely supplanted by synthetic organic pesti-

cides). In orchards in Ontario, the annual spray rate was as great as 8.7 kg ha^{-1} year^{-1} of lead, 2.7 of arsenic, 7.5 of zinc, and 3.0 of copper, depending on the crop, the pest, and the pesticide formulation (Frank *et al.*, 1976a,b).

Because these toxic elements are bound by soil organic matter and other ion exchange surfaces, they tended to accumulate in treated soil. Frank *et al.* (1976b) found concentrations as high as 890 ppm lead and 126 ppm arsenic in the surface soil of apple (*Malus pumila*) orchards in Ontario, compared with background levels of these elements of <25 ppm and <10 ppm, respectively. This accumulation in soil was caused by as many as 70 years of use of lead arsenate as a pesticide, particularly against the codling moth (*Laspeyresia pomonella*), which causes "wormy" apples. The progressive accumulation of Pb and As residues in soil is illustrated in Fig. 3.1. In a similar study, Veneman *et al.* (1983) reported that an abandoned apple orchard near Amherst, Massachusetts, had a surface soil contamination of up to 1400 ppm lead and 330 ppm arsenic, and a strong positive correlation between the concentrations of these elements.

Mercury can also contaminate agricultural soil.

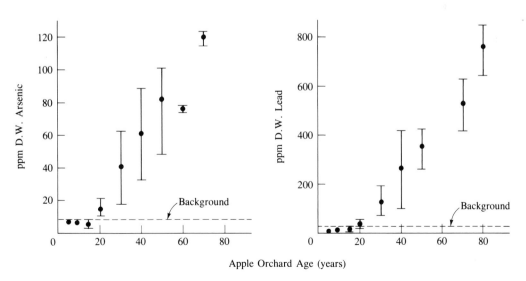

Fig. 3.1 The accumulation of arsenic and lead in the surface soil (0–15 cm) of 31 apple orchards in Ontario, Canada. Lead arsenate was applied at various rates to the orchards. Mean values and ranges are indicated. After Frank *et al.* (1976b).

The most important sources have been the use of organic mercurials as a seedcoat dressing to prevent fungal diseases of seeds before or just after germination, of mercury sulfate as a root dip for cruciferous crops, and of phenyl mercuric acetate for the treatment of apple scab (Frank *et al.*, 1976a,b). Mercurial compounds have also been used for the control of fungal turfgrass diseases and to reduce infestations of crabgrass (*Digitaria* spp.). MacLean *et al.* (1973) found a mercury concentration ranging from 24 to 120 ppm in the surface (0–5 cm) soil of golf course putting greens, where intense efforts are made to maintain a weed- and disease-free lawn.

Mercury compounds have been especially widely used as a seedcoat dressing. The intent of this practice is to prevent fungal diseases of newly germinated seedlings, especially damping-off, a fungal infection that initiates at the soil–air interface and that causes the weakened seedling to fall over and die. The planting of mercury-coated seed has had an effect beyond that of soil contamination. There have been observations of mercury accumulation and poisoning in wild birds and mammals that consume the planted seed, and in their predators (Tejning, 1967; Fimreite, 1970; Fimreite *et al.*, 1970; Johnels *et al.*, 1979). The use of alkyl mercury compounds such as methyl mercury was particularly damaging, since Hg in this form is very toxic and readily assimilated by animals from their food. Fimreite *et al.* (1970) examined the mercury concentration in tissues of rodents and birds, and found a large difference between alkyl mercury-sprayed and unsprayed areas of western Canada (Table 3.4).

The use of alkyl mercury compounds as a seed dressing was prohibited in most industrialized countries in the late 1960s, after the associated ecological problems of these chemicals became recognized. In Sweden their use was prohibited in 1966, and the much less toxic alkoxyl-alkyl mercury compounds were approved as a replacement. This led to an almost immediate decrease in the mercury contamination of previously affected wildlife such as raptorial birds (Figs 3.2a and b), and has contributed (along with banning of DDT; see Chapter 8) to an increased population of some species (Wallin, 1984).

Humans have also been poisoned by the inadvertent ingestion of mercury-treated seed that was intended for planting. One of the largest cases of a human population experiencing direct toxicity from exposure to a toxic metal occurred in Iraq in the 1960s, when about 6500 people were poisoned (about 500 people died) by the consumption of mercury-treated grain that had been supplied as crop seed by foreign aid. This episode happened in

Table 3.4   Mercury concentration in tissue of seed-eating rodents and birds and in their avian predators, in alkylmercury–treated and untreated agricultural areas of western Canada[a]

Organism	Mercury Concentration (ppm d.w., mean ± SD)	
	Treated Area	Untreated Area
Rodents[b]	1.25 ± 0.68 (n = 6)	0.18 ± 0.15 (n = 5)
Songbirds[b]	1.63 ± 1.00 (n = 10)	0.03 ± 0.01 (n = 3)
Upland game birds[b]	1.88 ± 0.44 (n = 19)	0.35 ± 0.22 (n = 12)
All seed eaters[b]	1.70 ± 0.38 (n = 35)	0.26 ± 0.14 (n = 20)
Predatory birds[c]	0.24 ± 0.04 (n = 89)	0.10 ± 0.02 (n = 34)
Seed-Eating Prey[d]	1.16 ± 0.23 (n = 61)	0.37 ± 0.15 (n = 32)

[a] After Fimreite *et al.* (1970).
[b] Mercury in liver; treated vs. untreated areas in Alberta.
[c] Mercury in egg; treated areas in Alberta vs. untreated areas in Saskatchewan.
[d] Mercury in liver; treated areas in Alberta vs. untreated areas in Saskatchewan.

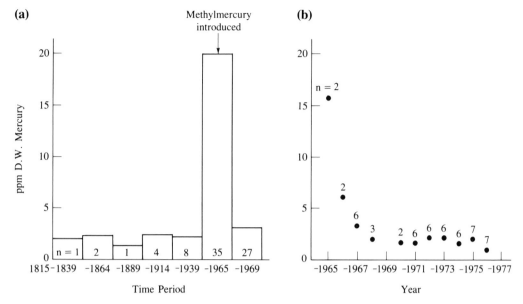

Fig. 3.2 (a) Mercury concentration in feathers taken from female goshawks (*Accipiter gentilis*) collected at nests in Sweden during April to June. (b) Mercury in tail feathers of young marsh harriers (*Circus aeruginosus*) from central Sweden. After Johnels *et al.* (1979).

spite of warnings on the bags in several languages, including Arabic. In some cases the warnings had actually been read, but were ignored by people who did not comprehend the concept that grain treated with a pesticide would be toxic to humans. Similar poisonings by the ingestion of mercury-treated grain took place in Iran, Pakistan, Guatemala, and elsewhere (Goldwater and Clarkson, 1972; Ehrlich *et al.*, 1977; Murphy, 1980; Ziff, 1985).

Sewage Sludge

The application of metal-containing sewage sludge can cause the contamination of agricultural soil and crops. Sewage sludge is a by-product of the secondary treatment of municipal sewage. Sewage sludge has a favorable soil conditioning property due to its high concentration of humified organic matter, and it contains a large concentration of the macronutrients nitrogen and phosphorus. As a result, it is frequently disposed of by application to agricultural land. For example, of the 5.6×10^6 tonnes/year of sludge produced in the United States, about 42% is

applied to farmland, as is about 40% of the 5.9×10^6 MT/year produced in western Europe (Brown and Jacobson, 1987). A similarly high rate of usage of sewage sludge is typical of most industrialized countries where wastewater treatment facilities are frequent.

Unfortunately, much of this sewage sludge, especially that having a significant industrial input, contains a considerable quantity of toxic elements. Table 3.5 summarizes the typical concentrations of toxic elements in sludge from several industrialized countries. Clearly, the metal concentrations in sludges vary tremendously. This variation largely reflects differences in the nature of the industrial inputs to particular wastewater systems. Of the various elements in Table 3.5, cadmium, copper, nickel, and zinc are most likely to cause phytotoxicity when the sludge is applied to agricultural land (Page, 1974; Keeney, 1975).

Only a portion of the "total" quantity of toxic elements present in sewage sludge is available for plant uptake. The uptake depends on such variables as soil cation exchange capacity and pH, amount of

Table 3.5 Average and/or range of elemental concentration (ppm, d.w.) in sewage sludge from a variety of locations

Element	Sweden[a]	Michigan[b]	Britain[c]	North America, Europe[d]	Ontario[e]
Ag			32 (5–150)	5–150	25 (4–60)
As		7.8 (1.6–18)		1–18	
B			70 (15–1000)		
Ba			1700 (150–4000)		
Cd	13 (2.3–171)	74 (2–1100)	<200 (<60–1500)	1–1500	29 (2–147)
Co	15 (2.0–113)		24 (2–260)	2–260	
Cr	872 (20–40,615)	2030 (22–30,000)	980 (40–8800)	20–40,000	4200 (16–16,000)
Cu	791 (52–3300)	1020 (84–10,400)	970 (200–8000)	52–11,700	1100 (162–3000)
Hg	6.0 (<0.1–55)	5.5 (0.1–56)		0.1–56	9 (1–24)
Mn	517 (73–3860)		500 (150–2500)	60–3860	310 (60–500)
Mo			7 (2–30)	2–1000	
Ni	121 (16–2120)	371 (12–2800)	510 (20–5300)	10–53,000	390 (7–1500)
Pb	281 (52–2910)	1380 (80–26,000)	820 (120–3000)	15–26,000	1200 (85–4000)
Sn			160 (40–700)		
V			75 (20–400)	40–700	
Zn	2060 (705–14,700)	3320 (72–16,400)	4100 (700–49,000)	74–49,000	4500 (610–19,000)

[a]93 Sewage works; after Berrgren and Oden (1972).
[b]57 Sewage works; after Blakeslee (1973).
[c]42 Sewage works; after Berrow and Webber (1972).
[d]300 Sewage works; after Page (1974).
[e]10 Sewage works; after Van Loon (1974).

Table 3.6 Influence of sludge addition on available metal concentration in three soil types[a]

| Soil Type | Treatment[b] | Available Metals (ppm d.w.)[c] | | | |
		Zn	Cu	Pb	Cd
Sandy loam (a)	Control	7.7	4.4	3.4	0.37
	+Sludge	52.8	11.5	6.2	0.70
Sandy loam (b)	Control	1.0	0.7	1.6	0.07
	+Sludge	31.1	4.9	3.5	0.25
Clay	Control	5.7	2.2	1.8	0.14
	+Sludge	74.6	12.8	5.5	0.72

[a] After Gaynor and Halstead (1976).
[b] 3.3 kg d.w. Sludge was added to 82 kg of soil, followed by an 8-week incubation. The metal concentration in the sludge was Cd, 30 ppm; Pb, 200 ppm; Ni, 360 ppm; Cu, 539 ppm; and Zn, 3200 ppm.
[c] The concentrations of available metals were determined after extraction with 0.005 M diethylenetriamine pentaacetic acid (DTPA).

sludge applied and its elemental composition, and plant species or variety. Gaynor and Halstead (1976) showed a large increase in "available" metals in soil treated with sewage sludge (Table 3.6), and other researchers have reported similar results (Page, 1974; Lagerwerff *et al.*, 1976; MacLean and Dekker, 1978). Crop plants grown on sludge-amended soil can have an increased metal con-

centration in their tissue, and metal toxicity has occasionally been reported (Table 3.7) (Cunningham *et al.*, 1975a,b; Webber and Beauchamp, 1975; Sidle *et al.*, 1976; Cohen *et al.*, 1978; MacLean and Dekker, 1978). The assimilated metals can enter the human food chain through the direct ingestion of produce, or indirectly if the crop is first fed to livestock. Clearly, the amount of toxic elements ap-

Table 3.7 Increase in the metal concentration of plants grown in soil containing sewage sludge at various rates of application: Average of four different sewage sludges, and two experiments using corn and one using rye[a]

| Application Rate (MT/ha) | Metal Concentration in Plant Tissue (ppm, d.w.) | | | | | |
	Cd	Cr	Cu	Mn	Ni	Zn
Control (0)	0.4	<3.0	7.4	33.3	<4.5	38
63	1.6	<3.0	14.4	44.1	<4.5	149
125	3.7	<3.0	15.8	110	<4.5	206
251	4.1	3.2	19.1	376	6.5	251
502	4.8	5.5	23.3	346	16.3	289
Average metal concentration in sludge (ppm, d.w.)	194	9200	1200	600	860	5100

[a] After Cunningham *et al.* (1975a).

plied to agricultural land in sewage sludge and the resulting contamination of crop plants require monitoring so that large tracts of otherwise fertile cropland are not contaminated or rendered toxic by this agricultural practice.

Pollution from Metal Mining and Processing

The Industrial Metal Cycle

Pollution by toxic elements is associated with various aspects of the metal mining and processing industry (Fig. 3.3). Mining produces ore as a product, along with a considerable amount of metal-contaminated waste. The ore is taken to a mill, which crushes the ore and separates mineral fractions, producing a metal-rich concentrate plus a large quantity of waste tailings. The tailings are usually discharged as a slurry into a contained dump, which is usually located in a natural basin or lake. Concentrate from the mill is processed in a primary smelter. Smelter wastes include a molten waste material called slag that is disposed on land, and atmospheric emissions of SO_2, other gases, and metal-containing particulates. Concentrate from the primary smelter is taken to a refinery, where a pure grade of metal is produced, along with a further emission of gases and metal-containing particulates to the atmosphere. The pure refined metal is used in diverse manufacturing processes, which may also emit pollution to the atmosphere as dust and gases. After their useful lifetime, manufactured products can be recycled through a secondary smelter and refinery, or they can be discarded in a solid waste disposal site. All of these processes produce environmental contamination, as is described below.

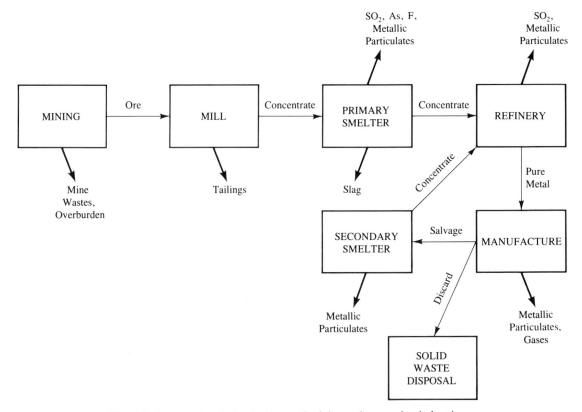

Fig. 3.3 Sources of pollution in the metal mining and processing industries.

Table 3.8 Metal concentrations in surface soil in studies of contaminated mine waste disposal sites

Study Site	Site	Metal Concentration in Soil (ppm, d.w.)			Reference[a]
		Cu	**Zn**	**Pb**	
Near operating mine,	Site 1	90	1500	500	(1)
British Columbia, Canada	Site 2	110	390	180	
Near abandoned mines,	Mine 1	680	120	2	(1)
British Columbia, Canada	Mine 2	120	130	40	
	Mine 3	160	390	300	
Near abandoned mine, Devonshire, U.K.		240	1750	960	(1)
Near abandoned mine, Wales, U.K.		—	1270	21,300	(2)
Near abandoned mines,	Mine 1	—	—	1100–1750	(3)
Wales, U.K.	Mine 2	—	4500–5000	—	
	Mine 3	680–2700	—	—	
Near abandoned mines, Wales, U.K.		90–2300	75–40,000	80–3600	(4)

[a]References: (1) Warren *et al.* (1966); (2) Williams *et al.* (1977); (3) Jain and Bradshaw (1966); (4) Gregory and Bradshaw (1965).

Mining Residues

Pollution around metal mine sites is caused by the dumping of contaminated overburden, excavation wastes, etc. (see Table 3.8 for examples). Because of toxicity from the large metal concentrations, the development of vegetation on the waste can be limited to an early successional grassland–herb community. In some cases, the toxicity can be so severe that even after a long period of time, only a few pioneer species ever establish themselves, as is the case with mine wastes from old Roman lead workings in Britain (Bradshaw and Chadwick, 1981).

Metal-Tolerant Plant Ecotypes

Sites that are contaminated by toxic elements from mine wastes or other sources are often dominated by genetically adapted plants, or metal-tolerant ecotypes, as they are often called. These individuals can survive and grow in a metal-contaminated environment, whereas nontolerant individuals are quickly eliminated by the toxic stress. Conversely, metal-tolerant ecotypes are usually poor competitors, and they are rare components of the vegetation of uncontaminated sites.

Several studies have shown that individual plants with a genetically based tolerance of toxic metals can be present at a small frequency (<1%) in populations growing in uncontaminated sites (Jowett, 1964; Bradshaw et al., 1965; Wu and Bradshaw, 1972; Cox and Hutchinson, 1980). Because of this preexisting tolerance within the population, the frequency of tolerant genotypes can increase very rapidly when toxic stress is caused by anthropogenic metal pollution (Bradshaw et al., 1965; Antonovics et al., 1971; Bradshaw, 1977). These population differences can be maintained over a short distance (<100 m) by a steep gradient of stress caused by metal pollution, since tolerant individuals are highly favored on contaminated soil but are poor competitors on uncontaminated soil (Jain and Bradshaw, 1966; Antonovics et al., 1971).

In the Sudbury region of Ontario, there is an extensive and severe contamination of soil by metals. Especially important pollutants are nickel and copper from mining and smelting activities. In addition, high concentrations of available soil aluminum have resulted from acidification of the soil. The vegetation on these sites largely consists of

Clones of a metal-tolerant ecotype of the grass *Deschampsia caespitosa*, growing in nickel- and copper-contaminated soil near the Coniston smelter at Sudbury, Ontario. (Photo courtesy of B. Freedman.)

metal-tolerant grasses, especially *Agrostis gigantea* (Hogan *et al.*, 1977a,b) and *Deschampsia caespitosa* (Cox and Hutchinson, 1979, 1980). Grasslands dominated by these species became much more extensive on barren, mesic-wet sites after a reduction in SO$_2$ pollution in 1972, even though the soil was still acidic and contaminated by Ni, Cu, and other metals (Cox and Hutchinson, 1980).

The likely source of seed of *Deschampsia caespitosa* was a site about 75 km from Sudbury where the grass grows in calcareous meadows beside Lake Huron. Huge piles of coal presently occur in the midst of one of these meadows at the port of Little Current, where coal was stored prior to its shipment to the Sudbury smelters. Mature seed-producing plants of *Deschampsia caespitosa* are present on this acidic coal substrate. Such seed could well have been transported with the coal to Sudbury. Metal tolerance occurs at a low level in the natural, calcareous population. Cox and Hutchinson (1980) grew seedlings from two populations of *Deschampsia caespitosa* on metal-contaminated soil collected

near a smelter (122 ppm Ni, 89 ppm Cu, pH 3.5) or near an old roast bed (2900 ppm Ni, 867 ppm Cu, pH 3.0). Seedlings from the uncontaminated Little Current site survived at a rate of 46% on the smelter soil and 2.6% on the roast bed soil, compared with rates for a smelter population of 94% and 14%, respectively. The above information suggests the following scenario regarding the history of *Deschampsia caespitosa* at Sudbury: seeds from a reference population having some individuals tolerant of acidity and metal stress were transported to the vicinity of the Sudbury smelters with the coal. The metal-tolerant individuals were positively selected at the metal-stressed sites near Sudbury, and tolerance became much more frequent in the population.

Cox and Hutchinson (1979, 1980) compared the metal tolerance of mature plants growing near Sudbury, with those from an uncontaminated reference site (Table 3.9). The Sudbury population was markedly more tolerant of Ni and Cu. These were the two most important metals in their soil environment, being present at an average concentration of

Table 3.9 Tolerance indices[a] to several metals, for the grass *Deschampsia caespitosa* collected from a metal-contaminated site near Sudbury, and from an uncontaminated reference site

Metal	Contaminated Sudbury Site	Uncontaminated Reference Site
Ni	0.96	1.97[b]
Cu	1.06	2.01[b]
Al	1.06	1.21[c]
Cd	1.24	1.36
Pb	1.37	1.90[b]
Zn	1.55	1.98[b]

[a] Note that the smaller the index, the greater the metal tolerance, since the tolerance index is $1 - \log(Rc/Rm)$, where Rc is the average length of the longest root of 20 tillers in a control nutrient solution and Rm is the average length of the longest root of 20 tillers in a metal-spiked solution. A tolerance index of 1.0 means that the growth of roots in the metal solution was the same as that in the reference solution without the metal. Modified from Cox and Hutchinson (1980).
[b] Statistically significant difference at $p < .001$.
[c] Indicates $p < .05$.

423 ppm Ni and 359 ppm Cu, compared with 22 ppm Ni and 19 ppm Cu at the reference site. The smelter population was also more tolerant of Al. This is explained by the acidic nature of soil near the smelter (pH 3.5–3.9, compared with pH 6.8–7.2 at the reference site), coupled with the fact that aluminum solubility and toxicity are much greater in acidic soil (Freedman and Hutchinson, 1986). The moderate but statistically significant increase in tolerance to Pb and Zn of the smelter population was unexpected, since the soil in which they grew was not contaminated by these metals. This observation of cotolerance implies that there may be similar physiological mechanisms of tolerance for several metals in *Deschampsia caespitosa*.

Metalliferous Tailings and Their Reclamation

Tailings are a waste by-product of milling. In this process, raw ore is ground to a fine powder, which is then separated either magnetically or by physicochemical flotation into (1) a metal-rich fraction that is roasted and then smelted and (2) waste tailings, which can still contain toxic concentrations of metals that make it difficult to establish plants for the eventual revegetation and stabilization of old tailings disposal sites. Additional toxicity can arise if the metal is present as a sulfide, since acidity can be generated when sulfides are oxidized by microbes. Table 3.10 lists metal analyses of tailings from a number of disposal sites in North America. All of these tailings contain a very large concentration of certain toxic elements, depending on the nature of the ore that had been fed to the mill. Toxicity of the acidic tailings is especially severe, since most metals have an increased solubility and bioavailability at acidic pHs.

Much research has been done on the establishment of vegetation on metal-contaminated tailings after they have been abandoned. Apart from the esthetic and economic problems associated with a large area of derelict land, tailings dumps can be an important source of water pollution and of windborne, metal-containing dust. These problems are greatly alleviated if the tailings are rehabilitated with a stable cover of vegetation. The establishment of vegetation on tailings usually involves some combination of liming to raise pH and reduce metal availability, fertilizing to alleviate nutrient deficiency, incorporation of organic matter to improve soil structure, and planting with a mixture of seeds of various plant species. Occasionally, a relatively novel approach is required, such as the use of acid- or metal-tolerant plant ecotypes, inoculation of the seed mix with metal-tolerant mycorrhizal fungi, or the overlaying of the entire tailings area with some sort of nontoxic, locally available overburden, which is then vegetated (Leroy and Keller, 1972; Dean *et al.*, 1973; Peterson and Nielson, 1973; Harris and Jurgensen, 1977; Moore and Zimmermann, 1977; Bradshaw and Chadwick, 1981; Peters, 1984).

As a case study, consider the rehabilitation of tailings from a large mill at Sudbury (after Peters, 1984). The mill can produce 54,000 MT/day of tailings, which are piped as a slurry for disposal in low-lying, contained depressions in the landscape. In 1983, there were ~1120 ha of active disposal area, 485 ha that were undergoing vegetation establishment, and 600 ha with an established grassland.

Table 3.10   Chemical analyses of a variety of metal-contaminated tailings

Site	pH	Concentration (ppm, d.w.)						S (%)	Refer-ence[b]
		As	Cd	Cu	Ni	Pb	Zn		
Yukon, Canada, gold mine tailings	1.9	5200	18	140	13	952	2400	1.1	(1)
Yukon, Canada, gold tailings	3.2	400	15	613	14	130	9600	44.4	(1)
Yukon, Canada, gold tailings	6.0	1350	102	172	15	3600	6200	5.7	(1)
Yukon, Canada, gold tailings	6.9	50,000	114	330	20	2300	18,000	13.5	(1)
Yukon, Canada, tungsten tailings	7.0	—	0.3	1420	22	12	288	6.5	(1)
Yukon, Canada, gold tailings	7.1	7	2.6	33	21	1130	1060	4.0	(1)
Yukon, Canada, copper tailings	9.0	15	0.7	1710	21	9	178	0.1	(1)
Ontario, Canada, nickel tailings	3.2	—	0.4	392	290	18	291	0.9	(1)
New Mexico, U.S.A., copper tailings[a]	2.2	—	—	>600	—	<5	21	—	(2)
Utah, U.S.A., uranium tailings[a]	2.9	—	—	475	—	16	103	—	(2)
Montana, U.S.A., copper tailings[a]	5.2	—	—	195	—	<5	24	—	(2)

[a] In ppm in saturation water extract.
[b] References: (1) after Kuja (1980); (2) after Peterson and Nielson (1973).

The mineral constituents of the Sudbury tailings are feldspar (>50%), chlorite (20%), quartz (10%), pyroxenes (7%), biotite (7%), and pyrites (7%). These are not toxic minerals. However, the acidity that is produced by oxidation of the pyrites can cause a pH lower than 3.7. In addition, there is a relatively large concentration of some metals, especially Cu (1–81 ppm in an "available" extract of the tailings measured with an acetic acid leachate), Ni (1–87 ppm), and Fe (59–441 ppm).

Over the years, various revegetation schemes were used to try to stabilize the Sudbury tailings, largely because they caused a severe dust problem during dry weather. The techniques that are currently used result in a well-established grassland, which is being secondarily colonized by seedlings of native trees and other plants. The methods used to achieve this include (1) the application of limestone ($CaCO_3$) at ~900 kg/ha to raise the tailings pH to 4.5–5.5; (2) the application of nitrogen and other macronutrients several times during establishment of the grassland; and (3) sowing with a mixture of pasture grasses (25% *Agrostis gigantea*, 25% *Poa compressa*, 15% *Phleum pratense*, 15% *Poa pratensis*, 10% *Festuca arundinacea*, and 10% *Festuca rubra*). Legumes have also been used to a limited extent; the most successful species have been *Medicago sativa* and *Lotus corniculatus*. An important innovation was the sowing of a short-lived nurse crop of annual rye (*Secale cereale*) to provide initial shade and reduce wind stress for the tender seedlings of the pasture grasses and legumes as they were becoming established. In addition, there has been a natural invasion by native herbaceous plants, especially of Asteraceae, and also by seedlings of tree species such as birch (*Betula papyrifera*), trembling aspen (*Populus tremuloides*), and willow (*Salix* spp.). The establishment of trees has been supplemented by the planting of various conifer species.

A reclaimed tailings area near the Copper Cliff mill at Sudbury. Most of the herbaceous vegetation was established using a mixture of pasture species, but the woody shrubs are seeding in from the nearby hardwood forest. The tailings pond, visible in the background, is used by ducks for breeding and during migration. (Photo courtesy of B. Freedman.)

As the ecosystem developed on the reclaimed tailings at Sudbury, animals began to invade. At least 90 bird species, most of which are migrants, have been observed on the vegetated tailings area and its central pond. Birds that regularly breed in the tailings habitat include mallard and black duck (*Anas platyrhynchos* and *A. rubripes*), American kestrel (*Falco sparverius*), killdeer (*Charadrius vociferus*), and savannah sparrow (*Passerculus sandwichensis*).

Primary Smelters

Several detailed studies have been made of the environmental consequences of pollution in the vicinity of metal smelters. Among the better-documented cases are the Sudbury nickel–copper smelters (Hutchinson and Whitby, 1974; Whitby and Hutchinson, 1974; Freedman and Hutchinson, 1980c; Amiro and Courtin, 1981; see also Chap-

ter 2), a zinc smelter at Palmerton, Pennsylvania (Buchauer, 1973; Beyer *et al.*, 1985; Jordan, 1975; Jordan and Lechevalier, 1975; Strojan, 1978a,b), a brassworks at Gusum, Sweden (Tyler, 1984, and below), and a lead–zinc smelter at Avenmouth, U.K. (Little and Martin, 1972; Hutton, 1984). Common features of these studies are (after Freedman and Hutchinson, 1981) (1) the occurrence of severe environmental contamination close to the point source, especially in surface soils; (2) an exponential decrease in the intensity of pollution with increasing distance from the point source; (3) the presence of damaged plant communities in the most polluted sites, characterized by a small biomass and productivity, short stature of the ecosystem dominants, altered species composition, and depauperate species richness and diversity; and (4) a disruption of nutrient cycling by toxic metals, including an impoverished abundance and species richness of soil-dwelling organisms, decreased litter decomposition, and slower mineralization and other transformations of nitrogen and phosphorus in the forest floor.

Sudbury. The total atmospheric emission of metal-containing particulates from the nickel–copper smelters near Sudbury averaged ~1.9×10^4 MT/year between 1973 and 1981, including 4.2×10^3 MT/year of iron, 6.7×10^2 MT/year of copper, 5.0×10^2 MT/year of nickel, 2.0×10^2 MT/year of lead, and 1.2×10^2 MT/year of arsenic (Chan and Lusis, 1985). The emitted particulates can be effectively removed from the atmosphere during wet precipitation events (~60–80% removal efficiency), but less so by dry deposition (<15%) Chan and Lusis, 1985). Overall, about 50% of the particulates emitted from the smelters are deposited nearby (Muller and Kramer 1977; Freedman and Hutchinson, 1980a; Chan and Lusis, 1985). The deposition rate is especially large close to the point sources, and it declines exponentially with increasing distance (Fig. 3.4). This pattern of deposition has caused a parallel pattern of environmental contamination, as is illustrated by the concentrations of nickel and copper in the forest floor of stands along a transect from the Copper Cliff smelter (Fig. 3.5). Concentrations of nickel and copper as large as 4900 ppm each were present in the forest floor of stands close to the smelter, while up to 370 ppm Ni and 260 ppm Cu occurred in plant foliage (Freedman and Hutchinson, 1980a). The large metal-binding capacity of the organic forest floor makes it a virtual sink for deposited metals. The concentrations of metals in mineral soil are much smaller because of its limited cation exchange capacity (Freedman and Hutchinson, 1980a).

Because sulfur dioxide has been such an important pollutant in the Sudbury area, it is impossible to attribute ecological damage directly or solely to toxic metals. However, the importance of metals has been demonstrated through experiments in which bioassay plants were grown in contaminated soil, but in the absence of sulfur dioxide. In a greenhouse experiment, Whitby and Hutchinson (1974) grew a variety of plant species in metal-contaminated soil collected in the vicinity of a Sudbury smelter. They

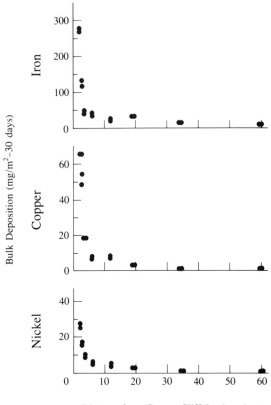

Fig. 3.4 Atmospheric deposition of metals along a transect from the Copper Cliff smelter at Sudbury. The sampling period was July 18 to August 16, 1977. Modified from Freedman and Hutchinson (1980a).

observed a large reduction in the productivity and root elongation of all bioassay species, compared with their growth in reference soil. This toxic effect was due to a high concentration of soluble nickel, copper, cobalt, and aluminum, and it persisted to a large degree when soil acidity was neutralized by liming. Similar toxic symptoms were observed when the bioassay plants were grown in a nutrient solution containing pure salts of these metals. In a field experiment, the concentration of toxic metals in limed soil precluded the establishment of test species by seed at sites within 12 km of the smelter.

Nickel- and copper-tolerant populations of the grasses *Deschampsia caespitosa* and *Agrostis gi-*

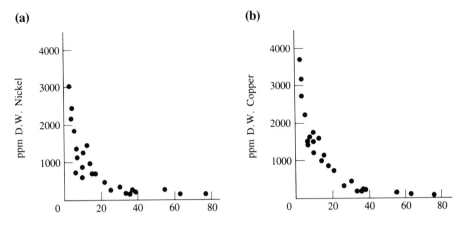

Fig. 3.5 (a) Total nickel and (b) total copper in the forest floor of stands in a transect from the Copper Cliff smelter. Each point is the average of 10 replicate determinations. Modified from Freedman and Hutchinson (1980a).

gantea were previously described. The presence of these ecotypes indicates that metal stress is important in sites close to the smelter. However, the rapid spread of a grassland dominated by these species plus *Agrostis scabra* only occurred after the local emissions of SO_2 were greatly reduced by building the very tall superstack in 1972 (Cox and Hutchinson, 1981; Chapter 2). This observation underscores the important role of that toxic gas in retarding the development of plant communities near the Sudbury smelters.

In addition to an effect on vegetation, toxic metals in soil near the Sudbury smelters have been shown to inhibit litter decomposition and other nutrient cycling processes (Freedman and Hutchinson, 1980b). In part, the effects on nutrient cycling explain the relatively small concentrations of some plant macronutrients in soil and foliage at contaminated sites near Sudbury (Lozano and Morrison, 1981).

Gusum. Gusum, Sweden, is another case where metals emitted from a smelter have damaged terrestrial vegetation. In this situation, the emission of SO_2 was unimportant, so that all ecological damage can be ascribed to toxic metals, primarily zinc and copper. The brassworks at Gusum has operated continuously since 1661, and there is now a large concentration of metals in soil near the point source (Fig. 3.6) (see also Tyler, 1974, 1975, 1976). Zinc is the most abundant metal, reaching 16,000–20,000 ppm (1.6–2.0%) in the surface organic matter of sites within 0.3 km of the source, compared with <200 ppm at reference sites 7–9 km away. Copper and lead have a similar pattern of contamination (copper was 11,000–17,000 ppm at <0.3 km, compared with <20 ppm at 7–9 km). The elevated soil pH at sites close to the source is due to the saturation of most of the cation exchange capacity by the large metal concentration.

The metal contamination has caused local ecological damage. The effects on vegetation (Folkeson, 1984; Tyler, 1984) include the death or deterioration of most pine (*Pinus sylvestris*) and birch (*Betula* spp.) trees close to the source, and a substantial reduction in the cover of lichens and the feather mosses *Hylocomium splendens*, *Pleurozium schreberi*, and *Dicranum* spp. (<1% cover, compared with 25–90% in reference stands). A decrease in the species richness of many groups of organisms is clearly related to the metal concentration of the surface soil (Fig. 3.7). Not unexpec-

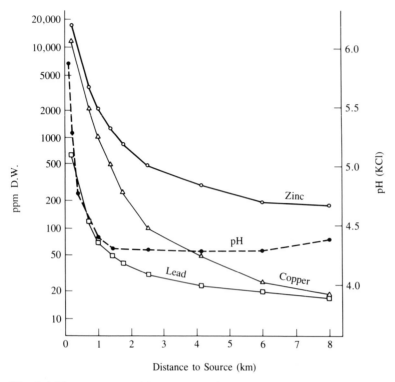

Fig. 3.6 Zinc, copper, and lead concentrations and pH of surface soil along a transect from the brassworks at Gusum. After Tyler (1984).

tedly, there are large differences in the tolerance of plant taxa to the metal contamination. For example, *Hylocomium splendens* is normally an abundant species of feather moss on the forest floor, but it is virtually eliminated from sites <1.5 km from the brassworks, whereas *Pohlia nutans* is relatively abundant in contaminated sites (*P. nutans* is also abundant in Ni- and Cu-contaminated sites near Sudbury). The grass *Deschampsia flexuosa* is more abundant close to the brassworks than in reference sites. This species may have experienced a competitive release after the death or deterioration of the overstory and most of the ground vegetation. Some fungi are relatively abundant close to the source, notably *Laccaria laccata* and *Paecilomyces farinosus*. However, some 25 species of macrofungi are less abundant in contaminated sites, while 11 species are indifferent (Ruhling *et al.*, 1984; see also Fig. 3.7). Soil dwelling invertebrates are also af-

fected; populations of earthworms are absent 175 m from the brass mill, compared with a density of about 25/m² at 8 km, and 50/m² at 20 km (Bengston *et al.*, 1983; see also Fig. 3.7).

Studies were also made of the effects of metal pollution on litter decomposition and nutrient cycling in the forest floor of sites near Gusum (Fig. 3.8; see also Ebregt and Boldewijn, 1977). The rates of decomposition, the activity of phosphatase enzymes, and nitrogen mineralization by ammonification are especially reduced by toxic metals.

Contamination from Automobile Emissions

Many studies have documented the distribution of heavy metals in soil, vegetation, and the atmosphere along transects perpendicular to busy highways. All have found a well-defined gradient of

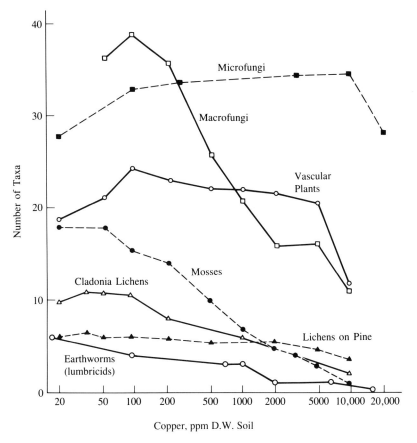

Fig. 3.7 Species richness of several groups of organisms, as related to the copper concentration of forest surface soil in the Gusum area. After Tyler (1984).

lead contamination, with the intensity of the effect being related to traffic volume. Some studies have also demonstrated a less well-defined contamination by other metals, including cadmium, chromium, copper, nickel, vanadium, and zinc (Lagerwerff and Specht, 1970; Page *et al.*, 1971; Hutchinson, 1972; Ward *et al.*, 1977; Peterson, 1978; Dale and Freedman, 1982). This pattern is illustrated in Table 3.11 for four roads of differing traffic density. A secondary contamination occurs in other biota living beside roads, for example, small mammals (Clark, 1979).

Lead emitted by automobiles is also a major contributor to the general lead pollution that occurs in cities, and that secondarily causes a contamination of urban wildlife with lead. For example, a high lead concentration has been measured in pigeons (*Columba livia*) living in London, England, and the most contaminated birds showed symptoms of acute lead poisoning (Hutton, 1980; Hutton and Goodman, 1980). A more important source of lead toxicity to birds is spent shotgun pellets, which are ingested during feeding and retained in the gizzard, where they are progessively abraded, and then dissolved by acidic gastric fluids. Waterfowl are especially widely affected by lead shot, with an estimated 2–3% of the North American autumn and winter population, or about 1 million individuals, dying each year (Bellrose, 1959, 1976). Upland game birds such as mourning doves (*Zenaida mac-*

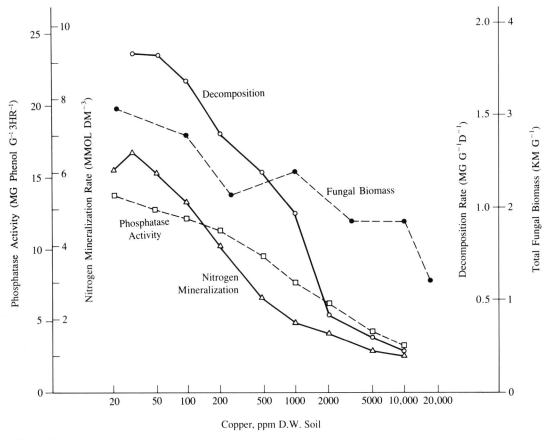

Fig. 3.8 Total fungal biomass, phosphatase activity, and decomposition and mineralization rates as related to the copper concentration of forest surface soil in the Gusum area. Modified from Tyler (1984).

roura) have also been poisoned by ingested lead shot (Kendall, 1982; Kendall and Scanlon, 1982).

The cause of lead emission from automobiles is the use of leaded gasoline. Since 1923 and especially since 1946, tetraethyl lead [Pb(CH₂CH₃)₄] has been added to gasoline as an antiknock compound at a rate of about 0.8 g lead/l. This has been done to increase engine efficiency and gasoline economy, while decreasing engine wear due to "knock" (Atkins, 1969). Most of the lead in the gasoline is emitted through the automobile tailpipe, at a rate of about 0.07 g/km, and accounting for up to 40% of automobile particulate emissions (Atkins, 1969; Cantwell *et al.*, 1972). It was estimated that 98% of the total 1970 U.S. lead emission of 1.6×10^5 MT/year originated with automobile exhaust

(NAS, 1972). Some 61% of the emitted lead particulates have a diameter >3 μm, and most of this emission settles gravitationally within about 50 m of the roadway. The remainder is well dispersed and contributes to regional lead pollution (Cantwell *et al.*, 1972).

The degree of urban lead contamination is being alleviated in many areas by the decreasing use of leaded gasoline. This trend is related in part to the increasing use of catalytic converters to reduce the emission of other automobile pollutants, especially carbon monoxide and hydrocarbons. Automobiles equipped with these devices require the use of unleaded gasoline, as the converter catalysts are inactivated by lead. Hoggan *et al.* (1978) reported a 50% decrease in atmospheric lead particulates in the

Table 3.11 Lead and zinc in surface (0–2 cm) soil collected at various distances from roads of different traffic density (ADT = average daily traffic) in Halifax, Canada[a]

Distance from Road Edge (m)	Lead Concentration (ppm, d.w.) and ADT				Zinc Concentration (ppm, d.w.) and ADT			
	Road A 50,000	Road B 18,500	Road C 16,000	Road D 3,000	Road A 50,000	Road B 18,500	Road C 16,000	Road D 3,000
0	3045	858	1075	465	880	422	272	106
1	2813	402	457	118	700	198	167	69
5	342	177	136	32	144	122	79	75
15	223	75	163	26	150	55	92	59
30	—	45	63	26	—	63	—	65
50	223	45	95	38	95	69	58	64
100	—	—	60	21	—	—	76	78
Background	14				60			

[a]Modified from Dale and Freedman (1982).

Los Angeles area between 1971 and 1976, a phenomenon that was attributed to the increasing use of unleaded gasoline.

The sources of the other metals that are roadside contaminants are less well defined than that of lead. They are presumably derived from the wear of metallic automobile parts containing these metals, from the erosion of tires, and from the use of metals such as nickel as a gasoline or oil additive (Lagerwerff and Specht, 1970; Hutchinson, 1972). For

Table 3.12 Distribution and movement of lead in a high-elevation forest of *Abies balsamea, Picea rubens,* and *Betula papyrifera* on Camel's Hump, Vermont[a]

	Amount (kg/ha)	Concentration (ppm)
Lead distribution		
Vegetation		
Foliage	0.045	3
Twigs	0.475	28
Bark	0.160	23
Wood	0.200	3
Herbs	0.005	16
Root wood	n.d.	10
Root bark	n.d.	33
Forest floor	20	219
Soil		
0–10 cm	14	26
10–20 cm	24	31
>20 cm	25	35
Lead flux	(kg/ha-year)	(ppm in water)
Rain water	0.300	0.017
Cloud water	0.400	0.051
Forest floor at 3 cm	0.085	0.004
at 12 cm	0.039	0.002
Mineral soil at 25 cm	0.023	0.001
at 40 cm	0.020	0.001
Streams	<0.012	<0.001

[a]Modified from Friedland and Johnson (1985); n.d. = no data.

example, it was estimated that in 1972 about 800 MT of zinc was released into the environment in Canada from tire erosion (zinc is used in the vulcanization of rubber) (Anonymous, 1973).

As noted earlier, a substantial fraction of the lead emitted from automobiles occurs as very small particulates that are dispersed far from the roadway. This dispersed lead (plus lead emitted from other sources, such as power plants) is largely removed from the atmosphere by precipitation. The rate of deposition can be especially great at high-elevation forested sites, because of a relatively large rate of wet deposition due to an orographic effect, and because of direct cloud-water deposition. The lead accumulation rate (i.e., atmospheric input minus stream-water export) in forests of the northeastern United States ranges from $\sim$ <0.1 to 0.7 kg ha^{-1} year^{-1}, with the largest recorded rate occurring at a 1000-m-elevation site on Camel's Hump, Vermont (Siccama and Smith, 1978; Andresen et al., 1980; Smith and Siccama, 1981; Friedland and Johnson, 1985). This accumulation has caused a large lead contamination of certain components of the forest, especially the organic forest floor (Andresen et al., 1980; Smith and Siccama, 1981; Johnson et al., 1982; Friedland et al., 1984, 1985). The forest floor acts as a chemical filter for lead and other metals, by the formation of relatively strong and immobile organo-metallic complexes with humic substances (Schnitzer, 1978; Friedland et al., 1984).

The distribution and movement of lead in a high elevation forest in Vermont is shown in Table 3.12

(Friedland and Johnson, 1985). Within the vegetation, the largest lead concentration (23–33 ppm) is present in tree bark and twigs (which have a large proportion of bark). However, the quantity of lead also depends on the standing crop of the biomass compartment; for example, wood only had a Pb concentration of 3 ppm, but because of the great mass of wood in the forest it contained 0.200 kg/ha of lead, next only to twigs at 0.475 kg/ha. By far the largest lead concentration was in the forest floor at 219 ppm, and this component contained 20 kg/ha of lead. However, because of its very large mass the quantity of lead in the mineral soil (63 kg/ha) was larger. The influx of lead from the atmosphere was 0.70 kg ha^{-1} year^{-1}, while stream-water export was <0.012 kg ha^{-1} year^{-1}. This yields a net accumulation rate of ~0.7 kg ha^{-1} year^{-1}. The average residence time of lead in the forest floor was ~500 yr, and lead was accumulating at a rate of 3.3% per year, for a doubling time of 21 years.

At present, we do not understand the ecological consequences of the increasing contamination of lead (and other metals) in remote forests. This is because of a lack of information on the toxic threshold of lead in foliage or in the soil solution, and because most of the soil lead is bound as relatively immobile and unavailable organometallic complexes. However, there is an important message in the observation of a progressive lead contamination of forests—even in fairly remote situations the trend is toward an increasing pollution, with an uncertain ecological effect.

4

ACIDIFICATION

4.1

INTRODUCTION

The environmental impacts of the deposition of acidifying substances from the atmosphere, popularly known as "acid rain," are currently a high-profile issue in both scientific and public forums. Only 15–20 years ago, acid rain research was essentially nonexistent, and few people were aware of or concerned about the potential effects of the acidification of natural ecosystems. This state of affairs changed quickly in the early 1970s, beginning with a catalytic conference held in 1970 in Stockholm: the United Nations Conference on the Human Environment. At this international scientific/environmental gathering, the Swedish government presented as a case study the results of an innovative research program, summarized in a report titled: "Air Pollution across National Boundaries. The Impact on the Environment of Sulfur in Air and Precipitation" [Royal Ministry of Foreign Affairs/Royal Ministry of Agriculture (RMFA/RMA), 1971]. Partly because of the attention that this report focused on the anthropogenic acidification of ecosystems, over the next 5 years or so this field of investigation metamorphosed to

a condition of vigorous and relatively well-funded scientific research, intense lobbying by nongovernment environmental organizations and by industry (of course, these groups lobbied in different directions), and even some pollution control legislation by various levels of government. Because of these activities, acid rain impacts are now one of the better understood ways by which humans can degrade their environment. Equally important, acid rain is one of the most identifiable environmental problems to the "person-on-the-street," at least in north temperate industrialized countries.

It is now known that the most important acidifying substances that are deposited from the atmosphere are weak solutions of sulfuric and nitric acids, which arrive in the form of acidic precipitation. In addition, the dry deposition of gaseous sulfur dioxide and oxides of nitrogen, and of certain particulates such as ammonium sulfate and ammonium nitrate, can contribute further acidifying potential to ecosystems. There is evidence that the area of North America and Europe that is experiencing acidic deposition has increased in the most recent few decades. Certain ecosystems are more susceptible than others to acidification by these atmospheric inputs. These ecosystems usually have a thin, often coarse-grained, noncalcareous soil overlying a mantle of hard, slowly weathering bedrock of such minerals as granite, gneiss, and quartzite. Although the evidence is still uncertain that terrestrial ecosystems have been degraded by acidic deposition, there is incontrovertible evidence for damage to fresh-water ecosystems. Data show that many lakes, streams, and rivers have acidified fairly recently, and that fish populations have declined or become extinct in many of these water bodies.

In this chapter, the chemical characteristics of acidic precipitation and dry deposition are examined, as are the effects of acidic deposition on the chemistry and biota of terrestrial and aquatic ecosystems, and possible techniques for use in the reclamation of acidic water bodies.

4.2

DEPOSITION OF ACIDIFYING SUBSTANCES FROM THE ATMOSPHERE

There are several ways by which acidifying substances can be deposited from the atmosphere to aquatic and terrestrial ecosystems. These are (1) the wet deposition of material entrained in rain, snow, and fog, that is, "acidic precipitation"; (2) the uptake of certain gases by vegetation, soil, and water surfaces; and (3) the dry deposition of particulates. These subjects are discussed below.

Chemistry of Precipitation

Acidic precipitation is functionally defined as having a pH less than 5.65. This is the degree of acidity that is produced by carbonic acid (H_2CO_3) at the equilibrium concentration that occurs when atmospheric CO_2 at ~340 ppm is in contact with distilled water (i.e., $CO_2 + H_2O \leftrightarrows H_2CO_3$). Therefore, the slightly acidic pH 5.65 is considered to be the acid rain cutoff, and not the pH corresponding to zero acidity, that is, pH 7.0 (Cogbill and Likens, 1974; Reuss, 1975). However, atmospheric moisture is not merely distilled water in a pH equilibrium with CO_2. Because of the neutralizing influence of other atmospheric gases, and

of soil-derived cations such as Ca^{2+} and Mg^{2+} that are naturally present in a trace concentration in precipitation, especially in agricultural and prairie landscapes, the pH of unacidified rainwater can be higher than pH 5.65 (Kramer and Tessier, 1982; Liljestrand, 1985). However, in remote areas where precipitation is not strongly influenced by either pollutants or calcareous dusts, for example, much of northern North America over Precambrian bedrock, the pH of unacidified precipitation may be closer to about 5.0 because of the presence of natural mineral acids (Schindler, 1988).

The most abundant cations in precipitation are usually H^+, NH_4^+, Ca^{2+}, Mg^{2+}, and Na^+, while the most abundant anions are SO_4^{2-}, Cl^-, and NO_3^-. Other ions are also present, but in relatively small concentrations. The acidity of precipitation is due to the presence of H^+ ions, and it is measured in units of pH. Since the pH scale is logarithmic to base 10, a one-unit difference in solution pH implies a 10–fold difference in the concentration of H^+. In addition, the quantity of H^+ in solution is directly related to the difference in concentration (in units of equivalents) of the sum of all anions and the sum of all cations other than H^+. If there are more equivalents of anions than of cations other than hydrogen ion, then H^+ goes into solution to balance the cation "deficit," that is (in microequivalents),

$$H^+ = SO_4^{2-} + NO_3^- + Cl^- - Na^+ - NH_4^+ - Ca^{2+} - Mg^{2+}$$

This follows from the principle of conservation of electrochemical neutrality of aqueous solutions: the total number of cation equivalents equals the total number of anion equivalents, so that the aqueous solution does not have a net electrical charge (Reuss, 1975). The above equation has been used to calculate the pH of precipitation samples that were collected and analyzed prior to about 1955, for which the reported pH value was suspected to be inaccurate, but there were reliable determinations of the concentrations of all other important ions (Cogbill, 1976).

Table 4.1 summarizes long-term data for precipitation chemistry at Hubbard Brook, New Hampshire, located in a region of the northeastern United States that is currently experiencing severe

Table 4.1 Precipitation chemistry at Hubbard Brook, New Hampshire: Average $\pm$ SE of the annual volume-weighted concentration of weekly bulk-collected precipitation over a 19-year study period (1963–1982)[a]

Constituent	μeq/l	%[b]
H	69.3 $\pm$ 2.1	71.1
NH$_4$	10.6 $\pm$ 0.6	10.9
Ca	6.5 $\pm$ 0.8	6.7
Na	4.8 $\pm$ 0.5	4.9
Mg	3.0 $\pm$ 0.5	3.1
Al	1.8 $\pm$ 0.2	1.8
K	1.5 $\pm$ 0.3	1.5
SO$_4$	54.0 $\pm$ 2.1	60.6
NO$_3$	23.5 $\pm$ 1.0	26.4
Cl	11.2 $\pm$ 1.2	12.6
PO$_4$	0.3 $\pm$ 0.1	0.3
HCO$_3$	0.1	0.1
Sum of cations	97.5 $\pm$ 2.5	
Sum of anions	89.1 $\pm$ 2.6	
pH	4.16	

[a]Modified from Likens *et al.* (1984).
[b]Percent of total cation or total anion equivalents.

acidic deposition with a mean annual pH of 4.1–4.2. This site also has the longest continuous record of precipitation chemistry in North America, with weekly collections having been made from 1963 to the present. The Hubbard Brook data set therefore represents a very important record of precipitation chemistry.

The average pH of precipitation at Hubbard Brook between 1963 and 1982 was 4.16, and hydrogen ion comprised 71% of the total cation equivalents (Likens *et al.*, 1984). The most important anions were sulfate and nitrate, which were present in a 2 : 1 equivalent ratio and together accounted for 87% of total anion equivalents. Therefore, most of the acidity in precipitation at Hubbard Brook is in the form of sulfuric and nitric acids. During the 19–year period between 1963 and 1982, there was no statistically significant trend in the mean annual volume-weighted concentration of either hydrogen ion or nitrate in precipitation, but there was a significant decrease of 34% for sulfate, 34% for ammonium, 63% for chloride, 79% for magnesium,

and 86% for calcium. The specific reasons for these changes in precipitation chemistry are not known.

Not surprisingly, the chemistry of precipitation events at Hubbard Brook is influenced by the direction of movement of the precipitation air mass. Munn *et al.* (1984) presented a trajectory analysis of 69 precipitation events from the period 1975 to 1978. They found that the largest concentrations of hydrogen ion, sulfate, and nitrate were associated with storms tracking from the SSE through SSW. These air masses had passed over the large metropolitan conurbations of Boston and New York. Approximately two-thirds of the wet deposition of H$^+$ was associated with storms from this sector. Air masses from the NNW–ESE sector had relatively small concentrations of H$^+$, NO$_3^-$, and SO$_4^{2-}$.

The chemistry of precipitation varies markedly on both regional and local scales (Fig. 4.1). In part this reflects the patterns of emission of SO$_2$ and NO$_x$, their degree of oxidation to sulfate and nitrate, and the prevailing direction traveled by the air mass containing these pollutants. Acid-neutralizing dusts in the atmosphere also affect precipitation chemistry. This is especially true where there is little vegetation cover, for example, where farming is important, since dust is relatively easily entrained into the atmosphere by wind blowing over bare fields. Dusts entrained by vehicles from dry, unpaved roads can also be important (Barnard, 1986; Gatz *et al.*, 1986).

These influences are illustrated in Table 4.2, which summarizes data for precipitation chemistry at four sites where wet-only precipitation was collected (i.e., the precipitation sampler was only open to the atmosphere during a precipitation event). The data for Dorset in southern Ontario are representative of a region where precipitation is very acidic, with an average pH of 4.1. Compared with the other stations, precipitation at Dorset had relatively large concentrations of H$^+$, SO$_4^{2-}$, and NO$_3^-$, implying that the acidity was largely composed of sulfuric and nitric acids. The large concentrations of sulfate and nitrate are believed to be partly anthropogenic, having been derived from oxidation in the atmosphere of SO$_2$ and NO$_x$. These gases are emitted from power plants, metal smelters, and automo-

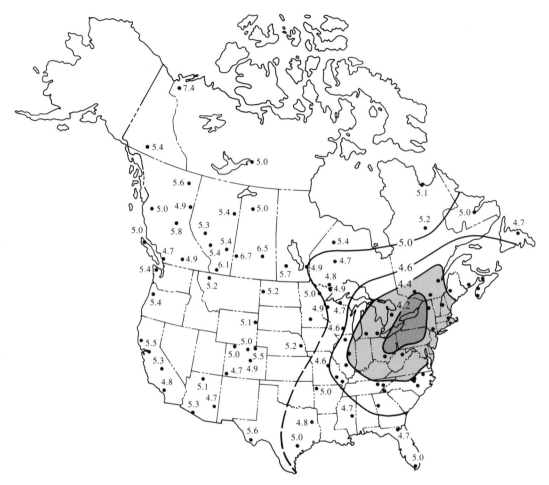

Fig. 4.1 The acidity of wet-only deposition in North America in 1982. • 5.4, pH at sample site; −5.0−, line of approximately equal pH value. From National Acid Precipitation Assessment Program (NAPAP) (1983).

biles, and are then dispersed over a long distance (see also Chapter 2).

Like Dorset, the Experimental Lakes Area (ELA) site in northwestern Ontario is on a Precambrian Shield landscape, with bedrock and soil mainly comprised of hard, plutonic minerals such as granite, gneiss, and quartzite. However, the ELA site is located in a remote area that is infrequently impacted by long-range transported air masses that are contaminated by anthropogenic pollution. Hence the precipitation at ELA is less acidic (mean

pH 4.7) than at Dorset, and it has smaller concentrations of nitrate and sulfate.

The Kejimkujik, Nova Scotia, site is also fairly remote from sources of emission. It frequently receives air masses from source areas in the northeastern United States and eastern Canada, but by the time these reach the Kejimkujik station much of their acidic constituents has rained out, and hence precipitation is only moderately acidic (mean pH 4.6). The influence of maritime Atlantic weather systems at Kejimkujik is apparent from the rela-

Table 4.2 Comparison of the average volume-weighted concentrations of chemical constituents in 1982–1983 precipitation collected at maritime, continental, and prairie sites in Canada [a]

Constituent		Maritime: Kejimkujik, N.S.	Industrial Continental: Dorset, Ont.	Remote Continental: ELA, Ont.	Prairie: Lethbridge, Alta.
Cations	H^+	25.1	73.6	18.6	1.0
	Ca^{2+}	4.3	10.0	12.0	112.8
	Mg^{2+}	2.9	2.4	2.4	25.5
	Na^+	26.1	3.9	4.3	9.6
	K^+	1.1	1.0	2.0	2.3
	NH_4^+	4.2	15.6	18.9	22.2
Anions	SO_4^{2-}	27.5	58.3	27.1	43.5
	NO_3^-	9.7	35.5	16.4	20.8
	Cl^-	29.5	4.2	5.4	9.9
pH		4.6	4.1	4.7	6.0

[a]Data are in μeq/l. Calculated from CANSAP (1983, 1984).

tively large concentrations of Na and Cl in precipitation. In addition, marine-derived aerosols help to neutralize some of the acidity of precipitation, since the pH of ocean saltwater is about 8.

The fourth site at Lethbridge, Alberta, is located in a prairie landscape. The precipitation is non-acidic (mean pH 6.0) because of the acid-neutralizing influence of calcareous dust that is entrained by wind into the atmosphere, where it affects the chemistry of precipitation. Note that the concentrations of calcium and magnesium are almost one order of magnitude larger at Lethbridge than at the other sites.

Calcareous dust can have a neutralizing influence over a narrow transition between prairie and boreal forest vegetation (Gorham, 1976; Eisenreich et al., 1980; Munger, 1982; Munger and Eisenreich, 1983). Eisenreich et al. (1980) collected snow and rain during precipitation events along a 560-km transect running from a calcareous prairie–agricultural landscape in southwest North Dakota, through an area of forest transition, to the granitic, conifer-forested terrain of northeastern Minnesota. The average volume-weighted pH of snow–rainfall ranged from 5.3 in the agriculture–prairie, to 4.6 in the conifer forest. This was paralleled by a change in the concentration of calcium in precipitation,

from 1.6 mg/l in the southwest, to 0.2 mg/l in the northeastern forest. Gorham (1976) sampled snow along a similar transect in 1975, after a windstorm that entrained a large quantity of prairie dust into the atmosphere. The average pH of precipitation of 10 sites at the prairie–agricultural end of the transect was 8.0 and there was 274 mg/l of particulates, while among nine boreal forest sites in the northeastern part of his transect the mean pH was 4.9 and there was only 17 mg/l of particulates. Another study done in southern Ontario compared the precipitation chemistry of eight sites in a largely forested Precambrian Shield landscape with a thin mantle of hard plutonic minerals to that of three sites in agricultural–forested terrain with calcareous soil, located just south of the Shield periphery (Dillon et al., 1977). The average volume-weighted pH of precipitation among the Shield sites ranged from 4.1 to 4.2, while for the sites south of the shield it was 4.8 to 5.8. The above studies clearly show that dust entrained into the atmosphere can strongly influence the chemistry of precipitation on both a local and a regional scale. This effect is strongest in terrain where agriculture is an important land-use.

Fog moisture can be especially acidic, because ionic constituents are less "diluted" by fog water than they are in precipitation. This phenomenon was first noted by Houghton (1955), in an analysis of fog-water samples collected at five locations in the northeastern United States. Fog water from Mount Washington, New Hampshire, had a pH as acidic as 3.0, and samples from Brooklin, Maine, had a pH as low as 3.5. More recently, pH values as acidic as 3.2–3.5 have been reported in the same study area (Hileman, 1983). In Ohio, 10 fog-water events in 1985–1986 ranged in pH from 2.85 to 4.06 (median 3.8), while rainwater pH ranged from 3.94 to 4.26 (median 4.2) (Muir *et al.*, 1986). Even in California, where rainwater pH is generally above 4.4, a fog-water pH of 2.3 has been recorded (Hileman, 1983).

At relatively high-elevation sites where fog is frequent and windy conditions impact small droplets of water vapor into the coniferous forest canopy, the direct deposition of cloud water and acidity can be considerable (Vogelmann *et al.*, 1968; Siccama, 1974; Lovett *et al.*, 1982). Lovett *et al.* (1982) estimated the total input of moisture to a balsam fir (*Abies balsamea*) stand at 1220 m in New Hampshire, and found that annual precipitation was about 180 cm/year, while direct cloud-water deposition was 84 cm/year (Table 4.3; fog was present 40% of the time). Because many chemical constituents had a relatively large concentration in the cloudwater, their subsequent rate of deposition was

large. Fog-water deposition accounted for 62% of the total deposition of H^+, and 81% of the deposition of SO_4^{2-} and NO_3^- (Table 4.3). Clearly, for stands at high altitude where fog is frequent, the direct cloud-water input of acidity and other atmospheric constituents can be very substantial.

Spatial Patterns of Acidic Deposition

The area that is currently experiencing acidic precipitation is widespread in eastern North America (Fig. 4.1), northwestern Europe, and elsewhere (Cogbill, 1976; Likens and Butler, 1981; Persson, 1982; Rodhe and Granat, 1984; Galloway *et al.*, 1987b). In eastern North America prior to 1955, the region that experienced precipitation with pH < 4.6 was relatively small, localized, and centered over New York, Pennsylvania, and southern New England. This area expanded in the next few decades, as did the area receiving precipitation less than pH 5.6, to the extent that most of the eastern United States and southeastern Canada is now receiving acidic precipitation (Cogbill and Likens, 1974; Cogbill, 1976) (Fig. 4.1). It seems that the general geographical pattern of acidic precipitation in eastern North America has existed since at least the early to mid 1950s. However, since that time the phenomenon has become more widespread, and to a

Table 4.3 Chemical composition of cloud water and deposition of selected chemical constituents to a balsam fir (*Abies balsamea*) forest at 1220 m on Mt. Moosilauke, New Hampshire[a]

Constituent	Concentration (µeq/l; mean ± SD)	Deposition (kg/ha-year)			Percent of Total from Clouds
		Cloud	Bulk	Total	
H^+	288 ± 193	2.4	1.5	3.9	62
NH_4^+	108 ± 89	16.3	4.2	20.5	80
Na^+	30 ± 29	5.8	1.7	7.5	77
K^+	10 ± 4	3.3	2.1	5.4	61
SO_4^{2-}	342 ± 234	275.8	64.8	340.6	81
NO_3^-	195 ± 175	101.5	23.4	124.9	81
H_2O (cm/year)		84	180	264	32

[a]Modified from Lovett *et al.* (1982).

degree its intensity has increased. Clearly, an important aspect of acidic precipitation is its regional character; it impacts a very large expanse of terrain.

Another notable characteristic of acidic precipitation is that it does not generally decrease in intensity with increasing distance from a large point-source emitter of acid anion precursors, such as a power plant or smelter that vents SO_2 to the atmosphere from a tall smokestack. This observation further reinforces the idea that acidic precipitation is regional in character. For example, several studies have reported an insignificant influence of distance from the world's largest point source emitter of SO_2 near Sudbury on the acidity of precipitation (Dillon *et al.*, 1977; Freedman and Hutchinson, 1980a; Chan *et al.*, 1982, 1984b). Furthermore, the acidity of precipitation did not change when that smelter was closed by a strike in 1978–1979 (Scheider *et al.*, 1980). During the 7-month strike period, the hydrogen ion concentration in precipitation averaged 33 μeq/l or pH 4.49, compared with 30 μeq/l or pH 4.52 in the comparable period prior to the shutdown when there were large emissions of SO_2.

In part, this remarkable lack of a distance effect at Sudbury is due to the fact that at any given time, only a relatively small sector of terrain (less than 10%) is directly beneath the smokestack plume. The rest of the area is solely under a regional influence. Millan *et al.* (1982) studied a rainfall event in 1980, and found that rain collected directly beneath the smelter plume had an average pH of 4.1, compared with 4.3 beyond the plume influence. In addition, sulfate was more concentrated by 60% beneath the plume, and the metals Cu, Ni, and Fe were enhanced by a factor of 5–20. Despite this measurable influence of the plume on the chemistry of precipitation that has passed through it, the average ground-level influence in the Sudbury area is minor because the plume sector is small and spatially variable. As a result, the long-range transport of air pollutants appears to have a larger influence than local sources on the chemistry of precipitation, even in the vicinity of the large SO_2 sources at Sudbury (Tang *et al.*, 1987).

Qualitatively similar observations have been made in the vicinity of power plants that emit SO_2,

that is, a relatively minor or no effect of distance from the point source on the acidity of precipitation (Granat and Rodhe, 1973; Hutcheson and Hall, 1974; Enger and Hogstrom, 1979; Kelly, 1984; however, see Li and Landsberg, 1975).

Dry Deposition of Acidifying Substances

Dry deposition of atmospheric constituents takes place in the interval between precipitation events. Dry deposition includes the direct uptake of gases such as SO_2 and NO_x by vegetation, soil, and water surfaces, and the gravitational settling and impaction/filtering of particulate aerosols.

These processes can result in a substantial input of certain atmospheric constituents, including some that can create acidity when they undergo chemical transformation in the receiving ecosystem [Reuss, 1975, 1976; International Electric Research Exchange (IERE), 1981; Glass *et al.*, 1981]. For example, gaseous SO_2 can dissolve directly into the surface water of a lake, or it can be absorbed at the moist surface of the spongy mesophyll after entering plant foliage through stomata, behaving similarly to gaseous CO_2 in this respect. The dry-deposited SO_2 is oxidized to the anion sulfite (SO_3^{2-}), which is then rapidly oxidized to the anion sulfate. The sulfate is electrochemically balanced by the same number of equivalents of hydrogen ion. Similarly, dry-deposited NO_x gas can be oxidized to the anion nitrate, which also generates an equivalent quantity of hydrogen ion. Even atmospheric chemicals that are not in themselves acidic can generate acidity when they are chemically transformed in soil or water. For example, the gas ammonia and the cation ammonium can be deposited from the atmosphere to soil, where if conditions are suitable they will be oxidized to nitrate plus two equivalent units of hydrogen ion (see pp. 93–96). Because inputs of these nonacidic substances can cause acidification when they are chemically transformed, the phrases "acid rain" and "acid deposition" are inaccurate (but shorter) replacements for "the deposition of acidifying substances from the atmosphere."

The dry deposition of sulfur and nitrogen compounds is greatest in areas with a large atmospheric concentration of gaseous NO_x and SO_2. Unfortunately, processes of dry deposition to ecosystems are not well understood, and the quantitative estimation of dry fluxes is difficult. This is especially true of sulfur and nitrogen compounds. The major reasons for the uncertainty are the paucity of data concerning (1) the spatial and temporal variations of concentration of atmospheric gases and particulates and (2) the rates of their deposition to natural ecosystems of great structural complexity, and under various weather conditions (Unsworth, 1979; Fowler, 1980a,b; Grennfelt et al., 1980). In spite of these problems, several estimates of dry deposition to natural ecosystems have been made, and they have helped to put into perspective the rates of dry deposition of S, N, and acidity, relative to their rates of wet input with precipitation.

A conceptually and mathematically simple model of dry deposition was computed by Fowler (1980a), using data from the literature on the average atmospheric concentrations of S- and N-containing gases and particulates, and their likely average deposition velocities to natural surfaces. These data were used to compute the dry-deposition fluxes to three landscapes located at different distances from sources of pollution. The computed dry inputs were then compared to the inputs via precipitation. Although this simple budget ignores a great deal of spatial and temporal variation by using average values for deposition velocities and the concentrations of pollutants, it is nevertheless useful for gaining an appreciation of the relative magnitudes of the various fluxes. Total (wet + dry) depositions of both S and N were much larger near to the sources than further away (Table 4.4). This pattern was almost entirely due to a difference in dry deposition (especially of gaseous S and N), since wet deposition was much less affected by distance from the emission sources. Dry deposition dominated in the relatively polluted environment close to the sources, where it was 8.5 times larger than wet deposition for S, and 4.9 times larger for N. Even at an intermediate distance, dry deposition was larger than wet, while at a relatively remote distance the

two input sources were similar in magnitude. These simple calculations indicate that the dry deposition of acidic substances and acid-forming precursors can be quantitatively important, and that it dominates the total input in a relatively contaminated environment close to sources of emission.

The determination of atmospheric deposition in the vicinity of large point sources also reveals this general pattern. Chan et al. (1984a) computed sulfur deposition within a 40-km radius of the largest smelter at Sudbury. They found that 55% of the total S deposition of 15.9×10^3 kg/day was due to dry deposition (Table 2.11, Chapter 2). Of this, 91% was dry deposition of SO_2, and the rest was dry input of sulfate aerosol. However, because the smokestack at Sudbury is so tall (380 m) and effective at dispersing emissions, the local ground-level pollution is only moderate. As a result, the total S deposition within a 40-km radius is only 16% larger than the background S deposition that would be expected if the smelter was not present. In fact, <1% of the total emission of SO_2 from the smelter is deposited within 40 km of the source; virtually all of the SO_2 emission is transported over a longer distance.

Several other studies have estimated dry and wet inputs to forests. Grennfelt et al. (1980) calculated atmospheric deposition to a coniferous forest in southern Sweden, and found that dry inputs accounted for 57% of the total input of 23 kg S ha^{-1} year^{-1}. Uptake of SO_2 accounted for 68% of the total dry input, and impaction of sulfate aerosol the remainder. The dry input of nitrogen accounted for 54% of the total N input of 13 kg ha^{-1} year^{-1}. At Hubbard Brook, New Hampshire, total S deposition to a northern hardwood forest was 19 kg ha^{-1} year^{-1}, of which 31% was dry uptake of SO_2, and <2% was dry particulate deposition (Eaton et al., 1980). Of the total deposition of S of 25.6 kg ha^{-1} year^{-1} to a hardwood forest in Tennessee, 56% was dry deposition, while for nitrogen dry inputs accounted for 63% of the total input of 7.6 kg NO_3-N ha^{-1} year^{-1} and 33% of the 2.5 kg NH_3-N ha^{-1} year^{-1} (Lindberg et al., 1986). Dry inputs were also important for a beech (Fagus sylvatica) forest at Solling in central West Germany, where they

Table 4.4 Deposition budgets for several important sulfur and nitrogen compounds (expressed as S or N) to terrestrial landscapes at various distances from emission sources[a]

	Deposition Velocity (cm/sec)	Distance from Emission Sources					
		3–30 km		30–300 km		300–3000 km	
		Concentration (µg/m³)	Flux (kg/ha-year)	Concentration (µg/m³)	Flux (kg/ha-year)	Concentration (µg/m³)	Flux (kg/ha-year)
Dry deposition							
Gas SO_2	0.8	100	126.2	20	25.2	5	6.3
Gas NO_2	0.5	30	14.4	5	2.4	2	1.0
Particulate SO_4^{2-}	0.1	20	2.1	10	1.1	10	1.1
Particulate NO_3^-	0.1	5	0.4	2	0.1	2	0.1
Wet deposition (90 cm/year precipitation)							
SO_4^{2-}		5	15.0	4	12.0	3	9.0
NO_3^-		1.5	3.0	0.7	1.4	0.5	1.0
Total deposition, S			143.3		38.3		16.4
Total deposition, N			17.8		3.9		2.1
Dry/wet, S			8.5		2.2		0.8
Dry/wet, N			4.9		1.2		1.1

[a]Modified from Fowler (1980a).

accounted for at least 38% of the total S input, >17% of NO_3-N, and >56% of total hydrogen ion. The dry inputs caused a much larger sulfate flux in precipitation reaching the forest floor of the beech stand (measured as throughfall + stemflow) than in incident precipitation (i.e., 48 versus 24 kg ha^{-1} year^{-1}). Because of its more complex surface architecture, larger foliage biomass, and longer foliage persistence, this enhancement effect was even more pronounced in a nearby spruce (*Picea abies*) forest, where throughfall plus stemflow was 80 kg ha^{-1} year^{-1} (Mayer and Ulrich, 1974, 1977; Heinrichs and Mayer, 1977).

4.3
CHEMICAL CHANGES
WITHIN THE WATERSHED

The chemistry of precipitation is greatly altered by interactions that take place within watersheds—especially in the terrestrial environment. Processes that are particularly important are ion exchange, uptake and leaching of nutrients by plants, and chemical transformations by microbial and inorganic reactions.

Canopy Effects

The first substrates that incoming precipitation encounters in a forest are the foliage and bark surfaces of trees. Most of the incoming water penetrates through the canopy and reaches the forest floor as a flux called throughfall (TF), while another relatively minor fraction runs down the trunks of trees as stemflow (SF). In addition, some of the incoming precipitation is intercepted by the canopy and evaporated back to the atmosphere (canopy evaporation, CE). During the growing season the incoming precipitation to four conifer stands in Nova Scotia was partitioned into an average of 34% CE, 66% TF, and <1% SF, while for four angiosperm stands it was 22% CE, 75% TF, and 3% SF (Freedman and Prager, 1986). Averaged over an entire year, Eaton *et al.* (1973) reported 14% CE, 81% TF, and 5% SF for a hardwood forest at Hubbard Brook, N.H. Be-

cause the evaporation of water back to the atmosphere as CE is essentially a distillation process, this should cause a proportionate concentrating of chemical constituents in TF + SF, compared with ambient precipitation.

However, the change in the chemistry of throughfall and stemflow is much larger than this distillation effect. Several major ions, particularly potassium, are relatively easily leached from foliage. Freedman and Prager (1986) reported an average volume-weighted concentration of potassium in TF + SF of 2.1 mg/l (2.0 mg/l in TF and 5.7 mg/l in SF) among eight stands that they studied, compared with <0.2 mg/l in the incident precipitation. The concentrations of calcium and magnesium were also enhanced in TF + SF (0.78 and 0.26 mg/l, respectively) compared with precipitation (0.26 mg/l and 0.06 mg/l, respectively). Experimental studies have shown that treatment with an acidic solution increases the rate of leaching of K, Ca, and Mg from foliage, even at an acidity that does not cause visible injury (Fairfax and Lepp, 1975; Lepp and Fairfax, 1976; Tukey, 1980; Tveite, 1980a). However, there is no evidence that increased leaching of these nutrients causes a physiologically important decrease in their foliar concentration (Tveite, 1980a,b; Morrison, 1984).

The average volume-weighted concentration of hydrogen ion in the ambient precipitation sampled by Freedman and Prager (1986) was 0.041 mg/l or pH 4.4. Acidity averaged slightly less in TF + SF of the four hardwood stands (0.018 mg/l or pH 4.7) and in two of four conifer stands (these averaged pH 4.5; the other two were pH 4.4). The decrease in acidity was probably caused by ion-exchange reactions taking place on foliage and bark surfaces, with H$^+$ largely exchanging for Ca^{2+}, Mg^{2+}, and K$^+$. Coupled with reduced water volume, the generally lesser acidity of TF + SF compared with incident precipitation resulted in an average canopy "consumption" of 66% of the precipitation input of hydrogen ion to the hardwood stands, and 42% for the conifer stands. A similar effect has been reported by others. Eaton *et al.* (1973) calculated an average annual H$^+$ consumption of 91% by a hardwood forest canopy at Hubbard Brook. Mahendrappa

(1983) found an H^+ consumption of 21–80% among seven stands in New Brunswick. Mollitor and Raynal (1983) reported 43% consumption in an Adirondack, N.Y., hardwood stand, and 14% in a conifer stand. Foster (1983) reported 38% consumption by a hardwood canopy in northern Ontario. Cronan and Reiners (1983) reported 31% for a New Hampshire hardwood stand, but an increase in H^+ flux of 16% under conifers. Finally, Abrahamsen *et al.* (1976) reported 47% consumption of H^+ flux (pH 4.5) under a *Betula* canopy in Norway compared with precipitation (pH 4.3), while flux increased by a factor of 2.2 times under *Pinus sylvestris* (pH 4.0), and was unchanged under *Picea abies*.

In an area where the atmospheric concentration of SO_2 and aerosol sulfate is small, the sulfate flux in TF + SF tends to be only slightly larger than that in incident precipitation. Freedman and Prager (1986) reported an average growing season TF + SF flux of 4.2 kg SO_4-S/ha (equivalent to 1.3 mg SO_4-S/l) in hardwood stands and 5.2 kg/ha (1.7 mg/l) under conifers, compared with 3.2 kg/ha (0.8 mg/l) in ambient precipitation, for an average enhancement of sulfate flux in TF + SF of 47%. In a more polluted environment, the enhancement of sulfate flux in TF + SF can be much larger, because of washoff of dry-deposited SO_2 and sulfate. Eaton *et al.* (1973) observed an average annual sulfate enhancement of 414% at Hubbard Brook, while in central West Germany Heinrichs and Mayer (1977) reported an enhancement of 200% in a beech forest, and 330% in a spruce forest. The study sites of Eaton *et al.* and Heinrichs and Mayer are located in relatively polluted regions, where there is an enhanced atmospheric concentration of aerosol sulfate, and to a lesser extent sulfur dioxide.

Chemical Changes in Soil

After precipitation reaches the forest floor, it begins to percolate through the soil. Important chemical changes take place as this water moves downward and interacts with mineral soil constituents, organic matter, microbes, and plant roots. Some examples are (1) the biota can selectively take up or release chemical ions; (2) ions can be selectively exchanged at non-living surfaces with exchange capability, such as those of clay and soil organic matter; (3) water-insoluble minerals can be made soluble by weathering processes; and (4) secondary minerals such as certain clays and sesquioxides of iron and aluminum can be formed by crystallization/precipitation reactions involving particular ions. These various chemical changes can contribute to acidification, to the leaching of basic cations such as calcium and magnesium, and to the mobilization of toxic cations of aluminum (especially Al^{3+}). These closely linked processes take place naturally in all sites where there is an excess of precipitation over evapotranspiration, and they may also be influenced by the nature of the vegetation. A potential influence of acidic deposition is to increase the rates of some of these processes, and to enhance the rates of leaching of toxic Al^{3+} and H^+ to surface waters.

Some of these chemical changes have been examined experimentally, by leaching soil monoliths isolated in plastic tubes with simulated rainwater of various pH. These lysimeter experiments have shown that a very acidic leaching solution can cause a number of effects on soil chemistry (Overrein, 1972; Abrahamsen *et al.*, 1976, 1977; Abrahamsen and Stuanes, 1980; Farrell *et al.*, 1980; Stuanes, 1980; Morrison, 1981, 1983; Tyler, 1981). These include (1) an increased rate of leaching of calcium, magnesium, and potassium, resulting in a decreased base saturation of the cation exchange capacity; (2) an increased concentration of toxic, cationic forms of metals such as Al, Fe, Mn, Pb, Ni, etc.; (3) acidification of the soil; and (4) saturation of sulfate adsorption capacity, after which sulfate leaches at about the rate of input, with a secondary effect on soil acidification if the mobile sulfate is accompanied by base cations and, more importantly, a secondary acidifying and toxic effect on surface waters if the leaching sulfate is accompanied by Al^{3+} and H^+ (Reuss *et al.*, 1987).

Some of these effects are illustrated in Tables 4.5 and 4.6. In the first experiment (Table 4.5), monolith profiles of two sandy soils were collected from jack pine (*Pinus banksiana*) stands in Ontario, and leached with an artificial "rainwater" adjusted with

Table 4.5 Concentration of selected chemical constituents in percolate from lysimeters containing a profile of sandy soil from jack pine (*Pinus banksiana*) stands in northern Ontario[a]

	pH of Leaching Solution	Concentration in Percolate (μeq/l)			
		Ca	Mg	All Bases	SO_4
Soil A	5.7	740	50	994	229
	4	670	49	896	224
	3	745	73	1098	152
	2	872	204	1348	527
Soil B	5.7	422	174	727	146
	4	446	183	765	152
	3	399	165	701	277
	2	3722	1474	5504	5427

[a]Data are for week 152; the leaching solution was added at 100 cm/year in equal weekly 1–2 hr events; average of three replicates per treatment. Modified from Morrison (1981, 1983).

mineral acid to pH 5.7, 4, 3, or 2. The percolate emerging from the bottom of the lysimeter was collected and analyzed chemically. The effects on soil acidity were minor. Even after 3 years of treatment with a solution as acidic as pH 2, the pH of the percolate was >6.5, compared with an initial pH of 7.3 (Morrison, 1981, 1983). An effect on the concentrations of calcium, magnesium, sum of basic cations (i.e., Ca + Mg + K + Na), and sulfate only occurred in the artificially very acidic pH 2 treatment. For all of these chemical constituents, no effects were observed until a time- (i.e., dose-) dependent threshold had passed, after which there were consistently larger concentrations in the percolate. For example, the sulfate concentration in the pH 5.7 treatment remained stable at less than 20

Table 4.6 Effect of irrigating a coniferous forest soil between 1972 and 1977 with several rates of acid $\pm$ NPK fertilizer: Average of eight replicate lysimeters per treatment[a]

				Quantity of Base Cations (keq/ha)			
	Acid Added (keq H+/ha)	pH of A_0	% Base Saturation of A_0	A_0	Mineral Soil		
Treatment[b]					0–5 cm	5–10 cm	Total
Control	0.0	4.0	16.0	5.2	1.8	0.7	7.7
Acid 1	6.1	3.9	12.2	3.2	1.9	0.6	5.7
Acid 2	13.3	3.8	9.0	2.1	0.9	0.5	3.5
NPK	0.0	4.2	20.1	5.7	—	—	—
NPK + acid 1	6.1	3.9	15.2	5.0	3.3	0.8	9.1
NPK + acid 2	13.3	3.8	11.5	3.5	2.7	0.8	7.0

[a]Modified from Farrell et al. (1980).

[b]Acid 1, annual application of H_2SO_4 at 15 kg S/ha-year; acid 2, annual application of H_2SO_4 at 33 kg S/ha-year; NPK, total application of 360 kg N/ha as NH_4NO_3, plus 80 kgP/ha and 150 kg K/ha as a compound PK fertilizer.

μeq/l for both soil types. In the pH 2 treatment, sulfate concentration remained at about 20 μeq/l for the first 100 and 150 weeks for the two soils, respectively, after which the concentration in the percolate rapidly increased to 400 μeq/l and 150 μeq/l, respectively. This threshold indicates that the sulfate adsorption capacity of the soil had become saturated, after which the added sulfate essentially flushed through the system, as a highly mobile anion.

The study of Farrell *et al.* (1980) involved the treatment of soil columns in the field with two rates of addition of sulfuric acid, which were equivalent to 15 and 33 kg S ha^{-1} year^{-1} (Table 4.6). These treatments simulated the rate of loading experienced with 100 cm/year of precipitation having a pH of 4.0 or 3.7, respectively. After 6 years of treatment the effect on pH of the surface organic layer was minor. However, there was a large decrease in the base saturation of cation exchange capacity, characterized by a decreased quantity of Ca + Mg + K + Na in the surface organic layer and in the top 0–5 cm of mineral soil, especially at the larger rate of acid loading. These effects are more marked than those observed by Morrison (1981, 1983) (Table 4.5), who only found a change in the concentration of these ions at pH 2. Farrell *et al.* (1980) found that the effect of acid treatment on base saturation could be largely overcome by the addition of NPK fertilizer; this was attributed to the presence of potassium in the fertilizer (Table 4.6). Farrell *et al.* (1980) also determined sulfur retention in their experiment. At their larger acid loading, the total input to the columns was equivalent to 241 kg S/ha; 28% of this was ambient bulk deposition, and 72% was the experimental addition. Of the total input, 66% was recovered as sulfate leached from the lysimeter; the other 34% was retained in the columns by various sulfate absorption mechanisms. The vertical distribution of the retained sulfur was investigated using a ^{35}S radiotracer technique. Three months after adding the radioactive sulfur, 24% of the retained ^{35}S was present in the organic strata (vegetation + forest floor). The rest was in the mineral soil, especially in the B horizon.

Soil Acidity

Factors Affecting Soil Acidity

The acidity of soil is influenced by a complex array of physical/biological chemical transformations and exchanges. Some of the more important chemical processes that affect soil acidity are summarized below. All of these processes operate naturally, but their rates can potentially be affected by acidic deposition.

Carbonic Acid.   In the forest floor and in soil rich in organic matter, there can be a large concentration of CO_2 (frequently >1%) in the interstitial soil atmosphere because of respiration by saprophytes and plant roots. As a result, there is a relatively large equilibrium concentration of H_2CO_3 in the soil solution, and the pH can be more acidic than the pH 5.65 realized from a CO_2 equilibrium with the above-ground atmosphere. This carbonic acid effect is most influential in soil having a pH >6 and with appreciable alkalinity—which usually means a calcareous substrate or a relatively young and unweathered, noncalcareous soil of glacial or other primary origin. In relatively acidic soil (i.e., pH < 5.0–5.5) with less or no alkalinity, the influence of partial pressure of CO_2 on acidity is relatively unimportant (Russell, 1973; Reuss, 1975, 1976; Hausenbuiller, 1978; Bache, 1980).

The Nitrogen Cycle.   Plant uptake/release and chemical transformations of nitrogen can have a large and complex effect on soil acidity (Kirkby, 1969; Russell, 1973; Reuss, 1975, 1976). Most plants take up inorganic nitrogen from the soil solution as one or both of NH_4^+ or NO_3^-. However, in acidic soil with a pH less than ~5.0–5.5, the dominant form of inorganic nitrogen is NH_4^+. The NH_4^+ may have originated from ammonification of organic nitrogen to form ammonia (NH_3), which then consumes one H^+ to form an ion of NH_4^+ (Fig. 4.2). If the NH_4^+ is taken up by a plant root, an equivalent quantity of H^+ is excreted to the soil solution in order to maintain electrochemical neutrality. In this case, there is no net change in the acidity of the soil.

(a) Nitrogen Transformations

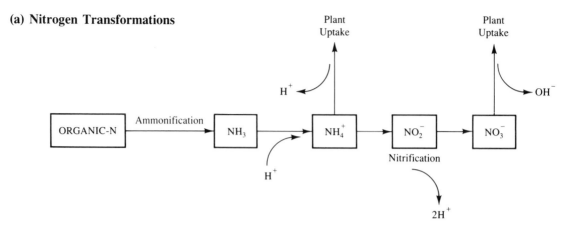

(b) Sulfur Transformations

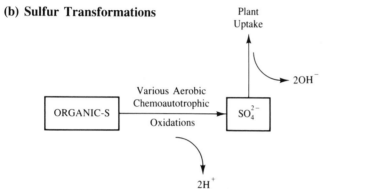

Fig. 4.2 A simplified flow diagram of the acidifying effect of some nitrogen and sulfur transformations in soil. See text for an explanation. Modified from Reuss (1975).

However, if there is a direct input of NH_4^+ to the soil, for example, from atmospheric deposition or by fertilization, then plant uptake of NH_4^+ with concomitant H^+ release will have a net acidifying effect.

In less acidic soil with pH > 5.5, most of the inorganic nitrogen occurs as NO_3^-. Much of this nitrate is produced by microbial oxidation of NH_4^+ via a NO_2^- intermediate in the acid-sensitive process of nitrification, carried out by bacteria in the genera *Nitrosomonas* and *Nitrobacter*. Overall, the oxidation of NH_4^+ to NO_3^- generates two H^+ for every NO_3^- produced. If the ammonium substrate was derived from ammonification of organic nitrogen, which consumes one H^+ per NH_4^+ produced,

then the net effect is production of one H^+ per NO_3^- ion produced ultimately from organic nitrogen. However, if the nitrate is subsequently taken up by a plant root, then to maintain elecrochemical neutrality one OH^- would be excreted, which is equivalent to the consumption of one H^+. Therefore, the net effect on soil acidity is zero.

If there is a direct input of NH_4^+, which then serves as a substrate for nitrification, then acidification will take place. Such a direct input could occur by the deposition of ammonia gas or ammonium ion from the atmosphere, or by the addition of an inorganic fertilizer such as ammonium nitrate, ammonium sulfate, or urea. The acidifying effect of the use of ammonium-containing fertilizer is a well-recognized phenomenon (Hausenbuiller, 1978; Bache, 1980; Jenny, 1980).

The Sulfur Cycle. The transformation and uptake of sulfur compounds can also influence soil acidity (Russell, 1973; Reuss, 1975, 1976). Much of the sulfur in soil is present in the reduced form of organic sulfur. This can be transformed by various microbial processes through incompletely oxidized forms such as sulfides, elemental sulfur, and thiosulfates, but in an aerobic environment all of these are ultimately oxidized to sulfate by aerobic chemoautotrophs. The oxidation of organic sulfur to SO_4^{2-} is accompanied by the release of an equivalent quantity of H^+ (i.e., two unit-weights of H^+ per unit of SO_4^{2-} produced). If the SO_4^{2-} is taken up by plant roots, then an equivalent quantity of OH^- is excreted to the soil solution to conserve electrochemical neutrality, and there is no net effect on soil acidity. However, if there is a direct input of sulfate by atmospheric deposition, followed by plant uptake, then the net effect would be a reduction of soil acidity. If the soil is anaerobic, for example, because of a high water table, then the sulfate would be reduced to sulfide or another reduced compound by anaerobic chemoautotrophs, accompanied by the consumption of an equivalent quantity of H^+ per equivalent of SO_4^{2-} reduced.

In most terrestrial ecosystems the sulfur cycle has a much smaller effect on soil acidity than the nitrogen cycle. However, there are certain situations in which the sulfur cycle dominates. Most notably, the drainage of wetlands causes the previously anaerobic soil to become aerobic. Under this condition, a large quantity of acidity can be generated by the chemoautotrophic oxidation of reduced forms of sulfur, especially sulfides. In agriculture these are called "acid sulfate soils," and pH <3 can be present for as long as 10 years following the drainage of wetlands that are susceptible to this syndrome (Dost, 1973; Rorison, 1980).

Leaching of Ions. In many well-drained terrestrial soils, sulfate and chloride leach readily to groundwater. For sulfate, this is especially true in the relatively young forest soils of formerly glaciated terrain of much of the northern hemisphere. In such soils sulfate adsorption capacity is small and easily saturated by enhanced atmospheric inputs caused by anthropogenic emissions; older, highly weathered soils of more southerly latitudes tend to have a considerably larger sulfate adsorption capacity (Reuss *et al.*, 1987). Nitrate may also leach, especially after the site is disturbed or fertilized. To preserve electrochemical neutrality, the mobile SO_4^{2-} and NO_3^- anions must be accompanied by an equivalent quantity of cations. In calcareous soil and in relatively young and unweathered soil, bases such as Ca^{2+} and Mg^{2+} comprise a large fraction of the leaching cations. The Ca^{2+} and Mg^{2+} originate from either the solubilization of primary carbonates and silicates, or by cation exchange with H^+. In soil with a small carbonate content, the loss of calcium and magnesium is important in reducing the acid-neutralizing ability of the soil, and in acidification via a reduction of base saturation of cation exchange capacity (Reuss, 1975, 1976; Bache, 1980; Tabatabai, 1987). In addition, aluminum and hydrogen ions may leach with mobile anions from base-poor soils and into surface waters, where they can cause acidification and toxic stress to the aquatic biota (Cronan and Schofield, 1979; Reuss *et al.*, 1987; see also Section 4). In many old, highly weathered soils of the humid tropics, intense leaching for a long period of time has removed much of the original content of silicate and base cations, leaving behind a matrix dominated by aluminum and iron hydroxides and silica-poor clays such as kaolinite (Ellenberg, 1986; Lavelle, 1987). Such soils can be very acidic, with pH values of about 4.0 or less, and there can be toxic concentrations of exchangeable aluminum (Uhl, 1983, 1987; Swift, 1984; dos Santos, 1987).

Soil Buffering Systems. Soil is usually strongly buffered, in comparison to precipitation or fresh water. Therefore, soil is relatively resistant to pH change, especially once it is already acidic. Different buffering systems come into play at particular ranges of soil pH. Carbonate minerals buffer soil within the pH range of >8 to 6.2; silicates from pH 6.2 to 5.0; cation exchange capacity from pH 5.0 to 4.2; aluminum from pH 5.0 to 3.5; iron from pH 3.8 to 2.4; and humic substances from pH 6 to 3 (Hausenbuiller, 1978; Bache, 1980; Seip and Freedman, 1980; Ulrich, 1983).

The role of aluminum buffering is especially important in strongly acidic soil (pH $<$ 5). In addition, exchangeable aluminum can contribute to total soil acidity through the ionization and hydrolysis of Al^{3+}, as follows (after Reuss, 1975; Bache, 1980; Reuss and Johnson, 1985):

$$Al^{3+} + H_2O \leftrightharpoons AlOH^{2+} + H^+$$

This reaction is most important at about pH 3.5. The $AlOH^{2+}$ can also hydrolyze to produce $Al(OH)_2^+ + H^+$ at about pH 5. However, since $Al(OH)_2^+$ is generally insoluble in acidic soil, it contributes little to the soil acidity.

Concentration of Soil Solutes.   Soil acidification is caused by the accumulation of soluble inorganic and organic acids, at a rate faster than their neutralization or removal by leaching. Ionization of these acids produces free H^+ in the soil solution. Exchangeable H^+ from cation exchange surfaces also contributes to acidity. The relative contribution of exchangeable H^+ is influenced by the total osmotic strength of the soil solution; the more concentrated the extracting solution, the greater the amount of H^+ that is exchanged. This fact is important in the measurement of soil pH. If distilled water is used as an extracting solution, the measured pH value is typically 0.5–1.0 pH units higher than that obtained when using a standardized salt solution as an extractant, usually $0.01\,M$ $CaCl_2$ (Russell, 1973; Allen *et al.*, 1974). Similarly, in a natural ecosystem the acidity of soil may be influenced by temporal and spatial variations in the total concentration of solutes. In addition, a location with a relatively large atmospheric input of total ions (e.g., a site near the ocean) may have a larger exchange acidity in soil than a more continental site (Rosenqvist, 1978a,b; Seip and Tollan, 1978).

Acidification of Soil

Soil acidification is a naturally occurring process. This can be illustrated by several classical studies of primary succession on newly exposed parent material. Crocker and Major (1955) examined a chronosequence of deglaciated sites at Glacier Bay, Alaska. The parent material that is initially available for biological colonization after the meltback of glacial ice is a fine morainic till with pH 8.0–8.4. The mineral composition is primarily of granite, gneiss, and schist, with as much as 7–10% carbonate mineral. As this primary substrate is progressively leached by precipitation water and colonized and modified by the developing vegetation, the acidity of the mineral soil increases progressively. The pH decreased to ~5.0 after about 70 years, when a spruce–hemlock (*Picea–Tsuga*) conifer forest began to develop. Acidity of the mineral soil eventually stabilized at pH 4.6–4.8 under a mature conifer forest, while the forest floor stabilized at pH 4.4. This acidification was accompanied by a large reduction of calcium in the surface mineral soil, from an initial average concentration of ~ 5–9%, to less than 1%. The decreased concentration of calcium in soil is a result of its leaching to below the rooting depth, and its uptake by vegetation. Other soil changes include a successional accumulation of organic matter and nitrogen, due to the biological fixation of atmospheric CO_2 and N_2.

Other studies of primary succession examined chronosequences of sand dunes on Lakes Michigan and Huron (Olson, 1958; Morrison and Yarranton, 1973). The initial development of vegetation on newly exposed sand consists of a dunegrass community dominated by *Ammophila breviligulata* and *Calamovilfa longifolia*. With time, a tallgrass prairie develops, dominated by various grasses and dicotyledonous herbs. The prairie is invaded by shade-intolerant shrub and tree species, which form forest nuclei. Eventually an edaphic climax develops that is dominated by oaks (*Quercus* spp.) and tulip-tree (*Liriodendron tulipifera*). Soil development in this sere is qualitatively similar to that observed at Glacier Bay (Olson, 1958). There is an increase in acidity, organic matter, and fixed nitrogen, and a decrease in carbonate minerals. Initially, $CaCO_3$ comprises about 1.5–2.5% of the sandy substrate. This decreases to less than 0.15% in the upper 10 cm after about 400 years of succession, because of the uptake of Ca by vegetation, and the solubilization and downward leaching of Ca, in conjunction with the development of a white, acidic, highly siliceous eluviated horizon through the process of podsolization. Concurrently, the soil acidified from an initial pH of 7.7, to ~pH 4.5 after 1000 years.

An important reason why successively aggrading vegetation causes the acidification of soil is that plant uptake of the bases Ca, Mg, and K is accompanied by the excretion of hydrogen ions into the soil solution. Over a long period of time (i.e., centuries) in natural ecosystems, there may not be a net acidifying effect via this process, because the bases are eventually returned to the soil surface by litterfall. If there is no net accumulation of organic matter on the soil surface, then the organically bound Ca, Mg, and K in litterfall is mineralized by microbes and returned to the soil. Alternatively, these cations may be returned to the soil in the form of base-rich ash following wildfire. However, in the shorter term, if the quantity of organic matter is increasing on the site in the form of successively aggrading plant biomass and/or an increasing quantity of organic matter in the forest floor, then soil acidification occurs. The reason for this is that the rate of incorporation of bases into organic matter exceeds their rates of input or recycling by such processes as mineralization, decomposition, weathering, and atmospheric deposition (Oden, 1976; Rosenvist, 1978a, 1980; Rosenqvist et al., 1980; Nilsson et al., 1982). Oden (1976) calculated that a typical aggrading pine forest could add 340 eq ha^{-1} year^{-1} of H$^+$ to the soil by the net uptake of basic cations, and 55 eq ha^{-1} year^{-1} by the accumulation of organic matter in the forest floor. This input of acid is similar in magnitude to the wet atmospheric input in many areas that are experiencing severe acidic precipitation. For example, the input of H$^+$ in 100 cm/year of precipitation with an average pH of 4.2 would be ~630 eq ha^{-1} year^{-1}. Therefore, an aggrading forest can cause the accumulation of a substantial quantity of acidity in the soil, and when this process combines with a large atmospheric deposition, the potential for acidification is great.

Additional acidification takes place if biomass is harvested from the site, since this removes a large quantity of bases that are incorporated into the biomass (Russell, 1973; Rosenqvist, 1978a, 1980; Rosenqvist et al., 1980; Nilsson et al., 1982). Freedman et al. (1986) predicted the nutrient removals by conventional bole-only and whole-tree clearcuts of eight mature stands of forest in Nova Scotia. On average, the bole-only clearcuts were predicted to remove 9300 eq Ca/ha, 1500 eq K/ha, and 1500 eq Mg/ha, for a total of 12,300 eq/ha. The whole-tree clearcuts were predicted to remove 18,000 eq Ca/ha, 3700 eq K/ha, and 3200 eq Mg/ha, for a total of 24,900 eq/ha. If a harvest rotation of 100 years is assumed, then acid-neutralizing capacity (by Ca, Mg, and K) of the site would be harvested at an average rate of 123 eq ha^{-1} year^{-1} by a conventional clearcut, and 249 eq ha^{-1} year^{-1} by a whole-tree clearcut. This removal of basic cations by harvesting would contribute to the acidification of the site, unless the cations were regenerated by the weathering of minerals, by atmospheric inputs, or by liming or fertilization as a management practice.

A number of studies have investigated whether the acidity of forest soils has increased as a result of the atmospheric input of acidifying substances. Conceptually, the more important soil effects are likely to be (1) a decrease in soil pH; (2) a decrease in base saturation of cation exchange capacity; (3) an increase in the saturation of cation exchange capacity by H$^+$ and aluminum ions, especially Al^{3+}; and (4) a saturation of sulfate adsorption capacity, leading to the leaching of mobile sulfate anions, accompanied by base cations and toxic Al^{3+} and H$^+$ (Oden and Andersson, 1971; Wiklander, 1975, 1980; Frink and Voigt, 1977; McFee et al., 1977; Bache, 1980; Morrison, 1984; Reuss et al., 1987). As described previously, these potential effects have been examined in experiments where soil-column lysimeters were treated with artificial rain-water solutions of various pH. In the short term, a notable effect only occurred when highly acidic leaching solutions were used, that is, pH ≤ 3 (Abrahamsen et al., 1976, 1977; Abrahamsen and Stuanes, 1980; Bjor and Tiegen, 1980; Stuanes, 1980; Morrison, 1981, 1983; however see Farrell et al., 1980, for an effect at higher pH). It could be argued that in terms of H^+ flux, 1 year of treatment with an artificial rain at pH 3 is equivalent to 10 years of treatment at pH 4. If this were true, then an effect observed with an artificially acidic pH treatment in a lysimeter experiment, could be extrapolated to an effect that might occur over a longer time period at a less acidic pH, for example, in a field situation exposed to ambient acidic precipitation. However, this is a controversial point, and there is no direct evidence that such an extrapolation is reasonable. There is a need for longer-term experiments using realistic rates of acid loading, so that the potential effects of atmospheric deposition on soil acidity can be better predicted.

In addition to the experimental approach, surveys of soil chemistry at the same location but conducted at different times can indicate whether acidification has taken place. Several studies have shown that the conversion of former agricultural land to forest can cause a substantial acidification of the soil, especially if conifers dominate the developing stand (Williams et al., 1979; Alban, 1982; Brand et al., 1986). For example, the afforestation of abandoned farmland in Ontario with red pine (Pinus resinosa,) Scotch pine (P. sylvestris), or white spruce (Picea glauca) caused an average acidification of about 1.1 pH units (from pH 5.7 to 4.7) after about 46 years (Brand et al., 1986). In general, it is well accepted that afforestation will cause an acidification of the site.

Of more relevance to the atmospheric deposition of acidifying substances is whether sites that are already forested will acidify further or more rapidly because of the enhanced atmospheric inputs. In part, this problem can be examined by resampling forest soils after widely spaced intervals of time. In one study, Linzon and Temple (1980) found no increase in soil acidity in a resurvey of six sites in Ontario after a 16-year interval, in an area where the mean annual pH of precipitation is 4.0–4.1. In another study, Troedsson (1980) compared analy-

ses of forest floor samples collected in Sweden in 1961–1963 to those collected in 1971–1973. He reported a general relationship between stand age and increasing soil acidity. A comparison of data from the two time intervals showed a minor change in pH and exchangeable aluminum, but a decreasing concentration of soil Ca, Mg, and K, implying that acidification may have taken place via a decrease in the base saturation of cation exchange capacity. However, it is important to note that the design of the above studies does not distinguish between the natural effects of increasing stand age on acidification, and changes caused by atmospheric deposition.

In another study, Oden and Andersson (1971) mapped the distribution of soil pH and base saturation of cation exchange capacity in Sweden. They found that a relatively acidic pH and low base saturation occurred in southern Sweden, where precipitation is more acidic than in the north. Because local sources of gaseous pollutants are relatively abundant in southern Sweden, the dry deposition of acidifying substances may also have influenced this spatial pattern, as could have various climatic and biological factors.

Another study in Sweden combined aspects of several of the studies just described, that is, (1) resampling soil in permanent plots after a widely spaced time interval and (2) a comparison of areas receiving a large or a small deposition of acidifying substances from the atmosphere (Hallbacken and Tamm, 1985; Tamm and Hallbacken, 1986, 1988). These researchers resampled soil in 1982–1983 at sites that had originally been sampled in 1927. They used a permanently staked grid in an experimental forest in Sweden, a design that allowed them to dig their new soil pits less than 1 m from the originals. The 1927 pH measurements were made with a quinhydrone/calomel electrode, and to avoid analytical bias the same method was used in 1982–1983 instead of the conventional glass electrode. The original landscape of the site in southern Sweden was an open Calluna vulgaris–Juniperus communis heathland that was planted to Picea abies beginning in the 1870s. In the mid-1920s there were middle-aged spruce plantations 37–50 years old, and there was also a natural hardwood forest dominated by beech, birch, and oak (*Fagus sylvatica*, *Betula* spp., and *Quercus robur*). In 1982–1983, there were mature spruce stands >100 years old, and a mature beech–oak forest. A large increase in acidity occurred throughout the soil profile between 1927 and 1982–1983 (Table 4.7). This change was especially large in the surface humus layer, which in the beech forest acidified from an initial average pH of 4.5, to 3.8 in the early 1980s, and from 4.6 to 3.6 in the spruce stands. To a lesser degree the mineral soil also acidified, with the change being somewhat smaller under beech than under spruce. The changes in the B and C horizons were much larger than those observed by Tamm and Hallbacken (1988) in a similar study done in spruce stands northern Sweden, where the rate of atmospheric deposition is considerably smaller. As a result, they interpreted their data to indicate that the atmospheric deposition of acidifying substances was the most important cause of acidification of the deeper soil horizons, whereas biological acidification was more important in the humus and upper A horizon. These are important studies; however, because the data are variously complicated by succession, local spatial heterogeneity, harvesting, and regional variations of climate, in addition to differences in acidic deposition, they do not provide conclusive evidence of soil acidification via acidic deposition—more data are still needed.

Surface Water Chemistry

Chemical Characteristics

We previously described changes that take place in water chemistry when incoming precipitation interacts with foliage and bark surfaces (throughfall and stemflow) and with the forest floor and mineral soil. These interactions result in a decreased concentra-

A view of terrain in the LaCloche highlands of Ontario that is hypersensitive to acidification caused by atmospheric depositions. The terrestrial part of the watershed is characterized by thin, slowly weathering, oligotrophic glacial deposits. There are frequent exposures of whitish, quartzitic bedrock on hilltops and slopes. The lakes in this terrain had very little acid-neutralizing capacity, and they were quickly acidified by a combination of acidic precipitation and the dry deposition of sulfur dioxide. (Photo courtesy of B. Freedman.)

Table 4.7 The change in average soil pH between 1927 and 1982–1983 in relocated pits in Swedish forests[a]

Horizon	Time	pH under *Fagus sylvatica*	pH under *Picea abies*
Humus	1927	4.5	4.6
	1982–1983	3.8	3.6
A$_2$	1927	4.5	4.7
	1982–1983	4.2	4.0
B	1927	4.9	4.9
	1982–1983	4.6	4.5
C	1927	5.3	5.2
	1982–1983	4.7	4.2

[a]Modified from Hallbacken and Tamm (1986).

Table 4.8   Volume-weighted mean annual concentration of chemical constituents in wet-only precipitation, and in two headwater oligotrophic lakes in Nova Scotia, Canada: Beaverskin is a clear-water lake, and Pebbleloggitch is a brown-water lake[a]

Constituent	Wet-Only Precipitation (1981–1983)	Beaverskin Lake (1979–1980)	Pebbleloggitch Lake (1979–1980)
Ca	4.3	20.0	18.0
Mg	2.9	32.0	30.0
Na	26.1	126.0	126.0
K	1.1	8.0	6.0
Fe	<0.1	1.0	4.0
Al	<0.1	2.2	23.4
NH_4	4.2	1.0	2.4
H	29.9	5.0	33.0
SO_4	27.5	48.7	57.9
Cl	29.5	124.0	111.0
NO_3	9.7	1.0	0.9
Organic anions	<0.1	32.0	66.0
Alkalinity	0.0	0.0	0.0
Sum cations	68.5	195.2	242.8
Sum anions	66.7	205.7	235.8
Anions/cations	0.97	1.05	0.97
DOC (mg/l)	0.0	3.9	10.0
TC (mg/l)	0.0	4.5	13.8
Color (Hazen units)	0	6	87

[a]Data are in µeq/l. Modified from Freedman and Clair (1987) and Kerekes and Freedman (1988).

tion of certain chemical constituents in the percolating solution, and an increased concentration of others. The net effect of these various chemical interactions in the terrestrial part of the watershed is reflected in the chemistry of surface waters such as streams, rivers, and lakes.

A comparison of the chemistry of precipitation with that of two remote, headwater, oligotrophic lakes in Nova Scotia is presented in Table 4.8. Compared with precipitation, the lake waters are a relatively concentrated solution—their sum of cations + anions averaged 440 µeq/l, versus 135 µeq/l in precipitation. This occurs in spite of the fact that these headwater, oligotrophic lakes are relatively dilute in comparison with most fresh water, especially with respect to calcium concentration (Hutchinson, 1967; Bowen, 1979; Kerekes *et al.*, 1982). In addition, the relative contributions of ionic constituents differ markedly—the lake waters

are relatively enriched in calcium, magnesium, sodium, potassium, iron, aluminum, sulfate, chloride, organic anions, and dissolved and total organic carbon. The increased concentrations of all of these constituents are largely due to their mobilization from the terrestrial part of the watershed. Because sodium, chloride, and sulfate have an important aerosol phase in the atmosphere, their dry depositon and subsequent solubilization also contribute to their larger concentrations in lake water. In addition, evapotranspiration has the effect of physically concentrating surface water in comparison with precipitation. However, this physical effect is fairly small, since evapotranspiration dissipates only about 33% of the annual atmospheric input of water to these watersheds (Kerekes and Freedman, 1988). In contrast, ammonium and nitrate have a markedly smaller concentration in the lakewater (Table 4.8), indicating that they have

Airphoto of the watershed of Pebbleloggitch Lake, Nova Scotia. About one-third of the terrestrial watershed of this lake is characterized by an organic-rich boggy substrate, visible as the low-texture, whitish area at the bottom of the photo. The rest of the terrestrial watershed is covered with coniferous forest. Because of the large inputs of dissolved organic substances from the bog, the lake has tea-colored, dark-brown water and it is naturally acidic, with an average pH of about 4.5. (Photo courtesy of J. Kerekes.)

been "consumed" by biological uptake and inorganic exchange reactions within the terrestrial part of the watershed.

The clear-water Beaverskin Lake is a typical oligotrophic, clear-water, slightly acidic lake. It is less acidic than incoming precipitation (average pH 5.3 versus 4.6, respectively), while the brown-water Pebbleloggitch Lake (pH 4.5) has a similar acidity to precipitation. These two oligotrophic lakes are situated only 1 km apart on a similar geological substrate. However, they differ in acidity because of the strong influence of organic acids in tea-colored Pebbleloggitch Lake, which has a *Sphagnum*–heath bog covering about one-third of its watershed (Kerekes and Freedman, 1988). Water with a large concentration of dissolved organic substances is usually naturally acidic, with a pH range of about 4–5 (Oliver *et al.*, 1983; Clymo,

1984; Gorham *et al.*, 1984; Krug *et al.*, 1985). In Pebbleloggitch Lake, organic anions comprise 28% of the total anions, further suggesting that they influence its acidity.

Seasonal variation in the concentration of chemical constituents is also important. In most places where a snowpack accumulates in winter, there is a tendency for winter and especially spring meltwater flows to be relatively acidic, particularly in streams, rivers, and surface lake waters (Haapala *et al.*, 1975; Johannessen and Henriksen, 1978; Jeffries *et al.*, 1976; Goodison *et al.*, 1986; Freedman and Clair, 1987). In part, this is caused by soil that is saturated and/or frozen at the time of snowmelt, so that there is relatively little neutralization of the acidity of precipitation by interactions with soil constituents. "Acid shock" events have also been linked to the initiation of snowmelt, since the first

A view of a set of tubular, clear polyethylene mesocosms in which columns of lake water were experimentally treated in order to examine the effects of acidification, liming, and fertilization on the open-water biota. The lake is located in a remote area of Nova Scotia, and it is oligotrophic and susceptible to acidification by atmospheric depositions of acidifying substances (the shallowness of the soil of its watershed can be appreciated from the many exposed boulders in shallow water along the shoreline). The largest changes in phytoplankton and zooplankton were caused by fertilization, but stresses associated with acidification and liming also caused changes in species richness, standing crop, and productivity. (Blouin *et al.* 1984; photo courtesy of P. Lane).

melt fraction of the snowpack is more acidic than are later fractions (Johannesen and Henriksen, 1978; Cadle *et al.*, 1984). In the Adirondack Mountains of New York State, acidic stream water in the springtime is believed to be due to (1) the occurrence of relatively small concentrations of the base cations Ca^{2+}, Mg^{2+}, Na^+, and K^+ and their associated alkalinity; (2) an increase in the concentration of nitrate; and (3) the constant high concentration of sulfate in spite of the lower concentrations of bases (Galloway *et al.*, 1987a).

The concentrations of other chemical constituents also vary seasonally in surface waters. Freedman and Clair (1987) examined streams in Nova Scotia, and found that the concentrations of hydrogen ion, dissolved organic carbon, sulfate, nitrate,

chloride, sodium, magnesium, calcium, and aluminum were all relatively high during the high-flow period of late autumn–winter–early spring (Fig. 4.3). Overall, however, they found that the seasonal variations in concentration were much smaller than the seasonal variation in water flow. As a result, water flow was by far the most significant determinant of the rate of export of chemicals from their study watersheds; variations of concentration were relatively unimportant (see also Lewis and Grant, 1979; Hall and Likens, 1984).

Another important geochemical consideration is that of the flux of chemicals (for a particular chemical constituent, annual flux is calculated as the product of its mean annual volume-weighted concentration, times water flux from the watershed).

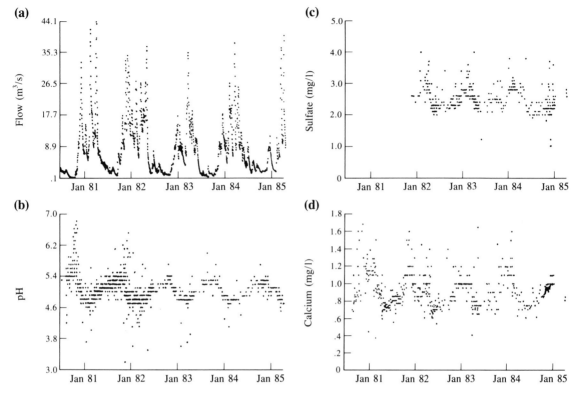

Fig. 4.3 Seasonal variations in water flow, pH, and the concentrations of sulfate and calcium in the Mersey River, Nova Scotia. Modified from Freedman and Clair (1987).

Because fluxes are standardized to time, area, and water volume, they allow a comparison of the total input and the total output of material. The input and output fluxes of certain elements for a forested watershed at Hubbard Brook, New Hampshire, are summarized in Table 4.9. For elements with a net flux (i.e., inputs minus outputs) that is positive, there is an accumulation in the watershed; if net flux is negative there is a net depletion of the watershed "capital" of the chemical.

Elements with a negative net flux include silicon, calcium, sodium, aluminum, magnesium, and potassium. The excess output of these derive from ion exchange, and from the weathering of minerals in the watershed (Schindler *et al.*, 1976; Likens *et al.*, 1977; N. M. Johnson, 1979).

Hydrogen ion accumulates in the watershed at a rate that is equivalent to the consumption of 90% of the atmospheric input of H^+ via precipitation (Table 4.9). Other studies in regions receiving acidic deposition have also reported H^+ consumption by the terrestrial part of the watershed. For example, Freedman and Clair (1987) observed 75–90% H^+ consumption among four forested watersheds in Nova Scotia. Within the terrestrial part of the watershed, H^+ consumption is largely due to ion exchange, the dissolution of aluminum hydroxide compounds, and weathering reactions (Likens *et al.*, 1977; N. M. Johnson, 1979). In part, these processes cause the loss of basic cations such as Ca^{2+} and Mg^{2+}, and they result in an acidification of the terrestrial watershed.

The net flux of fixed nitrogen was also strongly positive at Hubbard Brook ($+16.7$ kg N ha^{-1} year^{-1}; Table 4.9). The nitrogen accumulates in the terrestrial part of the watershed, as organic nitrogen

Table 4.9 Total inputs and outputs and net fluxes of chemical constituents for a forested watershed at Hubbard Brook, New Hampshire[a]

Constituent	Total Inputs[b]	Total Outputs	Net Fluxes
Si	<0.1	23.8	−23.8
Ca	2.2	13.9	−11.7
Na	1.6	7.5	−5.9
Al	<0.1	3.4	−3.4
Mg	0.6	3.3	−2.7
K	0.9	2.4	−1.5
Organic C	1484	12.3	+1472
N	20.7	4.0	+16.7
Cl	6.2	4.6	+1.6
S	18.8	17.6	+1.2
H	0.96	0.10	+0.86
P	0.036	0.019	+0.017

[a]Data are in kg/ha-year. Modified from Likens *et al.* (1977).
[b]Based on bulk-collected precipitation, except for C, N, and S, where inputs include aerosol deposition and gaseous uptake.

of the aggrading plant biomass and organic matter of the forest floor. The very large positive net flux of carbon is, of course, due to the net fixation of atmospheric CO_2 into organic carbon of the aggrading forest (Likens *et al.*, 1977).

Acidification of Surface Water

A widespread acidification of weakly buffered surface waters has been attributed to the deposition of acidifying substances from the atmosphere. Impacted regions include Scandinavia (RMFA/RMA, 1971; Brakke, 1976; Wright and Gjessing, 1976; Seip and Tollan, 1978; Nilssen, 1980), eastern Canada (Beamish and Harvey, 1972; Beamish *et al.*, 1975; Thompson *et al.*, 1980; Watt *et al.*, 1979, 1983), and the northeastern United States (Schofield, 1976a, 1982; Wright and Gjessing, 1976; A. H. Johnson, 1979; Glass *et al.*, 1981; Schindler, 1988). For example, the pH of 21 water bodies in central Norway decreased from an average of 7.5 in 1941, to 5.4–6.3 in the early 1970s (Brakke, 1976). In southwestern Sweden, the average pH of 14 surface waters decreased from 6.5–6.6 prior to 1950, to pH 5.4–5.6 in 1971 (Brakke, 1976). In the Ad-

irondack Mountains region of New York state, more than 51% of a sample of 217 high-altitude lakes had pH < 5 in 1975 (90% of these acidic lakes were also devoid of fish), compared with only 4% having pH < 5 in a sample of 320 lakes from the same area in the 1930s (Schofield, 1976a). In Nova Scotia, the average pH of 7 rivers was 4.9 in 1973, compared with 5.7 in 1954–1955 (Thompson *et al.*, 1980).

It is important to note that some of the historical comparisons that indicate acidification may suffer from systematic biases that cause erroneously high values in the earlier pH and alkalinity determinations (Kramer and Tessier, 1982). These errors can result from (1) the use of relatively soft-glass containers, which may have contributed alkalinity to early samples; (2) a difference in analytical technique (the pH meter with glass electrode came into common use in 1946–1960. Prior to this time the most frequent techniques for determining the concentration of H^+ involved the use of color-indicator comparators, which may have overestimated the pH of dilute, poorly buffered, oligotrophic water); and (3) a difference in the timing and frequency of sampling between early and recent studies (on an annual basis, pH tends to be lower in winter, while on a diurnal basis it tends to be higher at midday due to alkalinity generated by primary production). Therefore, pH determinations made prior to ~1960 should be regarded with some caution. However, in spite of these analytical problems, there is a broad concensus among researchers that there has been a recent acidification of many surface waters. Even if there is some doubt about the accuracy of some of the older pH data, the evidence for biological changes caused by the acidification (described later) is incontrovertible (Kramer and Tessier, 1982; Havas *et al.*, 1984b).

Few data are available that illustrate the changes that take place in the concentrations of hydrogen ion and other chemicals as fresh waters acidify. The most important reason for this lack is that there are few accurate data on the characteristics of water bodies prior to their acidification. There have been a few experimental studies of natural water bodies in which acidification was caused by the addition of concentrated mineral acid. A whole-lake experiment was performed in the Experimental Lakes Area (ELA) of northwestern Ontario (Schindler and Turner, 1982; Cook and Schindler, 1983; Mills, 1984; Schindler *et al.*, 1985; Rudd *et al.*, 1988). The experimental water body is Lake 223, a 27-ha oligotrophic lake located on a Precambrian Shield landscape, with an average depth of 7.1 m, and a volume of 19.5×10^5 m^3. The lake was studied for 2 years (1974, 1975) prior to its experimental manipulation, and then sulfuric acid was added to progressively acidify the system. Prior to its acidification, Lake 223 had a mean annual epilimnetic pH of 6.5, an average ice-free season pH of 6.8, and an alkalinity of 80 μeq/l (the mean annual pH of precipitation was 4.9–5.0). Beginning in 1976, sulfuric acid was added to Lake 223; by 1983 27,400 litres of 36 N H$_2$SO$_4$ had been added! The acidity of Lake 223 was increased progressively, from an initial mean annual epilimnetic pH of 6.49 in 1976, to 6.13 in 1977, 5.93 in 1978, 5.64 in 1979, 5.59 in 1980, and to 5.02–5.13 between 1981 and 1983, when the objective was to maintain pH rather than to further acidify the lake.

The addition of sulfuric acid caused various di-

rect and indirect chemical changes in Lake 223. As expected, both sulfate and hydrogen ions increased in concentration (sulfate averaged 35 μmol/l in 1975, compared with 115 μmol/l in 1979). There were also increases in the concentrations of manganese (a 980% increase in 1980, compared with the 1976 concentration), zinc (550%), aluminum (155%), and sodium (26%); these were probably solubilized from sediment by the acidifying fresh water. The wintertime concentration of ammonium also increased in Lake 223, after pH reached 5.4. The normal pattern under less acidic conditions is for a wintertime increase in the concentration of nitrate, but in Lake 223 ammonium increased instead because of the inhibition of bacterial nitrification. Decreased concentrations were observed throughout the year for alkalinity, dissolved nitrogen, iron, and chloride. An increase in the transparency of the water column was reflected by a change in the extinction coefficient from an initial 0.50–0.58, to 0.44–0.50 after acidification. The increased light penetration caused additional heating of the hypolimnion during the growing season and an increased depth of the thermocline. Many biological changes were also caused by the acidification of Lake 223; these are described later.

Interestingly, the observed pH decreases were substantially smaller than were predicted from calculations that assumed that Lake 223 would behave as an isolated, dilute, alkalinity-buffered system that was being titrated with sulfuric acid. In part, this was due to the consumption of H$^+$ by weathering and solubilization reactions at the water/sediment interface; these resulted in increased concentrations of metals such as Al and Mn. However, the most important acid-consuming mechanism was due to alkalinity produced by reactions that formed iron sulfide in the bottom 10% of the seasonally anoxic hypolimnion. Under this condition, Fe^{2+} is mobilized from the sediment. This ferric iron reacts with sulfide produced by the reduction of sulfate under anoxic conditions, to produce an FeS precipitate. This process is accompanied by the consumption of one equivalent of H$^+$ per equivalent of SO$_4^{2-}$ reduced. Overall, an estimated 66–81% of the added acid was neutralized by alkalinity, 75% of which was generated by sulfate reduction, iron reduction, and iron sulfide formation (Cook *et al.*, 1986). The exchange of H⁺ for other cations in the sediment accounted for a further 19% of the alkalinity, and inputs from the terrestrial watershed for about 5%. The remaining 19–34% of the H⁺ that was not neutralized by alkalinity served to decrease the pH of the water column and to weather minerals, or it was exported from the lake. A sulfur budget for Lake 223 showed that streamwater outflow accounted for 33–35% of the total input of sulfate, while FeS sedimentation consumed 10–15%. The remaining 50–57% remained in the water column as increased sulfate concentration, as was described earlier (Cook and Schindler, 1983).

Surface waters that are susceptible to acidification tend to exhibit a syndrome of physical and chemical characteristics. The most important of these are discussed below (after Hendry *et al.*, 1980b; Harvey *et al.*, 1981; Henriksen, 1982; Kramer and Tessier, 1982; Schindler, 1988).

One characteristic is that susceptible water tends to have a small alkalinity or acid-neutralizing capacity. Usually H⁺ is absorbed until a buffering threshold is exceeded, after which there is a rapid decrease in pH until another buffering system comes into play (Fig. 4.4). In the circumneutral pH range (pH 6–8), bicarbonate alkalinity is the critical buffering system that can be depleted by acidic deposition, causing acidification of the water body. In this pH range bicarbonate alkalinity (HCO_3^-) reacts with H⁺ to form H_2O + CO_2; by this mechanism, added H⁺ is neutralized, and there is little or no decrease in pH unless the supply of alkalinity is exhausted. In this sense, the acidification of surface water by the deposition of acidifying substances from the atmosphere can initially be viewed as a large-scale, whole-lake titration of alkalinity (Henriksen, 1980, 1982).

Even distilled water contains some bicarbonate as a result of the carbonic acid (H_2CO_3) equilibrium with atmospheric CO_2. At the equilibrium pH of 5.6, ~18% of the total inorganic carbon is present as HCO_3^-; the rest occurs as free CO_2 (Table 4.10). However, the concentration of bicarbonate in surface water is influenced by other geochemical fac-

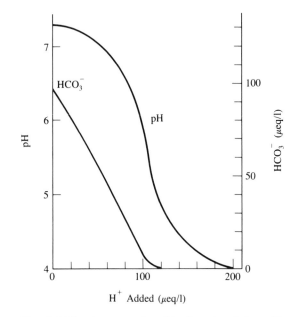

Fig. 4.4 Titration curve for a bicarbonate solution with an initial concentration of alkalinity (HCO_3^-) of 100 μeq/l. This curve is qualitatively similar to that for a clear, oligotrophic surface water. Modified from Henriksen (1980).

tors. When water comes in contact with mineral carbonates [especially calcite ($CaCO_3$) and dolomite ($Ca,MgCO_3$)] in the watershed soil, bedrock, or aquatic sediment, a large quantity of bicarbonate alkalinity is generated, and the acid-neutralizing capacity of the water is large. Alkalinity is produced by the dissolution of these minerals, for example,

$$CaCO_3 + H_2O + CO_2 \leftrightarrows Ca^{2+} + 2HCO_3^-$$

In most circumneutral, clear, dilute waters, the quantity of HCO_3^- (in equivalents) is strongly correlated with the sum of Ca^{2+} plus Mg^{2+}. The ratio of equivalents of HCO_3^- to (Ca^{2+} + Mg^{2+}) is close to 1:1, indicating that bicarbonate associated with these basic cations is the predominant source of alkalinity (Almer *et al.*, 1978; Harvey *et al.*, 1981; Schindler, 1988).

In a watershed with a large quantity of calcium and magnesium carbonate in its terrestrial soil and/or aquatic sediment, there is enough alkalinity-generating capacity to effectively neutralize the

Table 4.10 Percentage of inorganic carbon occurring as CO_2, HCO_3^-, and CO_3^{2-} in water at various pH levels[a]

pH	Free CO_2 (%)	HCO_3^- (%)	CO_3^{2-} (%)
4	99.6	0.4	1.3×10^{-7}
5	96.2	3.8	1.2×10^{-5}
6	72.5	27.5	0.0009
7	20.8	79.2	0.026
8	2.5	97.2	0.32
9	0.3	96.6	3.1
10	0.002	75.7	24.3

[a]Modified from Hutchinson (1957).

acid inputs from the atmosphere. A water body in such a watershed will not acidify at the ambient rate of acid input, even in a relatively polluted region with a large rate of atmospheric deposition. However, the situation is different where the bedrock, soil, and sediment are comprised of hard, slowly weathering, oligotrophic minerals such as granite, gneiss, and quartzite, which do not contain an appreciable quantity of carbonate minerals. In such a case, the alkalinity generating capacity of the terrestrial watershed is small, and alkalinity generation reactions within the lake become relatively important. This *in situ* production of alkalinity is largely due to the biological reduction of sulfate and nitrate, and the exchange of H^+ for Ca^{+2} in sediment (Schindler *et al.*, 1986). However, only a relatively small quantity of alkalinity is generated by these processes, so that the acid-neutralizing capacity is small and therefore acidification can take place relatively easily as a result of the atmospheric deposition of acidifying substances. Susceptible watersheds with minimal alkalinity are frequent in glaciated terrain of eastern Canada and Scandinavia, where a thin till soil typically overlies a hard plutonic bedrock of Precambrian granite and gneiss. Susceptible watersheds are also frequent at high elevation in older mountains where erosion has exposed crustal granite, for example, in the Adirondacks of New York and other parts of the mountain chains of eastern North America. Another impor-

tant characteristic of the soil of these susceptible watersheds is that they have a small sulfate adsorption capacity, which can be quickly overwhelmed by large sulfate inputs via atmospheric deposition. When this takes place, the additional sulfate inputs flush from the terrestrial part of the watershed as mobile sulfate, accompanied by H^+ and Al^{3+}, which can acidify surface waters and cause toxic stress to the aquatic biota (Reuss *et al.*, 1987).

Even within these susceptible regions, the presence of a pocket of calcareous soil or till in the watershed can confer sufficient alkalinity for the water body to be essentially nonsusceptible to acidification by atmospheric deposition. This spatial effect on susceptibility can even exist in an area where the till and bedrock are dominated by hard, plutonic minerals. Dillon *et al.* (1977) surveyed 15 lakes on a Precambrian Shield landscape in Ontario, in an area where the mean annual pH of precipitation was 4.0–4.4. Fourteen of the lakes had a relatively small alkalinity (range of 95–175 μeq/l during the nonstratified period) and were slightly acidic (pH 5.8–6.7); these lakes were considered to be highly to moderately sensitive to acidification. However, the other lake had some calcareous glacial till in its watershed. As a result, it had an anomalously high alkalinity (1200 μeq/l) and high pH (7.1), and because of its large acid-neutralizing capacity it would be unlikely to acidify in the future.

Watershed area is another consideration with respect to the susceptibility of surface water to acidification. In general, relatively high-altitude, headwater systems are more at risk, because they tend to have a small watershed area in comparison with their surface area of water, and because the soil in their watershed is often relatively thin and oligotrophic. In a small watershed with little soil, acidic rain water has less opportunity to interact with the soil and bedrock, and therefore relatively little of the acidity is neutralized before the percolating groundwater reaches surface water. We previously examined water chemistry data for two small, headwater lakes in Nova Scotia (Table 4.8). Both of these have a relatively small watershed; their ratio of watershed area to lake surface area is less than 4. These headwater lakes have an average calcium concentration of only 18–20 µeq/l and magnesium of 30–32 µeq/l. Nearby, a large downstream lake with a watershed to lake area ratio of more than 30 has 38 µeq/l of calcium and 39 µeq/l of magnesium (Kerekes and Freedman, 1988).

Therefore, the acidification of typical oligotrophic fresh waters is a process that is analogous to the titration of a weak bicarbonate solution with sulfuric and nitric acids deposited from the atmosphere. Where there is only a small supply of alkalinity in the watershed, and where atmospheric deposition results in a large flux of sulfate from the terrestrial part of the watershed in accompaniment with hydrogen and aluminum ions, the surface water is generally susceptible to acidification.

4.4

BIOLOGICAL EFFECTS OF ACIDIFICATION

Injury to Terrestrial Vegetation

Acidic deposition can potentially affect vegetation in several direct and indirect ways. As summarized by Tamm and Cowling (1976), the potential direct effects include:

1. Damage to the leaf cuticle, caused by an accelerated erosion of the waxy protective layer, or by direct injury to surface cells by acidic droplets or acidic particles, causing a micro- or macroscopic surface injury.
2. Interference with the functioning of stomatal guard cells. To a degree, the turgidity of these cells is influenced by cytoplasmic pH, which may be affected by precipitation pH. An effect on guard cells could result in decreased control over stomatal aperture, and thus over the rate of transpiration and the flux of CO_2, O_2, and other gases.
3. Acute injury to plant cells after the penetration of acidic substances through the cuticle or stomata.
4. A "hidden injury" metabolic effect that could result from a change in the rates of photosynthesis, respiration, or some other

metabolic function. This does not cause acute injury, but it may result in a growth decrement, developmental dysfunction, or premature senescence.
5. Alteration of the chemistry or quantity of root or foliar exudates, with potential secondary effects on the microflora and microfauna of leaf and root surfaces, including organisms that are important in nutrient cycling and nutrient uptake, and pathogens that cause disease.
6. Interference with reproduction, for example, by decreasing the viability of pollen, by interfering with stigmatic receptability, or by otherwise decreasing fruit set or viability.

In addition, Tamm and Cowling (1976) identified potential indirect effects of acidic deposition on vegetation, including:

1. An increased rate of leaching of substances from foliage, especially mineral nutrients and organic chemicals. This could be an indirect effect of damage to the cuticle or guard cells, or it could be due to acid-leaching effects.
2. Damage to the cuticle or guard cells or other physiological disruptions could indirectly make plants more susceptible to drought, air pollution, and other environmental stresses.
3. A metabolic disruption of plants could affect their symbiotic association with nitrogen-fixing microorganisms, mycorrhizal fungi, etc., for example, via changes in the nature of root exudates.
4. Susceptibility to parasites, pathogens, and insect damage may be indirectly affected by damage to the protective cuticle, by a change in the nature of root or foliar exudates, by predisposing stress caused by acidic input or by metal toxicity in soil, or by changes in competitive or other interactions occurring among the microbial flora.
5. Root damage or other effects of an increased concentration of available metals in

soil as a result of acidification. Aluminum toxicity would be most important in this respect.

6. A synergistic interaction with other environmental stresses such as gaseous air pollutants, drought, etc., in such a way as to increase their damage to plants.

A review of the scientific literature indicates that a case has not yet been clearly documented where terrestrial vegetation has suffered acute injury caused by a naturally occurring acidic precipitation event. In contrast, a number of laboratory and field-based experiments in which plants were treated with an artificial "acid rain" solution have demonstrated injury to natural and agricultural vegetation. However, the toxic thresholds in these experiments are usually at a level of pH that is artificially more acidic than is observed in precipitation in nature (Tamm and Cowling, 1976; Abrahamsen, 1980; Jacobson, 1980; Cohen et al., 1981; Evans, 1982; Inving, 1983; Amthor, 1984; Morrison, 1984; McLaughlin, 1985).

Abrahamsen and Stuanes (1980) and Tveite (1980a,b) reported the results of field experiments in Norway in which an artificial acid rain solution was sprayed on a young conifer plantation using an overhead apparatus. The effects on tree growth were small and sporadic. In lodgepole pine (*Pinus ponderosa*) watered for as long as 3 years at 50 mm H_2O/month, height growth was stimulated by 15–20% at pH values 4 and 3 compared with the pH 5.6–6.1 control treatment; at 25 mm/month there was no treatment effect at all. One year of treatment over the pH range 5.6 to 2.5 had no effect on the height growth of Norway spruce (*Picea abies*). The growth of Scotch pine (*Picea sylvestris*) was stimulated by as much as 15% by pH levels of 3.0 and 2.5 after 4 years of treatment, compared with pH 5.6–6.1, while pH 2.0 had a slight negative effect. Birch (*Betula verrucosa*) was generally stimulated by the acid treatments. The feather mosses *Pleurozium schreberi* and *Hylocomium splendens* dominated the ground vegetation of the plantations. These bryophytes were very sensitive to the acid treatments, and were greatly damaged at pH $\leq$ 3.

Similar results were reported by Tamm and Wiklander (1980) in studies in Sweden. Six years of treatment of an 18-year-old *Pinus sylvestris* stand with water containing the equivalent of 16, 33, or 49 kg H_2SO_4-S ha^{-1} year^{-1} had a moderate effect on tree growth (there was a 30% increase in growth of basal area at the largest acid loading without NPK fertilization; with NPK there was a 10% decrease at the highest acidity). They also noted a demise of the ground vegetation (dominated by feather mosses and the shrub heath *Calluna vulgaris*) when the most acidic treatment was combined with fertilization. Note that a negative effect on the ground vegetation of northern coniferous forest is usually observed after fertilization.

Because they are better controlled, laboratory experiments with artificial acid rain solutions are more satisfactory than field experiments for the examination of dose–response effects on vegetation. In general, among a wide range of plant species that have been examined, a growth reduction does not occur at a laboratory treatment pH >3.0, and in some cases a growth stimulation has been observed at pH levels more acidic than this (Jacobson, 1980). In a greenhouse experiment, Wood and Bormann (1976) observed an enhanced growth of white pine (*Pinus strobus*) seedlings that were treated with acidic solutions ranging from pH 2.3 to 4.0, compared with pH 5.6. They attributed this effect to nitrate fertilization, since their artificial rains contained that nutrient at a relative concentration of 24% of the total anion equivalents. The growth stimulation occurred in spite of an increase in the leaching of base cations from the pine foliage caused by the acid treatments [see also Fairfax and Lepp (1975), Wood and Bormann (1975), and Lepp and Fairfax (1976) for other studies of acid leaching of ions from foliage]. In another laboratory experiment, Percy (1983, 1986) exposed seedlings of 11 eastern North American tree species to acid rain treatments that ranged from pH 2.6 to 5.6. Acute injury in the form of macroscopic lesions on foliage of some species was only observed at pH 2.6. The lesions covered as much as 10% of the leaf surface, and 1 week of treatment with this very acidic pH was required to cause this effect. There were significant reductions of height growth, needle number, axillary meristems, and some other growth- or development-related variables of some species at pH ≤4.6. In general, conifer species were more sensitive than angiosperm species. Overall, however, these detrimental effects were not sufficient to cause a significant decrease in the yield of dry weight of the seedlings (except at pH 2.6).

In general, it appears that trees and other vascular plants may not be at risk of direct, short-term, acute injury from ambient acidic precipitation. However, there remains the possibility that hidden injury or unidentified interactions with other environmental stresses could cause a growth decrement (this is also discussed in Chapter 5). Because acidic precipitation is regional in character, such a yield decrease could occur over a very large area, and therefore it would have great economic implications. This potential problem is most relevant to forests and other natural vegetation, since agricultural land is regularly limed and has an intrinsically larger rate of endogenous acid production, via cropping and fertilization, than is caused by acidic deposition from the atmosphere.

Several studies carried out in western Europe and eastern North America have addressed the question of decreased forest productivity caused by acidic deposition. These studies examined the temporal pattern of tree ring width over a large geographic area, searching for a recent trend of declining productivity that might be related to the regional patterns of acidic deposition from the atmosphere (Johnsson and Sundberg, 1972; Abrahamsen et al., 1976, 1977; Cogbill, 1977; Johnson et al., 1981; Siccama et al., 1982; Strand, 1983; Hornbeck et al., 1987b). Many of these studies have demonstrated that productivity has recently declined in various tree species and in various regions. However, a progressively decreasing ring width is a natural phenomenon in an aging forest, because of (1) the intensification of competitive stress when the forest canopy closes during the course of succession and (2) the immobilization of a large fraction of the site nutrient capital within aggrading biomass and organic matter. So far, it has not been possible in these tree-ring studies to clearly separate successional effects and other background influences,

such as insect defoliation and climate change, from effects that might be related to acidic deposition or other air pollution stresses (this is also discussed in Chapter 5).

Acidification and Fresh-Water Biota

Phytoplankton

In general, the phytoplankton community of lakes is species-rich. Nonacidic, oligotrophic lakes in the temperate zone are typically dominated numerically by golden-brown algae and diatoms (Chrysophyceae and Bacilliarophyceae, respectively) (Schindler and Holmgren, 1971; Findlay and Kling, 1975; Ostrovsky and Duthie, 1975; Hendry et al., 1980a). Acidic, clear-water, oligotrophic lakes are somewhat different in terms of the dominant families of algae. Yan and Stokes (1978) studied a pH 5.0 clear-water lake in Ontario, and reported domination by dinoflagellates (Dinophyceae, especially Peridinium limbatum) and cryptomonads (Cryptophyceae, esp. Cryptomonas ovata). In a less acidic (pH 5.3) clear-water lake in Nova Scotia, Kerekes and Freedman (1988) reported domination by blue-green algae (Cyanophyceae, esp. Agmenellum thermale), green algae (Chlorophyceae, esp. Sphaerocystis schroeteri), and golden-brown algae. In two more acidic (pH 4.5), brown-water lakes nearby, the phytoplankton was dominated by golden-brown algae (esp. Mallomonas caudata), green algae (esp. S. schroeteri), diatoms (esp. Asterionella formosa), and yellow-green algae (Xanthophyceae); this assemblage is typical of brown-water lakes (Ostrovsky and Duthie, 1975; Ilmavirta, 1980).

In the experimental whole-lake acidification described earlier, the phytoplankton of Lake 223 shifted from a preacidification community dominated by golden-brown algae, to one dominated by chlorophytes (esp. Chlorella cf. mucosa). Species diversity and richness were not affected (Findlay and Saesura, 1980).

Unlike species composition, the standing crop and productivity of the phytoplankton of oligotrophic lakes are rather unresponsive to a change in acidity (Hendry et al., 1980a). In fact, if acidification is accompanied by a clarification of the upper water column, then standing crop and production can increase due to the greater depth of the euphotic zone. In Lake 223, the preacidification, whole-lake phytoplankton biomass was estimated as 1465 kg, compared with 2410 kg following acidification (Findlay and Saesura, 1980). The average pre-acidification annual production of Lake 223 ranged from 16 to 26 mg C/m^2·year over 3 years when the lakewater pH was 6.5 to 6.7, while it ranged from 28 to 47 mg C/m^2·year over the 4 years that the pH was 5.6 to 6.1, and from 31 to 60 mg C/m^2·year over the 3 years that the pH was 5.0 to 5.1 (Shearer *et al.*, 1987). These data indicate a rather small effect of the acidification of Lake 223 on primary production, in contrast to the relatively large effect on species composition.

A change in the trophic status of a water body has a much larger effect on the standing crop and productivity of phytoplankton. To specifically illustrate this for acidic water, consider the case of a pair of small, adjacent lakes in Nova Scotia (Kerekes *et al.*, 1984). These lakes were acidified by the oxidation of pyrite in slate within their watershed, after bedrock was exposed to the atmosphere during the construction of a highway. Little Springfield Lake (mean annual pH = 3.7) is oligotrophic, while Drain Lake (pH 4.0) receives an input of sewage and is eutrophic in spite of its extreme acidity. Drain Lake had a midsummer phytoplankton volume of 15.5 mm^3/l and a chlorophyll a concentration of 10.3 mg/l. These are much larger than in (1) oligotrophic Little Springfield Lake with 0.85 mm^3/l and 0.49 mg/l, respectively, or (2) the range of values among 15 nearby, less acidic (pH >4.5), oligotrophic lakes (0.21–3.31 mm^3/l and 0.23–2.76 mg/l, respectively) (Blouin, 1985).

Periphyton

Periphyton consist of algae that live on a substrate. Periphyton can be very species-rich, even in an acidic lake. Nakatsu (1983) studied three brown-water lakes in Ontario and identified 460 taxa of algae, most of which were littoral periphyton.

In some acidic lakes with very transparent water, periphyton can be quite prominent, and may even form a benthic cloud or felt-like mat. The dominant algal taxa in such oligotrophic water bodies tend to be *Mougeotia* spp., *Spirogyra* spp., and *Binuclearia tatrana* (all Chlorophyceae), and *Tabellaria flocculosa* and *Eunotia lunaris* (Bacilliariophyceae) (Hendry *et al.*, 1980a; Stokes, 1980).

In the Lake 223 whole-lake acidification, a mat of the filamentous green alga *Mougeotia* sp. began to develop in the littoral zone after the pH had decreased to less than 5.6 (Mills, 1984; Schindler *et al.*, 1985). Similar observations of increased abundance of *Mougeotia* plus other Chlorophyceae of the periphyton have been made in other experimentally acidified lakes in northwestern Ontario (Turner *et al.*, 1987). These filamentous green algae are probably (1) relatively tolerant of physiological stresses associated with acidity; (2) relatively efficient at obtaining dissolved inorganic carbon for photosynthesis (in acidic water, DIC is present in a relatively small concentration, and it largely occurs as CO_2, which most algae are not well adapted to utilize); and (3) they may experience less grazing pressure from invertebrates at low pH (Turner *et al.*, 1987).

Muller (1980) performed an acidification experiment in plastic tubes in Lake 223, and found no effect on the biomass or productivity of periphyton, with acidification to pH 3.7–4.7. At pH ≥6.3 diatoms dominated the periphyton biomass; at pH < 6 chlorophytes were dominant. At pH 4, *Mougeotia* sp. was the only dominant, similar to the observation in the whole-lake acidification.

Macrophytes

Macrophytic plants have apparently decreased in abundance in some acidified lakes [e.g., *Phragmites communis* (Almer *et al.*, 1974); *Lobelia dortmanna* and *Isoetes* spp. (Grahn *et al.*, 1974; Grahn, 1977)]. In some cases in Scandinavia, the decline of particular vascular species was accompanied and possibly caused by an increase in the abundance of acidophilous *Sphagnum* species and benthic filamentous fungi and algae (Grahn *et al.*, 1974; Hultberg and Grahn, 1975a; Grahn, 1977). Benthic *Sphagnum* mats have also been reported in acidic North American lakes, usually when there is an extreme clarity of the water column (Hendry and Vertucci, 1980; Kerekes and Freedman, 1988). Because of its apparent restriction to very clear lakes, benthic *Sphagnum* is not necessarily characteristic of acidic water bodies (Singer *et al.*, 1983; Wile and Miller, 1983; Wile *et al.*, 1985). *Sphagnum* and other mosses have also increased in abundance in water bodies that have been impacted by acid mine drainage (Harrison, 1958; Hargreaves *et al.*, 1975). Acidophilous mosses are also abundant in acidic volcanic lakes in Japan (*Leptodictyum* sp.; Yoshimura, 1935), and in acidic lakes near Sudbury and at the Smoking Hills (Chapter 2).

It has been hypothesized that the invasion of acidified lakes by bryophytes, particularly *Sphagnum* spp., may contribute to a "self-accelerating oligotrophication" (Grahn *et al.*, 1974). This process has been attributed to (1) the acid-generating potential of acidophilous *Sphagnum* spp., which have an efficient cation exchange ability that allows them to remove calcium, magnesium and other basic cations from water, in exchange for hydrogen ion (Skene, 1915; Clymo, 1963, 1964, 1984; Hutchinson, 1975; Grahn, 1977), and (2) the physical isolation of the sediment from the water column by the benthic *Sphagnum* mat, thereby interfering with acid neutralization and nutrient cycling processes that would otherwise take place at the sediment/water interface (Hultberg and Grahn, 1975a).

The macrophyte communities of brown-water and clear-water acidic lakes can be quite different. This can be illustrated by a comparison of the macrophytes of two small, oligotrophic, headwater lakes in Nova Scotia (Kerekes and Freedman, 1988). Beaverskin Lake has a mean annual pH of 5.3, very clear water, and a euphotic zone that extends to all of its bottom. Macrophytes are found as deep as 6.5 m, and 96% of the lake bottom is vegetated. The most important littoral community is dominated by *Eriocaulon septangulare*, *Eleocharis acicularis*, and *Lobelia dortmanna*, and it covers 25% of the lake bottom. The deeper community is dominated by *Sphagnum macrophyllum* and *Utricularia vulgaris*, and it covers 71% of the bottom. The average standing crop of macrophytes in Beaverskin Lake is 61 g dry weight (d.w.)/m².

In contrast, the nearby, relatively acidic (pH 4.5), brown-water Pebbleloggitch Lake has a much more restricted macrophyte vegetation. Plants are only found in a littoral fringe at a depth of less than 1 m, and only 15% of the bottom of the lake is vegetated. The most extensive macrophyte community is dominated by the floating-leaved *Nuphar variegatum*. A second community has relatively sparse *Eriocaulon*, *Lobelia*, and *Eleocharis*. The average standing crop of macrophytes is only 2.6 g/m², but 17 g/m² within the vegetated zone.

Clearly, a critical factor that determines the distribution and abundance of macrophytes in these oligotrophic lakes is water clarity; acidity *per se* is relatively unimportant. However, if a sample of lakes that covers a wider range of acidity is considered, then rather different conclusions can be reached—acidity and associated chemical variables such as calcium and alkalinity appear to have an important influence on macrophytes (Seddon, 1972; Hutchinson, 1975; Kadono, 1982; Catling *et al.*, 1986).

Aquatic macrophytes respond vigorously to fertilization (Lind and Cottam, 1969; Seddon, 1972; Hutchinson, 1975; Porcella, 1978). This is also true of acidic lakes. The previously described eutrophic Drain Lake (pH 4.0) has a lush growth and large productivity of macrophytes (Kerekes *et al.*, 1984). There are also some anomalous species present. Two species of *Potamogeton* occur in Drain Lake (*P. pusillus* and *P. oakesianus*); the next most acidic water from which that genus has been reported in northeastern North America is pH 5.0 (Hellquist and Crow, 1980; Catling *et al.*, 1986). Therefore, trophic status may have a more important direct effect than acidity on the distribution and productivity of aquatic macrophytes.

Crustacean Zooplankton

The response of zooplankton to the acidification of an oligotrophic water body is a complex function of (1) the toxicity of hydrogen ion and associated metals such as aluminum, (2) the effects of any change in primary production and the standing crop of the phytoplankton food of zooplankton, and (3) the effect of any change in the nature and intensity of predation, particularly if acidification causes the extinction of planktivorous fish. Toxic effects can act directly to eliminate relatively susceptible taxa of zooplankton, while all three of the above factors can act indirectly, especially by changing competitive interactions within the zooplankton community, and by affecting the nature of predation (Sprules, 1975a; Hendry *et al.*, 1980a; Hobaek and Raddum, 1980).

Regional surveys have documented the typical crustacean zooplankton community of oligotrophic lakes, including acidic ones. Sprules (1975b) surveyed the midsummer zooplankton community of 47 lakes in Ontario with a pH range of 3.8 to 7.0. Acidity-indicator species that were primarily in lakes with pH < 5 were *Polyphemus pediculus*, *Daphnia catawba*, and *D. pulicaria*. Intolerant species that were only at pH > 5 included *Tropocyclops prasinus mexicanus*, *Epischura lacustris*, *Diaptomus oregonensis*, *Leptodora kindtii*, *Daphnia galeata mendotae*, *D. retrocurva*, *D. ambigua*, and *D. longiremis*. Apparently indifferent species that were present over the entire pH range were *Mesocyclops edax*, *Cyclops bicuspidatus thomasi*, *Diaptomus minutus*, *Holopedium gibberum*, *Diaphanosoma leuchtenbergianum*, and *Bosmina* sp. The most frequent species was *Diaptomus minutus*; this was also the most acid-tolerant species, and was the sole zooplankter in the most acidic (pH 3.8) water body that was sampled. In lakes with high pH, the zooplankton community was relatively species-rich and had a relatively even distribution of dominant species. Above pH 5 there were 9–16 species with three or four dominants; below this pH there were 1–7 species with one or two dominants.

Sprules (1977) performed a multivariate principal components analysis (PCA) on his data matrix of lakes × zooplankton species (Table 4.11). The first PCA axis accounted for 50% of the variance, and it separated small, acidic, clear lakes dominated by *Diaptomus minutus*, from relatively large, neutral, low-clarity lakes dominated by four acid-indifferent species. The second PCA axis accounted for 15% of the variance, and separated acidic, high-sulfate, clear lakes from neutral, low-sulfate, low-

clarity lakes. Of the various physical and chemical limnological variables that Sprules correlated with the PCA axes, pH showed the strongest relationship. The correlation of pH with axis 1 was 0.63, and it was −0.54 for axis 2 (both $p<.001$). Note, however, that these are not strong correlations, since they only account for 40% and 29% of the variance, respectively. This indicates that a complex of environmental factors that includes, but is not necessarily overwhelmed by, pH is important in structuring the zooplankton community. Other important (but unmeasured in this study) environmental factors that may be affected by acidity and that influence the zooplankton community include biological interactions such as predation and competition.

In another survey, of 27 lakes in Norway, Hobaek and Raddum (1980) found that acidic lakes had a zooplankton fauna that was dominated by *Bosmina longispina*, *Eudiaptomus gracilis*, and *Kellicottia longispina*. These, along with the less abundant species *Heterocope saliens*, *Holopedium gibberum*, *Keratella serrulata*, and *Polyarthra* spp., were regularly present in acidic lakes, but were not necessarily restricted to them. Hobaek and Raddum noted that *Daphnia* spp. and cyclopoid copepods were generally absent from acidic lakes, in contrast to Sprules's study, where two of the three acid-indicator taxa were *Daphnia* species. The acid-lake community in Norway was characterized by a low species richness and diversity, and by a tendency to have only a few, strongly dominant taxa.

Another study compared the crustacean zooplankton of a pH 5.3 clear-water lake and a nearby pH 4.5 brown-water lake (Blouin, 1985; Kerekes and Freedman, 1988). The clear-water lake was dominated by the copepod *Diaptomus minutus* and by the rotifer *Keratella cochlearis*. The brown-water lake was dominated by the rotifers *Keratella cochlearis* and *Conochilis unicornis*, and the copepod *Diaptomus minutus*. Zooplankton were somewhat more abundant in the brown-water lake, with a biomass of 330 mg/m^3 and a density of 222/m^3, compared with 200 mg/m^3 and 124/m^3, respectively, in the clear-water lake. The greater abun-

Table 4.11 Principal components analysis of the distribution matrix of 23 crustacean zooplankton species in 60 lakes in the Killarney area of Ontario[a]

Species	Principal Component 1 (50.1%)		Principal Component 2 (14.9%)	
	Eigenvector	Correlation	Eigenvector	Correlation
Diaphanosoma leuchtenbergianum W	0.215	0.66[b]	−0.142	−0.24
Bosmina longirostris W	0.346	0.64[b]	0.605	0.84[b]
Mesocyclops edax W	0.193	0.60[b]	−0.249	−0.27[c]
Cyclops bicuspidatus thomasi W	0.275	0.54[b]	−0.593	−0.63[b]
Daphnia retrocurva I	0.067	0.45[b]	−0.094	−0.35[d]
Tropocyclops prasinus mexicanus I	0.077	0.37[d]	−0.079	−0.20
Diaptomus oregonensis I	0.086	0.37[d]	−0.190	−0.45[b]
Leptodora kindtii I	0.007	0.34[d]	−0.013	−0.34[d]
Daphnia galeata mendotae I	0.064	0.34[d]	−0.151	−0.44[b]
Diaptomus reinhardi	0.057	0.33[d]	0.017	0.05
Ceriodaphnia reticulata	0.021	0.30[c]	0.008	0.06
Holopedium gibberum W	0.119	0.28[c]	0.330	0.42[b]
Daphnia longiremus I	0.039	0.27[c]	−0.077	−0.29[c]
Daphnia ambigua I	0.021	0.26[c]	−0.022	−0.15
Senecella calanoides	0.005	0.25	−0.008	−0.22
Cyclops scutifer	0.012	0.17	−0.021	−0.16
Epischura lacustris I	0.030	0.13	−0.062	−0.14
Polyphemus pediculus A	0.003	0.08	0.033	0.50[b]
Daphnia sp.	0.001	0.04	−0.008	−0.17
Daphnia catawba A	−0.002	−0.01	−0.007	−0.02
Daphnia pulicaria A	−0.007	−0.01	0.055	0.19
Orthocyclops modestus	−0.017	−0.10	−0.020	−0.07
Diaptomus minutus W	−0.823	−0.99[b]	−0.044	−0.03

[a] A, acid indicator; I, acid intolerant; W, acid independent (see text). Modified from Sprules (1977).
[b] Probability (*p*) <.001 of obtaining a product–moment correlation if true correlation is zero (two-tailed test).
[c] *p* <.05.
[d] *p* <.01.

dance in the brown-water lake could have been due to several factors. In brown water, a high concentration of dissolved and suspended organic matter from allochthonous sources could be partly responsible for sustaining a relatively large productivity of zooplankton (Nauwerck, 1963; Schindler and Noven, 1971; Ostrovsky and Duthie, 1975). Predation could also be important. Although fish were equally abundant in the two lakes (Kerekes and Freedman, 1988), the feeding efficiency of visual predators is undoubtedly less in a brown-water lake.

The above studies describe the crustacean zooplankton community of acidic lakes, but they do not necessarily shed light on the changes that take place while acidification is proceeding. For such information, we can consider the Lake 223 whole-lake acid-

ification, since the dynamics of zooplankton were monitored in this experiment (Malley *et al.*, 1982; Nero and Schindler, 1983). Overall, acidification caused an increase in the abundance of zooplankton. The total density of cladocerans was larger by 66% in 1980 (pH 5.4) compared with 1974 (pH 6.6), while the density of copepods was 93% greater. Throughout the study, the numerically dominant zooplankters were the copepods *Diaptomus minutus* and *Cyclops bicuspidatus*; the cladocerans *Bosmina longirostris*, *Daphnia galeata*, *Holopedium gibberum*, and *Diaphanosoma brachyurum* were also present throughout, but at a relatively low density. An important extinction was of the nocturnal predator *Mysis relicta*, which declined from a whole-lake abundance of 6.78 × 10⁶ in August

1978 (pH 5.9), to 0.27 × 10⁶ in August 1979 (pH 5.6), and then to zero. Two other relatively minor zooplankters (*Epischura lacustris* and *Diaptomus silicilis*) also became extinct. *Daphnia catawba × schroederi*, a species that was not recorded prior to acidification, appeared suddenly, and by 1980 it comprised 12% of the total cladoceran abundance. The overall increase in abundance of the zooplankton was attributed to the previously described increase in phytoplankton productivity and biomass. Toxicity of acidic water and changes in the nature and/or intensity of predation and competition were also believed to have had an important influence on the zooplankton of Lake 223.

Benthic Invertebrates

Various surveys have demonstrated that benthic invertebrates are generally less species-rich in acidic oligotrophic water, compared with less acidic water (Sutcliffe and Carrick, 1973; Conroy *et al.*, 1976; Hendry and Wright, 1976; Leivestad *et al.*, 1976; Almer *et al.*, 1978; Hendry *et al.*, 1980a). However, benthic invertebrates can be abundant in an acidic lake, especially if predatory fish have disappeared or have been reduced in numbers. The benthos of acidic lakes is generally dominated by (1) Crustacea—various Copepoda, Cladocera, Amphipoda, and Isopoda—and (2) Insecta—especially Notonectidae, Corixidae, Chrironomidae, and Megaloptera; Trichoptera, Ephemeroptera, and Plecoptera may also be present. However, all of the above groups also have species that are intolerant of acidity. For example, Raddum (1978) found that while many Plecoptera species were indifferent to pH in Scandinavian fresh waters, there were also several sensitive taxa, including *Amphinemura sulcicollis*, *Brachyptera risi*, and *Leuctra hippopus*. The Ephemeropteran *Baetis rhodani* tends to dominate its order in circumneutral streams, but it disappears with acidification. In northern Europe, acid-indicating mayflies include *Baetis rhodani*, *B. lapponicus*, and *B. macani* (Raddum and Fjellheim, 1984). The benthic amphipods *Gammarus lacustris* and *Hyallela azteca* and the snails *Valvata macrostoma* and *Ancyclus fluviatilis* are also very sensitive to acidification (Raddum and Fjellheim, 1984; Stephenson and Mackie, 1986).

Hall and Ide (1987) resampled two oligotrophic, low-alkalinity streams in Ontario in which benthic insects had been thoroughly studied 48 years previously. In one stream without severe acid pulses in the spring (pH 6.4–6.1) there were few differences in insect taxa between the two samplings. The other stream currently experiences severe acid pulses (pH 6.4–4.9) and there were some notable changes in the community of benthic insects. Four species of Ephemeroptera (mayflies) disappeared since 1942, but seven acid-tolerant species were newly recorded, while two species of Plecoptera (stoneflies) disappeared and two new species were added.

As an example of the benthos of typical acidic water bodies, consider the fauna of two previously described oligotrophic lakes in Nova Scotia (Kerekes and Freedman, 1988). Insects were the most abundant group in both the clear-water pH 5.3 and the brown-water pH 4.5 lake, comprising 56% and 78% of the total density of benthic invertebrates, respectively. The insects were dominated by Diptera (45% and 68% of total insect density, respectively), especially Chironomidae (96% and 92% of dipteran density). Dominance by Chironomidae appears to be a general characteristic of the benthos of acidic lakes (Roff and Kwiatkowski, 1977; Raddum, 1978; Havas and Hutchinson, 1982; Kerekes *et al.*, 1984). Other important insects in the Nova Scotia lakes were Trichoptera (5% of total invertebrate abundance in both lakes), Ephemeroptera (1% and 3%, respectively), and Odonata (both 2%). Crustacea comprised 18–19% of the benthic invertebrate density. The most abundant orders were Copepoda, Cladocera, Amphipoda, and Isopoda. The oligochaete families Naididae and Enchytraeidae accounted for 13% of the total invertebrate abundance in the clear-water lake, and 3% in the brown-water lake. Overall, there was a rather minor difference in the benthic invertebrates of these two lakes, in spite of large differences in their water chemistry.

It is notable that the acidity of surface sediment did not change during the Lake 223 experiment, even after 8 years of acidification (Kelly *et al.*,

1984). Microbial processes such as sulfate reduction (described earlier) kept pH > 6 within a few millimeters of the sediment–water interface. In 1981, the pH at and just above the sediment surface was 5.3–5.4, while at a depth of 0.5 cm in the sediment it was >6.0, and at 2 cm it was 6.7–6.8. After the acidification of Lake 223, there was an increased emergence of adult dipterans, especially of chironomids (Mills, 1984). The population of the crayfish *Orconectes virilis* suffered a demise as a result of the acidification (Schindler and Turner, 1982; Mills, 1984; France and Graham, 1985; France, 1987). This was largely caused by reproductive failure, along with an inhibition of carapace hardening following moult (which makes the animals susceptible to predation, including cannibalism), and the effects of a microsporidian para-

site. Direct toxicity to mature crayfish did not seem to be a major problem, even at pH 5.1.

Fish

Arguably the most important and highest-profile victims of acidification are populations of susceptible species of fish. The extinction of fish populations, particularly of commercially important Salmonidae, has been reported from various locations in the northern hemisphere where there is severe acidification of aquatic systems. For example, in Scandinavia, many surface waters have lost sport and commercial salmonid fisheries, especially in the southern areas where acidic deposition is relatively severe (Fig. 4.5) (Jensen and Snekvik, 1972; Wright and Snekvik, 1978; Rosseland *et al.*, 1986).

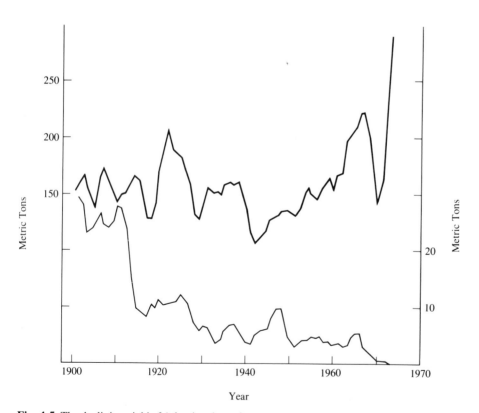

Fig. 4.5 The declining yield of Atlantic salmon from seven acidified rivers in southernmost Norway, compared with the rest of the country (upper curve, left vertical axis). Modified from Leivestad *et al.* (1976).

A survey of 700 small lakes done in 1974–1975 showed that fish populations, primarily brown trout (*Salmo trutta*), were absent from 40% of the lakes, and that fish were sparse in another 40%. Prior to the 1950s most of these lakes had sustained healthy fish populations. Another survey of more than 2000 lakes in southern Norway showed that one-third of them had lost their fish population since the 1940s. Most of the fishless lakes had a pH < 5.0. The important salmonid fish species differed in their sensitivity to acidification. Atlantic salmon (*Salmo salar*) required a pH greater than 5.0–5.5 for successful hatching and development of fry; sea trout (a distinct anadromous form of brown trout) also required a pH >5.0–5.5; stationary brown trout were more tolerant and required a pH >4.5. In general, younger life-history stages were more sensitive than were adult fish. However sporadic kills of adults were documented in the springtime, when water pH is lowest.

In the Adirondack Mountains of New York state, many high-altitude, oligotrophic lakes have lost their fish populations as a result of acidification (Schofield, 1976b, 1982; Haines and Baker, 1986). In sport fishery surveys done in the 1930s, brook trout (*Salvelinus fontinalis*) was the dominant species, and it was present in 82% of the lakes that had fish. In the same area in the mid-1970s, brook trout were extinct in at least 26 of the previously surveyed lakes. In nearby Ontario, brook trout become extinct when the average pH decreases below about 5.0 (Beggs and Gunn, 1986). In total, fish were extinct in 93 of the 215 New York lakes surveyed by Schofield (1982) in the mid-1970s. In another survey of fish status in Adirondack lakes, Haines and Baker (1986) reported that 42% of 707 brook trout populations had been lost, as had 39% of 111 lake trout (*Salvelinus namaycush*) populations, 36% of 90 rainbow trout (*Salmo gairdneri*) populations, 18% of 412 white sucker (*Catostomus commersoni*) populations, 19% of 520 brown bullhead (*Ictalurus nebulosus*) populations, 32% of 351 pumpkinseed sunfish (*Lepomis gibbosus*) populations, 34% of 326 golden shiner (*Notemigonus crysoleucas*) populations, and 42% of 254 creek chub (*Semotilus atromaculatus*) populations, among others. These

dramatic impacts on previously vigorous sport fisheries and on the fish community in general have paralleled the documented, concurrent acidification of lakes in the area.

Losses of sport-fish populations have also occurred in acidified lakes and rivers in Canada. Watt *et al.* (1983) reported the extinction of Atlantic salmon from seven acidic (pH < 4.7) rivers in Nova Scotia that had previously supported this species. Atlantic salmon were declining in other rivers with pH 4.7–5.0, but were stable at pH > 5.0 (Fig. 4.6). Lacroix and Townsend (1987) penned juvenile Atlantic salmon in four acidic streams in Nova Scotia, and reported no survival in the two streams with pH < 4.7, but complete survival where pH stayed >4.8.

Other studies documented the loss of fish populations from the Killarney region of Ontario (Beamish and Harvey, 1972; Beamish *et al.*, 1975; Beamish, 1974; Harvey and Lee, 1982). This area receives acidic precipitation (pH 4.0–4.5), and because it is fairly close to the Sudbury smelters it has episodes of dry deposition of acidifying SO_2. In a survey done in the early 1970s, 33 of 150 lakes had a pH < 4.5. The extinctions of several species of fish were actually monitored for two lakes (Lumsden and George Lakes). There was also circumstantial evidence for the extinction of fish in other lakes, in the form of eyewitness accounts of an historical sport fishery in presently fishless lakes. In total, there are known extinctions of lake trout in 17 lakes in the Killarney area. Lake trout is the most important sport fish in this region. In general, this species fails to recruit at pH < 5.5, and it is extirpated from Ontario lakes with an average pH of about <5.2 (Beggs and Gunn, 1986). In addition, smallmouth bass (*Micropterus dolomieui*) have disappeared from 12 lakes in the Killarney area, largemouth bass (*M. salmoides*) and walleye (*Stizostedion vitreum*) from four, and yellow perch (*Perca flavescens*) and rock bass (*Ambloplites rupestris*) from two (Harvey and Lee, 1982). Therefore, it seems that yellow perch and rock bass are among the most acid-tolerant fish species in the Killarney lakes. These species are also relatively tolerant of acidification elsewhere in eastern North America, as are the central mudminnow (*Umbra limni*), largemouth bass,

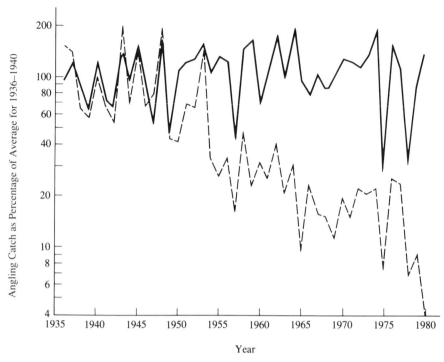

Fig. 4.6 The catch of Atlantic salmon by sport fishing in Nova Scotia rivers. The data were standardized to facilitate the comparison of acidic and less acidic rivers. ——— Mean for 12 rivers with pH > 5.0 (1980). — — — Mean for 10 rivers with pH ≤ 5.0 (1980). From Watt *et al.* (1983).

bluegill (*Lepomis macrochirus*), black bullhead (*Ictalurus melas*), and American eel (*Anguilla rostrata*), all of which are known to be present in water bodies with pH ≤4.5 (Rahel and Magnuson, 1983; Kerekes and Freedman, 1988).

In Lumsden Lake, the water pH decreased from 6.8 in 1961, to 4.4 by 1971 (Beamish and Harvey, 1972). This acidification was accompanied by a reproductive failure and loss of population of several fish species, including lake trout, lake herring (*Coregonus artedii*), and white sucker. In George Lake with a pH of 4.8–5.3, lake trout, walleye, burbot (*Lota lota*), and smallmouth bass disappeared. In 1973, unsuccessful reproduction was observed in other species that were not yet extinct. These were northern pike (*Esox lucius*), rock bass, pumpkinseed sunfish, brown bullhead, and white sucker. Clearly, an important cause of the extinction of fish in these lakes was a persistent failure to reproduce.

Fish populations were carefully monitored dur-

ing the Lake 223 acidification, and consideration of that study gives insight into the dynamics of the response of the fish community (Schindler and Turner, 1982; Mills, 1984; Mills *et al.*, 1987). Initially, there were five species of fish in Lake 223: lake trout, white sucker, fathead minnow (*Pimephales promelas*), pearl dace (*Semotilus margarita*), and slimy sculpin (*Cottus cognatus*). These species were all abundant, except for the relatively uncommon pearl dace. As acidification progressed, there was a marked change in the fish community. The most sensitive species was the fathead minnow, which declined precipitously in 1979 when the lake pH reached 5.6 and became extinct in 1980. The first of many year-class failures of lake trout was in 1980 (pH 5.4), and of white sucker in 1981 (pH 5.1). Slimy sculpins declined throughout the experiment. The pearl dace began to increase markedly in abundance in 1980, and it quickly became the most abundant small fish species. This species may have experienced a competitive release following the demise of the ecologically similar fathead minnow. However, the pearl dace also began to decline, in 1982 when the pH of Lake 223 reached 5.1. In fact, by 1982 no fish species were reproducing in Lake 223. Although lake trout and white sucker were still quite abundant in 1983, in the absence of successful reproduction in water at pH ~5.0–5.5, they will of course become extinct. The Lake 223 experience indicates a general sensitivity of the fish community to lake-water acidification. However, within limits set by the ultimate pH that is reached, there can be a replacement of sensitive species by relatively tolerant ones.

Many studies have examined the physiological effects of acidity and associated chemical stresses on fish. An important generalization is that early life-history stages are more susceptible to acidity than are adults. As was described above, most extinctions of fish populations have been attributed to reproductive failure, rather than to the death of adult fish (although episodic mortality of adult fish has been caused by events of acid shock; Leivestad and Muniz, 1976, Leivestad et al., 1976).

Laboratory studies of egg and embryonic development of Atlantic salmon indicated a toxic threshold (measured as LL_{50}, the point at which 50% of the exposed eggs fail to develop succesfully) at pH 3.9, while alevins were affected at pH 4.3 (Daye and Garside, 1979). In another laboratory study, Peterson et al. (1980) found that the hatching of Atlantic salmon eggs was inhibited at pH 4.0. This effect was eliminated by transfering the eggs to pH 6.8 water, providing the pH 4.0 exposure was less than 7–10 days. Lacroix (1985) incubated Atlantic salmon eggs in five streams in Nova Scotia with pH 4.6–6.5, and found an LL_{50} of pH 4.7 in the interstitial water of gravel in the spawning bed. The toxic threshold reported in this field study was at a higher pH than would have been predicted from the results of the two previously described laboratory experiments. The discrepancy could be due to some unknown influence of environmental variables that are controlled in the laboratory, but that are important in the field, such as oxygen tension or water temperature. For example, Kwain (1975) examined the effect of temperature and acidity on the em-

bryonic development of rainbow trout, and found greater toxicity at lower temperature. At 15°C there was a toxic effect at pH 4.5; at 10°C the threshold was pH 4.8; at 5°C it was at pH 5.5. The pattern of toxicity differed between yearling and fingerling trout (for fingerlings, pH 4.5 was toxic at 20°C; pH 4.2 at 15°C; pH 4.1 at 10°C), but both of these stages were more tolerant of acidity than were embryos.

Certain chemical constituents increase in concentration at low pH. From the perspective of toxicity to fish and other aquatic biota, the most notable example of this effect is aluminum (although most other metals also increase in concentration in acidic water). In fact, aluminum concentration in many acidic waters is sufficient to cause fish mortality, irrespective of any direct effect of hydrogen ion or other toxic stressors. Baker and Schofield (1982) reported that over the pH range 4.2–5.6, there was a reduced survival and growth of larvae and older life history stages of white sucker at 0.1 mg/l of aluminum, and at 0.2 mg/l for brook trout. An aluminum concentration of this magnitude is regularly exceeded in acidic waters. The most toxic forms of aluminum in acidic clear water are the ions Al^{3+} and $AlOH^{2+}$ (Havas, 1986). It is important to note that in brown water with a large concentration of dissolved organic matter, most of the soluble aluminum and other toxic metals is present in a complexed form. In this state metals are relatively nonexchangeable and less bioavailable than are free ionic forms, and aluminum in this bound form is less toxic than an equivalent concentration in clear water (Driscoll et al., 1980; Baker and Schofield, 1982; Campbell et al., 1983; Lazerte, 1984). In four acidic brown-water streams in Nova Scotia (pH 4.6–5.6; dissolved organic carbon 6.6–18.1 mg/l), virtually all of the aluminum dissolved in the water (0.08–0.19 mg/l) was organically bound (Clair and Komadina, 1984).

Amphibians

Most amphibian species are either completely aquatic, or they enter the water seasonally to breed. Several studies have suggested that acidification of their aquatic habitat could reduce the population size or restrict the distribution of amphibians (reviewed in Freda, 1986). For example, Gosner and Black (1957) found that the distribution of frog species in the New Jersey pine barrens is strongly influenced by acidity (pH range 3.6–5.2; the acidity is due to drainage from *Sphagnum*-dominated peatlands). The carpenter frog (*Rana virgatipes*) and the pine barrens tree frog (*Hyla andersoni*) are restricted to this area, and were the most acid-tolerant of the species that were tested, surviving pH 3.8 in a laboratory bioassay. Other species with a more widespread distribution were less tolerant of acidity. Saber and Dunson (1978) showed that the species richness of amphibians was smaller in an acidic bog in Pennsylvania than in less acidic water. For five of the eight species that were present in the bog, reproductive and adult stages were both present. For the other three species only adults were present, possibly indicating a reproductive failure. In England, Cooke and Frazer (1976) found that the smooth newt (*Triturus vulgaris*) rarely bred in ponds with pH < 6.0, whereas the palmate newt (*T. helveticus*) was present in water as acidic as pH 3.9, but not at greater acidity. In the Netherlands, Strijbosch (1979) found that six frog species avoided water with pH < 4.5, and that there was a relatively large frequency of dead egg masses in acidic habitat. In Nova Scotia, the distributions of 11 species of amphibian were not clearly related to acidity among 159 sites with a pH range of 3.9 to 9.0 (Dale et al., 1985). The bullfrog (*Rana catesbeiana*) was present between pH 4.0 and 9.0; the spring peeper (*Hyla crucifer*) and yellow-spotted salamander (*Ambystoma maculatum*) between pH 3.9 and 7.8; the green frog (*R. clamitans*) between pH 3.9 and 7.3; and the wood frog (*R. sylvatica*) between pH 4.3 and 7.8. Habitat structure, predation, and competition may have had a more important influence than acidity in the distribution of these amphibians. Successful reproduction was indicated at pH 4 by the presence of eggs, developing larvae, and adults of five species: the yellow-spotted salamander, and green, bull, wood, and pickerel (*R. palustris*) frogs. In contrast, the experimental acidification (to pH 4) of a small stream in New Hampshire caused salamanders to leave the treatment area (Hall and Likens, 1980).

In addition to the above field studies, laboratory experiments have revealed that among 14 species of amphibian, exposure to a pH of 3.7–3.9 during embryonic development caused mortality that exceeded 85%, and a prolonged exposure to pH 4.0 caused mortality that exceeded 50% (for reviews, see Tome and Pough, 1982; Dale et al., 1986; Freda, 1986). However, pH levels this acidic are uncommon in nature, and are not generally associated with an acidification of fresh water that was caused by acidic precipitation (which has a threshold of about pH 4.5). It seems that at least some amphibian species may be at lesser risk than fish for population changes from this source of acidification.

Waterfowl

A direct effect of acidification on aquatic birds is unlikely, and has not been documented. However, if acidification were to cause an important change in the habitat of waterfowl, then an indirect effect on the bird population would be anticipated (Eriksson, 1984). For example, a reduction or extinction of the fish population would be detrimental to piscivorous water birds. In contrast, an increased abundance of aquatic insects or zooplankton, possibly resulting from decreased predation caused by an extinction of predatory fish, could be beneficial to birds that eat these arthropods (Hunter et al., 1985; McNeil et al., 1987). Piscivorous waterfowl that breed on oligotrophic lakes of the northern hemisphere include common and red-throated loon (*Gavia immer* and *G. stellata*, respectively) and mergansers (*Mergus merganser* and *M. serrator*); the osprey (*Pandion haliaetus*) may also be at risk in certain locations. Planktivorous birds that might benefit from an increased abundance of invertebrates include diving ducks such as the common goldeneye (*Bucephala clangula*), ring-necked duck (*Aythya collaris*), and hooded merganser (*Lophodytes cucullatus*), and dabbling ducks such as mallard and black duck (*Anas platyrhynchos* and *A. rubripes*).

Interestingly, one of the most productive lakes in Nova Scotia for black and ring-necked ducks is the previously described, very acidic (pH 4.0), eutrophic Drain Lake (Kerekes et al., 1984). In spite of its extreme acidity, this lake sustains a very high rate of duck productivity of 0.5 broods/ha. Comparable productivity is only observed on other anthropogenically fertilized, but nonacidic, lakes in Nova Scotia. In Drain Lake, the large duck productivity is sustained by vigorous macrophyte and invertebrate communities. This observation suggests that there may not be an important direct toxicity to waterfowl in acidic water bodies.

4.5
RECLAMATION OF ACIDIFIED WATER BODIES

Since at least the 1950s, limnologists have been interested in the liming of brown-water lakes as a potential management practice to reduce acidity,

clarify the water, improve productivity, and create sport-fish habitat (Hasler *et al.*, 1951; Waters, 1956; Waters and Ball, 1957; Stross and Hasler, 1960; Stross *et al.*, 1961). More recently, much work has been done on the neutralization of acidified clear water by the addition of bases, usually limestone ($CaCO_3$) or lime [$Ca(OH)_2$] (Flick *et al.*, 1982). This treatment can be viewed as a whole-lake titration to raise pH. In some parts of Scandinavia liming is now used in a wide-scale remedial program to combat the biological symptoms of acidification. For example, by 1985 about 3000 lakes and 100 streams had been operationally limed in Sweden, mostly with limestone [Ministry of Agriculture and Environment '82 Committee (MAEC), 1983; Lessmark and Thornelof, 1986].

The effect of base addition on lake-water pH is illustrated in Fig. 4.7 for three limed and one refer-

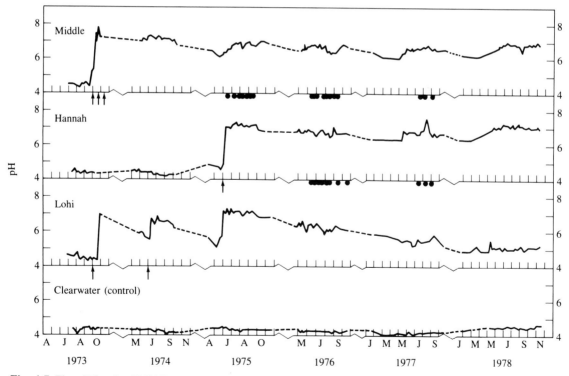

Fig. 4.7 The pH levels of Middle, Hannah, Lohi, and Clearwater Lakes, Ontario. Treatment with neutralizing agents [$CaCO_3$, $Ca(OH)_2$] is indicated by arrows; and addition of phosphorus by solid dots. From Dillon *et al.* (1979).

ence lake in Ontario. Prior to treatment, the lake waters had a pH of 4–5, which after liming increased to pH 7–8. The pH of Middle and Hannah Lakes remained fairly stable after treatment, while that of Lohi Lake quickly drifted back toward an acidic condition. This difference largely reflects the sizes of the drainage basin of these lakes; that of Lohi is relatively large, the lake flushes rapidly because of a large water input, and therefore neutralization was relatively short-lived (Dillon *et al.*, 1979). Similar liming experiments have been done in Scandinavia (Wilander and Ahl, 1972; Hultberg and Grahn, 1975b; Bengtsson *et al.*, 1980; Hultberg and Grennfelt, 1986).

Initially in the Ontario liming experiment, there was a drastic decline in the abundance and productivity of phytoplankton and zooplankton (Scheider *et al.*, 1975, 1976; Dillon *et al.*, 1979). Phytoplankton abundance returned to the preneutralization condition fairly quickly, but there was a shift in dominance from Dinophyceae to Chrysophyceae. Zooplankton were slower to recover, and 3 years after the first addition of lime to Lohi Lake, they had not returned to their preneutralization abundance. In addition, fish caged in the neutralized lakes suffered a high rate of mortality (Yan *et al.*, 1979), contrary to the expectation based on liming experiments elsewhere (Hultberg and Grahn, 1975b). The results in Ontario were influenced to some degree by metal toxicity, since the experimental lakes are close to the Sudbury smelters. Although the concentrations of Cu, Ni, Zn, and Al all decreased after the lakes were limed, they still remained large and probably continued to exert a toxic stress; this was especially true of copper (Yan *et al.*, 1979).

These neutralization experiments clearly show that acidified oligotrophic water bodies can be neutralized. Not unexpectedly, liming causes a severe shock to the acid-adapted biota, resulting in shifts in species dominance until a new steady-state community is achieved. It is important to realize that liming is not a long-term, permanent solution to the acidification of fresh water, since the acidic waters have to be retreated periodically, as the neutralizing substance is exhausted or flushed from the system.

Another possible way to treat acidified water is by fertilization. Even in very acidic water, there is a strong trophic response to fertilization (Dillon *et al.*, 1979; DeCosta *et al.*, 1983; Kerekes *et al.*, 1984), as was described earlier for Drain Lake. Fertilized but still acidic lakes could sustain a large rate of primary and secondary productivity, nurture waterfowl and other wildlife, and have other positive ecological attributes. However, the creation of a large number of mesotrophic or eutrophic lakes may not be an appropriate management objective in many situations, especially where nonconsumptive recreation is an important resource use. Nevertheless, fertilization may be an effective management practice for selected acidified water bodies.

Of course, it should be stressed that the liming of acidified ecosystems treats the symptoms but not the causes of acidification. Moreover, liming in a sense transforms the water body from one polluted state, to another still polluted but less toxic one. Clearly, a large reduction in the anthropogenic emission of acid-forming gases will be the ultimate solution to this widespread environmental problem. However, there is a great deal of controversy about the amount of reduction of emissions of SO_2 and other acid-precursor gases that will be required to reduce the impacts of acidic deposition, and about the various emission reduction strategies to be pursued (e.g., whether to target large point sources such as power plants and smelters, while paying less attention to smaller individual sources such as automobiles and oil-burning furnaces in homes). Not surprisingly, industries and political jurisdictions that are large emitters of acid precursors are strongly lobbying against effective emission controls, for which they argue the scientific justifications are not yet adequate. In addition, there is controversy about how small the rate of sulfur and nitrogen deposition should be in order to avoid further acidification of sensitive fresh waters, or to allow their recovery. For example, for the purposes of bilateral negotiations about the transboundary transport of acidic precipitation precursors between the United States and Canada, the Canadian government is advocating a desirable rate of sulfate deposition of 20 kg/ha-year. However, it is awkward or impossible to rigidly justify this desired level of sulfate deposition using the available scientific data. In fact, it appears that many sensitive fresh waters could acidify and suffer biological damage at sulfate loadings as small as 10 kg/ha-year (Schindler, 1988). However, in spite of all of the uncertainty about the specific causes and magnitude of the damage caused by the deposition of acidifying substances from the atmosphere, it seems intuitively clear that what goes up (i.e., the acid precursor gases) must come down (i.e., as acidic deposition).

5

FOREST
DECLINE

5.1
INTRODUCTION

Around large point sources of SO_2 pollution, a predictable zonation of vegetation damage can often be observed. In parallel with the pollution stress, plant damage decreases geometrically with increasing distance from the source. In contrast to these patterns near point sources, the effects of regional air pollution are not usually obvious, and they are always difficult to measure.

In recent years, there has been alarm over the widespread decline in vigor and dieback of mature forests in many parts of the world. It has been suggested that where the regional airshed is contaminated by various combinations of acidic deposition, ozone, sulfur dioxide, nitrogen compounds, and toxic elements, pollution may be a factor contributing to forest decline. However, such decline also appears to take place in areas where pollution is unimportant, and in these cases it has been hypothesized that dieback may be a natural process that could variously involve climatic change, the synchronous senescence of a cohort of overmature trees, or the effects of pathogens. In this chapter, the phenomenon of forest decline will be examined,

and some of the current hypotheses for its occurrence in landscapes both subject to and free from the effects of pollution will be discussed.

5.2
THE NATURE OF FOREST DECLINE

Forest decline is characterized by a progressive, often rapid deterioration in vigor of trees of one or several species, sometimes resulting in mass mortality (or dieback) within stands over a large area. Decline often affects mature individuals and is thought to be triggered by stresses or a combination of stresses, such as severe weather, nutrient deficiency, toxic substances in the soil, and air pollution. Stressed trees suffer a marked decline in vigor. In this weakened condition, trees are relatively susceptible to attack by insects and microbial pathogens. Usually, such secondary agents are not very harmful to vigorous individuals, but they may cause the death of severely stressed trees. It is important to note that although the occurrence and characteristics of forest decline can be well documented, the primary environmental variable(s) that trigger the decline disease are not known. As a result, the etiology of the decline syndrome is often attributed to a vague concatenation of biotic and abiotic factors (Manion, 1985; Wargo, 1985; Mueller-Dombois, 1986, 1987a; Klein and Perkins, 1987; Last, 1987).

Decline symptoms vary markedly among tree species. Frequently observed effects include (1) decreased net production, (2) chlorosis, abnormal size or shape, and premature abscission of foliage, (3) progressive death of branches, beginning at the extremities and often causing a "stag-headed" appearance, (4) root dieback, (5) a high frequency of attack by secondary agents such as fungal pathogens and insects, and ultimately (6) mortality, often as a stand-level dieback (Hepting, 1971; Cowling, 1985; Mueller-Dombois, 1986, 1987a; Klein and Perkins, 1987).

5.3
CASE STUDIES OF FOREST DECLINE
Naturally Occurring Decline

Periodic occurrences of a widespread decline of various tree species have been recognized for at least a century, and in these older cases the phenomenon may be unrelated to human activity. Some of these cases are described below.

Ancient Declines

Davis (1978, 1981) reported what appears to have been a decline of eastern hemlock (*Tsuga canadensis*) in North America. She documented a decrease of about 90% in the influx of hemlock pollen to lake sediment, taking place during a 50-year period about 4800 years ago. This phenomenon was synchronous at a number of sites throughout eastern North America. A similar widespread decline of eastern hemlock has not been observed in modern times. The hemlock decline was probably not due to climatic change, since other tree species that presently co-occur with hemlock, and that presumably have a similar habitat requirement, were not affected. In fact, during and after the hemlock decline these other species apparently increased in abundance to compensate for the decreased prominence of hemlock in the forest. The reason for the putative hemlock decline is unknown, but the triggering stress could have been a fungal disease pandemic or some other environmental factor.

There is similar palynological evidence for a widespread decline of elm (*Ulmus* spp.) in northwestern Europe about 5000 years ago (Smith and Pilcher, 1973). It has been speculated that this phenomenon could have been caused by a widespread clearing of the forest by Neolithic humans, but the influence of an unknown disease pandemic is an alternate hypothesis (Perry and Moore, 1987).

Birch Dieback

Because they took place so long ago, little is known about the hemlock and elm declines. There are other, more recent examples of forest decline that do not appear to have been caused by an anthropogenic influence. In North America, the best example is the widespread birch dieback that caused severe mortality throughout the northeastern United States and eastern Canada, especially from the 1930s to the 1950s (Fowells, 1965; Hepting, 1971; Auclair, 1987). The most susceptible species were yellow birch (*Betula alleghaniensis*), paper birch (*B. papyrifera*), and grey birch (*B. populifolia*). Birches over a vast area were affected by this condition, often with extensive mortality. For example, in 1951 at about the time of peak dieback damage in Maine, an estimated 67% of the birch trees had been killed. Birch dieback is less important today, but it still causes some mortality. In spite of a considerable research effort, a single primary cause was not determined for birch dieback. It is known that a heavy mortality of fine roots often preceded the deterioration of the above-ground tree, but the environmental cause(s) of this below-ground effect are unknown. No biological agent was identified as a primary predisposing factor, although a number of secondary fungal pathogens were observed to attack weakened trees and to cause their death, as did the bronze birch borer beetle (*Agrilis anxius*).

By definition, the birch dieback syndrome only affects birch trees in a natural forest environment.

However, the symptoms of the disease are virtually identical to "another" decline called "postlogging decadence" that affects the same species of birch on logged sites. The primary cause of this latter disease syndrome is also unknown, but it is suspected that stresses associated with the large microclimatic changes that occur after removal of much of the stand by logging are involved. Such environmental changes could predispose birch trees to damage or death caused by disease or defoliating insects that are normally tolerated by more vigorous individuals. Other apparently "natural" tree declines in North America have affected black willow (*Salix nigra*), oaks (*Quercus* spp.), ashes (*Fraxinus* spp.), balsam fir (*Abies balsamea*), pines (*Pinus* spp.), and others (Staley, 1965; Hepting, 1971; Sprugel, 1976; McCracken, 1985a,b; Auclair, 1987; Mueller-Dombois, 1987a; Weiss and Rizzo, 1987).

Ohia Decline

Other forest declines have been observed on Pacific islands. One of the best-known is that of the ohia tree (*Metrosideros polymorpha*), an endemic species of the Hawaiian archipelago (see below). Similar declines have occurred elsewhere in the Pacific, for example, in New Zealand forests of *Metrosideros umbellata–Weinmannia racemosa* and of *Nothofagus* (Batcheler, 1983; Stewart and Veblen, 1983; Wardle and Allen, 1983; Mueller-Dombois, 1987b).

In the Hawaiian Islands, *Metrosideros polymorpha* is the dominant tree species of the native forest (Mueller-Dombois *et al.*, 1983; Hodges *et al.*, 1986; Mueller-Dombois, 1980, 1986, 1987b). It usually occurs in monospecific stands, which comprise about 62% of the total forest area of the islands. From anecdotal accounts, events of widespread mortality of ohia are known to have occurred for at least a century, but the phenomenon is probably more ancient than this. The most recent decline began in the late 1960s. A survey of 76,900 ha of the island of Hawaii in 1982 found that about 50,000 ha exhibited symptoms of ohia decline. On about 35% of that area the damage was slight (i.e., less than 10% of the trees were declining or were dead), while 24% exhibited a moderate decline (11–50%), and in 41% the damage was severe (>50%). It is important to note that in most declining stands only the canopy individuals were affected. Understory saplings and seedlings were not in decline, and in fact they were released from competitive stress by the deterioration of the overstory.

As with all declines, the etiology of the ohia decline is unknown. Soil waterlogging may be an important predisposing environmental stress. An important role of nutrient deficiency has been ruled out by fertilization experiments, which did not reduce the mortality rate in declining stands (Gerrish *et al.*, 1988). A number of secondary biological stresses become important in many declining ohia stands. These include the fungal root pathogens *Phytophthora cinnamomi* and *Armillaria mellea*, and the endemic borer beetle *Plagithmysus bilineatus*. Often it is these secondary agents that kill the weakened trees.

Mueller-Dombois (1980, 1986) and Mueller-Dombois *et al.* (1983) have developed an interesting hypothesis to explain the etiology of ohia decline in Hawaii, and they have suggested the extension of their theory to forest declines elsewhere. They envisage stand-level dieback as being due to the phenomenon of "cohort senescence." This is a stage of the tree life history that is characterized by simultaneously decreasing vigor in many individuals, occurring in old-growth stands. The development of senescence in individuals is primarily governed by intrinsic genetic factors, but the timing of its onset can be influenced by extrinsic stress. The decline-susceptible life-history stage follows a more vigorous, younger, mature stage in an even-aged stand of individuals of the same generation (i.e., a cohort) that had initially established on the site following a severe disturbance. In Hawaii, lava flows, volcanic ash deposition, and hurricanes are disturbances that initiate primary succession, and sites disturbed in this way are colonized by a cohort of ohia individuals, which produce an even-aged stand. In the absence of an intervening catastrophic disturbance, the ohia forest eventually matures, and then becomes senescent and enters a decline phase. The original senescent stand is then replaced by

Stand-level dieback of *Metrosideros polymorpha* on a dry upland site in Hawaii. Note the dense advance regeneration of younger Ohia trees in the understory. (Photo courtesy of D. Mueller-Dombois.)

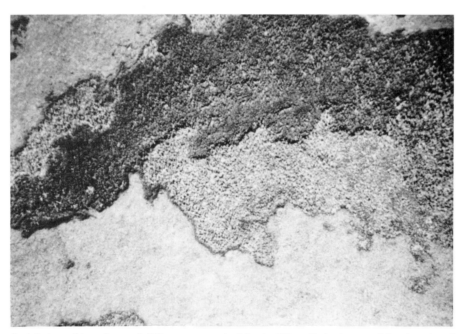

Aerial view of lava flows of different age on the island of Hawaii. At the top of the photo is a dark, healthy, even-aged forest of *Metrosideros polymorpha*. Below is a lighter-colored older forest that is suffering dieback. These are surrounded by an immature forest on a younger lava flow. (Photo courtesy of J. Jacobi.)

another ohia forest comprised of an advance regeneration of individuals released from the understory.

Therefore, according to the cohort senescence theory, the ohia decline should be considered to be a characteristic of the natural population dynamics of the species *Metrosideros polymorpha*. This theory appears to be most reasonable for cases of forest decline involving relatively short-lived tree species that are intolerant of the stressful conditions of a mature, closed forest. As such, it may be useful in explaining the declines that have variously occurred in species of birch, willow, ash, oak, and some other taxa. However, the cohort senescence model may be less tenable as an explanation for declines that affect relatively tolerant, longer-lived species such as sugar maple, and red and Norway spruce (described below). Declines of these latter species appear to involve stands that are younger than the potential longevity of these trees, which can exceed several centuries.

Forest Decline Possibly Triggered by Air Pollution

In recent years, declines of uncertain etiology have taken place in forests located in regions with polluted air. In western Europe the tree species that have been most severely affected are Norway spruce (*Picea abies*) and, to a lesser extent, beech (*Fagus sylvatica*). In North America the most prominent declines have occurred in red spruce (*Picea rubens*) and sugar maple (*Acer saccharum*). These forest declines are described below.

The "New" Type of Forest Decline in Europe

The recent widespread forest damage in Europe has been described as a "new" decline syndrome that may be in some way triggered by stress caused by air pollution. Although the symptoms appear to be similar, the "new" decline is believed to be different from the forest diebacks that are known to have occurred historically, and that are believed to have been natural (Prinz, 1984; Blank, 1985; Hinrichsen, 1986; Huettl, 1986a; Krause *et al.*, 1986). The

modern decline syndrome was first noted in fir (*Abies alba*) in West Germany in the early 1970s. In the early 1980s a larger-scale decline was apparent in Norway spruce (*Picea abies*), the commercially most important tree species in the region. In the mid-1980s decline also became apparent in beech (*Fagus sylvatica*), and to a lesser extent oak (*Quercus* spp.).

Decline symptoms in the conifers include (1) chlorosis or yellowing of needles, especially of older foliage; (2) premature shedding of foliage, starting from the base of the crown and the inner parts of branches; (3) a decrease or elimination of the net production of woody tissue, usually beginning after the loss of older needle biomass reaches 30–50%; (4) death of branches, often from the top of the tree and causing a "stag-headed" appearance; (5) a diagnostic nutrient deficiency in foliage, especially of magnesium, and sometimes of calcium, potassium, or zinc; (6) root dieback; and ultimately (7) death of trees, often as a synchronous, stand-level dieback. Symptoms of decline in beech include (1) abnormal coloration of foliage, (2) premature leaf drop, (3) dieback of lateral branches, (4) reduced net production, and (5) death.

Decline of this type has been observed in many countries of Europe, including Austria, Britain, Czechoslovakia, France, Germany, Italy, Yugoslavia, Poland, Scandinavia, Switzerland, and elsewhere. In total, in 1986 more than 19 million ha or 14% of the total forest area was damaged in Europe west of the Soviet Union, comprising about 15% of the coniferous forest volume and 17% of the angiosperm forest volume (Nilsson and Duinker, 1987; Postel, 1987). The spatial extent of the decline has been relatively well documented in West Germany (Blank, 1985; Krause *et al.*, 1986). In 1984, some degree of decline was observed in some 3.7 million ha, or about 50% of the West German forest area (33% of the forest was "lightly damaged" with 11–25% foliage loss; 16% was "moderately damaged" with 26–60% loss; and 1.5% was "severely damaged" with >60% loss). The area of damage had increased substantially from the condition in 1983, when 34% of the total forest area was classified as suffering damage, and from 1982 when the total was only 8%. [Note, however, that this apparent large increase in the area of damaged forest may in part be an artifact of different methodologies that were used to assess damage at different times (Blank *et al.*, 1988).] However, by 1984 a widespread forest dieback had not occurred, although there were many severely damaged stands, indicating that the former could yet occur.

Decline symptoms were variable in the German stands. However, in general, (1) mature stands older than about 60 years tended to be more severely affected; (2) dominant individuals within the stand were relatively vulnerable; and (3) individuals located at or near the edge of the stand were more severely affected by decline symptoms, suggesting that a shielding effect may protect trees situated further into the interior. Interestingly, epiphytic lichens often flourish in badly damaged stands, probably because of a greater availability of light and other resources caused by the diminished cover of tree foliage. In some respects this is a paradoxical observation, since lichens as a group are usually hypersensitive to air pollution, especially toxic gases (see Chapter 2).

It has been difficult to generalize about the role of site characteristics in predisposing stands to decline injury. The disease can occur in stands growing on either calcareous or acidic soil. However, it appears that stands on impoverished sites with oligotrophic soil or on sites that experience drought during the growing season are relatively susceptible to decline damage.

From the information that is now available, it appears that the "new" type of forest decline in Europe is triggered by a variable concatenation of environmental stresses. The weakened trees then decline rapidly, and may die as a result of attack by secondary agents such as fungal disease or defoliating or boring insects. A number of suggestions have been made regarding the identity of the primary inducing factor. These include gaseous air pollution, acidification, and toxic metals in soil. Other proposals suggest a natural climatic effect, in particular drought. However, there is not yet a concensus as to which of these interacting factors is the primary trigger that induces forest decline in Eu-

rope. It is quite possible that no single natural or air pollution–related stress will be proven as the primary cause of the new forest decline (Hinrichsen, 1986, 1987; Last, 1987). In fact, there may be several "different" declines occurring simultaneously, but in different areas (Blank *et al.*, 1988).

In order to actually prove that a specific environmental factor is the primary predisposing mechanism of forest decline, several or all of the following items of evidence would have to be demonstrated: (1) the environmental stressor would have to be shown to be present at an intensity that is known to be sufficient to cause acute or at least chronic toxicity; (2) experimental amelioration of the environmental factor should arrest or reverse decline symptoms; (3) experimental intensification of the environmental factor should predispose the stand to decline; and (4) in the case of a pathogenic biological agent, Koch's postulates must be demonstrated, that is, the pathogen must be isolated from declining trees, and it must be shown to induce decline in healthy trees that are injected or otherwise treated with an axenic culture or suspension of the putative pathogen (Agrios, 1969; Last, 1987). So far, these requirements have not been met for any of the proposed agents of forest decline. This indeterminate state of affairs with respect to the causes of forest decline is not unlike that for some human diseases, for example, many types of cancer, in which the predisposing or causal agent(s) have not been identified.

The major hypotheses put forward as mechanisms for the initiation of the European forest decline are summarized briefly below. Note that they are not listed in any particular order.

1. *Aluminum toxicity in acidified soil* was one of the first suggestions for the triggering stress in the "new" forest decline. In this scenario, aluminum is thought to be mobilized from naturally occurring minerals as a consequence of the acidification of forest soil, with the latter possibly caused by the deposition of acidifying substances from the atmosphere (Ulrich *et al.*, 1980). The aluminum hypothesis is consistent with (1)

the presence of large amounts of extractable aluminum in soils at many sites where decline has occurred and (2) frequently observed damaged and dead roots, exhibiting symptoms that resemble those of typical aluminum toxicity. In an experiment carried out in a mature *Picea abies* forest in West Germany, Matzner *et al.* (1986) limed and fertilized soil and then measured the *in situ* net production of root biomass after 7 months. At the initial soil pH of 3.9–4.0, there was a total root production of 180 kg/ha. However, liming to pH 4.3 allowed the production of 630 kg/ha, while a fertilized pH 4.5 soil produced 1820 kg/ha. This experiment suggests that root productivity was limited by acidity and associated aluminum toxicity, along with nutrient deficiency. However, as noted previously, stands are also declining on calcareous sites where aluminum availability in soil is usually low. Furthermore, laboratory bioassay studies have shown that many tree species have a considerable tolerance to plant-available aluminum (McCormick and Steiner, 1978; Ogner and Tiegen, 1980; Steiner *et al.*, 1980; Hutchinson *et al.*, 1986; Thornton *et al.*, 1986). This tolerance is a result of selection by the naturally high concentrations of soluble aluminum in acidic forest soils.

2. *Ozone* pollution has also been implicated as a potential contributor to the European forest decline. During sunny periods in the growing season, a peak ozone concentration of up to 400 μg/m^3 has been measured in West Germany. This is well above the toxic threshhold for sensitive plants (100–200 μg/m^3; Hinrichsen, 1986, 1987), and ozone damage has occasionally been observed in declining stands. However, the foliar damage that most often occurs in such stands does not resemble the diagnostic acute injury caused by ozone. Although this may mean that acute damage by ozone is not a primary predisposing agent, it does not rule out secondary effects (Blank, 1985; Krause *et al.*, 1986; Last, 1987).

3. *Sulfur dioxide and oxides of nitrogen* are present in regionally elevated concentrations throughout the area where forest decline is taking place (Hinrichsen, 1986). There have only been infrequent observations of the characteristic acute foliar injuries caused by exposure to these air pollutants in the European forests where decline has been observed. However, it is possible that chronic injury, occurring without obvious symptoms of damage (i.e. "hidden injury"; see Chapter 2) could be stressing these forests, thereby contributing to decline. On the other hand, the previously noted flourishing community of epiphytic lichens in the badly damaged forest suggests that SO_2 and other gaseous pollutants are not present in a phytotoxic concentration, since lichens are well known for their sensitivity to these toxic gases.

4. A *climatic effect* or extremes of weather during the growing season have also been considered as possible agents in the etiology of the "new" forest decline. The decline was first noticed in the mid-1970s, following several droughty growing seasons. The intensification of symptoms in the early 1980s may also be related to relatively dry growing conditions at that time (Blank, 1985; Andersson, 1986; Hauhs and Wright, 1986). Drought could have caused severe stress to trees, predisposing them to damage from other agents.

5. *Nutrient imbalance* has also been implicated in the decline disease in Europe. Two primary mechanisms have been proposed: (1) nitrogen fertilization, especially by the deposition of ammonium from the atmosphere, and (2) nutrient deficiency, especially of magnesium.

a. *Nitrogen fertilization.* Many of the declining forests are subject to large rates of atmospheric deposition of inorganic nitrogen compounds, largely originating from anthropogenic emissions. A typical background deposition of fixed nitrogen from the atmosphere is about 2–4 kg N ha^{-1} year^{-1}. However, in central Europe the rate of wet plus dry deposition is much larger than this, typically 10–25 kg ha^{-1} year^{-1}, and reaching 60–75 kg ha^{-1} year^{-1} at some sites (Nihlgard, 1985; Grennfelt and Hultberg, 1986). Anthropogenic sources of inorganic nitrogen include (1) the emission of NO_x from automobiles, power plants, and home heating, (2) ammonia volatilization from fertilized agricultural fields and animal manure, and (3) emission of N_2O by microbial denitrification in fertilized agricultural fields. These sources can cause regionally elevated rates of wet and dry deposition of inorganic nitrogen. This is especially true of forests, which because of their complex architecture are relatively efficient at "filtering" gases and particulates from the atmosphere. A relatively high rate of fertilization of forests with nitrogen can have various effects (after Nihlgard, 1985): (1) an increased net productivity of forest biomass, though often accompanied by relatively little resource allocation to the root system, causing small root/shoot ratios; (2) a possible accumulation of toxic concentrations of nitrogen-containing metabolites in foliage, leading to premature abscission as a detoxification mechanism; (3) decreased or delayed frost hardiness, a frequent observation after the fertilization of conifer forest (Soikkeli and Karenlampi, 1984); (4) an increased susceptibility to pathogens and insect attack; and (5) nutritional imbalance caused by the abnormally large supply of fixed nitrogen (see below).

b. *Nutrient deficiency.* In addition to a nutritional imbalance, trees may become deficient in magnesium, calcium, or potassium because of increased leaching from foliage or soil by acidic precipitation (see Chapter 4). Some German researchers have claimed that nutrient deficiency is the dominant predisposing stress in the "new" type of forest decline, particularly in their study area in southwestern West Germany. To support this hypothesis, they have presented data from field trials indicating that declining stands can be revitalized with an appropriate fertilization regime (Huettl, 1986a,b; Huettl and Wisniewski, 1987; Zoettl and Huettl, 1986). This is illustrated in Table 5.1 for a young stand of *Picea abies*, originally having a deficient foliar concentration of K and Mg. This stand responded favorably to fertilization with these nutrients, the symptoms of deficiency disappeared, and the stand appeared to be revitalized.

Decline of Red Spruce and Sugar Maple in Eastern North America

Red spruce (*Picea rubens*) and sugar maple (*Acer saccharum*) are both long-lived, shade-tolerant

Table 5.1 Effects of fertilizing a 12-year-old *Picea abies* stand in southern Germany on foliar nutrient concentration: The stand had symptoms of deficiency and decline[a]

Nutrient	Foliar Nutrient Concentration		Optimal Foliar Concentration
	May 1984 (Before Fertilization)	Oct. 1984 (After Fertilization)	
N (mg/g, d.w.)	15.0	14.5	13.0
P	1.4	1.8	1.3
K	2.6	5.6	4.5
Mg	0.4	0.9	0.7
Ca	1.3	2.6	1.0
Zn (μg/g, d.w.)	11	17	15
Mn	80	70	15

[a]The fertilizer treatment was 800 kg/ha of $MgSO_4$ + K_2SO_4. Modified from Huettl and Wisniewski (1987).

Dieback of red spruce (*Picea rubens*) at high altitude on Camel's Hump Mountain in Vermont. Between 1965 and the early 1980s about one-half of the spruce individuals died. Although the etiology of the dieback has not yet been demonstrated, it is suspected that the deposition of acidifying substances from the atmosphere may play a key role. (Photo courtesy of H. W. Vogelmann.)

trees, but they can be shallow-rooted and susceptible to drought (Fowells, 1965).

The modern epidemic of decline in sugar maple began in the late 1970s and early 1980s. It has been most prominent in Quebec, Ontario, New York, and parts of New England (Bordeleau, 1986; Carrier, 1986; McIlveen *et al.*, 1986). The symptoms are similar to those described for an earlier dieback (Westing, 1966), and include abnormal coloration, size, shape, and premature abscission of foliage, death of branches from the top of the tree downward, reduced productivity, and death of trees. There is a frequent association with the pathogenic fungus *Armillaria mellea*, but this is believed to be a secondary agent that attacks weakened trees (Hibber, 1964; Wargo and Houston, 1974). Many declining stands had recently been severely defoliated by the forest tent caterpillar (*Malacosoma disstria*), and many were commercial sugar bushes in which trees were tapped each spring for sap to produce maple sugar. As for red spruce (below), the declining maple stands are located in a region that is subject to a high rate of deposition of acidifying substances from the atmosphere, and this has been suggested as a possible predisposing factor, along with soil acidification and mobilization of available aluminum. However, little is known about the modern sugar maple decline, apart from the facts that it is occurring and the affected area appears to be increasing. No conclusive statements can yet be made about its causal factor(s).

The occurrence of stand-level decline in red spruce has been most frequent in high-elevation sites of the northeastern United States, especially in upstate New York, New England, and the mid- and southern Appalachian states (Siccama *et al.*, 1982; Johnson, 1983; Johnson and Siccama, 1984; Johnson *et al.*, 1986; Scott *et al.* 1984; Adams *et al.*, 1985; Bruck, 1985; Friedland and Battles, 1987). These sites are variously subject to acidic precipitation (mean annual pH about 4.0–4.1), to very acidic fog water (pH as low as 3.2–3.5), to a large rate of

Another view of damage in a red spruce dieback stand on Camel's Hump. (Photo courtesy of H. W. Vogelmann.)

deposition of sulfur and nitrogen from the atmosphere, and to stress from metal toxicity in acidic soil (Lovett *et al.*, 1982; see also Chapter 4).

The modern decline of red spruce can involve all mature size classes. Symptoms include (after Bruck, 1985; Johnson *et al.*, 1986) (1) discoloration, and premature death and abscission of foliage, starting from the top of the tree and working downward, and from the outside of the crown and progressing inward; (2) death of branches, starting from the top of the tree; (3) a general thinning of the crown; (4) decreases or zero production of radial growth by affected trees; and ultimately (5) death of trees singly or in patches of various size. The pathogenic root fungus *Armillaria mellea* frequently plays a secondary role and may kill weakened trees,

especially at relatively low altitude sites (Carey *et al.*, 1984). Secondary attack by the bark beetle *Dendroctonus rufipennis* and other insects is also frequent (Johnson *et al.*, 1986).

Where red spruce dieback is severe, there may be a large change in the species composition of stands. Siccama *et al.* (1982) observed such a change between 1964 and 1979 in permanent plots in high-elevation, mixed-conifer stands at Camel's Hump in the Green Mountains of Vermont. The total live basal area of the stand was relatively unaffected by mortality caused by spruce decline (29.6 m²/ha in 1965 versus 26.5 m²/ha in 1979). However, over that time period the basal area of red spruce decreased by about 45% (from 6.7 m²/ha to 3.7 m²/ha), and its relative dominance decreased from 23% to 14%. Similarly, Scott *et al.* (1984) remeasured permanent plots on Whiteface Mountain in New York. Between 1964–1969 and 1982, stands of red spruce at an elevation of less than 900 m suffered decreased basal areas of 40–60%, while higher elevation stands decreased by 60–70%. This effect was due to both a high rate of mortality of red spruce and decreased growth of surviving trees. Data from several other studies illustrate the latter effect. Siccama *et al.* (1982) found that the average ring width of red spruce at Camel's Hump had decreased from 1.33 mm/year between 1965 and 1969, to 0.92 mm/year between 1967 and 1971, and 0.63 mm/year between 1975 and 1979. In a larger-scale survey of 3000 dominant and codominant red spruce trees from New England and upstate New York, Hornbeck *et al.* (1986, 1987b) found that the average ring width was 13–40% smaller in 1980 compared with 1960 (note that 1960 was a time of peak productivity of red spruce in their study region). Hornbeck *et al.* also noted a decrease of 7–37% in the average ring width of balsam fir (*Abies balsamea*) over the same time period, even though that species was not apparently suffering from decline. These observations suggest that a climatic effect or a natural successional influence, related to the crown closure of developing stands, could have been a cause of the decreased net production of both species. However, note that in more southern mature forests of red spruce and fir (*Abies fraseri*) in the Appalachian Mountains, reduced productivity and dieback of both species is taking place (Weiss and Rizzo, 1987).

Declines of red spruce are known from anecdotal accounts to have taken place in the past. A major episode of dieback occurred in the 1870s and 1880s in the same general area where the "new" decline is occurring. An estimated one-third to one-half of the mature red spruce in the Adirondacks of New York was lost during that early decline episode, and there was also extensive damage in New England (Johnson et al., 1986; Hamburg and Cogbill, 1988). As in the case of the European forest decline, the "old" and "new" episodes appear to have similar symptoms, and it is possible that both occurrences are examples of the same kind of disease.

The hypotheses that have been suggested to explain the initiation of red spruce decline are similar to those proposed for the European forest decline. They include the acidification of soil and concomitant aluminum toxicity, acidic deposition, drought, winter injury exacerbated by insufficient hardiness due to nitrogen fertilization, heavy metals in soil, calcium deficiency caused secondarily by a high concentration of aluminum in soil, which may interfere with calcium uptake by tree roots, and gaseous air pollution (Friedland et al., 1985; Evans, 1986; Johnson et al., 1986; Shortle and Smith, 1988). An additional, novel hypothesis is that decline of spruces may be related to nitrogen deficiency that occurs in late-successional stands, caused by the immobilization of this primary limiting nutrient in aggrading biomass and spruce litter (Pastor et al., 1987). Climatic change, in particular a longer-term warming that has occurred subsequent to the end of the Little Ice Age in the early 1800s, may also be important in the decreased prominence of red spruce in the northeastern United States since that period (Hamburg and Cogbill, 1988).

At present, so little is known about the etiology of red spruce decline that the possible role(s) of anthropogenic air pollution and of natural environmental factors remain quite uncertain. This does not necessarily mean that air pollution is not involved. Rather, it stresses that more information is required before any conclusive statements can be made regarding the cause and effect of the phenomenon of red spruce decline.

6

OIL
POLLUTION

OIL
POLLUTION

6.1
INTRODUCTION

Petroleum is a critically important but nonrenewable natural resource. In its refined forms it is used for the production of energy and for the manufacture of synthetic materials such as plastics, while its asphaltic residues are used in heating and in construction. Energy production is by far the largest of the uses of petroleum, globally accounting for an energy equivalent of about 119×10^{18} J in 1983, or 45% of the total production of energy by all means (World Resources Institute, 1986). About 50% of the 1983 global petroleum use was in North America and Europe, even though these only comprise about 14% of the human population. The global production of petroleum had a value of more than $600 billion in 1983, in constant 1980 U.S. dollars (WRI, 1986).

Because most petroleum is extracted in locations that are remote from places where consumption occurs, it is a commodity that must be transported in a very large quantity. The most important methods of transportation are by oceanic tanker and overland pipeline. These transportation methods can pollute the environment by accidental oil spills and by operational discharges (i.e., the cleaning of storage and ballast tanks). There have been spectacular accidental spills, involving the loss of very large quantities of crude oil from disabled supertankers and offshore platforms. In addition, the large stationary facilities that are used for the refining of petroleum have the potential to cause chronic pollution by the discharge of hydrocarbon-laden waste waters, and by frequent small spills. The purpose of this chapter is to examine the ecological effects of oil pollution from these various sources.

6.2
CHARACTERISTICS OF PETROLEUM AND ITS REFINED PRODUCTS

Petroleum is a complex, naturally occurring mixture of organic compounds that is produced by the incomplete decomposition of biomass over a geologically long period of time. Petroleum compounds can occur in a gaseous form that is often called natural gas, as a liquid called crude oil, and as a solid or semisolid asphalt or tar associated with oil sands and shales. In aggregate, these materials comprise hundreds of molecular species. The molecules range in complexity from the gaseous hydrocarbon methane with a molecular weight of only 16 g/mole, to substances having a molecular weight greater than 20,000 g/mole (Clark and Brown, 1977; Kornberg, 1981).

Hydrocarbons are the quantitatively most important constituent of petroleum. Hydrocarbons can be classified into three broad groups, each with various subclasses (Clark and Brown, 1977; Kornberg, 1981).

1. *Aliphatic hydrocarbons* are open-chain compounds. If there is only a single bond between all adjacent carbon atoms, the molecule is said to be saturated. Unsaturated molecules have at least one double or triple bond. This is illustrated by the following series of two-carbon aliphatics: ethane, H_3C-CH_3 ; ethylene, $H_2C=CH_2$; and acetylene, $HC \equiv CH$. Saturated aliphatics are known as paraffins or alkanes, and they are chemically more stable than unsaturated aliphatics. The latter are not present in crude oil, but they can be produced secondarily during an industrial refining process, or photochemically after crude oil is spilled.
2. *Alicyclic hydrocarbons* have some or all of their carbon atoms arranged in a ring structure, and they can be saturated or unsaturated.
3. *Aromatics* are hydrocarbons that contain at least one six-carbon ring in the molecular structure. The basic C_6H_6 ring is known as benzene.

Crude oils from different locations vary greatly in their hydrocarbon composition (Table 6.1). On average, the three most important groups of hydrocarbons in petroleum are paraffin molecules, ranging from one carbon to >78 carbons, saturated and unsaturated five- and six-carbon alicyclics or naphthenes, and a great variety of aromatics. Other elements that are present in crude oil include sulfur at a concentration of from <0.1% to 5–6% by weight, and nitrogen at <0.1% to 0.9%. Both of these are typically present in an organically bound form. Oxygen is also present, at up to 2%. The most important trace elements in petroleum are vanadium and nickel, both at concentrations of up to 300 ppm, and present as organometallic complexes (Clark and Brown, 1977).

During the refining of crude oil, the various hydrocarbon products are separated by fractional distillation at specific temperatures (see the boiling-

Table 6.1 Chemical characteristics of several crude oils[a]

Component	Source of Crude Oil		
	Prudhoe Bay	South Louisiana	Kuwait
Sulfur (wt. %)	0.94	0.25	2.44
Nitrogen (wt. %)	0.23	0.69	0.14
Nickel (ppm)	10	2.2	7.7
Vanadium (ppm)	20	1.9	28
Naphtha (20–205°C) (wt. %)	23.2	18.6	22.7
Paraffins	12.5	8.8	16.2
Naphthenes	7.4	7.7	4.1
Aromatics	3.2	2.1	2.4
High-boiling fraction (>205°C)	76.8	81.4	77.3
Saturates	14.4	56.3	34.0
Aromatics	25.0	16.5	21.9
Polar materials	2.9	8.4	17.9
Insolubles	1.2	0.2	3.5

[a]Modified from Clark and Brown (1977).

Table 6.2 Typical analysis of the components of a crude oil from Prudhoe Bay, Alaska[a]

	Crude Oil	Natural Gas	Naphtha		Middle Distillate	Wide-Cut Gas Oils	Residuum
			Gasoline	Kerosene			
Boiling-point range (°C)	—	<20	20–190	190–205	205–343	343–565	565+
Yield: crude oil (vol. %)	100	3.1	18.0	2.1	24.6	35.0	17.6
Paraffins	27.3	100	47.3	41.9	8.9	9.3	9.3
Naphthenes	36.8	0	36.8	38.1	14.4	22.8	22.8
Aromatics	25.3	0	15.9	20.0	76.7[b]	67.9[b]	67.9[b]
Others	10.6	0	0	0			
Composition: sulfur (wt. %)	0.94		0.011	0.04	0.34	1.05	2.30
Nitrogen (wt. %)	0.23	—	0.02	0.02	0.04	0.16	0.68
Vanadium (ppm)	18	0	0	0	0	<1	93
Nickel (ppm)	10	0	0	0	0	<1	46
Iron (ppm)	4	0	0	0	0	<1	25

[a]Modified from Clark and Brown (1977).
[b]Numbers indicate a percentage derived from a mixture of aromatics and others.

point range for major refinery products in Table 6.2). A typical yield of products by the refining of petroleum from the Prudhoe Bay, Alaska field is: natural gas, 3%; gasoline, 18%; kerosene, 2%; middle distillates, 25% (including heating oil, and jet and rocket fuels); wide-cut gas oil, 35% (lubricating oils, waxes, feedstock for catalytic cracking); and residual fuel oil, 18% (bunker fuel for ships, and fuel for electrical ultilities) (Table 6.2). In addition, the process of catalytic cracking is often used to convert some of the heavier fractions to lighter, more valuable products. This secondary process can increase the yield of gasoline to more than 50%.

6.3

OIL POLLUTION

Oil in the Environment

Oil pollution can be caused by the spillage of crude oil or any of its refined products. However, the largest and most damaging pollution events usually involve a spill of petroleum or heavy bunker fuel from a disabled tanker or drill platform at sea, or to a lesser extent from a blowout of a well or a broken pipeline on land.

A spill on land can occur in many ways, but the largest events generally involve a pipeline or a blowout. As of 1982, the global tally was 64,500 km of pipeline in place or being constructed for the transportation of liquid petroleum, and another 136,100 km of natural gas pipeline (Gilroy, 1983). The causes of pipeline rupture are diverse. They include faulty pumping equipment and pipe seam welds, earthquakes, sabotage, and target practice by hunters in the case of aboveground pipelines. The total quantity of oil spilled from pipelines is not well quantified in many parts of the world. However, because of the widespread use of spill sensors and mechanisms for shutting down sections of pipeline, individual events are usually much smaller than can potentially be spilled by oceanic supertankers or by blowouts of offshore platforms. Because the spread of spilled oil is much more restricted on land than on water, terrestrial spills usually impact a relatively localized area (unless the spilled oil reaches a water-

course). The characteristics and ecological effects of terrestrial spills will be described in more detail later, in the context of oil spills in the Arctic.

Much information is available about the magnitude, behavior, and effects of oil spilled at sea. Recent estimates of the input of petroleum hydrocarbons to the world's oceans range from about 3.2 to 6.1 million metric tons/year (Table 6.3). This is equivalent to about 0.2–0.3% of the total quantity of petroleum transported by tankers in 1980, and 0.1–0.2% of the global petroleum production (Cormack, 1983). In comparison, the natural production of nonpetroleum hydrocarbons by marine plankton has been estimated at about 26 million metric tons/year, or about 4–8 times the total input of petroleum hydrocarbons. These biogenic hydrocarbons are an important component of the background concentration of hydrocarbons in the marine environment, but they are, of course, very well dispersed and should not be considered to be an important source of marine pollution. An estimate of the emission of petroleum hydrocarbons from natural submarine and coastal oil seeps is about $200–600 \times 10^3$ MT/year, or about 6–13% of the total petroleum input to the oceans. The remaining input is anthropogenic, except for an unknown fraction of the atmospheric deposition of hydrocarbons, which could have originated with emissions from terrestrial vegetation and from other natural sources. Of the anthropogenic sources, discharges contaminated by urban runoff, refineries, and other coastal effluents are most important in causing local, chronic pollution in the vicinity of harbors and coastal cities around the world. The discharges from tankers, other ships, and offshore exploration and production platforms are essentially episodic, and they occur as accidental spills and deliberate discharges of various size.

Of course, it is impossible to predict the location or magnitude of an accidental spill of petroleum. As might be expected, spills from tankers are most frequent in coastal areas of the most heavily traveled sea lanes (Fig. 6.1). In magnitude, marine spills range from frequent losses of trivial amounts, to very infrequent but massive events. Some examples of the latter include supertanker accidents such as

Table 6.3   Estimates of inputs of petroleum hydrocarbons to the world's oceans[a]

Source	1973 Ref. (1)[b]	1979 Ref. (2)	1981 Ref. (3)	1983 Ref. (4)
Natural seeps	600	600	300 (30–2600)	200 (20–2000)
Atmospheric deposition	600	600	300 (50–500)	300 (50–500)
Urban runoff and discharges	2500	2100	1430 (700–2800)	1080 (500–2500)
Coastal refineries	200	60	—	100 (60–600)
Other coastal effluents	—	150	50 (30–80)	50 (50–200)
Accidents from tankers at sea	300	300	390 (350–430)	400 (300–400)
Operational discharges from tankers	1080	600	710 (440–1450)	700 (400–1500)
Losses from nontanker shipping	750	200	340 (160–640	320 (200–600)
Offshore production losses	80	60	50 (40–70)	50 (40–60)
Total discharges	6110	4670	3570	3200

[a]Data are in units of 10^3 MT/year.
[b]References: (1) National Academy of Sciences (NAS) (1975d); (2) Kornberg (1981); (3) Baker (1983), best estimate, range in parentheses; (4) Koons (1984), best estimate, range in parentheses.

Fig. 6.1 Major oceanic transportation routes for petroleum. From NAS (1985).

the 1978 *Amoco Cadiz* incident in the English Channel (about 230,000 metric tons spilled; Anonymous, 1982), the *Torrey Canyon* in 1967 off southern England (117,000 MT; Smith, 1968), the *Argo Merchant* in 1976 off Massachusetts (26,000 MT; Grose and Mattson, 1977), and the *Arrow* in 1970 off Nova Scotia (11,000 MT; Thomas, 1977). Massive spills have also occurred from offshore platforms, such as the 1979 IXTOC I blowout in the Gulf of Mexico [more than 400,000 MT of petroleum was lost, making it the largest spill event to date; Advisory Committee on Oil Pollution of the Sea (ACOPS), 1980], the 1977 Ekofisk blowout in the North Sea (30,000 MT; Cormack, 1983), and the 1969 Santa Barbara spill off the coast of southern California (an estimated 10,000 MT; Foster and Holmes, 1977). As will be described later in the case studies of large marine oil spills, some of these events caused considerable ecological damage.

It is important to stress that in itself, the size of the spill does not necessarily tell much about its potential to cause damage. Even a small spill can wreak havoc in an ecologically sensitive environment. For example, near Norway in 1981 a small operational discharge of oily bilge washings from the tanker *Stylis* killed an estimated 30,000 seabirds, because it impacted a location where they were seasonally abundant (Kornberg, 1981). In addition, the type of petroleum product can influence the severity of the ecological damage. The most important considerations in this respect are the relative toxicity and environmental persistence of the spilled material (discussed later).

The operational discharge of oily tank washings from tankers is an important and frequent source of marine spills. This source of pollution is caused by the practice of filling the holding tanks with seawater after the delivery of a load of petroleum or a refined product, and later discharging the bilge water into the sea as the ship travels to pick up its next load (Clark and MacLeod, 1977; Koons, 1984). The dicharge typically contains a hydrocarbon residue that is equivalent to about 0.35% of the tanker's capacity (range of 0.1% for light refined oil, to 1.5% for heavy bunker fuel). Beginning in the early 1970s, this operational source of oil pollution was

decreased substantially by a simple change in the tank-cleaning procedure called the load-on-top (LOT) process. In LOT, the oily wash water is retained on board for some time as ballast. This allows most of the oil to separate from the seawater. The oily residue is combined with the next cargo, while the relatively clean seawater is discharged. The LOT technique has a potential recovery efficiency of 99%, but in operational use it is usually 90% or less, depending on how turbulent the sea is during the separation phase (Clark and MacLeod, 1977). As a result of the widespread (but not universal) adoption of LOT by tankers, oil pollution from this source was reduced from about 1.1 million metric tons/year in 1973, to 0.7×10^6 MT/year in 1983 (Table 6.3).

Partitioning of the Spilled Oil

After oil is spilled it is partitioned in the environment in a number of ways (Fig. 6.2).

1. *Spreading* is the process by which spilled oil physically dilutes itself over the surface of the water. The surface slick can then be transported by a water current, or it can be moved by wind at a rate of about 3–4% of the wind speed (Kornberg, 1981). The degree of spreading is directly influenced by the viscosity of the spilled oil, and by such environmental conditions as wind, turbulence, and the presence of ice on the water surface. A surface oil slick can be visually detected as a sheen when only 0.0003 mm thick. After an experimental spill of 1 m^3 of a Middle Eastern crude oil, a slick 0.5 mm thick and 48 m in diameter formed in 10 min, and after 100 min it was 100 m wide and 0.1 mm thick (Kornberg, 1981).

2. *Evaporation* is initially important in reducing the volume of spillage that remains in the aqueous environment. Evaporation is most important in the dissipation of relatively light and volatile hydrocarbon fractions, and it is aided by high ambient temperatures and wind speed, and by rough

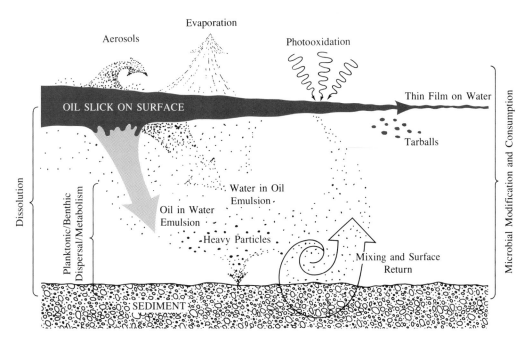

Fig. 6.2 Diagrammatic representation of the processes involved in the dissipation of spilled petroleum at sea. From Clark and MacLeod (1977).

seas that move spilled oil into the atmosphere by the formation of a fine spray at wave crests. In the marine environment, evaporation typically accounts for as much as 100% of spilled gasoline, 75% of number 2 fuel oil, 30–50% of spilled crude oil, and 10% of bunker C (Clark and MacLeod, 1977; Kornberg, 1981). Since the lighter-molecular-weight fractions of petroleum are evaporated preferentially, the relative concentration of heavier molecules increases greatly in the residual spill volume. For example, after a spill of an Alaskan oil there was a 15–20% loss of mass by evaporation. This caused the relative concentration of the heavy, nondistillable molecular fraction to increase from an initial 34% of the mass to more than 50% (Payne and McNabb, 1984).

3. *Solubilization* is the process of dissolution of oil fractions into the water column. This causes a contamination of water in the vicinity of the oil spill. For example, the concentration of total hydrocarbons dissolved in water directly beneath (at a 2-m depth) a petroleum slick in the North Sea was up to 4 g/m^3 (ppm), compared with about 1 mg/m^3 (ppb) or less in uncontaminated seawater (Cormack, 1983). In general, lighter fractions are more soluble in water than are heavier ones, and aromatics are much more soluble than alkanes. Among the alkanes, the solubility of the gases C_1–C_4 ranges from 24 to 62 g/m^3 in fresh water, while the light liquids C_5–C_9 are 0.05–39 g/m^3, species in the kerosene fraction C_{10}–C_{17} are 1–2 × 10^{-4} g/m^3, gas oils C_{16}–C_{25} are 6 × 10^{-4} to 10^{-8} g/m^3, lubricating oils C_{23}–C_{37} are 10^{-7} to 10^{-14} g/m^3, and residual hydrocarbons >C_{37} are less than 10^{-14} g/m^3. Benzene, the lightest aromatic, has a solubility of

1780 g/m³, while toluene is 515 g/m³, and naphthalene is 31 g/m³ (Parker *et al.*, 1971; Clark and MacLeod, 1977);

4. *Residual material* that remains after most of the evaporation and solubilization of lighter fractions has taken place forms a rather stable and gelatinous water-in-oil emulsion (although it can contain up to 70–80% water), known as "mousse" because of its superficial resemblance to the whipped desert with that name. It is in this partially weathered form that most oil that has been spilled offshore actually impacts the shoreline. Further weathering by biological oxidation and photooxidation of relatively light components creates lumps of a very dense, semisolid, asphaltic residuum, known as "tar balls."

Tar balls can be important in the chronic pollution of beaches and some pelagic environments. In most cases, the primary source of tarballs is thought to be tank washings, rather than the weathered residue of accidental spills. The presence of a large quantity of floating tar was first reported in the mid-Atlantic gyre known as the Sargasso Sea by Horn *et al.* (1970). However, the phenomenon did not receive much attention from the popular media until a dramatic report by the anthropologist Thor Heyerdahl (1971) of "shocking" and "terrible" pollution by "floating asphalt-like material" along the edge of the Sargasso Sea. He made this observation at sea level during his two east–west crossings of the Atlantic Ocean in 1969 and 1970, in small papyrus raft boats. Other studies of pelagic tar in the northwest Atlantic Ocean estimated a total quantity of >86,000 metric tons, 75% of which was in the Sargasso Sea (Butler *et al.*, 1973). The tar balls accumulate in the Sargasso Sea quite efficiently, because of the natural ability of this gyre to collect floating material, including the locally dominant seaweed *Sargassum* spp., as well as human detritus. Because the tar balls are typically comprised of C_{30} to C_{40} hydrocarbons, and are highly weathered of their relatively toxic light hydrocarbon fractions, they can serve as an ecological substrate for

some marine life. Typical organisms of these tiny pelagic ecosystems include the isopod *Idothea metallica*, the goose barnacle *Lepas pectinata*, and the epipelagic fish *Scomberesox saurus* (Butler *et al.*, 1973).

6.4
BIOLOGICAL EFFECTS OF HYDROCARBONS

Petroleum and hydrocarbon toxicity is a well known and well studied phenomenon, with a large amount of both field- and laboratory-based bioassay data. Comprehensive reviews of specific toxicity data for a wide variety of organisms are available in various sources (e.g., Currier and Peoples, 1954; Baker, 1971f, 1983; O'Brien and Dixon, 1976; Malins, 1977; Corner, 1978; Boyles, 1980; Nounou, 1980; Neff and Anderson, 1981; Wells, 1984; NRC, 1985b; Wells and Percy, 1985; Vandermeulen, 1987). This subject will not be detailed here.

Several recent syntheses have attempted to derive an empirical relationship between the physical/chemical properties of hydrocarbons and their toxicity. The value of such a relationship lies in its potential ability to predict the likely toxicity of the multitude of hydrocarbon species and mixtures that have not been directly bioassayed (Birge, 1983; Miller, 1984). An important observation that has emerged from these studies is that the toxicity of particular hydrocarbons is strongly related (in a statistical sense) to their chemical structure and hydrophobicity (Hutchinson *et al.*, 1979; Veith *et al.*, 1979; Bobra *et al.*, 1985; Hermens *et al.*, 1985). To state the relationship in another way, those hydrocarbons that are most soluble in water are least toxic on a molar basis. This statistical relationship has been found to be remarkably strong and consistent across a wide range of hydrocarbons and organisms. The biophysical mechanism of the hydrophobicity effect is that the rate of transport of hydrocarbons into organisms depends on their solubility in the lipid phases of the cellular membrane. Therefore, lipid solubility is a major controlling factor of the rate and degree of bioconcentration of specific hydrocarbons from the aqueous environment (i.e., organism–water partitioning). Furthermore, in cases of acute exposure, lipid solubility affects the degree of membrane disruption that is caused (loss of integrity of the plasma membrane is a frequently observed toxic effect of an acute exposure to hydrocarbons). From these observations, a toxicity index has been proposed that is based on the partitioning of particular hydrocarbons between water and octanol phases (octanol chemically mimics partitioning into a generalized lipid phase) (Veith *et al.*, 1979; Bobra *et al.*, 1985; Hermens *et al.*, 1985; M. Miller *et al.*, 1985; Abernethy *et al.*, 1986). This index is also useful for modeling the toxicity and food web accumulation of chlorinated hydrocarbons (described in Chapter 8). However, it should be stressed that even though large, water-insoluble hydrocarbon species have a greater toxicity *on a per mole basis* than do relatively light and water-soluble ones, the greater part of the ecological damage after an actual petroleum spill is often attributed to the lighter fractions. The reason for this apparent contradiction is that the light hydrocarbons typically make up a very large fraction of the spill volume, and since they are also of relatively small molecular weight, they comprise a much larger fraction of the total number of moles of hydrocarbons in the spill than do the heavy species.

The bioconcentration of hydrocarbons from the aqueous environment has been determined in many situations of chronic exposure and after spill events [e.g., Clark and MacLeod, 1977; Lee, 1977; Kornberg, 1981; National Oceanic and Atmospheric Administration (NOAA-CNEXO), 1982; NRC, 1985b]. A few selected examples of hydrocarbon concentration in organisms of polluted marine environments are as follows (from Clark and MacLeod, 1977; Ohlendorf *et al.*, 1978).

1. Plants: the littoral macroalga *Enteromorpha clathrata*, 429 ppm after a fuel oil spill; the salt marsh grass *Spartina alterniflora*, 15 ppm after a fuel oil spill; the estuarine dicot *Zostera marina*, 17 ppm after a fuel oil spill.
2. Invertebrates: the snail *Littorina littorea*,

27–604 ppm after a bunker C spill; the mussel *Modiolus edulis*, 21–372 ppm after a bunker C spill; the mussel *Mytilus edulis*, 77–103 ppm after a bunker C spill; the oyster *Crassostrea virginica*, 38–126 ppm after a fuel oil spill, 236 ppm in a chronically polluted estuary, 160 ppm in a chronically polluted harbor; the starfish *Asterias vulgaris*, 14–400 ppm after a bunker C spill; the lobster *Homarus americanus*, 103–130 ppm after a bunker C spill.
3. Birds: the herring gull *Larus argentatus*, 584 ppm in brain tissue after a fuel oil spill; the common murre *Uria aalge*, 8820 ppm in a composite organ sample after a fuel oil spill; the western grebe *Aechmophorus occidentalis*, 9100 ppm in liver after a fuel oil spill.

Of course, the toxic effect of an exposure to a particular concentration of hydrocarbons in the environment is influenced by many variables. Some of the more important of these are (after Mackay and Wells, 1981; Baker, 1983) (1) the amount of oil; (2) the type of oil, and the relative toxicity of its component hydrocarbons; (3) the frequency and timing of the exposure event (chronic versus episodic pollution), including the persistence of residues under particular environmental conditions; (4) the state of the oil, for example, thickness of the slick, nature of the emulsion, degree of weathering, etc.; (5) environmental variables that affect toxicity, such as weather and climate, oxygen status, presence of other pollutants, etc.; (6) the use of chemical dispersants to create an oil-in-water emulsion, including toxicity of the dispersant itself; and (7) the sensitivity of the specific biota of the impacted ecosystem to the toxic effects of hydrocarbons.

6.5

ECOLOGICAL EFFECTS OF OIL POLLUTION

A number of case studies will be considered to illustrate the ecological effects of oil pollution in various environments. The marine cases will describe the effects of (1) massive spills from wrecked supertankers, (2) spills from offshore drilling rigs, (3) experimental oiling of salt-marsh vegetation, and (4) chronic oil pollution. To illustrate the impacts of oil spilled on terrestrial ecosystems, we will examine the effects of petroleum spilled on northern terrain.

Oil Spills from Wrecked Tankers

The Torrey Canyon

One of the largest and best studied cases of oil pollution caused by the wreckage of a tanker was the *Torrey Canyon* incident of 1967 (Smith, 1968; Michael, 1977; Nelson-Smith, 1977). The *Torrey Canyon* was a supertanker bound for Milford Haven in southern England, when it ran aground on the Seven Stones Rocks. This caused the eventual loss of its entire cargo of 117,000 metric tons of Kuwait crude oil. Pollution from this wreck contaminated about 225 km of the Cornish coastline, and to a lesser degree the Brittany coast of France. Most of the oil that washed on shore was a viscous mousse of about 80% water-in-oil.

As with many marine oil spills, birds were the most obvious victims of the *Torrey Canyon* incident. In the perception of the popular press and most people, birds are also the most tragic victims of oil pollution. The *Torrey Canyon* spill was estimated to have caused the death of at least 30,000 birds, of which 97% were common murres (*Uria aalge*) and razorbills (*Alca torda*) (Ohlendorf *et al.*, 1978). About 7850 oiled birds were captured and cleaned. However, partly because of inexperience in the cleaning and handling of a large number of oiled seabirds, the rehabilitation program was not very successful. Only 6% of the cleaned birds survived for at least 1 month. The total mortality was suffi-

Aquatic birds are among the most evocative victims of oil spills. This blue-winged teal (*Anas discors*) was impacted by a spill of heavy fuel oil in the St. Lawrence River near Alexandria Bay, New York. (Photo courtesy of B. Freedman.)

A rocky shore on the Brittany coast of France that was heavily contaminated by an oil-in-water emulsion after the wreck of the *Amoco Cadiz*. The dominant macroalgae are *Fucus serratus* and *Ascophyllum nodosum*. When these were directly oiled, their fronds were killed, but most of their holdfasts survived, so that regeneration was fairly rapid. (Photo courtesy of J. Vandermeulen.)

cient to cause a population decrease of 80–88% among breeding puffins (*Fratercula arctica*), murres, and razorbills on the Brittany coast in the year following the *Torrey Canyon* spill (Michael, 1977; Evans and Nettleship, 1986).

Seabirds that spend much of their time on the surface of the ocean are especially vulnerable to oil, and pelagic species such as sea ducks, alcids, and penguins that aggregate in large numbers as surface "rafts" can suffer tremendous mortality (Brown *et al.*, 1973; Bourne, 1976; Holmes and Cronshaw, 1977; Ohlendorf *et al.*, 1978; Clark, 1984; Evans and Nettleship, 1986; Leighton *et al.*, 1985; Cairns and Elliot, 1987). In addition, the alcids and penguins have a relatively small reproductive potential, and as a result it can take a considerable time for their populations to recover from an event of mass mortality. For example, murres (*Uria* spp.) do not breed until they are at least 3 years old, not all adults breed in a given year, they lay only one egg per

clutch, and they fledge an average of only about one young per five breeding adults (Ohlendorf *et al.*, 1978).

The most important toxic effect of petroleum on birds is the fouling of feathers. This causes a loss of the critical feather properties of insulation and buoyancy, so that the animal may die from drowning or from an excessive heat loss leading to hypothermia. There is also a direct toxic effect of oil ingested during attempts to clean feathers by preening, and the developing embryo may suffer mortality from even a light oiling of the egg by the feathers of a contaminated adult (Bourne, 1976; Ohlendorf *et al.*, 1978; Clark, 1984; Holmes, 1984; Leighton *et al.*, 1985; Vandermeulen, 1987).

An important aspect of the *Torrey Canyon* case is that an intensive effort at beach clean-up was mounted that emphasized the use of detergent and first-generation dispersants to create a milky, oil-in-water emulsion that washed out to sea for dispersal.

In total, about 10,000 metric tons of detergent and dispersant was used to clean up the 14,000 MT of crude oil residue that had washed up on the Cornish beaches. These chemicals greatly exacerbated the toxic effect of the oil on the shoreline and nearshore ecosystems.

The ecological effects of the *Torrey Canyon* spill were well described (Smith, 1968). A rocky intertidal habitat was common in the area and it was heavily damaged, especially where detergent and dispersants were used in the clean-up. In the absence of such cleaning, algae were damaged by oil but they generally survived (including species that proved to be quite sensitive to detergent; see below). In addition, limpets (*Patella* spp.) often survived a complete covering by oil, and they were even observed grazing on oil-covered rocks. There was a much more intensive damage to seaweeds on oiled-and-cleaned shores. Relatively sensitive species of macroalgae included *Fucus serratus*, *F. spiralis*, *F. vesiculosus*, *Pelvetia canaliculata*, and *Ulva lactuca*. *Laminaria digitata* suffered widespread damage to its fronds, but it was able to regenerate vigorously from this acute injury. The algae that were most resistant to oil plus detergent were *Chondrus crispus* and *Ascophyllum nodosum*.

The most resistant invertebrates were the beadlet *Actinia equina* and the dahlia anemone *Tealia felina*. These survived in tidal pools that were devoid of all other animals as a result of the combined toxicities of oil and detergent/dispersant. Among the more notable crustacea that were wiped out on many rocky near-shores were the lobster *Homarus vulgaris*, and the crabs *Cancer pagurus*, *Carcinus maenas*, *Pilumnus hirtellus*, *Porcellana platycheles*, *Portunus puber*, and *Xantho incisus*. The limpets *Patella vulgata*, *P. intermedia*, and *P. asper* were also devastated, as were other molluscs such as the previously abundant saddle oyster *Anomia ephippium*. There was also a complete demise of echinoderms, including the starfish *Marthasterias glacialis*, the cushion-star *Asterina gibbosa*, and the urchin *Psammechinus miliaris*. A widespread mortality occurred among fish of the rocky nearshore, especially the blenny *Blennius pholis*, the Cornish sucker-fish *Lepadogaster lepadogaster*, the eel *Conger conger*, and others. Virtually no fish were present in the vicinity of oiled and cleaned beaches.

Because such a large fraction of the invertebrate herbivore population was killed on the oiled and cleaned rocky beaches, the regenerating intertidal habitat was characterized by an atypical plant community dominated by the green alga *Enteromorpha* spp. As herbivores gradually recovered in abundance, the *Enteromorpha* was replaced by species of brown algae that are more typical of the rocky intertidal. However, during the initial years of the postoiling succession, the zonation of algal species was somewhat abnormal, an artifact that was probably caused by the disruption of grazing by limpets on oiled and cleaned shores. For example, the upper limit of the distribution of *Laminaria digitata* and *Himanthalia elongata* was higher by as much as 2 m during the first few years of succession. As limpets progressively recolonized, the distributions of seaweed species returned to their more typical limits (Southward and Southward, 1978).

Another important consequence of the massive use of detergent and dispersants during the *Torrey Canyon* cleanup was the damage caused in submersed habitat as far as 0.4 km offshore. Usually, a marine oil spill only causes severe damage to biota that are directly contacted by petroleum at the surface or on beaches. However, the impaction of submersed habitat by water with a large concentration of dissolved detergent, dispersant, and emulsified petroleum caused mortality of many marine species. Especially susceptible decapods were the squat lobster *Galathea strigosa*, the swimming crab *Portunus puber*, and the edible crab *Cancer pagurus*. Serious mortality occurred in the bivalve molluscs *Ensis siliqua* and *Mactra corallina* at a depth as great as 14.5 m. Susceptible echinoderms included the starfish *Martasterias glacialis*, the heart urchin *Echinocardium cordatum*, and the sea urchin *Echinus esculentus*. However, some invertebrates tolerated an exposure that was fatal to other species in their phylum, notably the spider crab *Maia squinado*, the starfish *Asterias rubens*, and the brittle star *Acrocnida brachiata*. In addition, there was relatively little mortality of macroalgae in the subtidal habitat.

The unexpectedly great ecological damage caused by the oil and detergent/dispersants during the *Torrey Canyon* cleanup proved to be an important lesson. This effect resulted in a much more judicial use of detergent and dispersants during subsequent oil spill cleanups, with the practice largely being restricted to offshore locations and to sites that are of high value for industrial or recreational purposes. For the latter purpose, small concentrations of dispersant are often used in a seawater diluent to remove oil residues. Except for birds, the ecological damage caused by the *Torrey Canyon* oil alone proved to be relatively small, and recovery was fairly vigorous. In contrast, some sites treated with detergent and/or dispersants took as long as a decade to recover an ecosystem similar to that present before the spill (Southward and Southward, 1978).

The Amoco Cadiz

In 1978, another supertanker broke up in the same general area as the *Torrey Canyon*, but closer to France. This was the *Amoco Cadiz* accident, in which about 233,000 metric tons of crude oil were spilled off the Brittany Coast. The accident was caused by a steering failure of the ship. The vessel drifted helplessly for several critical hours while attempting repairs. At the same time the vessel's captain and a marine salvage company negotiated assistance terms under conditions complicated by different native languages, long-distance communications between the captain and owners of the vessel, and other factors. In any event, the *Amoco Cadiz* went aground on a nearshore reef.

Of the total oil spilled, about 32% was estimated to have evaporated, 11% sank to the bottom, 11% was dispersed at sea, 10% was recovered and disposed of, and the remaining 36% contaminated beaches, rocky shores and other parts of the marine environment (Ganning *et al.*, 1984). About 62,000 MT of oil impacted more than 360 km of shoreline with some degree of oiling. This included about 140 km of heavily oiled Brittany coast, which was polluted by a ~50% water-in-oil mousse (Berne *et al.*,

1980; Finkelstein and Gundlach, 1981; Anonymous, 1982). There was a large-scale, intensive cleanup of much of the contaminated shoreline, an effort that cost an estimated $117 million (in 1978 U.S. dollars; Berne and Bodennec, 1984). The cleanup largely involved the physical removal of oily residues and contaminated sand and sediment. The use of dispersants and detergents was restricted to sites of high economic value, especially harbors. In addition, the specific dispersants used were much less toxic than those used in the *Torrey Canyon* incident, and the dispersants were mostly applied as a dilute fraction of the washwater.

Much effort went into monitoring the degree of environmental pollution by hydrocarbons, and their subsequent dissipation by biodegradation and physical weathering. For example, 14–28 days after the spill, a very polluted nearshore site had an average hydrocarbon concentration in water of 122 ppb. This decreased to 39 ppb after 28–39 d, 10 ppb after 54–63 d, 2–5 ppb after 87–112 d, and 1 ppb after 1 year (the latter is the approximate background concentration of hydrocarbons in seawater) (Berne *et al.*, 1980). Nearshore sediment contained as much as 1100 ppm of hydrocarbons, but the degree of contamination decreased rapidly in seaward transects. The anaerobic sediment of an estuary had a concentration of up to 50,000 ppm (5%) of hydrocarbons (Berne *et al.*, 1980). The oyster *Crassostrea gigas* accumulated up to 2700 ppm d.w. of total petroleum hydrocarbons in its flesh, and even after 2 years some individuals contained as much as 100 ppm (Boehm, 1982). The initial contamination of some commercial fish species in the vicinity of oiled areas was up to 154–170 ppm (Boehm, 1982).

The presence of a large concentration of hydrocarbons in the contaminated sediment of various habitats stimulated a strong numerical response by microorganisms (Ward *et al.*, 1980; Atlas, 1982). Interestingly, the general, indigenous microbial community proved to be quite capable of utilizing hydrocarbons as a metabolic substrate; the response of taxa that are metabolically specific to hydrocarbons comprised only a small proportion of the total microbial response to the oil pollution (Table 6.4). The average rate of biodegradation was about 0.5

Table 6.4   Average abundance of bacteria in various Brittany sediments 1 month after the *Amoco Cadiz* oil spill[a]

	Total Number of Bacteria (10^3/g sediment)	Hydrocarbon-Utilizing Bacteria (10^3/g sediment)
Oiled sites		
Salt marsh	630,000	14,000
Estuary—high intertidal	190,000	18
Estuary—low intertidal	690,000	390
Sandy beach	100,000	350
Reference sites		
Salt marsh	96,000	0.52
Beach	45,000	0.38

[a]Modified from Ward *et al.* (1980).

μg hydrocarbon/g d.w. sediment-day, and in an aerobic, nutrient-rich environment as much as 80% of the mass of aliphatic and aromatic hydrocarbon residues was metabolized by microorganisms in the first seven postspill months (Ward *et al.*, 1980).

The ecological consequences of oil pollution and the subsequent cleanup after the *Amoco Cadiz* incident were less severe than those of the *Torrey Canyon* spill. In addition, recovery was more rapid, and was largely complete within several years (NRC, 1985b). The principal reason for this smaller ecological effect was the effective use of relatively low-toxicity dispersants rather than industrial detergent and first-generation dispersants to clean up the *Amoco Cadiz* residues.

Seaweeds of the rocky coast were most affected in the intertidal zone (Floc'h and Diouris, 1980). The red alga *Rhodothamniella floridula* is present on rock surfaces that are exposed except at high tide. Where this species was directly exposed to oily redidues, it died within a few days. The brown alga *Pelvetia canaliculata* occupies rocks that are only exposed at low tide. Where this species was directly oiled it suffered an initial damage, and then for as long as 8 months it experienced a progressive decline that resulted in a virtual denudation of previously vegetated substrates. About 18 months after the spill regeneration of *Pelvetia* began by the establishment of germlings, and this led to a progressive

and rapid recovery. Lower down in the intertidal zone, the dominant alga *Fucus spiralis* suffered some damage, but it was not as intensively damaged as were the preceding two algae of more xeric habitat. However, the encrusting intertidal alga *Hildenbrandia* was drastically affected. In the mid- and low-tide zone the dominant algae *Ascophyllum nodosum* and *Fucus serratus* were little affected, except for relatively minor injury occurring where there was direct contact with oil. As in the *Torrey Canyon* incident, the opportunistic green alga *Enteromorpha* spp. formed a thick but ephemeral carpet on denuded rock substrate in the intertidal zone, wherever herbivores were eliminated or reduced greatly in density by oil pollution. Overall, injury to seaweeds was largely restricted to situations where there was direct contact with oil. Compared with the *Torrey Canyon* experience, much less long-term damage was caused to intertidal habitat by the *Amoco Cadiz* oil spill.

Similarly, the overall effect was not great in a small eelgrass (*Zostera marina*) community that was impacted by *Amoco Cadiz* oil (Zieman *et al.*, 1984). The eelgrass (an angiosperm) was only damaged by direct contact of foliage with oil. However, there was very little plant mortality since the perennating tissues of this plant are in the sediment, where they were not directly oiled. There was great mortality of some of the fauna of this ecosystem, particularly among the amphipods. Prior to the spill there were 26 species of amphipods, but even 1 year after the spill only five species were present in the oiled area. In contrast, gastropod molluscs were little affected in the seagrass ecosystem.

There was some damage to salt-marsh vegetation, much of which was caused by the physical removal and trampling of vegetation during cleanup of the oily residue. However, this cleanup activity was clearly needed in order to remove the thick deposits of mousse from the salt marsh, and the methods used were generally effective in spite of some initial blunders. Plant regeneration was fairly vigorous, and involved the resprouting of unkilled perennials and the establishment of seedlings of salt-marsh annuals (Levasseur and Jory, 1982; Seneca and Broome, 1982).

Biological effects in the sublittoral zone were important, but smaller than after the *Torrey Canyon* spill (Cabioch, 1980; Glemarec and Hussenot, 1982). In this zone the biota was affected by water contaminated by dissolved and finely dispersed hydrocarbons. In general, severe damage was restricted to nearshore situations, usually less than 30 m from shore. The most heavily damaged community was characterized by the bivalve *Abra alba* and the amphipod *Amplelisca tenuicornis*. This community suffered mortality equivalent to about 40% of its total invertebrate biomass. Overall, the most sensitive taxa of the nearshore benthos were molluscs, amphipods, and the sea urchin *Echinocardium cordatum* (millions of dead urchins washed up along the polluted beaches).

Commercially important oyster (*Crassostrea gigas*) beds were contaminated by oil residues, which caused a discernible "tainting" of the animals. However, the oysters themselves did not exhibit any histopathological injury, and they appeared to be unaffected (Neff and Haensly, 1982). In contrast, a flatfish (plaice, *Pleuronectes platessa*) was affected by serious and progressive histopathological and biochemical injuries, even though it was much less contaminated by petroleum residues than were the oysters (Neff and Haensly, 1982).

At least 4600 seabirds died as a result of contamination by the *Amoco Cadiz* oil spill (Clark,

1984). A larger number of deaths was prevented by the fact that the avian population was at a seasonally small abundance at the time of the oil spill.

Pelagic Oil—The Argo Merchant

Another case of oil pollution from a stricken tanker concerns the *Argo Merchant* wreck in 1976 off the coast of Nantucket Island, Massachusetts (Grose and Matton, 1977; NRC, 1985b). About 26,000 MT of fuel oil was lost at sea. However, because the spilled oil was mainly comprised of light hydrocarbons, virtually all of the spill evaporated to the atmosphere. Furthermore, the prevailing wind kept the oil slicks at sea until they dissipated, so that there was no contamination of relatively susceptible littoral and other shallow-water ecosystems. Relatively minor effects on the pelagic biota were documented. For example, up to 55% of sampled individuals of the zooplankter *Calanus finmarchicus* were visibly contaminated by oil, 61% of *Centropages typicus*, and 34% of *Pseudocalanus* sp., largely in the form of oil-containing fecal pellets. However, this contamination did not cause a measurable change in the zooplankton community. There was no demonstrable effect on the phytoplankton of oiled compared with reference areas. Of the six fish species that were common in the oiled area, only the sand lance *Ammodytes americanus* was apparently decreased in abundance. Although the mortality of seabirds was not determined, it likely was not large since bird density in the spill area was small. The most frequently oiled species were the gulls *Larus argentatus* and *L. marinus*. Therefore, the *Argo Merchant* oil spill caused a relatively minor ecological damage, but this was largely because the oil did not wash ashore. In addition, seabirds were not abundant in the vicinity of the spill.

Oil Spills from Offshore Drill Platforms

The world's largest oil spill occurred in 1979 from the IXTOC I, an exploration semisubmersible platform well located 80 km off the eastern coast of Mexico (ACOPS, 1980; Kornberg, 1981). The rate of flow was as large as 6400 m³/day, and over the more than 9-month blowout period an estimated 476,000 MT of crude oil was spilled. Within a week of the beginning of the spill, a slick 180 km long and as much as 80 km wide had formed. An estimated 50% of the spilled oil evaporated to the atmosphere, 25% sank to the bottom of the Gulf of Mexico as weathered residue, 12% was degraded by microorganisms and photochemical processes, and 6% was mechanically removed or burned at the well site, while 6% reached and contaminated about 600 km of Mexican shoreline, and another 1% landed on beaches in Texas (Ganning *et al.*, 1984). This massive spill caused great disruption of tourism in the Gulf of Mexico, and affected the fishing industry by fouling boats and gear, by tainting commercial species, and by eliminating fishing in the large area where slicks were present. However, there was relatively little documentation of the effects of this spill on offshore and coastal ecosystems (NRC, 1985b).

Better ecological information is available about a blowout that took place in 1969 in the Santa Barbara Channel of southern California (Straughan and Abbott, 1971; Steinhart and Steinhart, 1972; Foster and Holmes, 1977). In total, approximately 10,000 MT of crude oil was spilled, resulting in contamination of the entire channel and more than 230 km of coastline. The average pollution of beaches by oil residue was 115 MT/km, compared with 10.5 MT/km in a nearby area that was only affected by natural petroleum seepages (see below), and 0.03 MT/km for all California beaches.

Interestingly, the Santa Barbara area has long been known for its naturally occurring underwater and inland petroleum seepages. These have historically caused slicks on the ocean surface, deposits of oil and tar on beaches, and tar pits inland. The seepages come from shallow underground reservoirs of petroleum, which leak to the surface through fractured or porous bedrock. One estimate of the rate of marine seepage was 3000–4000 MT/year (Allan *et al.*, 1970). Because of the long-term occurrence of natural oil seepages in the Santa Barbara area, it has been speculated that some of the local biota may have evolved a physiological toler-

ance to the toxic effects of hydrocarbons. However, studies of marine invertebrates have not generally shown this effect (NRC, 1985b).

As in many cases of marine pollution by oil spills, the most evocative victims of the Santa Barbara spill were birds (Steinhart and Steinhart, 1972; Foster and Holmes, 1977). An estimated 9000 birds were killed, or about 45% of the estimated population that was present at the time of the spill (the area is a wintering ground for many marine bird species). About 60% of the deaths was comprised of three species of loon (*Gavia immer*, *G. stellata*, and *G. arctica*) and western grebes (*Aechmophorus occidentalis*). However, in spite of this mortality the number of birds in the channel in subsequent winters was not significantly reduced. An attempt was made to clean and rehabilitate many of the oiled birds that made it to beaches, but the techniques were not as advanced as they are today, and the effort was not very successful. Of more than 1600 individuals that were treated, only 246 were still alive after 1 month.

The habitat type that was most affected by the Santa Barbara spill was the rocky intertidal, which comprised about 13% of the impacted coastline (Foster and Holmes, 1977). There was a great mortality of barnacles, especially *Chthamalus fissus* and *Balanus glandula*. Mortality ranged from 60–90% under "very heavily" oiled conditions, to 20% in "moderately" oiled situations, and 1–10% after "light" oiling. Relatively fresh oil was especially toxic to barnacles. Weathered residues caused much less mortality; in some cases, *Chthamalus fissus* survived a complete covering by weathered residue. The rocky intertidal surfgrasses *Phyllospadix souleri* and *P. torreyi* were also badly damaged, suffering 50–100% damage after heavy oiling, and 30–50% after light exposure. These were the most prominent short-term effects of the Santa Barbara oil spill on the rocky intertidal, but most other species of this habitat were also damaged. However, the recovery of most species was fairly rapid. Within a year of the spill barnacles began resettling, even on rocks that were still covered with asphaltic residues.

In recreationally important areas of the Santa

Barbara Channel there was a cleanup of the oil that washed onto beaches. Various techniques were used, including the physical removal of oil and oily sand, blast-cleaning with steam, sand, or water, and spraying with a light naphtha-like mixture. When these cleanup techniques were used in natural habitats they usually exacerbated the ecological damage, because they physically removed all biota along with the oil residues.

Effects of Oil on Salt Marshes

Salt marshes occur in coastal situations that may be frequently impacted by oil from a coastal refinery or terminal, or by oil spilled offshore. Because salt marshes are a relatively accessible natural habitat for experimental research, they have been better studied with respect to the effects of oil pollution than any other type of marine ecosystem. In the following, a series of experiments done in salt marshes in southern Britain will be briefly examined (similar research has also been done in North America; e.g., DeLaune *et al.*, 1979; Getter *et al.*, 1984; Webb *et al.*, 1985).

Baker (1971a) reported the effects of a single experimental treatment with crude oil on three graminoid-dominated communities along a salt-marsh gradient in southern England (Table 6.5). The oil caused initial damage to all species. However, since the regenerative meristematic tissue of most plants was not killed, there was a progressive refoliation during the growing season. The five most prominent plant species of these habitats displayed a wide range of tolerance to crude oil (Table 6.5). The rush *Juncus maritimus* was very sensitive, and suffered a large decrease in cover even in the lightly oiled treatment. The grass *Agrostis stolonifera* was most tolerant, and it actually increased in abundance in the lightly oiled treatment. An apparent growth stimulation after a light oiling was also observed with the grasses *Puccinellia maritima* and *Festuca rubra* in another field experiment (Baker, 1971e). The reasons for such a positive response are not clear, but the stimulation could have been due to (1) an indirect effect on nutrient cycling caused by the vigorous microbial activity that takes place in oiled soil, (2) a microclimatic effect related to the relatively dark and warm oiled soil, or (3) a hormone-like action of particular hydrocarbons.

Baker (1971d) also carried out an experiment that was designed to examine the effects of cleaning oiled salt-marsh vegetation by (1) cutting and removing oiled biomass, (2) burning, and (3) spraying with detergent to emulsify the oil residues. At best, none of these treatments provided a better recovery than was achieved if the oiled vegetation was left alone. Moreover, some of the clean-up procedures greatly exacerbated the initial damage caused by the oil, notably the use of detergent or burning. The conclusion was that it was best to leave salt-marsh vegetation to recover naturally after the relatively light experimental oilings. However, in situations contaminated by thick deposits of

Table 6.5 Effects of a single experimental treatment of salt-marsh vegetation with crude oil: The plots were oiled in June, and then sampled 3 months later in September[a]

Treatment	Low Marsh: *Spartina anglica* density (number/0.025 m²)	Mid Marsh (% cover)		High Marsh (% cover)	
		Puccinellia martima	*Festuca rubra*	*Agrostis stolonifera*	*Juncus maritimus*
Unoiled	19.9	84	60	10	17
Lightly oiled (9 l/36 m²)	8.1	47	41	54	3
Heavily oiled (54 l/36 m²)	1.4	10	9	10	5

[a]Modified from Baker (1971a).

mousse or other residues, an active cleanup may be required.

In another set of experiments, Baker (1971c,g) investigated the seasonality of the effects of a crude oil spill on salt-marsh vegetation. She found that most plant species were relatively sensitive to oiling when their buds were expanding early in the growing season. Plants were less sensitive when they either were growing less actively or were senescent, and when their stomata were closed because of drought. The annual *Suaeda maritima* was very sensitive to oil treatment at any time in the growing season, and once oiled it was unable to recover. In contrast, although the leaf tissue of perennial plants was killed by contact with oil, meristematic tissue

did not always die and therefore regrowth of new foliage could take place.

Baker (1971b, 1973) also investigated the effects of successive oilings of salt-marsh vegetation. This was done in order to model the effects of frequent exposures in a chronically polluted situation, for example, near a refinery or a marine oil terminal. At the higher frequencies of oiling, all of the important graminoid species suffered a relatively large initial damage (Table 6.6). Afterward, most species exhibited a progressive recovery of abundance on the oiled plots. The recovery was largely by vegetative incursion from outside the experimental plots, rather than by the establishment of seedlings. Of course, there were large differences

Table 6.6 Response of salt-marsh graminoids to as many as 12 treatments of crude oil at 4.5 1/10 m², followed by 5 years of regeneration[a]

Species	Number of Oil Treatments	Prespray	Postspray Years of Recovery			
			1	2	3	4
Spartina anglica	0	400	240	256	336	256
(shoots/m²)	2	352	160	256	240	288
	4	288	80	240	208	272
	8	336	32	112	160	272
	12	288	0	0	48	160
Puccinellia martima	0	86	85	90	86	78
(% cover)	2	95	78	84	91	74
	4	91	71	80	75	65
	8	90	0	0	0	6
	12	92	0	0	0	0
Juncus martimus	0	22	16	20	25	18
(% cover)	2	28	2	4	8	7
	4	21	0	0	2	5
	8	28	0	0	0	0
	12	30	0	0	0	0
Festuca rubra	0	72	30	40	35	41
(% cover)	2	75	30	52	45	40
	4	73	35	40	50	32
	8	68	0	0	5	8
	12	62	0	0	10	8
Agrostis stolonifera	0	10	13	13	15	12
(% cover)	2	16	22	40	30	12
	4	14	18	32	15	8
	8	18	17	43	36	46
	12	16	0	15	35	42

[a]Modified from Baker (1973).

among species in their relative tolerance to the crude oil treatment. The most sensitive taxa were the annuals *Suaeda maritima* and *Salicornia* spp., while the most tolerant species was the umbellifer *Oenanthe lachenalii*, which survived 12 oilings in one growing season. Among the dominant graminoids, *Juncus maritimus* and *Puccinellia maritima* were initially most sensitive, and were slow or unable to recover after 4 postoiling years. The most tolerant grass was *Agrostis stolonifera*, which was initially stimulated by as many as eight oiling treatments, and even by up to 12 oilings after several years of regeneration.

Overall, it appears that some species of the salt marsh are very sensitive to oiling, but the ecosystem itself is resilient. The plant community responds to the experimental pollution stress by a change in species composition, but not necessarily by a decrease in productivity and biomass.

Effects of Chronic Oil Pollution

The environment near a petroleum refinery or a tanker terminal can be subject to chronic oil pollution from frequent spills, and from the continuous discharge of contaminated process water. Chronic oil pollution of the coastal environment is also caused by cities that allow oil to be discharged into storm or sanitary sewers. In some cases these can cause longer-term ecological damage.

Dicks (1977) studied changes in a *Spartina anglica*-dominated salt marsh in the vicinity of a large oil refinery at Southampton Water, England. He observed a progressive deterioration of the salt marsh, which began when the refinery began operating in 1951. After 1970, the ecosystem started to recover as a result of a decreased frequency of oiling and an improved chemical quality of the refinery outfall. In 1970, the ecological damage was severe but local, and was characterized by death of the dominant *Spartina* grass over an area of about 60 ha, resulting in a devegetated area of bare mud. In the primary succession that began in 1970 in the denuded area, the initial colonists were the annuals *Salicornia* spp. and *Suaeda maritima*. These were followed by the perennial dicot herbs *Aster tri-*

polium and *Halimione portulacoides*, and ultimately by *Spartina anglica*.

Baker (1976) reported qualitatively similar changes in a *Spartina anglica* salt marsh subjected to chronic oil pollution in the vicinity of a large terminal and refinery complex at Milford Haven in southern England. However, the damage was restricted to a smaller area. She also described short-term effects of small oil spills on the intertidal biota, and observed an increased abundance of the opportunistic green alga *Enteromorpha* in a small area of frequently impacted rocky shore. Overall, Baker concluded that the petroleum industry at Milford Haven had not caused important damage to the salt marsh or the marine biota.

A chronic exposure to a diverse array of hydrocarbons and other pollutants can occur in many situations where effluents from industrial and municipal sources impact water. In some cases a high frequency of both cancerous and noncancerous diseases of fish and/or shellfish has been observed in chronically polluted situations (e.g., E. R. Brown *et al.*, 1977; Mearns and Sherwood, 1977; Sherwood and Mearns, 1977; Sonstegard, 1977; Mallins *et al.*, 1983; Fabacher *et al.*, 1986). Although the precise etiology of these diseases is unknown, it is suspected that they may be somehow caused by chronic pollution, and that they could represent an important bellwether of environmental degradation. For example, Sonstegard (1977) found a large incidence of gonadal tumors (up to 100% among older male fish) in goldfish × carp hybrids (*Carassius auratus × Cyprinus carpio*) collected in 1977 from the River Rouge, a polluted river in Detroit, Michigan. Examination of a large museum sample of hybrids collected in the same location in 1952 indicated that there were no tumors at that time. More generally, however, it has proven difficult to demonstrate a consistent linkage of animal diseases with chronic water pollution. Yevich and Barszcz (1977) examined biota that were exposed to hydrocarbons at 16 oil-spill sites on the east coast of the United States. They found cancerous neoplasia in only one of the 18 species of invertebrate that they surveyed (the soft-shell clam, *Mya arenaria*), but only in apparent conjunction with two of the 16 spill incidents. R. S. Brown *et al.* (1977) also investigated the frequency of neoplasia in the soft-shell clam in a survey of the New England coast. They found that a high incidence of this cancerous disease was widespread in their study area, but the neoplasia was not obviously linked to pollution (some polluted sites had a large frequency of neoplasia while other polluted sites did not, and some nonpolluted sites also had an epizootic). Clearly, the role of the environment (including pollution) in the etiology of neoplasia and other diseases is complex and uncertain (see Mix, 1986, for a review).

Oil Spills in the Arctic

In this section, some of the effects of oil spills on terrestrial arctic ecosystems are described. In a very general way, this section also serves to illustrate the effects of terrestrial oil spills in other biomes.

Natural Oil Seeps

Some insight into the effects of petroleum and its residues can be gained from consideration of some natural oil seeps on the North Slope of Alaska. Fresh seepage ranges in quality from a heavy fluid oil to tar. The seepage weathers progressively to a very thick tar, and ultimately to a dry, oxidized, crumbly asphalt (McCown *et al.*, 1973b). Associated with the fresh seepage material is a vigorous bacterial community that is adapted to the oxidation of petroleum hydrocarbons, with the most prominent genera being *Pseudomonas*, *Serratio*, *Vibrio*, *Cytophaga*, and *Flavobacterium* (Agosti and Agosti, 1973). Bacteria are also abundant in shallow tundra ponds that are affected by oil seepage (Barsdate *et al.*, 1973). In both the terrestrial and aquatic environments, these microorganisms are responsible for much of the oxidative weathering of seepage materials. In laboratory studies these microbes are

A 1-year-old spill of crude oil in a freshwater marsh near Norman Wells in the Canadian subarctic. The dominant plant is the sedge *Carex aquatilis*. The foliage of this plant was killed by oiling, but its perennating rhizomes in the sediment were little affected. As a result, after only 1 year, there was a vigorous regeneration, which at this location has penetrated a weathered residue of crude oil without suffering much damage. (Photo courtesy of B. Freedman.)

A 1-year-old experimental spill of crude oil in dwarf shrub tundra near Tuktoyaktuk in the Canadian low arctic. Initially, the crude oil acted as a contact herbicide and killed all foliage and active woody stem buds that were directly contacted. However, not all individuals of the dominant shrubs [willow (*Salix glauca*) and birch (*Betula glandulosa*)] were killed. In this photo, regeneration can be seen issuing from lateral stem buds, which were stimulated to break dormancy by the death of the terminal stem bud, or by suckering from the root crown. Other plant species of this ecosystem were less resilient to the effects of the oil spill. (Photo courtesy of B. Freedman.)

capable of oxidizing hydrocarbons at a temperature as low as 4°C (Agosti and Agosti, 1973).

Plants of wet meadow tundra can grow in an intimate contact with the seepages. Plants suffer foliar damage when they contact a fresh seepage, but they can tolerate weathered seepage without obvious injury. In fact, some species are relatively lush and vigorous when growing in tar, apparently because of the warm microclimate that is associated with the dark surface. This is especially true of the sedge *Carex aquatilis*, the cottongrass *Eriophorum scheuchzeri*, and the grass *Arctagrostis latifolia* (McCown *et al.*, 1973b; Deneke *et al.*, 1975).

Experimental Oil Spills on Land

Since the early 1970s, several research projects have examined the environmental impacts of oil and

gas development in the Arctic of North America. Of course, this research paralleled the vigorous hydrocarbon exploration and resource development activities in the north.

Most of the research on the potential effects of oil spills on terrestrial vegetation involved the perturbation of permanent plots, which were then monitored over time (e.g., McCown *et al.*, 1973a; Freedman and Hutchinson, 1976; Hutchinson and Freedman, 1978). As in experimental studies of the effects of oil on vegetation elsewhere, there was an initial contact damage that killed foliage and some exposed woody tissue. However, in many species not all of the perennating tissue was killed. Vegetative growth originating from unkilled meristematic tissue was the most important mechanism of the progressive, postoiling regeneration. This general pattern of effect can be illustrated by the results of field experiments done in the western Canadian Arctic (Table 6.7). In two study sites in black spruce (*Picea mariana*) boreal forest, the treatment of vegetation with crude oil caused a rapid defoliation, to 21–37% of the prespray plant cover after 1 month. Subsequently, most of the plants that survived the initial oiling succumbed to normally tolerated winter stress. As a result, plant cover 1 year after the spill was only about 5% of the prespray value. A protracted period of postspray mortality was especially marked in black spruce, which dominates that boreal ecosystem. Individuals of this tree species took as long as 4 years to die after being oiled at ground level. After the initial period of mortality, a slow recovery of many species began. An exception was black spruce, for which no seedlings were observed in the 5 years of the study.

There was a similar pattern of vegetation damage in the tundra communities that were experimentally treated with crude oil. Both the cottongrass- (*Eriophorum vaginatum*-) dominated wet meadow and the mesic, dwarf shrub (*Salix glauca* and *Betula glandulosa*) community showed a large initial defoliation. There was a further winter-kill in the wet meadow, and then both vegetation types began to recover from the perturbation. In both of these tundra communities, the dominant plant species

were able to recover from the oiling treatment. The regeneration emerged from unkilled basal meristematic tissue of the cottongrass, and from previously dormant lateral stem buds of the dwarf willow and birch. As a result, the initial community dominants were also prominent in the postdisturbance secondary succession.

Lichens and bryophytes were notably susceptible to the experimental oiling of tundra and boreal forest vegetation. These lower plants suffered almost complete mortality, and had little ability to recolonize the oiled plots in the first few postoiling years.

Several studies in Alaska examined the ability of the natural, terrestrial microflora to oxidize petroleum residues. The most frequent genera of oil-degrading bacteria were *Arthrobacter*, *Brevibacterium*, *Pseudomonas*, *Spirillum*, and *Xanthomonas*, while the most frequent microfungi were *Beauveria bassiana*, *Mortierella*, *Penicillium*, *Phoma*, and *Verticillium*. These and many other taxa are more-or-less ubiquitous in northern soils. After oiling they respond variously, depending on their relative competitive ability in the presence of a large quantity of hydrocarbon substrate (Scarborough and Flanagan, 1973; Linkins *et al.*, 1984). Other studies on the North Slope of Alaska showed

Table 6.7   Effect of experimental crude oil spills (9 l/m²) on live vegetation cover of four arctic plant communities: (1) mature *Picea mariana* boreal forest, (2) 40-year-old *P. mariana* boreal forest, (3) cottongrass wet meadow tundra, and (4) dwarf shrub tundra[a]

| | | | Total Foliage Cover (%) of Green Plants | | | | |
| | | | Postspill | | | | |
	Treatment	Prespill	1 Month	1 Year	2 Years	3 Years	5 Years
Mature forest	Reference	195	198	215	255	210	240
	Oil spill	350	130	18	10	35	20
40-Year-old forest	Reference	355	350	360	260	210	235
	Oil spill	420	90	20	20	23	95
Cottongrass tundra	Reference	339	342	284	268	—	—
	Oil spill	358	56	26	34	—	—
Dwarf shrub tundra	Reference	339	342	338	292	—	—
	Oil spill	322	57	55	82	—	—

[a]Modified from Freedman and Hutchinson (1976) and Hutchinson and Freedman (1978).

that an experimental oiling elicits a vigorous but short-term numerical response of heterotrophic microbes. For example, Scarborough and Flanagan (1973) found that microfungal propagules were 17 times more abundant in an oiled soil than in a reference soil at Prudhoe Bay, while yeasts were 20 times as abundant. Similarly, Campbell *et al.* (1973) found that the respiratory activity (i.e., efflux of CO_2 from soil) of experimentally oiled soil near Barrow was about two times larger than that of reference soil, while bacteria were about five times as abundant.

Overview of Arctic Oil Spills

Overall, the potential effects of oil spills from pipelines onto northern terrain appear to be relatively moderate. Although there is vegetation damage, it is relatively restricted spatially. (That is, except for a very large spill, the area of land that is impacted is fairly small because of the great absorptive capacity of the terrain, and because much of the spilled oil would accumulate in low spots and therefore not spread widely. This is a notable difference between aquatic and terrestrial oil spills.) Furthermore, many arctic plants have the ability to survive an exposure to oil and to invade habitat damaged by an oil spill. In addition, even at a low ambient temperature the microflora is capable of oxidizing spilled hydrocarbons. Along with the evaporation that takes place just after a spill, this microbial activity would progressively remove most of the spilled oil from the environment. However, because of the relatively short growing season and low ambient temperature, this process could take several decades (Atlas, 1985). The postspill recovery of terrestrial ecosystems can be sped up by fertilization and by artificial revegetation with arctic grasses (Linkins *et al.*, 1984). Of course, oil that reaches a watercourse could cause damage there, especially in relatively stagnant water. (Depending on the volume of oil that is spilled, rivers and streams might be less affected because of a rapid dilution and dissipation of the oil.) Overall, the ecological damage caused by an isolated, accidental terrestrial spill of oil in the Arctic might be considered to be environmentally acceptable by resource managers, in view of the economic and strategic importance of hydrocarbon resource development in the north (Hanley *et al.*, 1981; Alexander and Van Cleve, 1983).

Of course, the environmental consequences of northern oil development are much broader than oil spills *per se*. They include a socioeconomic impact (Freeman, 1985), the various environmental consequences of building roads and pipelines in remote and inhospitable terrain, the effects on wildlife, and many other problems that are not considered here. However, some ecologists have concluded that hydrocarbon resource development in the North Slope of Alaska has been an ecological "success story," even after considering the environmental impacts of exploration, oil extraction, the building and operating of an oil pipeline across the state, and the operation of a large tanker terminal at Valdez (Alexander and Van Cleve, 1983). In 1988, the U.S. Republican presidential candidate George Bush (subsequently elected) was quoted as saying, "the caribou love the pipeline, they rub up against it and have babies, so now there are more caribou in Alaska than you can shake a stick at" (Anonymous, 1988).

In contrast to the terrestrial situation, the potential environmental consequences of oil spills resulting from offshore exploration and production activity in the Arctic are much more problematic, and potentially catastrophic. Some of the more prominent reasons for this conclusion are as follows (after Pimlott *et al.*, 1976; Milne and Herlinveaux, 1977; Ross *et al.*, 1977; Percy and Wells, 1984; Engelhardt, 1985):

1. Because of the severe physical and climatic operating conditions in the Arctic, there is a relatively large risk of an oil spill caused by equipment failure or by human error.
2. A blowout from a drilling or production platform in the Arctic sea could potentially remain uncontained for as long as several years, because of the difficulties in quickly drilling a relief well.
3. The cleanup of spilled oil could be extraordinarily difficult in the rigorous Arctic marine environment.

4. Oil trapped under sea ice would not weather appreciably by evaporation and dissolution, and biodegradation would be very slow. As a result, the quantity of spilled petroleum would not decrease very much, and the oil would retain most of its initial toxicity for a long time.

5. When they return to their northern breeding habitat in the spring, arctic marine seabirds and mammals often congregate in large numbers in ice-free water such as leads and polynya. Spilled oil would accumulate in these open-water sites and cause a widespread mortality of these animals.

6. Arctic marine foodwebs are relatively simple, and therefore the elimination or great reduction in abundance of a few species as a result of an oil spill could cause relatively great ecological damage.

7. The successional recovery of ecosystems devastated by an offshore oil spill in the Arctic would be very slow.

To date, there have been no large oil spills caused by offshore drilling activity in the North American Arctic. However, it appears likely that when a severe spill incident does occur, it could cause great ecological damage, from which recovery would be slow.

7

EUTROPHICATION
OF FRESH WATER

7.1
INTRODUCTION

A eutrophic water body is characterized by a high rate of productivity, as a result of a good nutrient supply. This contrasts with oligotrophic water, which is unproductive because of a restricted availability of nutrients, and mesotrophic water, which is intermediate between these two states (Wetzel, 1975).

The most conspicuous symptom of eutrophication is a large increase in the standing crop of phytoplankton, known as an algal "bloom." Usually, there is also a change in algal species composition. In a shallow water body, there may also be a vigorous growth of vascular plants. These primary responses are generally accompanied by secondary changes at higher trophic levels, in response to greater food availability and other habitat changes such as a poor oxygen status of deep water. In the extreme case of very productive hypertrophic water, there can be noxious algal blooms, an off-flavor of drinking water, the production of toxic substances by algae and other microorganisms, periods of hypolimnetic oxygen depletion causing kills of fish and other biota, and the evolution of noxious gases such as hydrogen sulfide (Vallentyne, 1974;

Wetzel, 1975; Barica, 1980; Vollenweider and Kerekes, 1982). These characteristics adversely affect the multiple use of a hypertrophic water body for such purposes as drinking water, a fishery, recreation, and esthetics.

In this chapter, the causes, ecological consequences, and control of the eutrophication of fresh water are examined.

7.2
CAUSES OF
EUTROPHICATION

Background

Cultural eutrophication is most frequently caused by the fertilization of water with nutrients in (1) sewage that contains detergents, human wastes, and animal wastes; and (2) agricultural runoff contaminated by fertilizers. This anthropogenic influence has affected lakes and other surface waters wherever the human population density is large.

The term eutrophication has also been used to describe the slow, natural process by which a geologically young and unproductive water body (such as a lake newly exposed after deglaciation) gradually increases in productivity as nutrients are accumulated over time, and as the lake basin becomes shallow due to sedimentation (Hutchinson, 1969; Vallentyne, 1974; Wetzel, 1975). In contrast, a condition of decreased productivity or oligotrophication often accompanies the natural process of paludification of the landscape in a cool, wet climate. The decreased productivity is caused by the formation of ombrotrophic, raised bogs on an initially minerotrophic, relatively productive landscape (Hutchinson, 1969).

The most widely accepted cause of cultural eutrophication is excessive nutrient loading, particularly with phosphorus. This process of fertilization stimulates primary productivity, which causes secondary effects at higher trophic levels. Another factor that can affect primary production and algal standing crop in some situations is a change in trophic structure. For example, an increase in the population of planktivorous fish can cause a decreased abundance of herbivorous zooplankton. Decreased grazing can then result in an increased standing crop of phytoplankton, assuming that nutrient supply is sufficient to sustain an increased abundance of autotrophs. Environmental factors that influence the primary productivity of fresh water are described in more detail below.

Nutrient Loading

According to the Principle of Limiting Factors, the rate of an ecological process is controlled by the metabolically essential environmental factor that is present in least supply relative to demand. If we exclude environmental factors other than nutrients from consideration (e.g., temperature, oxygen supply), then biological productivity is limited by whichever nutrient is present in least supply relative to demand. Note that, in general, the continuous or steady-state rate of supply of the limiting nutrient is of particular importance; a pulse of availability tends only to cause a short-term change in productivity (Wetzel, 1975; Odum, 1983).

Of the various nutrients that can potentially affect the rate of primary productivity, phosphorus is the one that is most frequently limiting in fresh water, especially in the form of ionic orthophosphate (PO_4^{3-}). The typical concentration of phosphorus in fresh water (an index of "supply") is small compared with the concentration of P in plants (an index of "demand"). Moreover, for P the ratio of supply to demand is much smaller than is observed for other important inorganic nutrients (Table 7.1). Following from Table 7.1, the next most frequent chemical limiting factor for primary production in fresh water would be inorganic nitrogen (i.e., in the form of ammonium or nitrate).

A large amount of these plant nutrients can be delivered to a water body by natural processes and by the activities of people. The average rate of anthropogenic supply of P is about 2.1 kg/person-year in the United States, or about one-sixth the rate of supply of N (Table 7.2). The relatively smaller rate of anthropogenic supply of P compared with N is

Table 7.1 Demand and supply of selected essential elements in fresh water[a]

Element	Concentration in Plants (%)	Concentration in Water (%)	Ratio of Plants to Water (Approx.)
C	6.5	0.0012	5,000
Si	1.3	0.00065	2,000
N	0.7	0.000023	30,000
K	0.3	0.00023	1,300
P	0.08	0.000001	80,000

[a]"Demand" is indexed as the typical concentration in aquatic plants, while "supply" is average river-water concentration. Modified from Vallentyne (1974).

Table 7.2 Per capita output of N and P in the United States, 1965–1970[a]

	Output (kg/person-year)	
Source	**N**	**P**
Sewage		
Physiological waste	4.5	0.6
Detergents	0.0	1.1
Industry	0.5	0.1
Total	5.0	1.8
Delivered to water	4.5	1.6
Agriculture		
Animal wastes	45	6
Fertilizers	20	8
Total	65	14
Delivered to water	8	0.3

[a]From Vallentyne (1974).

another reason why P is the primary limiting factor in the cultural eutrophication of fresh water.

Studies that have analyzed data for surface water in many parts of the world have provided empirical evidence that demonstrates the importance of phosphorus as the most typically limiting nutrient in eutrophication. The intent of these studies was to identify the nutrient that most strongly correlated with primary productivity and algal standing crop, thereby indicating a possible controlling mechanism for the rates or amounts of these biological variables (e.g., Sakamoto, 1966; Dillon and Rigler, 1974; Schindler, 1978; Smith, 1980; Canfield and Bachman, 1981; Vollenweider and Kerekes, 1981, 1982). In all cases, the correlation with total phosphorus concentration was strongest, indicating that it is the most likely limiting nutrient. For example, Vollenweider and Kerekes (1982) examined a data matrix of the aqueous concentrations of chlorophyll (an index of the standing crop of phytoplankton) and nutrients among a large number of lakes in the northern hemisphere. A log–log plot of the mean annual total P concentration (this includes phosphate and organically bound P) and the mean annual concentration of chlorophyll yielded a highly significant correlation coefficient of .88 ($n = 77$ lakes), while the correlation of total P with peak chlorophyll concentration (which occurs during the summer "bloom" of phytoplankton) was $r = .90$ ($n = 50$ lakes). The log–log correlations with total N were also statistically significant, but much weaker (for mean annual chlorophyll, $r = .61$ among $n = 41$ lakes; for peak chlorophyll, $r = .66$ among 40 lakes). Largely because of these statistical relationships, Vollenweider and Kerekes concluded that P was the most generally limiting nutrient for primary productivity in fresh water, and that N did not play an important role except in situations of exceptional supply of P. They were so confident in the predictive capability of phosphorus that they used its concentration in a scheme of boundary values for the trophic status of lakes (Table 7.3). They also proposed a probability distribution for these trophic categories around mean annual concentrations of P

Table 7.3 Proposed boundary values for trophic categories of inland lakes and reservoirs[a]

Trophic category	Mean Annual Total P (mg/m³)	Mean Annual Chlorophyll (mg/m³)	Maximum Chlorophyll (mg/m³)	Mean Annual Secchi Transparency (m)	Minimum Annual Transparency (m)
Ultraoligotrophic	<4.0	<1.0	<2.5	>12	>6
Oligotrophic	<10	<2.5	<8	>6	>3
Mesotrophic	10–35	2.5–8	8–25	3–6	1.5–3
Eutrophic	35–100	8–25	25–75	1.5–3	0.7–1.5
Hypertrophic	>100	>25	>75	<1.5	<0.7

[a]Note that the concentration of chlorophyll and secchi transparency are indices of the standing crop of phytoplankton, and that maximum values correspond to the summer "bloom" condition. Modified from Vollenweider and Kerekes (1982).

(Fig. 7.1). Trophic status can, of course, be measured directly via the biological characteristics of the water body, for example, algal standing crop or chlorophyll in Table 7.3 and Figure 7.1. The importance of the strong statistical relationship between phosphorus and trophic status is that it suggests a mechanism for eutrophication via P fertilization. A corollary of this relationship is that cultural eutrophication should be controllable or reversible, by a reduction of the anthropogenic inputs of phosphorus.

A few studies of fresh water have demonstrated a

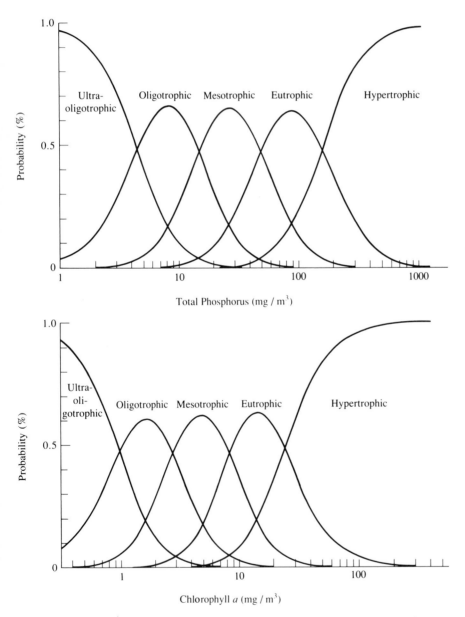

Fig. 7.1. The probability that a lake will fall within a given trophic category at a particular average annual concentration of total P or chlorophyll. Modified from Vollenweider and Kerekes (1982).

limitation to primary productivity by a nutrient other than phosphorus. These cases most frequently involve a limitation by inorganic nitrogen. For example, a bioassay study of 238 Scandinavian lakes indicated that 22 of them were primarily limited by nitrogen, and 216 by phosphorus (Vollenweider and Kerekes, 1982).

Very rarely, a micronutrient deficiency can limit primary productivity. Goldman (1961, 1965, 1972) demonstrated a stimulation of phytoplankton productivity in Castle Lake, California, after fertilization with molybdenum. The chrysophycean *Dinobryon sertularia* was especially limited by Mo. In a laboratory bioassay this alga responded to fertilization with 100 ppb Mo by a 2.6-fold increase in net productivity. In plants, Mo is necessary for the production of the enzyme nitrate reductase, which reduces absorbed nitrate to ammonium, and as a cofactor in the enzymatic fixation of atmospheric dinitrogen to ammonium. The Mo deficiency in Castle Lake was apparently caused by the development of a stand of *Alnus tenuifolia* around the fringe of the water body. This N_2-fixing alder tied up so much of the watershed's molybdenum that it became limiting to the phytoplankton.

Other studies have shown that silica can limit phytoplankton productivity in fresh water. This is an especially frequent observation with diatoms, which require this element in a relatively large quantity for the construction of their siliceous cell wall or frustule (Lund, 1950; Lund *et al.*, 1963; Heron, 1961; Schelske and Stoermer, 1972).

Because algal species are variously affected by particular micronutrient limitations, this environmental feature can influence the species composition and seasonal dynamics of the phytoplankton community. However, the influence of micronutrients on the productivity of the phytoplankton community considered as a whole is much smaller than is observed with a macronutrient such as phosphorus.

In very productive water bodies, it can be difficult to demonstrate a relationship between primary productivity and the concentrations of major nutrients such as P and N. Such a situation has been shown to exist in certain shallow, hypertrophic prairie lakes of central North America (Hammer, 1969;

Haertl, 1976; Bierhuizen and Prepas, 1985). These lakes are generally shallow and saucer-shaped, and located in a watershed with fertilized or naturally fertile soil. Because the shallow water column is mixed by summer winds it does not stratify and the productive surface water is not cut off from nutrient-rich deep water, as it is in lakes that do stratify during the growing season. Nutrient supply in this sort of hypertrophic system is sustained at a high rate throughout the year. This allows the development of a protracted bloom of blue-green algae, which persists from about mid-June to autumn. Because this situation is so fertile, the eutrophic condition does not usually diminish quickly when anthropogenic nutrient loading is decreased. This is largely because internal phosphorus loading takes place at such a high rate in these lakes (i.e., P release from anaerobic sediment; Jacoby *et al.*, 1982; Ryding, 1985). Light, temperature, and wind that causes mixing of the water column appear to be important physical factors in the initiation of conditions appropriate for an algal bloom (Haertl, 1976).

Sources and Control of Phosphorus Loading

There are many natural sources of phosphorus input to fresh water. These include wet and dry deposition from the atmosphere, leaching from soil of the watershed, and even biological transport (e.g., an input of marine phosphorus to the upper reaches of a watershed in the bodies of anadromous fish such as salmon, which may die after spawning and release nutrients via decomposition). Typically, the natural inputs of P total considerably less than 1 kg/ha-year, averaged over the entire watershed (Freedman, 1981). The anthropogenic input of phosphorus can be much larger than this in a watershed with a large population of humans or domestic animals. In the United States, the average per capita discharge to water was about 1.9 kg P/person-year between 1965 and 1970 (Table 2). About 84% of this total came from municipal sources and the rest from agriculture (the total nitrogen discharge was 12.5 kg/person-year: 36% from municipal sources and 64% from agriculture). For Lake Erie, the most eutrophic of the Great Lakes, the anthropogenic input of phosphorus comprised at least 87% of the total P input in 1976 (about 2 kg/ha-year), compared with only 11% for oligotrophic Lake Superior (0.2 kg/ha-year) (Gregor and Johnson, 1980).

Because phosphorus is so clearly the limiting nutrient in most fresh-water systems, there has been a great effort to reduce its anthropogenic input so as to alleviate eutrophication in waterbodies that receive wastewater.

Detergents were a very large contributor to municipal phosphorus loading in the 1960s and early 1970s. A typical chemical formulation of domestic detergents used at that time included about 65% by weight (16% as P) of complex phosphates, especially sodium tripolyphosphate. These so-called "builders" were used to chelate ionic Ca, Mg, Fe, and Mn in the wash water, so that the chemical surfactants, the actual cleaning agents in the detergent, were not immobilized by complexation (Duthie, 1972a,b). At about that time, the total production and use of detergents in North America was about 3 million kg/year, and much of this quantity was eventually flushed into fresh water with municipal sewage effluent (most of the rest was discharged into the ocean by coastal cities). Detergent P comprised about one-third to one-half of the total P in waste water discharge at that time (Duthie, 1972a; Rohlich and Uttormark, 1972; Vallentyne, 1974). Because detergent use is a discrete activity sector, and there were substitutes available for phosphorus in the builder function, detergents were a relatively easy and inexpensive target for the rapid reduction of phosphorus loading to fresh water. As a result, between 1970 and 1973 the maximum permissible concentration of P in detergents was reduced from 16% to 2.2% in Canada. Larger reductions were legislated in other places, for example, to less than 0.5% in Dade County, Florida (Duthie, 1972b; Prakash, 1976).

Another strategy for reducing the rate of phosphorus loading to a particular water body is the diversion of sewage effluent to some other sink (see page 179). If diversion is not possible, then sewage can be treated to reduce the concentration of phosphorus in the effluent. Unfortunately, this practice has not been actively pursued in many places, because it can involve an expensive investment in facility, technology, and operating costs. The chemical characteristics of treated sewage effluent vary tremendously, depending on the nature of the domestic and industrial inputs, and on the technology used in the treatment facility. Typical data for U.S. municipal water that receives no more than primary or secondary treatment are 5–15 mg/l of total P, 25–35 mg/l total N, 150–250 mg/l of biochemical oxygen demand (BOD), and 200–300 mg/l of suspended solids (Rohlich and Uttormark, 1972). An effluent with this sort of chemical quality is a rich growth medium for algae and microorganisms, but it is somewhat unbalanced nutritionally because the P : N ratio (about 1 : 2) is large relative to the optimal condition for the growth of freshwater plants (about 1 : 9; Vallentyne, 1974).

The concentrations of nutrients and other material in sewage can be greatly reduced by treatment of the waste water (Rohlich and Uttormark, 1972):

1. Primary sewage treatment involves the screening or settling of the raw sewage to remove large materials. The effluent may also be disinfected to kill pathogenic microorganisms. This process has a nutrient removal efficiency of 5–15%.

2. Primary treatment may be followed by a secondary treatment process that is intended to reduce BOD and form soluble nutrients and colloidal organic material. This is done by the use of an activated sludge or a trickling filter, in either of which microorganisms are used to aerobically degrade the organic waste. A waste product of this process is a humus-like sewage sludge that is often disposed of by application to agricultural land (Chapter 3), by solid waste disposal, or by incineration. Secondary treatment can have a nutrient removal efficiency as large as 30–50%.

3. Tertiary treatment usually involves the use of a suite of processes that is designed to remove nutrients from the sewage effluent. Phosphorus removal can be achieved by the flocculation or precipitation of phosphate

compounds by aluminum, iron, calcium, or other agents, or by the use of algae or macrophytes to incorporate phosphate into aggrading biomass that is later harvested. The removal efficiency of these processes is typically 90% or larger. Tertiary treatment can also include nitrogen removal, using biological uptake and harvesting, denitrification, ammonia stripping, ion exchange, or some other process.

Effects of Changes in Trophic Structure on Eutrophication

Several studies have shown that in some circumstances, a change in trophic structure can affect the standing crop of phytoplankton. This effect can have a strong influence on the symptoms of eutrophication, but it only occurs if the nutrient supply is sufficient to support an enhanced productivity of algae.

In a microcosm experiment involving 2-m^3 plastic enclosures in a pond, Shapiro (1980) found that the addition of planktivorous bluegill sunfish (*Lepomis macrochirus*) caused a large increase in the standing crop of algae, by reducing the abundance of grazing cladoceran zooplankton. In two enclosures without fish the concentration of chlorophyll was <5 μg/l and secchi depth extended 2.3 m to the bottom of the pond, while in enclosures with four or five fish chlorophyll was 42–50 μg/l and secchi depth only 0.9–1.0 m. The concentration of total P in the water was unaffected by the experimental treatment. In another microcosm experiment, five enclosures with planktivorous fathead minnows (*Pimephales promelas*) had a small abundance of cladoceran zooplankton and an average late-summer phytoplankton biomass of 60 × 10^5 μm^3/ml, while five enclosures without fish averaged 24 × 10^5 μm^3/ml (Lynch and Shapiro, 1981).

In a larger-scale experiment, Spencer and King (1984) examined four eutrophic ponds that varied in trophic structure (Table 7.4). One of the ponds was fishless and had a relatively large population of zooplankton, a small standing crop of phytoplankton because of intense grazing, and, since the water was very transparent, a great abundance of macrophytes. Another pond had piscivorous largemouth bass (*Micropterus salmoides*), which caused a small density of planktivorous minnows, and a consequent large density of zooplankton, a small standing crop of phytoplankton, and a large standing crop of macrophytes. Two other experimental ponds had a large population of planktivorous fathead minnows and brook stickleback (*Culaea inconstans*), a small density of zooplankton, a large standing crop of phytoplankton, and a small standing crop of macrophytes.

Predation may also play a role in relieving the symptoms of eutrophication in much larger water-

Table 7.4 Role of predation and grazing in some small (3.3–5.0 ha), experimental, eutrophic ponds in Michigan: Pond 1 had no fish, pond 2 only had piscivorous largemouth bass, ponds 3 and 4 had planktivorous minnows but no bass[a]

	Pond 1	Pond 2	Pond 3	Pond 4
Fish community	No fish	Bass, 3000/ha	Minnows, 25,000/ha Sticklebacks, 1,000/ha	Minnows, 13,000/ha Sticklebacks, 58,000/ha
Zooplankton (number/l)	24.7	35.0	0.06	0.03
Phytoplankton (mm^3/l)	2.6	7.4	20.1	36.2
Light penetration (% of surface at bottom)	17.4	10.9	0.45	0.25
Macrophytes (g d.w./m^2)	196	264	3	75

[a]See text for discussion. Modified from Spencer and King (1984).

bodies. It has been speculated that predation by the burgeoning population of introduced salmon species in Lake Michigan has caused a large decrease in abundance of the planktivorous alewife (*Alosa pseudoharengus*), resulting in an increased abundance of herbivorous zooplankters such as *Daphnia* spp., which then caused a decreased standing crop of phytoplankton and an increased clarity of the water column (Scavia *et al.*, 1986). These trophic effects may have helped to relieve the symptoms of eutrophication in this very large lake, in conjunction with a decrease in nutrient loading since about the mid-1970s.

Clearly, trophic structure can influence the characteristics of a eutrophic water body. This effect can potentially be manipulated to ameliorate some of the symptoms of eutrophication, particularly the standing crop of phytoplankton (Uhlmann, 1980; Shapiro and Wright, 1984). It is likely that this technique would be especially useful in relatively small eutrophic lakes or ponds. Because these small water bodies have a large internal rate of P loading from their sediment, they do not respond quickly to a decrease in the external supply of nutrients.

7.3
CASE STUDIES OF EUTROPHICATION

The first case that will be considered involves several whole-lake experiments that investigated the role of nutrients in limiting the primary productivity of remote oligotrophic lakes. The next case describes the response of an ecologically simple, High Arctic lake to nutrient loading via sewage. We will then consider the relatively complex case of Lake Erie, the most productive of the Great Lakes of North America. This lake has been affected by nutrient loading, extensive modification of its watershed that caused severe siltation, a vigorous commercial fishery, toxic chemicals, and other anthropogenic stresses. Finally, we will discuss the recovery of eutrophied water, by reference to the case of Lake Washington near Seattle.

Whole-Lake Fertilization Experiments

Much has been learned about the mechanisms and dynamics of the eutrophication of oligotrophic lakes from a series of whole-lake experiments done in the Experimental Lakes Area (ELA) of northwestern Ontario by D. Schindler and his coworkers.

In a relatively long-term experiment, Lake 227 was fertilized from 1969 to 1983 at an average rate of 5.5 kg PO_4-P/ha-year [from 1969 to 1974 it also received 6.3 kg NO_3-N/ha-year, and then 2.2 kg NO_3-N/ha-year from 1975 to 1983 (Schindler, 1985)]. Observations were made of phosphorus dynamics in Lake 227. Typically, 1–5% of the mass of phosphorus in the euphotic zone sedimented to deeper water each day. If the lake was thermally stratified, the sedimented phosphorus became unavailable to sustain primary productivity in surface waters. The residence times of phosphorus (0.6 years) and nitrogen (2.5 years) within the water column of Lake 227 were considerably shorter than that of water (6.1 years), because of the rapid rate of sedimentation of these nutrients.

Studies using radioactive ^{32}P as a tracer in Lake 227 showed that within minutes of its addition to the surface water, more than 90% of the phosphorus was incorporated into the smallest plankton fraction (comprised of bacteria and microphytoplankton <10 μm in diameter), and the ^{32}P tended to remain in that fraction (Levine *et al.*, 1986). Less than 5% of the added ^{32}P was present in larger phytoplankton and zooplankton.

The ELA observations on the dynamics and size fractionation of phosphorus have been generally confirmed in research on P-limited lakes elsewhere (Currie and Kalff, 1984; Currie *et al.*, 1986; Heath, 1986). Collectively, these studies indicate that relatively large phytoplankton are inefficient competitors for phosphate, and they raise important questions about the mechanisms by which the larger species of phytoplankton acquire this critical nutrient.

Because phosphate is so rapidly immobilized by

bacteria and microphytoplankton, its concentration in surface water is very small during the growing season. However, it is important to note that the tiny pool of dissolved phosphorus is quite labile. For example, the turnover time of P in the epilimnetic water of several ELA lakes is only 8–14 min, and 15–77 min in the hypolimnion (Planas and Heckey, 1984). Because of the dynamic character of the pool of dissolved phosphorus, a relatively small concentration can drive a large rate of planktonic primary productivity.

As noted earlier, Lake 227 was fertilized with both phosphate and nitrate. It subsequently responded with a large increase in primary productivity, but because of the experimental design it was not possible to determine which of these two key nutrients had acted as the primary limiting factor. However, observations from other experimental treatments in ELA lakes clearly indicate that phosphorus is the primary limiting nutrient in these oligotrophic water bodies (Schindler and Fee, 1974), as follows.

1. Lake 304 was fertilized for 2 years with phosphorus, nitrogen, and carbon, and it became eutrophic. It recovered rapidly and became oligotrophic again when the P fertilization was stopped, even though N and C fertilization was continued.
2. Lake 226 is an hourglass-shaped lake that was partitioned with a heavy vinyl curtain into two experimental basins. One basin was fertilized with C + N, and the other with P + C + N. Only the latter treatment caused an algal bloom.
3. Lake 302 received an injection of P + C + N directly into its hypolimnion during the summer. Because the lake was thermally stratified at that time, the hypolimnetic nutrients were not available to fertilize phytoplankton in the epilimnetic euphotic zone, and no algal bloom resulted from the fertilization.

It is notable that even though N and C were not directly added to the lakes that were only fertilized

with P, their rate of supply nevertheless increased markedly. In the case of N this took place by an increase in the rate of fixation of atmospheric dinitrogen by bluegreen algae, while for C it occurred via an enhanced rate of diffusion of atmospheric CO_2 into the water column (Schindler *et al.*, 1972; Schindler, 1977; Smith, 1983). Therefore, the secondary emergence of N and C limitation after fertilization with phosphorus was compensated by natural mechanisms that increased the rate of supply of these nutrients.

Detailed observations were made of the response of the phytoplankton community in the Lake 227 eutrophication experiment, and these are briefly summarized below (Schindler *et al.*, 1973).

1. Prior to fertilization, the phytoplankton biomass averaged about 1 g/m^3 in winter and 2 g/m^3 during the ice-free season, it had a peak summer value of 4–5 g/m^3, and the chlorophyll concentration was 1–5 mg/m^3. The dominant taxa during the spring and summer were *Dinobryon* spp., *Chrysochromulina parva*, *Mallomonas pumilo*, *Chrysoikos skujai*, *Pseudokepherion* spp., and *Kepherion* spp., and the overall dominant group within the phytoplankton was the Chrysophyta (golden-brown algae).
2. In the first year of fertilization (1969), the peak summer biomass increased more than threefold to 16 g/m^3, while chlorophyll increased to 50 mg/m^3. The dominant group of phytoplankton in summer changed to the Chlorophyta (green algae).
3. In the second year of fertilization, the peak summer biomass increased further to 35 g/m^3, and chlorophyll increased to 90 mg/m^3. The peak summer phytoplankton in this and subsequent years of the experiment was dominated by bluegreen algae (Cyanobacteria), especially *Oscillatoria geminata*, *O. amphigranulata*, *Pseudoanabaena articulata*, and *Lyngbya lauterbornii*.
4. In the third year of fertilization, the peak biomass was only 12.0 g/m^3 and chloro-

phyll 30 mg/m³. The decreased standing crop of phytoplankton compared with the previous year was attributed to grazing pressure from an exceptionally great abundance of rotifer zooplankton.

5. In the fourth year of fertilization, the peak biomass increased again to 30.0 g/m³, and chlorophyll to 180 mg/m³.

Observations were also made of the responses of other trophic levels to some of the experimental fertilizations in the ELA. In the hourglass-shaped Lake 226, the abundance and growth rate of whitefish (*Coregonus clupeaformis*) were measured in each of the oligotrophic and eutrophic basins (Mills, 1985; Mills and Chalanchuk, 1987). In the first year of fertilization, the growth rate of fish in the basin receiving P averaged 21% larger than in the oligotrophic basin. In the second year of fertilization this increased to 106% larger, then to 141% in year three, and 171% in year four. Therefore, the increased primary production that was caused by experimental fertilization with phosphorus resulted in increased productivity and biomass at a higher trophic level.

Eutrophication of a High Arctic Lake

Meretta and Char Lakes are small water bodies located in the Canadian High Arctic at about 75°N latitude (Kalff and Welch, 1974; Kalff *et al.*, 1975; Schindler *et al.*, 1974). They are relatively simple ecosystems because of severe climatic limitations on ecological processes. Char Lake is an ultra-oligotrophic polar lake, while Meretta Lake receives sewage from a small community and is moderately eutrophic. The loading rate of P was about 13 times larger in Meretta Lake than in Char Lake, and that of N was 19 times larger (Table 7.5). The fertilized lake had a summer chlorophyll concentration that averaged about 12 times larger than the reference lake, while the maximum summer biomass was 40 times greater, and annual primary production 2.8 times greater. Both lakes had a similar richness of phytoplankton species, and the phytoplankton biomass of each was dominated by only a few species, but these were generally different in the two water bodies (Table 7.5). Because of the large demand for oxygen to sustain the decomposition of organic materials from sewage and the pri-

Table 7.5 A comparison of two High Arctic lakes at 75°N on Cornwallis Island, Canada: Char Lake is a typical, ultraoligotrophic polar lake, while Meretta Lake receives sewage and is eutrophic[a]

Characteristics	Char Lake	Meretta Lake
Nutrient inputs (kg ha⁻¹ year⁻¹)		
-P	0.16	2.0
-N	0.32	6.0
Phytoplankton		
Summer chlorophyll (mg/m³)	0.13–0.69	3–5
Maximum summer biomass (g/m³)	0.20–0.55	14–16
Primary production (gC.m⁻².yr⁻¹)	4.1	11.3
Dominant phytoplankton species		
	Gymnodinium helveticum	*Gymnodinium mirable*
	Rhodomonas minuta	*G. veris*
	Chromulina sp.	*G. lacustre*
	Two Chrysophycean spp.	Two Chrysophycean spp.
	Chrysococcus sp. (episodic)	*Cyclotella glomerata* (episodic)
	Cyclotella comensis (episodic)	

[a]Data are from Kalff and Welch (1974), Kalff *et al.* (1975), and Schindler *et al.* (1974).

Aerial view of Lake 226 in the Experimental Lakes Area of northwestern Ontario. This remote lake was partitioned into two experimental basins using a heavy vinyl curtain. The upper basin in the photograph received fertilizer containing phosphorus, nitrogen, and carbon, while the lower basin received nitrogen and carbon. The role of phosphorus as the limiting nutrient for primary productivity was clearly demonstrated, since only the basin that was fertilized with phosphorus became eutrophic and developed a large standing crop of phytoplankton. Because of the bloom of phytoplankton, the water of the eutrophic basin developed a greenish color, represented by a whitish hue in this photograph. (Photo courtesy of D. Schindler.)

Gull Lake, and part of its watershed, in southwestern Michigan. The lake area is 827 ha and, as can be clearly seen in the airphoto, the predominant land-use in its 6960-ha terrestrial watershed is agriculture [mainly cultivation of corn (*Zea mays*)], which results in large input of fertilizer nutrients to the lake. The lake itself is used for recreation (the many white spots on the water are sailboats and motorboats), and there are more than 500 cottages and other recreational facilities along the shore, which also contribute nutrients via sewage, lawn fertilizers, and other sources. As a result of the large input of nutrients, Gull Lake has changed from an initially oligotrophic condition to a meso-trophic/eutrophic state. The consequent biological changes have included (after P. Lane, unpublished observations): (1) a change in phytoplankton from an assemblage dominated by diatoms of the genera *Asterionella*, *Fragilaria*, and *Melosira*, to a community dominated by blue-green algal genera such as *Aphanizomenon*, *Anabaena*, and *Oscillatoria*; (2) the development of nuisance growths in the littoral zone of the macrophytic alga *Cladophora* and various aquatic angiosperms; (3) a change in zoo-plankton from an assemblage dominated by *Daphnia galeata*, *D. retrocurva*, and *Diaptomus ore-gonensis* to one dominated by relatively small species such as *Bosmina longirostris, Cyclops bicuspidatus*, and *Tropocyclops parsinus*; and (4) a change from a relatively species-rich fish commu-nity dominated by piscivorous lake trout (*Salvelinus namaycush*) and planktivorous lake herring (*Coregonus artedii*) to one dominated by an introduced planktivore, the rainbow smelt (*Osmerus mordax*). (Photo courtesy of W. K. Kellogg Biological Station, Michigan State University.)

mary production in Meretta Lake, there was a hypolimnetic oxygen deficit. The anoxia may have been responsible for the disappearance of the zooplankter *Limnocalanus macrurus*, and it caused recruitment problems in arctic char (*Salvelinus alpinus*), a salmonid fish.

Consideration of the case of these polar lakes shows that even simple ecosystems that are under severe climatic stress can exhibit a profound eutrophication response to fertilization.

Lake Erie—Effects of Eutrophication and Other Stresses

Lake Erie is the most productive of the Great Lakes of North America. Lake Erie has been impacted by several anthropogenic stresses in addition to nutrient loading, especially disturbance of its watershed, an intensive fishery, and pollution by toxic chemicals. Nevertheless, Lake Erie is an important case study of eutrophication because of the large scale of the changes that have been caused by the fertilization of this lake.

The watershed of Lake Erie is much more agricultural and urban in character than are those of the other Great Lakes (Table 7.6). Consequently, the dominant sources of phosphorus to Lake Erie are agricultural runoff and municipal point sources (Table 7.7). The total input of P to Lake Erie (standardized to watershed area) is about 1.3 times larger

than to Lake Ontario, and more than five times larger than to the other Great Lakes (Tables 7.6 and 7.7).

During the late 1960s and early 1970s, when the eutrophication of Lake Erie was at a maximum, the concentrations of total P and inorganic N (i.e., nitrate + ammonium) were larger than in any of the other Great Lakes (Table 7.8). The supply of both of these nutrients, but primarily P, is critical in the limitation of primary productivity in the Great Lakes, along with silica for diatoms under a condition of P fertilization (Thomas *et al.*, 1980; Schelske *et al.*, 1986). During this period, the eutrophic western basin of Lake Erie (which is relatively shallow and warm, and which directly receives large inputs of sewage and agricultural runoff) had a spring P concentration that averaged about eight times larger than that of oligotrophic Lake Superior, and a concentration of inorganic N that was almost three times as large (Table 7.8).

As a consequence of the large loading rate and concentration of nutrients in Lake Erie, it is more productive and has a larger standing crop of phytoplankton than do the other Great Lakes (Table 7.8). During the late 1960s and early 1970s, the eutrophic western basin of Lake Erie had a summer chlorophyll concentration that averaged about twice as large as in Lake Ontario, and 11 times larger than in oligotrophic Lake Superior. Transparency showed a similar pattern. However, since the late 1960s and early 1970s the eutrophication of Lake

Table 7.6 Size and watershed characteristics of the Great Lakes[a]

	Superior	Michigan	Huron	Erie	Ontario
Lake surface area (km^2)	82,414	58,068	59,596	25,719	19,477
Watershed area (km^2)	138,586	117,408	128,863	78,769	75,272
Land use in watershed (%)					
Forest	95	50	66	17	56
Agriculture	1	23	22	59	32
Urban	<1	4	2	9	4
Brush, wetland, other	4	23	10	15	8
Average P export (kg P/km^2-year)	19	40	24	198	86

[a]Modified from Gregor and Johnson (1980).

Table 7.7 Importance of various source categories to the total 1978 phosphorus load to the Great Lakes[a]

| Lake | Inputs of P (MT/year) | | | Relative Contribution of Source Categories (% of Total) | | | Nonpoint Inputs via Tributaries | | |
	Total	Point Sources	Nonpoint Sources	Industrial Point	Municipal Point	Atmospheric Inputs	Agriculture	Urban	Forest
Superior	4200	161	4039	2	1	40	4	4	49
Huron	6350	155	6195	1	2	64	23	4	6
Michigan	4850	1072	3778	1	21	28	34	6	10
Erie	17,450	6010	11,440	2	33	5	40	12	8
Ontario	11,750	2204	9546	1	25	41	23	6	5

[a]Data modified from Zar (1980) and Gregor and Johnson (1980).

Table 7.8 Average values for nutrients and other water quality variables in the Great Lakes during the late 1960s and early 1970s[a]

| Lake | Spring Values | | | Summer Values | | Trophic Status[b] |
	Total P (μg P/l)	Inorganic N (μg N/l)	Reactive Si (mg SiO$_2$/l)	Chlorophyll a (μg/l)	Secchi Depth (m)	
Superior	4.6	275	2.25	1.0	8.8	0
Huron	5.2	259	1.36	1.2	8.3	0
Michigan	9.0	200	1.50	2.0	6.0	0–M
Erie						
Western	39.5	631	1.32	11.1	1.5	E
Central	21.2	133	0.33	3.9	4.4	E–M
Eastern	23.8	180	0.30	4.3	4.5	E–M
Ontario	24.0	279	0.42	5.3	2.5	E–M

[a] Modified from Thomas *et al.* (1980).
[b] O = oligotrophic, M = mesotrophic, E = eutrophic.

Erie has been alleviated somewhat, in direct response to a decrease in phosphorus loading (Table 7.9).

A consequence of the eutrophic state of Lake Erie was the development of anoxia in its hypolimnion during the summer stratification. In the summer of 1953 there was a period of 10 days of hot calm weather, which caused a very stable thermal stratification of the water column. Because of a large demand for oxygen for the decomposition of organic material in the hypolimnion, a widespread anoxia developed in deep water, especially in the western end of the lake (Hartman, 1972). This deoxygenation had a great effect on benthic animals.

Prior to this time, the benthos was dominated by mayfly larvae, especially *Hexagenia rigida* and *H. limbata*. In 1929 their density was about 397/m², and in 1942–1943 they averaged 422/m². Just prior to the severe stratification in 1953, the density of *Hexagenia* was 300/m², but this collapsed to only 44/m² in September. Density remained small until 1957 when it averaged 37/m², but by 1961 these insects had almost disappeared, as density was then less than 1/m². The collapse of the population of benthic mayflies was widely reported in the popular press, which interpreted the phenomenon to indicate that Lake Erie was "dead." After this effect on the initially mayfly-dominated benthos, a low-

Table 7.9 Recent reductions in algal standing crop and phosphorus concentration in Lake Erie[a]

Time Period	Western Basin	Central Basin	Eastern Basin
Total phosphorus in water (μg/l; spring maximum)			
1970–1975	39 ± 5	19 ± 2	24 ± 6
1978–1980	31 ± 4	14 ± 1	12 ± 2
Chlorophyll a (μg/l; summer values)			
1970–1975	11.6 ± 2.4	4.8 ± 0.8	4.3 ± 1.0
1979–1980	10.0 ± 2.2	4.1 ± 1.4	2.7 ± 0.9

[a] Modified from Rapport (1983).

oxygen-tolerant benthic community was established. This was dominated by the tubificid worms *Limnodrilus hoffmeisteri* and *L. cervix*, by chironomid midges, and by gastropod and sphaeriid molluscs (Beeton and Edmondson, 1972; Hartman, 1972; Wetzel, 1975).

It has been speculated that the western and central basins of Lake Erie have long been subject to periodic incidents of oxygen depletion, even prior to their cultural eutrophication (Charlton, 1979, 1980; Delorme, 1982). This effect apparently resulted from the morphometry of the basins, which causes them to be relatively susceptible to the formation of a stable thermal stratification. In addition, the deep water of western Lake Erie has always been subject to a large input of oxygen-demanding organic matter from biological productivity. Nevertheless, it is likely that the frequency and intensity of the events of deoxygenation in Lake Erie have been exacerbated by the modern increase in biological productivity, and by the input of organic matter from sewage outfalls (Charlton, 1979, 1980).

The most conspicous change caused by the eutrophication of Lake Erie has been an increased productivity and standing crop of phytoplankton and macrophytes. The standing crop of phytoplankton (indexed by chlorophyll concentration and secchi depth) was previously shown to be larger than in the other Great Lakes (Table 7.8). However, studies of phytoplankton in Lake Erie have suggested that there may not have been a large change in species composition in recent years compared with the previous century, in spite of the apparently more severe eutrophication of the lake. Harris and Vollenweider (1982) reviewed historical data on algal species composition, using the records of a water treatment facility at Buffalo, New York. They also examined siliceous diatom frustules in a sediment core taken from the central basin of Lake Erie. These authors concluded that the lake had been characterized by a meso- to eutrophic condition since at least 1850: "the phytoplankton assemblages present in the last century (from historical data) resemble those at the present time." In the earliest years of their sediment core the most frequent diatoms were *Melosira dis-*

tans and *Stephanodiscus niagarae*, while more recently *Fragilaria capucina* was most frequent. However, large fluctuations in the abundance of these dominants made it impossible to generalize a clear trend in long-term dominance. In the data from the Buffalo water works, the earliest appearance of a phytoplankton species that is considered to be a reliable indicator of a eutrophic condition was in the late 1880s, when *Cyclotella bodanica* was present. By 1900 there were other indicators of eutrophication, including the blue-green algae *Aphanizomenon*, *Anabaena*, and *Oscillatoria*.

The historical mesotrophic to eutrophic condition of Lake Erie was likely partly natural, especially in the western basin, which was well known for its lush growth of macrophytes and its large fish production. In addition, events in the mid-1800s such as the draining at the southwestern end of the lake of a million-hectare wetland known as the Great Black Swamp, the general clearing of land in the Lake Erie watershed for agriculture, and sewage discharges from developing cities must have caused a degree of cultural eutrophication (Regier and Hartman, 1973). Nevertheless, the rate and intensity of eutrophication have undoubtedly increased in modern times, in response to an accelerated nutrient loading caused by the rapidly increasing human population in the Lake Erie basin, and the introduction of phosphorus-containing detergents in the late 1940s.

The phytoplankton of Lake Erie differ markedly among its three basins, and between its nearshore and offshore water. The eastern and central basins are considered to be mesotrophic to eutrophic, while the western basin is eutrophic (Table 7.8), and in all basins the relatively shallow nearshore water is more productive than the offshore. In general, the spring algal bloom in relatively eutrophic situations is dominated by the diatom *Melosira*, while the late summer–autumn bloom is dominated by the blue-greens *Anabaena*, *Microcycstis*, and *Aphanizomenon*, along with the diatom *Fragilaria* and the green alga *Pediastrum* (Hartman, 1972, 1973).

A notable development that has taken place since about 1940 in rocky, nearshore habitat has been a profuse growth of the filamentous green alga *Cladophora glomerata*. Mats of this plant frequently separate from their rocky substrate and drift on the surface of the lake, eventually washing ashore as a rotting, smelly nuisance, or sinking to the bottom where they create a large oxygen demand and contribute to the anaerobic condition during thermal stratification (Hartman, 1972, 1973). In Lake Erie and in localized eutrophic situations in some of the other Great Lakes, *Cladophora* appears to become established when the total P concentration in the spring exceeds 15 μg/l (Thomas *et al.*, 1980).

There have also been changes in the standing crop, species composition, and size spectrum of the zooplankton of Lake Erie. In 1939–1940 the zooplankton density of the western basin never exeeded about 7000/m^3 during July and August, compared with 10,000–22,000/m^3 in 1949, and 26,000–110,000/m^3 in 1959 (Brooks, 1969; Hartman, 1972, 1973). Historical collections of zooplankton were dominated by relatively large species, such as *Limnocalanus macrurus* and *Daphnia* spp. More recently, these taxa have declined or disappeared and they have been replaced by previously rare species, especially the eutrophic indicator *Diaptomus siciloides*. This change in the zooplankton was probably caused in part by the increasing productivity of Lake Erie. However, at about the same time that Lake Erie was undergoing cultural eutrophication, a vigorous commercial fishery caused an increasing dominance of the fish community by relatively small planktivorous species, which replaced the largely piscivorous taxa that were dominant earlier (described below). This change in predation must have influenced the size spectrum and species composition of the zooplankton of Lake Erie (Brooks, 1969; Hartman, 1972, 1973).

The changes in zooplankton were most marked in the shallow, western basin of Lake Erie. In the deeper eastern basin, some taxa characteristic of an oligotrophic condition have managed to survive even to the present. The most notable examples of these oligotrophic taxa are the opossum shrimp *Mysis relicta* and the amphipod *Pontoporeia affinis* (Regier and Hartman, 1973).

Tremendous changes took place in the fish community of Lake Erie during the time that it was becoming more eutrophic. It is possible that these effects were partly due to habitat changes caused by eutrophication. However, it is more likely that the changes in the fish community were caused by its intensive exploitation by a vigorous fishery, and by other habitat effects such as the damming of streams required for spawning by anadromous fish, and the sedimentation of shallow water habitat by silt eroded from deforested parts of the watershed (Hartman, 1972, 1973; Regier and Hartman, 1973).

In terms of fishery landings, Lake Erie has always been the most productive of the Great Lakes. Over the past 150 years, the average yield of Lake Erie's commercial fishery has exceeded the combined landings of all the other Great Lakes (Regier and Hartman, 1973). The peak years of the commercial fishery in Lake Erie were in 1935 (28.5 million kg) and 1956 (28.3 million kg), while the minima were in 1929 (11.2 million kg) and 1941 (11.6 million kg). Overall, the catch by the gross commercial fishery has been remarkably stable (Figure 7.2f), in spite of large changes in fish species, in fishery effort, in the degree of eutrophication and pollution by toxic chemicals, and other habitat effects.

There has been a dramatic change in the species composition of the fish community of Lake Erie during the past 150 or so years, and this reflects a deterioration of the quality of that natural resource. The historical pattern of development of the Lake Erie fishery is similar to that of most previously unexploited natural resources. The most desirable and valuable species were exploited first. As these species declined in abundance because of an unsustainable fishing pressure, coupled with deterioration of the habitat, the industry diverted to a progression of less desirable species of fish. This pattern of fishery resource development has been labeled "fishing up" by Regier and Loftus (1972).

The history of the Lake Erie fishery has been described by various researchers (Hartman, 1972, 1973; Regier and Hartman, 1973; Smith, 1972a,b; Sonzogni *et al.*, 1983). The initial fishery exploited nearshore, and then offshore, populations of lake

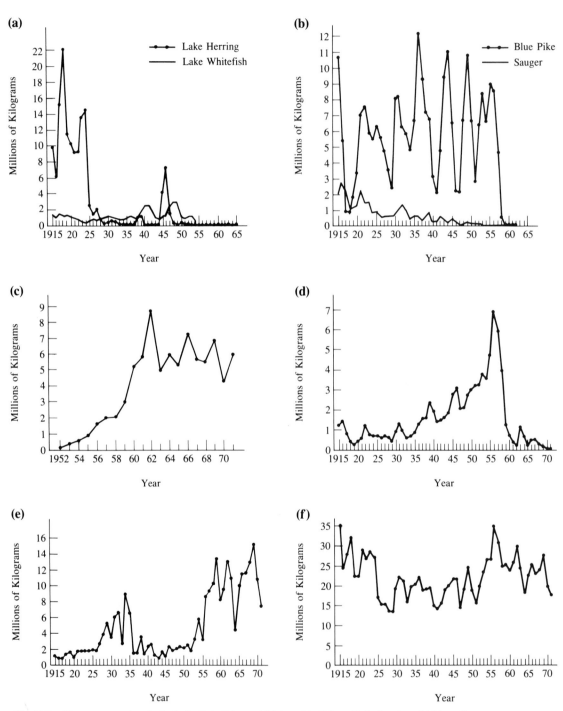

Fig. 7.2. Changes over time in the Lake Erie fishery: (a) lake whitefish and lake herring; (b) blue pike and sauger; (c) rainbow smelt; (d) walleye; (e) yellow perch; and (f) total commercial catch of all species. Modified from Hartman (1973).

Even very acidic lakes can exhibit a strong trophic response to nutrient input. Drain Lake in Nova Scotia became very acidic (pH 4.0) when pyritic minerals in its watershed were oxidized after their exposure to the atmosphere by construction activity. However, the lake also receives input of sewage, and the nutrients in this waste caused Drain Lake to become eutrophic. For such an acidic body of water, there are remarkably large standing crops and productivity of phytoplankton and macrophytes, a great abundance of crustacean and insect invertebrates, and a large population of breeding waterfowl and muskrats. Obvious in the photo are a productive fringing community dominated by the grass *Calamagrostis canadensis* and dense beds of the floating-leaved macrophyte *Nuphar variegatum*, which are visible as a stippling of the water surface (see also Chapter 4). (Photo courtesy of B. Freedman.)

whitefish (*Coregonus clupeaformis*) and lake trout (*Salvelinus namaycush*; this species was almost exclusively in the relatively deep eastern basin), along with lake herring (*Coregonus artedi*). Lake sturgeon (*Acipenser fulvescens*) were also fished at this time, but not for food; they were killed because they damaged nets. The species of the initial resource were rapidly overfished, and they declined to a small abundance or extinction (Figure 7.2a). The initially most-desired species were then replaced as the target of the fishery by "second choice" percid species, such as blue pike (*Stizostedion vitreum glaucum*), walleye (*S. v. vitreum*), sauger (*S. canadense*), and yellow perch (*Perca flavescens*). Today, the *Stizostedion* species are ex-

tinct or rare (Figures 7.2b and d), and the fishery is dominated by species that were initially of lowest value, such as yellow perch, rainbow smelt (*Osmerus mordax*; Figure 7.2c), freshwater drum (*Aplodinotus grunniens*), and carp (*Cyprinus carpio*).

In summary, Lake Erie is an important example of the effects of cultural eutrophication on the ecological structure and function of a very large lake. Its case is also representative of the detrimental effects of other anthropogenic stresses, especially overexploitation of a potentially renewable natural resource (the fishery), deterioration of aquatic habitat caused by the clearing of forest for agricultural purposes, and pollution by oxygen-demanding sewage and toxic chemicals.

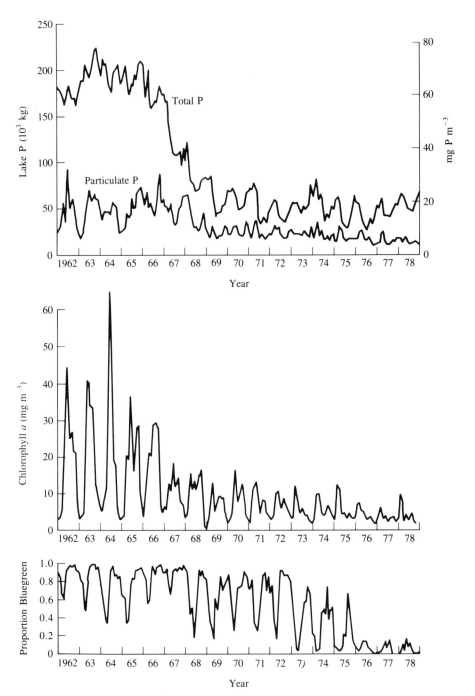

Fig. 7.3. The oligotrophication of Lake Washington as the result of a diversion of sewage effluents that was completed in 1968. From Edmondson and Lehman (1981).

Recovery of Culturally Eutrophied Water Bodies

In almost all situations where fresh water has become culturally eutrophic, the primary cause has been fertilization with phosphorus. It is therefore reasonable to expect that a reduction of the cultural input of P and other nutrients would result in a reversal of eutrophication.

A well-known case of the reversal of cultural eutrophication is the recovery of Lake Washington (Edmondson, 1972, 1977; Edmondson and Lehman, 1981; Edmondson and Litt, 1982). In the early 1900s, Lake Washington received raw sewage from Seattle, then a town of 50,000 people. Because of the ensuing eutrophication and public health problems caused by the sewage dumping, the city diverted its effluents to Puget Sound, and Lake Washington quickly recovered its oligotrophic character.

A second and more intense episode of eutrophication of Lake Washington took place between 1941 and 1963, when there were as many as 10 secondary sewage treatment plants with outfalls into the lake, serving a population of more than 64,300. In addition, private septic-field systems were used by about 12,000 people, and there was a storm-flow input that was equivalent to the annual sewage output of another 4,500 people. In 1964, phosphorus loading to Lake Washington was at its maximum, and sewage accounted for 72% of the total P input of about 204 MT P/year (the input of N was 1,400 MT/year).

This large nutrient loading caused summer blooms of phytoplankton. When the nutrient loading was at its maximum in the mid-1960s, about 98% of the phytoplankton volume during the summer bloom was comprised of colonial bluegreen algae, especially *Oscillatoria rubescens* and *O. agardhii*.

In 1968, a system was completed to divert the sewage from Lake Washington, and the total input of P was decreased to 21% of its peak loading in 1964 (the loading of N was decreased to 52% of the 1964 rate). This large decrease in loading led to (1) a smaller nutrient concentration in the euphotic zone, (2) a decrease in the summer standing crop of algae, and (3) a diminished dominance of *Oscillatoria* in the summer bloom (Figure 7.3). In addition, the average secchi transparency in summer increased from only 1.0–1.1 m in the mid-1960s, to 6.4–7.8 m between 1977 and 1979.

It is believed that the oligotrophication of Lake Washington was primarily achieved by the reduction of the loading of phosphorus, and that the reduction of nitrogen input had a smaller influence. This conclusion was based in part on the observation that the summer biomass of phytoplankton was most strongly correlated with the maximum concentration of phosphorus in water (which occurs in winter). Over the 12-year period of 1950 to 1972, (1) the correlation of summer chlorophyll concentration × maximum (winter) phosphate concentration yielded a correlation coefficient (r) of 0.91, while the correlation with total P was 0.93, and (2) maximum phytoplankton biomass × maximum PO_4 gave $r = 0.93$, while (3) summer chlorophyll × maximum nitrate concentration only gave $r = 0.60$.

In summary, the cultural eutrophication of Lake Washington was rapidly alleviated by a reduction of the loading of phosphorus. From this and other observations of oligotrophication following a reduction of nutrient loading, it is clear that the process of cultural eutrophication can be reliably alleviated by actions that reduce the supply of phosphorus (Bjork, 1977; Ahlgren, 1978; Loehr et al., 1980; Vollenweider and Kerekes, 1982). This mitigation can be accomplished by the diversion of sewage effluents (as in Lake Washington), by the replacement of phosphorus-containing detergents with nonfertilizing alternatives, and by the installation of tertiary sewage treatment processes to remove phosphorus from sewage effluents.

8

_____ PESTICIDES

8.1 _____

INTRODUCTION

Pesticides are substances that are used to protect humans against the insect vectors of disease-causing pathogens, to protect crop plants from competition from weeds, and to protect crops, plants, and livestock from disease and depredation by fungi, insects, mites, and rodents. To these various pesticidal ends, a diverse array of natural biochemicals extracted from plants and of inorganic and synthetic organic chemicals have been used. Our reliance on pesticides has increased greatly in recent decades, and the practice of chemical pest control is now a firmly entrenched component of the technological culture of modern humanity.

There have been many tangible benefits of the use of pesticides. The most important of these have been (1) an increased production of food and fiber because of the protection of crop plants from pathogens, competition from weeds, defoliation by insects, and parasitism by nematodes; (2) the prevention of spoilage of harvested, stored foods; and (3) the saving of many millions of human lives by the prevention of certain diseases.

Unfortunately, the considerable benefits of the use of pesticides are partly offset by some important environmental detriments. There have been rare but spectacular incidents of toxicity to humans, as occurred in 1984 at Bhopal, India, where more than 2800 people were killed and more than 20,000 seriously injured by a large emission (about 40 tons) of poisonous methyl isocyanate vapor, a chemical intermediate in the production of an agricultural pesticide (Rozencranz, 1988). An ecologically more pervasive problem is a widespread environmental contamination by persistent pesticides, including the presence of chemical residues in wildlife, in well waters, and even in humans. Ecological damages have included the poisoning of wildlife by some pesticides, and the disruption of such ecological processes as productivity and nutrient cycling. Many of the worst cases of environmental damage caused by pesticides have been caused by the use of relatively persistent chemicals such as DDT. Most modern pesticide use involves less persistent chemicals. However, the global use of pesticides is expanding in scale and intensity, and although we know a great deal about certain of the environmental consequences of this technological practice, not all of the potential impacts are understood.

Because of the wide dichotomy between the benefits and risks of pesticide use, a polarization of society has been caused by differing perceptions of the effects of the use of these chemicals. This state of affairs can be illustrated by quoting two highly influential persons (from McEwen and Stephenson, 1979). First, here is a phrase from a speech by Winston Churchill after the successful use of DDT to control a potentially deadly plague of typhus among Allied troops in Naples during the Second World War: "that miraculous DDT powder." In contrast, in her seminal book "Silent Spring," which was the first high-profile chronicle of the environmental consequences of the widespread use of persistent pesticides, especially DDT, Rachael Carson (1962) referred to that same chemical as the "elixir of death" because of her concern for the ecological damages that were caused by its use.

This chapter describes some of the ecological impacts of the use of pesticides. The most prominent uses of pesticides are considered first, and then the characteristics of selected groups of these chemicals are described. This is followed by a consideration of some notable examples of the detrimental effects of pesticides on nontarget, offsite wildlife, especially birds. Then, to give an appreciation of the ecological impacts of large-scale pesticide spraying in a complex natural ecosystem, several case studies are examined. The first case deals with the extensive spraying of conifer forest infested with spruce budworm. The second case examines the use of herbicides to manipulate plant regeneration on forestry clear-cuts.

8.2
CLASSIFICATION OF PESTICIDES BY THEIR USE AND CHEMICAL CHARACTERISTICS

Pesticides comprise a diverse group of chemicals, which can be classified according to (1) the pest organisms against which they are targetted, (2) the use sector, such as in agriculture, around the home, or in forestry, and (3) their similarity of chemical structure.

Classification of Pesticides by Their Biological Target

First, a classification of pesticides based on their intended pest target will be considered, with some prominent examples of each group [after Metcalf, 1971; Eto, 1974; McEwen and Stephenson, 1979; Entomological Society of America (ESA), 1980].

1. *Fungicides* are used to protect crop plants and animals from fungal pathogens. Fungicides include (a) inorganic chemicals such as elemental sulfur, and copper compounds such as Bordeaux mixture (pp. 186–189 give a more detailed description of these and other pesticides used as examples in this section); (b) organometallic compounds of mercury and tin; (c) chlorophenolics such as tri-, tetra-, and pentachlorophenol; (d) antibiotics such as penicillin and streptomycin; and (e) synthetic organics such as the dithiocarbamates and captans.

2. *Herbicides* are used to kill weedy plants, so as to release desired crop plants from competition. Herbicides include (a) inorganics such as various arsenicals, cyanates, and chlorates; (b) chlorophenoxy acids such as 2,4-D and 2,4,5-T; (c) chloroaliphatics such as dalapon and trichloroacetate; (d) triazines such as atrazine and simazine; and (e) organophosphates such as glyphosate.

3. *Insecticides* are used to kill insect pests and vectors of deadly human diseases such as malaria, yellow fever, trypanosomiasis, plague, and typhus. Some prominent insecticides are (a) inorganic arsenicals and fluorides; (b) "natural" plant-derived chemicals and their synthetic analogs, such as nicotine, pyrethroids, and rotenoids; (c) the DDT group of chlorinated hydrocarbons, including DDT, DDD, and methoxychlor; (d) lindane, an insecticidal isomer of benzene hexachloride; (e) highly chlorinated cyclodienes such as chlordane, heptachlor, mirex, aldrin, and dieldrin; (f) chlorinated terpenes such as toxaphene; (g) organophosphorus esters such as parathion, di-

azinon, fenitrothion, malathion, and phosphamidon; (h) carbamates such as carbaryl and aminocarb; and (i) microbial agents such as *Bacillus thuringiensis* or B.t.

4. *Acaricides* are used to kill mites, which are pests in agriculture, and ticks, which can carry encephalitis of humans and domestic animals. Most insecticides are effective against these arachnid arthropods, but there are also some specific acaricides with a relatively specialized use.

5. *Molluscicides* are used against snails and slugs, which can be important pests of citrus groves and vegetable and flower gardens. In addition, aquatic snails are the vector of certain human diseases, most prominently schistosomiasis. Important molluscicides include copper sulfate, the carbamates isolan and zectran, the organophosphate guthion, and others.

6. *Nematicides* are used to kill nematodes, which can be important parasites of the roots of crop plants. Chemical control involves fumigation of the soil with halogenated organics such as ethylene dibromide, dichloropropane, dichloropropene, dibromochloropropane, or with certain carbamates and organophosphorus chemicals.

7. *Rodenticides* are used to control rats, mice, gophers, and other rodent pests of human habitation and agriculture. Important rodenticides include plant-derived chemicals such as the alkaloid strychnine and the cardiac glycoside red squill, the hydroxycoumarin compound warfarin, and others.

Production and Use of Pesticides by Their Target

The total production of pesticides in the United States in 1977 was 630.9 million kg, while total imports were 23.5 million kg, and total exports were 273.6 million kg, leaving 380.8 million kg that was used domestically (ESA, 1980). About 10% of the land area of the continental United States is treated annually with pesticides (ESA, 1980).

The largest quantities of pesticides are used against weeds, insects, and fungi. Of the 380.8×10^6 kg of synthetic pesticides that was used in the United States in 1977, herbicides accounted for 46% of the total, insecticides 43%, and fungicides 11%. These target categories accounted for 58%, 35%, and 7%, respectively, of the $2.8 billion value of pesticides used (in 1977 dollars; ESA, 1980). Note that these totals do not include the use of creosote to protect railroad ties from fungal-caused rot, a use which in itself totals $> 500 \times 10^6$ kg/year (McEwen and Stephenson, 1979). The global proportions of 1980 pesticide sales were 40% herbicides, 34% insecticides, 21% fungicides, and 5% other uses (OECD, 1985).

The use of pesticides has increased tremendously since prior to 1950, when their application was insignificant by today's scale. In 1970, the global sales of pesticides totalled some $3 billion (in constant 1980 U.S. dollars). This increased to $12 billion in 1980, and will increase to a projected $50 billion by 1990 (OECD, 1985).

Classification of Pesticides by Their Use Category

Pesticides can also be considered from the perspective of use category. The most important of these are human health, agriculture, and forestry, as described below (after McEwen and Stephenson, 1979).

Human Health

In various parts of the world, species of insects and ticks play a critical role as vectors in the transmission of certain disease-causing pathogens of humans, livestock, and wild animals. Worldwide, the most important of these human diseases and their vectors are (1) malaria, caused by the protozoan *Plasmodium* and spread to humans by a mosquito vector, especially *Anopheles* spp.; (2) yellow fever and related viral diseases such as encephalitis, which are spread by mosquito vectors, especially *Aedes aegypti* and *Culex* spp. in the case of yellow fever; (3) trypanosomiasis or sleeping sickness, caused by the flagellated protozoans *Trypanosoma gambiense* and *T. rhodesiense* and spread by the tsetse fly *Glossina* spp.; (4) plague or black death, caused by the bacterium *Pasteurella pestis* and transmitted to people by the oriental rat flea *Xenopsylla cheops*, a parasite of various species of rat that live in association with humans; and (5) typhoid fever, caused by the bacterium *Rickettsia prowazeki* and transmitted to humans by the body louse *Pediculus humanus*.

The incidence of each of these diseases can be greatly reduced by the judicious use of pesticides to control the abundance of their arthropod vectors (McEwen and Stephenson, 1979). For example, (1) there are many cases where the local abundance of mosquito vectors has been reduced by the application of insecticide to their aquatic breeding habitat, or by the application of a persistent insecticide to walls and ceilings, which serve as a resting place for these insects; (2) infestations of the human body louse have been treated by dusting people with insecticide; and (3) plague has been controlled by the reduction of rat populations, using rodenticides in conjunction with manipulation of their habitat via sanitation programs.

The use of insecticides to reduce the abundance of the mosquito vectors of malaria has been especially successful, although in many areas this disease is now reemerging because of the evolution of tolerance by mosquitoes to various insecticides. The use of DDT to control mosquitoes during and just after the Second World War was so successful that predictions were made of the eradication of this debilitating and often fatal disease. Malaria has always been an important disease in the tropics and subtropics. During the 1950s, it was estimated that each year more than 5% of the world's population was infected with malaria. A large reduction in the incidence of malaria was rapidly achieved through the use of insecticides. For example, between 1933 and 1935 India annually recorded about 100 million cases of malaria, and 0.75 million deaths. However, there were only 0.15 million cases and 1500 deaths in 1966, largely because of the vigorous use of DDT. [In 1985, India utilized about 40–50% of its public health budget for anti-malarial insecticide

spraying, and about 15×10^6 kg of insecticide were used for that purpose (Hubendick, 1987).] In 1962, during a vigorous campaign to control malaria in the tropics, about 59.1×10^6 kg of DDT was used, as were 3.6×10^6 kg of dieldrin and 0.45×10^6 kg of lindane. Most of the pesticide was sprayed inside of homes and on other likely resting sites for mosquitoes, rather than in their breeding habitat (McEwen and Stephenson, 1979).

Agriculture

Modern, technological agriculture uses pesticides for the control of weeds, arthropods, and plant diseases. Worldwide, pests and diseases cause a loss that is equivalent to about 24% of the potential crop of wheat (*Triticum aestivum*), 46% of rice (*Oryza sativa*), 35% of corn or maize (*Zea mays*), 55% of sugar cane (*Saccharum officinale*), 37% of grapes (*Vitis vinifera*), and 28% of vegetables. Of the 230×10^6 kg of pesticides that was used in agriculture in the United States in 1971, 104×10^6 kg was herbicides, 77×10^6 kg was insecticides, and 19×10^6 kg was fungicides (McEwen and Stephenson, 1979).

In agriculture, arthropod pests can be regarded as animals that compete with humans for a common food resource. From the human perspective, that competition is direct when insects cause a large reduction in agricultural yield by the defoliation of crops in the field, or when stored foods are attacked. In a few exceptional cases, defoliation can cause a total loss of the economically harvestable agricultural yield, as in the case of acute infestations by "locusts" (in North America, epidemic locusts are comprised of four species of spur-throated grasshopper, *Melanoplus* spp.; in Eurasia various genera are important, including the desert locust, *Schistocerca gregaria*). More usually, insect defoliation causes a reduction in the yield of crops. For example, in the United States between 1963 and 1973, defoliation by the European corn borer (*Ostrinia nubilalis*) resulted in an average loss of maize yield of 9% (range of 3% to 17%; McEwen and Stephenson, 1979). In many cases insects may cause only trivial damage in terms of the quantity of biomass that they consume, but by causing cosmetic damage they can greatly reduce the economic value of the crop. For example, in an unsprayed orchard of apples (*Malus pumila*) the frequency of infestation of the fruit by codling moth (*Carpocapsa pomonella*) can be 20–90% (McEwen and Stephenson, 1979). This can render the harvest of the crop impractical in an economical sense. Although codling moth larvae do not consume much of the fruit that they infest, they cause great psychological damage to any consumer who finds a half worm in a just-bitten apple. Therefore, even seemingly "minor" insect (or fungal) damage to a fruit or vegetable crop can render it unsalable.

In agriculture, a weed can be considered to be any plant that interferes with the productivity of a crop plant (even though in other contexts weed species may have ecological and economic values). Weeds exert this effect by competing with the crop for light, water, and nutrients. Studies in Illinois demonstrated an average reduction of yield of corn of 81% in unweeded plots, while an average 51% reduction was reported in Minnesota. Competition from weeds can reduce the yield of small grains such as wheat and barley by 25–50% (McEwen and Stephenson, 1979). To reduce the influence of weeds on agricultural productivity, fields may be sprayed with a herbicide that is toxic to the weeds, but to which the crop plant is insensitive.

There are several herbicides that are toxic to dicotyledonous weeds, but not to grasses. As a result, herbicides are most intensively used in grain crops of the Gramineae. In North America, more than 83% of the acreage of maize is treated with herbicides (McEwen and Stephenson, 1979). This practice is especially important in maize agriculture because of the widespread use of no-tillage cultivation, a system that reduces erosion and saves fuel. Since an important purpose of plowing is to reduce the abundance of weeds, the no-tillage system would be impracticable if it were not accompanied by the use of herbicides. The most prominent herbicides used in corn cultivation in the United States are atrazine (accounting for about half the area treated), propachlor, alachlor, 2,4-D, and butylate. In

aggregate, these amount to about 39×10^6 kg/year (NAS, 1975c). Most of the area planted to other agricultural grass crops such as wheat, rice, and barley is also treated with herbicide. About 50–80% of the small-grain area in North America is treated with the phenoxy herbicides 2,4-D or MCPA. In 1975, about 28×10^6 kg of these herbicides was used for that purpose (McEwen and Stephenson, 1979).

There are many diseases of agricultural plants that can be controlled by the use of pesticides. In some cases insecticides can be used to control the insect vectors of viral, bacterial, and fungal diseases of plants (Borror *et al.*, 1976). More importantly, fungicides are used to control diseases caused by fungal pathogens that could otherwise reduce or totally destroy the production of crops. Examples of important fungal diseases of crop plants include (after Agrios, 1969) (1) late blight of potato (*Phytophthora infestans* on *Solanum tuberosum*), (2) apple scab (*Venturia inequalis* on *Malus pumila*), (3) powdery mildew of peach and other rosaceous crops (*Sphaerotheca pannosa* on *Prunus persica*), and (4) Pythium seedrot, damping-off, and root rot of many agricultural species (caused by *Pythium* spp.). These diseases can be controlled by the use of fungicides, usually in conjunction with the cultivation of resistant plant varieties and with particular cultural practices that help to reduce the incidence and severity of the disease.

Forestry

In forestry, the most important uses of pesticides are for the control of defoliation by epidemic insects, and the reduction of weeds in plantations. If left uncontrolled, these pest problems could result in a large decrease in the yield of merchantable timber. In the case of some insect infestations, particularly spruce budworm (*Choristoneura fumiferana*), repeated defoliation can cause the death of a large area of forest (Johnson and Lawrence, 1977; see pp. 198–213).

The quantity of pesticide that is used in forestry is much smaller than that used in agriculture. In spite of this fact, pesticide use in forestry has attracted a disproportionate amount of high-profile attention from environmental advocates and the media. The reasons for this phenomenon may be related to:

1. The fact that in forestry very large tracts of natural and seminatural ecosystems are sprayed. These are often perceived as pristine "wilderness," in contrast to the intensely and frequently disturbed technological agroecosystems that are treated in agriculture.
2. In intensive agriculture, wildlife populations are relatively sparse and tend not to be directly affected by pesticide spraying (although there can be important indirect effects, especially when persistent pesticides such as DDT are used; see pp. 191–198).
3. In forestry, spraying is aften aerial from a relatively high tree-top altitude. Such an application greatly increases the risk of drift of the spray, and therefore there is a much greater risk of direct off-target ecological impacts.
4. Some earlier forestry programs caused widely publicized toxicity to high-profile nontarget organisms, particularly birds and sportfish.
5. The government agencies and multinational forestry companies that are most involved in silvicultural pesticide use are large, immobile, and impersonal targets for environmental activists. In contrast, farmers are a pesticide-using group for which the public is much more empathetic and supportive.

In North American forestry, by far the largest insecticide spray program is carried out against several species of spruce budworm (*Choristoneura* spp.), which defoliate several conifer species (see pp. 198–213). In the United States between 1945 and 1974, about 6.4 million hectares of budworm-infested forest were sprayed with insecticide (including repeatedly sprayed areas), accounting for 52% of all forest insecticide spraying (NAS,

1975a). In Canada between 1952 and 1986, about 45.4 million hectares were sprayed (see Table 8.5), with the most prominent chemicals being DDT, fenitrothion, phosphamidon, and aminocarb.

DDT is one of the most important chemicals that have been used in the United States for forest insecticide spraying (NAS, 1975a). DDT was sprayed on 84.4% of the total area of forest treated between 1952 and 1974. The peak year of DDT use was 1957, when 2.3 million kg was used; however, its use for this purpose was discontinued in 1972. Also important in the United States have been Carbaryl (used over 7.5% of the sprayed area), Zectran (5.0%), and Malathion (1.2%) (NAS, 1975a).

The second most important forest insecticide program in the United States is targeted against the gypsy moth (*Lymantria dispar*), a defoliator of many tree species. Between 1945 and 1974 spraying against this pest totaled about 5.1 million hectares, and accounted for 41% of total insecticide spraying in forestry (NAS, 1975a). The gypsy moth was accidentally introduced to North America in 1869 at Medford, Massachusetts, by M. L. Trouvelet. He had hoped to develop a commercial silkworm industry by crossing the gypsy moth with the silkworm moth (*Bombyx mori*). Instead he released North America's second most important forest defoliator. The continental area of infestation is more than 1 million ha, almost entirely in the eastern United States and southeastern Canada (Gerardi and Grimm, 1979; Doane and McManus, 1981).

Other insect infestations that have been treated by the broadcast spraying of forest include those of Douglas-fir tussock moth (*Hemerocampa pseudotsugata*), hemlock looper (*Lambdina fiscellaria*), tent caterpillars (*Malacosoma disstria* and *M. americanum*), and bark beetles (especially *Ambrosia* spp.) (Johnson and Lawrence, 1977). In addition, spot spraying has been used to control infestations of urban elm trees (mainly *Ulmus americana*) by Dutch elm disease, which is caused by an introduced fungal pathogen (*Ceratocystis ulmi*) that is spread by two species of elm bark beetle (the introduced *Scolytus multistriatus*, and the native *Hylur-*

gopinus rufipes) (Agrios, 1969). The practice of spraying urban elm trees has largely been discontinued, mainly because of controversy caused by the spraying of insectides close to homes. Today, particularly valuable elm trees can be treated using injections of a systemic fungicide, a relatively expensive process.

A widespread use of herbicides in forestry began in the 1950s, and it is progressively becoming a routine silvicultural practice. Most herbicide use is for the release of desired conifer species from the effects of competition with angiosperm herbs and shrubs. The topic of silvicultural herbicide use is discussed in detail in pp. 213–223.

Chemical Classification of Pesticides

Pesticides can be classified according to their similarity of chemical structure, as is outlined in the following section (after Metcalf, 1971; Eto, 1974; McEwen and Stephenson, 1979; ESA, 1980; Newton and Knight, 1981) (see Table 8.1 for chemical names). Note that this brief treatment only gives a sampling of the great variety of pesticides that is currently available for use.

Inorganic Pesticides

This group includes compounds of various toxic elements, predominantly arsenic, copper, lead, and mercury. Compounds of these elements do not degrade in the conventional sense, and when used as a pesticide they have a long persistence as toxic substances. However, a degree of environmental detoxification may take place by changes in molecular structure caused by organic and inorganic chemical reactions. In addition, the persistence of inorganic chemicals in soil is affected by dissipative processes that physically remove residues, for example, leaching, and soil erosion by wind and water. Chapter 3 gave some examples in which the use of inorganic pesticides has caused environmental contamination. Some prominent inorganic pesticides include:

1. Bordeaux mixture, a complex pesticide with several copper-based active ingredients, including tetracupric sulfate and pentacupric sulfate. Bordeaux mixture is used as a foliar fungicide for fruit and vegetable crops. It acts by inhibiting a variety of fungal enzymes.

2. Various arsenicals, including arsenic trioxide, sodium arsenite, and calcium arsenate, which are nonselective herbicides and soil sterilants. Insecticides in this group include Paris green, lead arsenate, and calcium arsenate.

Organic Pesticides

Organic pesticides are a chemically diverse group of chemicals. Some are produced naturally by certain plants, but the great majority of organic pesticides have been synthesized by chemists. Some prominent organic pesticides are:

1. *Natural organic pesticides* that have been extracted from various species of plant. An important insecticide is the alkaloid nicotine and other nicotinoids, largely extracted from tobacco (*Nicotiana tabacum*), and often applied as the salt nicotine sulfate. Another insecticide is pyrethrum, a complex of five chemicals (pyrethrin I and II, cinerin I and II, and jasmolin II) extracted from the daisy-like *Chrysanthemum cinerariaefolium* and *C. coccinium*. A third chemical complex that can be used as an insecticide, piscicide, and rodenticide is the rotenoids, especially rotenone, which is extracted from the tropical plants *Derris elliptica*, *D. malaccensis*, *Lonchocarpus utilis*, and *L. urucu*. Another example is the rodenticide red squill, comprised of cardiac glycosides extracted from the lily *Scilla maritima*. A final example is the rodenticide strychnine, an alkaloid extracted from the tropical plant *Strychnos nux-vomica*.

2. *Synthetic organometallic pesticides* have

Table 8.1   Common and chemical names of selected pesticides mentioned in the text[a]

Class	Common Name	Chemical Name
1. Inorganic pesticides		
a. Bordeaux mixture	Tetracupric sulfate + pentacupric sulfate	$4CuO \cdot SO_3 + 5CuO \cdot SO_3$
b. arsenicals	Arsenic trioxide	As_2O_3
	Sodium arsenite	$NaAsO_2$ and Na_2HAsO_3
	Calcium arsenate	$Ca_3(AsO_4)_2$
	Paris green	$Cu(C_2H_3O_2)_2 \cdot 3Cu(AsO_2)_2$
	Lead arsenate	$PbHAsO_4$
2. Organic pesticides		
a. Natural organics	Nicotine	l-1-Methyl-2-(3′-pyridyl)-pyrrolidine
	Nicotine sulfate	$(C_{10}H_{14}N_2)_2 \cdot H_2SO_4$
	Pyrethroids	Pyrethrins I,II, cinerins I,II, jasmoline II
	Rotenone	1,2,12,12a-Tetrahydro-2-isopropenyl-8,9-dimethoxy-[1]benzo-pyrano[3,4-b]furo-[2,3-h][1]-benzopyran-6(6aH)-one
	Red squill	Various cardiac glycosides
	Strichnine	Complex alkaloid
b. Organomercurials	Phenyl Hg acetate	$C_6H_5HgOCOCH_3$
	Methylmercury	CH_3Hg
	Methoxyethyl Hg chloride	$CH_3OCH_2CH_2HgCl$
c. Phenols		2,4,5-Trichlorophenol
		Pentachlorophenol
d. Chlorinated hydrocarbons		
1. DDT and relatives	DDT	2,2-Bis-(p-chlorophenyl)-1,1,1-trichloroethane
	DDD or TDE	2,2-Bis-(p-chlorophenyl)-1,1-dichloroethane
	Methoxychlor	2,2-Bis-(p-methoxphenyl)-1,1,1-trichloroethane
	DDE	2,2-Bis-(p-chlorophenyl)-1,1-dichloroethylene
2. lindane	Lindane	1,2,3,4,5,6-Hexachlorocyclohexane
3. cyclodienes	Chlordane	2,3,4,5,6,7,8,8-octochloro-2,3,3a,4,7,7a-hexahydro-4,7-methanodiene
	Heptachlor	1,4,5,6,7,8,8-Heptachloro-3a,4,7,7a-tetrahydro-4,7-methanoindene
	Aldrin	1,2,3,4,10,10-Hexachloro-1,4,4a,5,8,8a-hexahydro-1,4-$endo,exo$-5,8-dimethanonaphthalene
	Dieldrin	1,2,3,4,10,10-Hexachloro-6,7-epoxy-1,4,4a,5,6,7,8,8a-octahydro-1,4-$endo,exo$-5,8-dimethanonaphthalene
4. Chlorophenoxy acids	2,4-D	2,4-Dichlorophenoxyacetic acid
	2,4,5-T	2,4,5-Trichlorophenoxyacetic acid
	MCPA	2-Methyl-4-chlorophenoxyacetic acid
	Silvex	2-(2,4,5-trichlorophenoxy)-propionic acid
e. Organophosphorus compounds	Parathion	O,O-Diethyl O-p-nitrophenyl phosphorothionate
	Methyl parathion	O,O-Dimethyl O-p-nitrophenylphosphorothionate
	Fenitrothion	O,O-Dimethyl O-3-methyl-4-nitrophenylphosphorothionate
	Malathion	O,O-Dimethyl S-(1,2- dicarboxyethyl)-phosphorodithioate
	Phosphamidon	Dimethyl 2-chloro-2-diethylcarbamyl-1-methyl vinyl phosphate
	Glyphosate	N-Phosphonomethylglycine

(continued)

Table 8.1   (*Continued*)

Class	Common Name	Chemical Name
f. Carbamate insecticides	Carbaryl	1-Naphthyl *N*-methylcarbamate
	Aminocarb	4-Dimethylamino 3-tolyl *N*-methylcarbamate
	Carbofuran	2,2-Dimethylbenzofuran-7-yl *N*-methylcarbamate
g. Triazine herbicides	Simazine	2-Chloro-4,6-bis-(ethylamino)-*s*-triazine
	Atrazine	2-Chloro-4-(ethylamino)-6-(isopropylamino)-*s*-triazine
	Hexazinone	3-Cyclohexyl-6-(dimethylamino)-1-methyl-1,3,5-triazine-2,4(1*H*,3*H*)-dione

*a*Modified from Metcalf (1971) and Entomological Society of America (ESA) (1980).

been widely used, almost entirely as fungicides. Most important in this category are the organomercurials. Examples include phenylmercuric acetate, methylmercury, methoxyethylmercuric chloride, and others. Examples of environmental contamination and human toxicity caused by the use of organomercurials were given in Chapter 3.

3. *Phenols* are fungicides used for the preservation of wood and other organic substrates. Prominent examples are the trichlorophenols, tetrachlorophenol, and pentachlorophenol.

4. *Chlorinated hydrocarbons* are a diverse group of synthetic pesticides. Prominent subgroups are:

a. *DDT and its insecticidal relatives*, including DDD and methoxychlor. The related chemical DDE is noninsecticidal, but it is an important and persistent metabolic breakdown product of DDT. Residues of DDT and its relatives are persistent, and they typically have a half-life of about 10 years in the environment. Their persistence, coupled with an ability to weakly codistill with water, has caused a global contamination with these compounds. In addition, their selective partitioning into lipids has caused their accumulation at the top of food webs, as is described in pp. 191–198.

b. *Lindane*, or the gamma isomer of 1,2,3,4,5,6-hexachlorocyclohexane. This is the active insecticidal constituent of benzene hexachloride.

c. *Cyclodienes* are a group of highly chlorinated cyclic hydrocarbons that are used as insecticides. Prominent examples are the alpha-cis and beta-trans isomers of chlordane, heptachlor, aldrin, and dieldrin.

d. *Chlorophenoxy acid herbicides*, which have an auxin-like growth regulating property, and are selective for broad-leaved angiosperm plants. The parent compound is 2,4-D. Other important chemicals are 2,4,5-T (which is more effective than 2,4-D for the control of many angiosperm shrubs), MCPA, and silvex.

5. *Organophosphorus pesticides* are a diverse group of chemicals, most of which are used as insecticides, acaricides, or nematicides. They generally have a high acute toxicity to arthropods but a short persistence in the environment, and some of the insecticides are highly toxic to nontarget organisms such as fish, birds, and mammals. Some prominent examples are the insecticides parathion, methyl parathion, fenitrothion, malathion, and phosphamidon. An important herbicide is the phosphonoalkyl compound glyphosate.

6. *Carbamate pesticides* generally have a high acute toxicity to arthropods but a moderate

environmental persistence. Important examples are carbaryl, aminocarb, and carbofuran.

7. *Triazine herbicides* are used in corn monoculture and for some other crops, and as soil sterilants. Prominent examples are simazine, atrazine, and hexazinone.

8.3
ENVIRONMENTAL IMPACTS
OF
PESTICIDE USE
Overview

The intended ecological effect of a pesticide application is to control a pest species, usually by reducing its abundance to an economically acceptable level. In a few situations, this objective can be attained without any important nontarget damage. For example, the judicious use of a rodenticide in the local environment of a human habitation can cause a selective kill of rats and mice. If care is taken in the placement of the poison, any exposure to nontarget mammals such as cats, dogs, and children can be minimized.

Of course, most situations where pesticides are used are more complex and less well controlled than this. Whenever a pesticide is broadcast-sprayed over a field or forest, a wide variety of onsite, nontarget organisms is impacted. In addition, much of the sprayed pesticide invariably drifts away from the intended site of deposition, and it deposits onto nontarget organisms and ecosystems. The ecotoxicological effects of nontarget exposures are influenced by a complex of variables, including (1) the biological sensitivity of specific organisms to the dose of the pesticide that is received and (2) environmental variables that can increase or decrease this sensitivity under the particular conditions of exposure. (Table 8.2 ranks the toxicity to rats of some pesticides and other chemicals, using LD_{50} measured under standardized laboratory conditions.) The ecological importance of any damage caused to nontarget, pesticide-sensitive organisms

largely depends on their role in maintaining the integrity of the structure and function of their ecosystem. However, from the human perspective, the importance of a nontarget pesticide effect is also influenced by specific economic and esthetic considerations.

There is such a great variety of pesticides available today, and such a diversity of uses, that it is impractical to specifically discuss the above principles for more than a few. The chemical complexity of environmental exposures can be illustrated by the observations of Frank *et al.* (1982), who studied pesticide use in 1975 on 11 agricultural watersheds in Ontario. At least 81 different pesticides had been applied in agriculture or along right-of-ways, and undocumented other pesticides were used in and around homes (note that in 1970, 19% of the pesticide sold in North America was for home use; McEwen and Stephenson, 1979). On average, 39% of the land surface had been treated at an average rate of 8.3 kg/ha-year. The rate of application in agriculture ranged from 0.005 kg/ha-year for hayfields and pasture, to an average of 51 kg/ha-year for potato, tomato, and tobacco crops. The intensive use of pesticides contaminated surface water in the study area. The herbicide atrazine accounted for 93% of the total pesticide flux in stream water of 2.18 g/ha-year. Although DDT has not been used since 1972, it was present in 41% of the 1975 water samples (as DDT or DDE), while PCBs or polychlorinated biphenyls (a group of chlorinated hydrocarbons that are largely used in electrical tranformers) were present in 78% of the samples.

In view of the extreme diversity of pesticides and their uses, a limited approach will be taken in this chapter. Initially, the ecological impacts of pesticide use will be characterized by reference to DDT and its chlorinated hydrocarbon relatives. Although the use of DDT has been almost eliminated in industrialized countries since the early 1970s, its use continues in less developed countries of warmer latitudes. Moreover, DDT is one of the best-studied pesticides from an ecotoxicological perspective. Consideration of its case gives insight into the movement, fate, and ecological damages that can be caused by contamination of the natural environ-

Table 8.2   Acute toxicity of selected chemicals to rats[a]

Chemical Substance	Oral LD$_{50}$ for Rats (mg/kg)
Sucrose (table sugar)	30,000
Fosamine (H)	24,000
Ethanol (drinking alcohol)	13,700
Benomyl (F)	10,000
Captan (F)	9,000
Picloram (H)	8,200
Maneb (F)	6,500
Simazine (H)	5,000
Glyphosate (H)	4,300
DDD (I)	4,000
Sodium chloride (table salt)	3,750
Sodium bicarbonate (baking soda)	3,500
Sodium hypochlorite (household bleach)	2,000
Dicamba (H)	2,000
Malathion (I)	2,000
Atrazine (H)	1,750
Acetylsalicylic acid (aspirin)	1,700
Hexazinone (H)	1,690
Dalapon (H)	1,000
DDE (I)	880
2,4-DP (H)	800
Mirex (I)	740
MSMA (H)	700
2,4,5-TP (H)	650
Triclopyr (H)	650
Carbaryl (I)	500
2,4,5-T (H)	500
Chlordane (I)	400
2,4-D (H)	370
Fenitrothion (I)	250
DDT (I)	200
Caffeine (alkaloid in coffee, tea, other foods)	200
Paraquat (H)	150
Diazinon (I)	108
Lindane (I)	88
Endosulfan (I)	75
Nicotine (alkaloid in tobacco)	50
Dinoseb (H, I)	40
Aminocarb (I)	39
Strychnine (R)	30
Phosphamidon (I)	24
Methylparathion (I)	14
Parathion (I)	13
Carbofuran (I)	10
TEPP (I)	6.8
Aldicarb (I)	0.8
Saxitoxin (paralytic shellfish neuropoison)	0.26
Tetrodotoxin (Japanese globe fish toxin, *Spheroides rubripes*)	0.01
TCDD (dioxin isomer)	0.01

[a](H), Herbicide; (I), insecticide; (F), fungicide; (R), rodenticide. Modified from Murphy (1980), Windholz (1983), and Walstad and Dost (1984).

ment with a persistent, toxic chemical. Therefore, DDT provides an excellent case study of the ecological impacts that can accompany the use of pesticides. Following the consideration of DDT, case studies will be examined of the ecological impacts of two large-scale pesticide spray programs in forestry. The first is the use of insecticides against spruce budworm in mature conifer forest, and the second concerns the silvicultural use of herbicides. These cases were chosen specifically because their spray programs impact complex, natural or semi-natural, managed ecosystems, and because there is a broad base of information about their various ecological impacts. As such, they allow a good appreciation of the effects on complex ecosystems of large-scale pesticide spraying. It would be much more difficult to comprehensively consider these effects in highly disturbed and relatively simple agroecosystems, and that is why nonagricultural pesticide programs were chosen as case studies in this chapter.

Environmental Impacts of the Use of DDT and Its Relatives

Background and Use History

DDT was first synthesized in 1874. In 1939 its insecticidal quality was recognized by Mueller in Switzerland, who won a Nobel Prize in 1948 for that discovery and his subsequent research on the uses of DDT. The first important application of DDT was in human health programs during and after World War II, and at that time its use for agricultural and forestry purposes also began. The global production of DDT peaked in 1970, when 175 million kg was manufactured. The peak of production in the United States was 90 million kg, in 1964 (Edwards, 1975). Because of the recognition of a widespread and persistent environmental contamination by DDT and its breakdown products, their ability to biomagnify, and the important non-target damages that they caused, most industrialized countries banned the use of DDT after the early 1970s (with a minor exception: it can still be

prescribed by North American physicians for control of the human body louse).

However, the use of DDT has continued in less developed countries of warmer latitudes, primarily for the control of the mosquito vectors of disease. Although the use of DDT was banned in the United States in 1972, its manufacture continued for the export market, which was almost entirely in less developed countries. The export of DDT from the United States was 26 million kg in 1974, 21 million kg in 1975, and 12 million kg in 1976, but beginning in 1977 (0.23 million kg) exports were greatly reduced (ESA, 1980). Today, there is no export of DDT from the United States, but its manufacture and use continue in less developed countries and elsewhere. However, largely because of the widespread evolution of resistance to DDT by many insect pests, it is becoming a decreasingly effective pesticide. There has even been a resurgence of previously well-controlled diseases such as malaria (Chapin and Wasserstrom, 1981). Because of the emerging ineffectiveness of DDT, its use will eventually be curtailed and it will be totally replaced by other insecticides.

DDT was the first pesticide to which a large number of insect pests developed resistance. This evolutionary process occurs because of the intense selection for resistant genotypes that takes place when a population of organisms is exposed to a toxic pesticide. Resistant genotypes may be present at a small frequency in an unsprayed population, but because they are not killed by the pesticide they become numerically dominant after treatment. According to NRC (1986), at least 447 taxa of insects and mites are known to be resistant to at least one insecticide, and there are at least 100 cases of resistant plant pathogens, 48 resistant weeds, and two resistant nematodes. Among arthropods, resistance is most frequent in the Diptera, with 156 resistant species. There are 51 resistant species of malaria-carrying *Anopheles* mosquitoe, including 47 that are resistant to dieldrin, 34 to DDT, 10 to organophosphates, and 4 to carbamates. The progressive evolution of resistance by *Anopheles* has been an important cause of the recent resurgence of malaria in warmer latitudes (NRC, 1986).

Residues and Biological Uptake

DDT has several chemical and physical properties that profoundly influence the nature of its ecological impact. First, DDT is persistent in the environment. It is not easily degraded to other, less toxic chemicals by microorganisms or by physical agents such as sunlight and heat. The typical persistence in soil of DDT and some other organochlorine insecticides is given in Table 8.3. DDE is the primary breakdown product of DDT, and it is produced by dechlorination reactions that take place in an alkaline environment, or enzymatically in organisms. Unfortunately, the persistence of DDE is similar to that of DDT. Therefore, once it is released into the environment, DDT and its breakdown products persist for many years.

Another important characteristic of DDT is its small solubility in water (less than 0.1 ppm). Its sparse aqueous solubility means that DDT cannot be physically "diluted" into this ubiquitous solvent, which is so abundant on the surface of the earth and in organisms. On the other hand, DDT is very soluble in organic solvents such as xylene (60%) and kerosene (80%). More important is its great solubility in fats or lipids, a characteristic that is shared with other chlorinated hydrocarbons. In the environment, most lipids are present in organisms. Therefore, because of its high lipid solubility DDT

has a great affinity for organisms, and it tends to bioconcentrate by a factor of several or more orders of magnitude. Furthermore, because organisms at the top of a food web are effective at accumulating DDT from their food, they tend to have an especially large concentration of DDT in their lipids.

The bioconcentration and food-web accumulation effects of DDT are illustrated in Fig. 8.1, which shows the typical concentrations of DDT in a variety of atmospheric, terrestrial, aquatic, and biotic compartments of the environment. Note that the residues of DDT in air, water, and nonagricultural soil are relatively small, compared with the concentrations in organisms. Note also that residues in plants are smaller than in herbivores, and that residues are largest in animals that are at or near the top of the food web, such as humans and predatory birds.

A similar pattern is seen in Fig. 8.2, which summarizes the pattern of residues in an estuarine ecosystem on Long Island, New York. The sources of the DDT were diffuse, and they included the spraying of salt marshes to control mosquitoes. The data indicate that there is a general concentrating of residues at the top of the food web. The largest concentration (75.5 ppm) was in a ring-billed gull (*Larus delawarensis*), an opportunistic species that often feeds on small fish. Large residues were also present in other piscivorous birds, for example, 26

Table 8.3   Persistence of some organochlorine insecticides in soil[a]

Chemical	Typical Annual Dose (kg/ha)	Half-life (years)	Average Time for 95% Disappearance (years)
Aldrin	1.1–3.4	0.3	3
Isobenzan	0.3–1.1	0.4	4
Heptachlor	1.1–3.4	0.8	3.5
Chlordane	1.1–2.2	1.0	4
Lindane	1.1–2.8	1.2	6.5
Endrin	1.1–3.4	2.2	7
Dieldrin	1.1–3.4	2.5	8
DDT	1.1–2.8	2.8	10

[a]Modified from Edwards (1975).

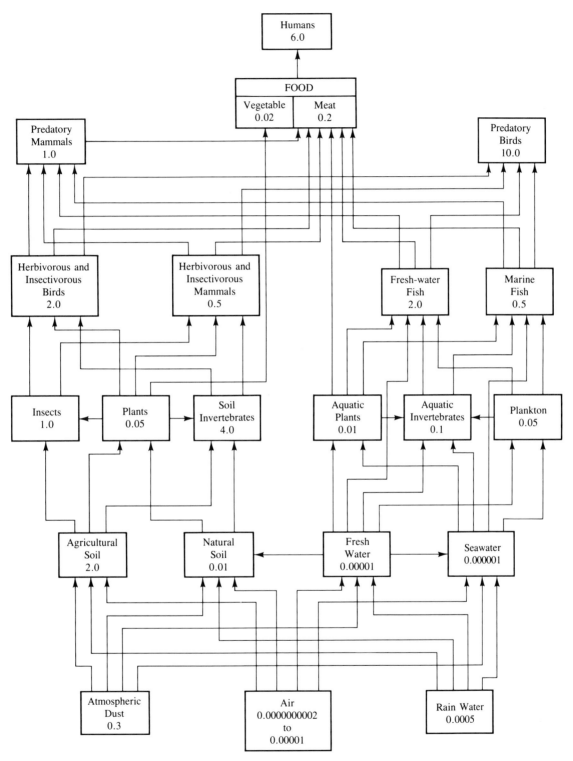

Fig. 8.1. Typical concentrations of DDT in the environment (ppm). Data were derived from a literature review of DDT residues. From Edwards (1975).

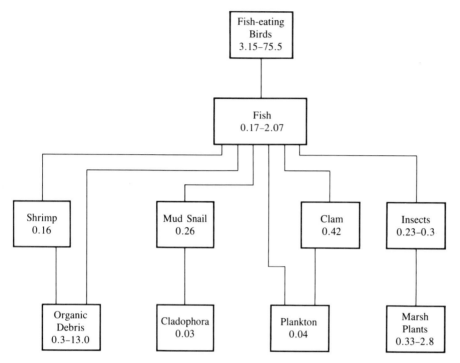

Fig. 8.2. Residues of total DDT (DDT + DDD + DDE; ppm) in various trophic levels of an estuary and salt marsh on Long Island, New York (Edwards, 1975, adapted from Woodwell *et al.*, 1967).

ppm in double-crested cormorant (*Phalacrocorax auritus*), 23 ppm in red-breasted merganser (*Mergus serrator*), and 19 ppm in herring gull (*Larus argentatus*) (Woodwell *et al.*, 1967).

Another environmental feature of DDT is its ubiquity—it is present throughout the biosphere in at least a trace concentration. The remarkably widespread environmental contamination with DDT and some related chlorinated hydrocarbons occurs because they can enter the global atmospheric cycle. This results from (1) the ability of DDT to codistill with water in a very small concentration; (2) the volatilization of DDT from treated surfaces (this is a relatively slow process, since its vapor pressure at 20°C is only 1.9×10^{-7} atm); (3) off-target drift of sprayed pesticide; and (4) the entrainment of contaminated dust into the atmosphere by wind (Edwards, 1975; Taylor, 1978; McEwen and Stephenson, 1979). Because of these processes, DDT and other pesticides can be present in a trace at-

mospheric concentration in remote locations. For example, atmospheric sampling for DDT between 1976 and 1978 found a trace concentration that ranged from 75 to 132×10^{-12} g DDT/m³ at three locations in the Indian Ocean, while at three sites in the North Atlantic it was $2.5–5.0 \times 10^{-12}$ g/m³ (Bidleman and Leonard, 1982). This difference reflected the contemporary pattern of DDT use, which had been banned in the temperate parts of the northern hemisphere but was still permitted in warmer latitudes.

Data on the residue levels of DDT in the biosphere were reviewed by Edwards (1975). The ubiquity of the contamination with DDT can be illustrated by its concentration in Antarctic wildlife, which live far remote from sites where DDT has been used. Norheim *et al.* (1982) reported residues of total DDT (i.e., DDT + DDE, most of which is DDE) in some Antarctic birds. The concentration in fat averaged largest in the southern polar skua

(*Catharacta maccormicki*) at 4.6 μg/g wet weight, compared with ≤ 0.43 μg/g in avian predators lower in the oceanic food web, such as the southern fulmar (*Fulmarus glacialoides*), cape pigeon (*Daption capense*), snow petrel (*Pagodroma nivea*), macaroni penguin (*Eudyptes chrysolophus*), and chinstrap penguin (*Pygoscelis antarctica*).

A much larger concentration of DDT and its related compounds is present in wildlife that live closer to the areas where the pesticides are manufactured and sprayed. Marine mammals are at or near the top of the marine food web, and they frequently have large residues of chlorinated hydrocarbons. The concentration of DDT in the fat of seals from the coast of California was as high as 158 μg/g during the late 1960s (a period of great use of DDT in that area), while in the Baltic Sea of northwestern Europe the residues were as great as 150 μg/g, and off eastern Canada they were up to 35 μg/g. Even larger residues were present in the fat of eastern Canadian porpoises (*Phocoena phocoena*), at up to 520 μg/g (Edwards, 1975).

Very large residues of DDT and other organochlorines have also been measured in birds, especially raptors. Residues as large as 356 μg/g (average of 12 μg/g) occurred in a sample of 69 U.S. bald eagles (*Haliaeetus leucocephalus*), up to 460 μg/g among 11 western grebes (*Aechmophorus occidentalis*), and up to 131 μg/g among 13 herring gulls (*Larus argentatus*) (Edwards, 1975). In some cases, remarkably large residues were measured. The white-tailed eagle (*Haliaeetus albicilla*) population of the Baltic Sea had individuals with as much as 36,000 μg/g of total DDT and 17,000 μg/g polychlorinated biphenyls (PCBs) in their fat, and addled eggs had up to 1900 μg/g of total DDT and 2600 μg/g of PCB (Koivusaari *et al.*, 1980).

Effects on Birds

The large residues of DDT and related chlorinated hydrocarbons have had a number of important ecological effects. Among the most prominent of these has been a widescale poisoning of birds. For example, there were a number of incidents in which dying or dead birds were found within a short time of the spraying of DDT, especially after urban sprays to kill the beetle vectors of Dutch elm disease. Wurster *et al.* (1965) recovered 117 dead birds of various species in a 6-ha spray area; undoubtedly many more individuals were killed by the spray, but were not found because they were hidden by vegetation or scavenged. It was not uncommon to have a greatly reduced abundance of songbirds in such a situation—hence the title of Rachael Carson's (1962) book, "Silent Spring." Avian mortality was also sometimes reported after the large-scale spraying of DDT for spruce budworm control (McEwen and Stephenson, 1979; see also pp. 198–213).

However, the effects of DDT and its related chlorinated hydrocarbons were usually more insidious than direct mortality. In some situations mortality was caused by chronic toxicity, and it was therefore less immediate and a relatively detailed investigation was required to link the decline of a bird population to the use of the pesticide. A graphic example of this took place at Clear Lake, California (Hunt and Bischoff, 1960). This is an important water body for recreation, but there were many complaints about the nuisance associated with events of great abundance of a nonbiting midge (*Chaoborus asticopus*). This pest was dealt with in 1949 by the application of DDD to the lake at about 1 kg/ha. Short-term bioassays were done prior to this treatment, and had indicated that this dose of DDD would adequately control the midges, but would have no immediate effect on fish. However, the unexpected happened. After a second application of DDD in 1954 to deal with a rebounding population of midges, about 100 western grebes (*Aechmophorus occidentalis*) were found dead on Clear Lake. The cause of mortality was not immediately identified, but an infectious disease was ruled out by autopsies. Sick and dead grebes were also found after later DDD treatments. Overall, the breeding population of western grebes on Clear Lake decreased from a prespray abundance of more than 1000 pairs, to none in 1960 (although 30 nonbreeding adults were present in 1960). The catastrophic decline of grebes was linked to DDD when, in 1957, an analysis of fat taken from dead birds indicated a residue as large as 1600 µg/g. Fish also proved to be heavily contaminated. Therefore, the treatment of Clear Lake with DDD was an essentially unsuccessful attempt to combat a relatively trivial pest problem. The failure was caused by (1) unexpected nontarget damage and (2) the development of resistance to the insecticide by the pest midges. However, it must be appreciated that in the early 1950s it was impossible to predict such an environmental "surprise" because of (1) the limited available information on the toxicological effect of pesticides on birds and other wildlife and (2) society's limited experience with the insidious, chronic effects of persistent and bioaccumulating pesticides.

Chronic ecotoxicological damage to birds also occurred in habitats that were remote from sprayed sites. This was especially true of raptorial birds of many species, because they were at the top of their food web, and they bioaccumulated chlorinated hydrocarbons to a very large concentration. In some species, effects were sufficiently severe to cause a decline in abundance beginning around the early 1950s, and resulting in local or regional extinction of breeding populations. Prominent examples of birds that suffered a population decline because of exposure to DDT, its metabolic breakdown product DDE, and other chlorinated hydrocarbons, include the bald eagle (*Haliaeetus leucocephalus*), golden eagle (*Aquila chrysaetos*), peregrine falcon (*Falco peregrinus*), osprey (*Pandion haliaetus*), brown pelican (*Pelecanus occidentalis*), and double-crested cormorant (*Phalacrocorax auritus*) (Ames, 1966; Hickey, 1969; Lockie *et al.*, 1969; Peterson, 1969; Cade and Fyfe, 1970; Ratcliffe, 1970; Cromartie *et al.*, 1975; Fyfe *et al.*, 1976). Fortunately, there are now encouraging signs of a recovery of abundance of these and other species since the use of DDT was banned in temperate countries around 1972 (Spitzer *et al.*, 1978; Spitzer and Poole, 1980; Grier, 1982; Wallin, 1984). Unfortunately, there are indications that the continued use of DDT and related insecticides in the tropics is causing reproductive damage to raptors there (Tannock *et al.*, 1983).

The damage to predatory birds was largely caused by the chronic effects of chlorinated hydrocarbons on reproduction, as opposed to direct toxicity to adults. The reproductive effects of chlorinated hydrocarbons included (1) a decrease in clutch size; (2) the production of a thin eggshell (Table 8.4), so that the egg could break under the weight of an incubating parent; (3) a high death rate of embryos, unhatched and pipping chicks, and nestlings; and (4) aberrant adult behavior while incubating or raising hatchlings, causing a decrease in fledging success. Because of the reproductive pathology of DDT and its chemical relatives, many populations of raptors had a small frequency of juvenile individuals, and the abundance of those species decreased because of inadequate recruitment (Nelson, 1976; Peakall, 1976; Cooke, 1979; McEwen and Stephenson, 1979).

The syndrome of chronic toxicity to birds of DDT and other chlorinated hydrocarbons can be illustrated by the peregrine falcon, the decline of which attracted high-profile attention in North America and Europe. A decreased reproductive success and declining population of this falcon were first noticed in the early 1950s in western Europe, and soon after in North America. In 1970, a North American census reported that there was virtually no successful reproduction by the eastern population of the *anatum* subspecies of the peregrine (*F. peregrinum anatum*), while the arctic *tundrius* subspecies was declining in abundance. Only the *peali* subspecies of the Queen Charlotte Islands of western Canada had a stable population and normal breeding success (Cade and Fyfe, 1970). The *peali* race is nonmigratory, it lives in an area where pesticides are not used, and it feeds on an essentially nonmigratory food resource of seabirds. In contrast, the eastern *anatum* race bred in a region where chlorinated hydrocarbon pesticides were widely used, and their prey was generally contaminated. Although the northern *tundrius* race breeds in a region that is remote from situations where pesticides are used, it winters in sprayed areas in Central and South America, and its migratory prey of waterfowl is contaminated (Cade and Fyfe, 1970; Fyfe *et al.*, 1976).

The implications of this latter characteristic of the *tundrius* peregrine can be illustrated by the study

Table 8.4 Changes in eggshell thickness of some populations of North American bird species: Change is percent change comparing post-1945 with pre-1945 data, and shell thickness is indexed as shell weight (mg)/[length (mm) × width (mm)][a]

Species	Location	Change in Thickness Index (%)
Bald eagle (*Haliaeetus leucocephalus*)	Texas	−30
Double-crested cormorant (*Phalacrocorax auritus*)	Wisconsin	−30
Prairie falcon (*Falco mexicanus*)	New Mexico	−28
Peregrine falcon (*Falco peregrinus*)	California	−26
Brown pelican (*Pelecanus occidentalis*)	California	−25
Marsh hawk (*Circus cyaneus*)	Oregon, Alberta	−24
Osprey (*Pandion haliaetus*)	Northeastern U.S.	−21
Cooper's hawk (*Accipiter cooperi*)	Western Canada	−20
Black-crowned night heron (*Nycticorax nycticorax*)	New Jersey	−18
Great horned owl (*Bubo virginianus*)	Florida	−17
Common loon (*Gavia immer*)	Ontario	−15
White pelican (*Pelecanus erythrorhynchos*)	British Columbia	−14
Golden eagle (*Aquila chrysaetos*)	California	−11
Herring gull (*Larus argentatus*)	Great Lakes	−10

[a]Modified from Anderson and Hickey (1970).

of Lincer *et al.* (1970), who compared the pesticide residues in Alaskan peregrine falcons and rough-legged hawks (*Buteo lagopus*). On a dry weight basis, three peregrines averaged 114 ppm DDE in muscle, 752 ppm in fat, and 131 ppm in eggs, while three rough-legged hawks averaged only 1.2 ppm in muscle, 13.3 ppm in fat, and 7.1 ppm in eggs. These differences were related to the residues in the prey of these raptors. The peregrines fed on migratory ducks, which had 10–20 ppm DDE in their fat, compared with less than 1 ppm in the nonmigratory small mammal prey of the rough-legged hawks.

Other studies done at about the same time showed that large residues of chlorinated hydrocarbons were widespread in North American peregrines (except *peali*). These residues were associated with eggshells that were thinner than the pre-DDT condition by 15–20%, and with a generally impaired reproductive ability of the adults (Hickey and Anderson, 1968; Ratcliffe, 1967, 1970; Berger *et al.*, 1970; Risebrough *et al.*, 1970).

In 1975, the North American peregrine survey was repeated. Again the *peali* subspecies had a stable population. In contrast, the eastern *anatum* pop-

ulation was virtually, if not entirely, extinct, and the *tundrius* subspecies had suffered a further decline in abundance and was clearly in trouble (Fyfe *et al.*, 1976). However, as with other raptors that have suffered from the chronic effects of chlorinated hydrocarbons, a recovery of peregrine populations has begun since the 1972 banning of DDT use in North America and Europe. In 1985, northern North American populations of peregrines were stable or increasing compared with 1975, as were some southern populations, although they remained very small (Murphy, 1987; Cade, 1988). In part, this recovery has been enhanced by a captive-breeding and release program over much of the former range of the eastern *anatum* race (Fyfe, 1976; Anonymous, 1987; Cade, 1988). In the United States, a falcon breeding facility at Cornell University began a program of experimental releases in 1974. This transformed to an operational program, and by the end of 1986 more than 850 peregrines had been released (Cade, 1988). In Canada, three captive breeding facilities were annually producing more than 100 young peregrine falcons for wild release during the mid-1980s, a total that is expected to increase to about 150/year by 1990, and eventually to 270/year until a satisfactory recovery of this species has taken place (Anonymous, 1987). In 1986, at least 43 pairs of peregrine falcons occupied breeding territories in eastern North America, a region from which it had been virtually extirpated as a breeding species prior to the operational release of captive-reared individuals (Cade, 1988).

Spraying Forests Infested with Spruce Budworm

The Setting

Spruce budworms are important lepidopteran defoliators of conifers in north temperate and boreal forests. In North America, the economically most important species are the spruce budworm (*Choristoneura fumiferana*) and the western spruce budworm (*C. occidentalis*), both of which feed on fir and spruce. A much larger amount of forest damage is caused by the spruce budworm. This is because of the tremendous area of its recent infestations, and because it causes extensive mortality in stands that have been continuously defoliated for a number of years. Defoliation by the western spruce budworm does not generally cause the death of trees, but it does cause an important loss of forest productivity. In this case study, we will focus on the spruce budworm (see Brooks *et al.*, 1985, and Sanders *et al.*, 1985, for information on the other budworm species).

Biology of Spruce Budworm

The spruce budworm produces a single generation each year (MacLean, 1984; Blais, 1985). Moths emerge from mid-July to early August, and each female lays an average of 10 egg masses, containing about 20 eggs each, on the foliage of host conifer

A moribund forest of balsam fir, which has suffered at least 7 years of severe defoliation by spruce budworm. This Cape Breton forest was not sprayed with insecticide to reduce the abundance of spruce budworm. The grey cast of the forest indicates a great deal of mortality of canopy trees. (Photo courtesy of B. Freedman.)

A closer view of a moribund balsam fir stand on Cape Breton. Note the dense advance regeneration of balsam fir under the dead overstory. In about 40–50 years, this regeneration will provide the next harvestable balsam fir forest for spruce budworm or for humans. (Photo courtesy of B. Freedman.)

species. The eggs hatch after about 10 days, and the first-instar larvae disperse locally on silken threads, or sometimes beyond the stand by "ballooning" with the wind. Larvae that survive the high-risk dispersal manufacture a silken hibernaculum on a host tree, and moult to the second instar. This stage overwinters, breaks its diapause in late April to early May, and emerges and begins to feed on 1-year-old needles or, if available, on seed and male flowers. Because of its relatively large size, the sixth instar does most of the damage, accounting for about 87% of the total defoliation. This instar prefers current-year foliage, but if this has been depleted it will feed on older needles. Pupation takes

place in early July, and the moths emerge after 8–12 days. Adult moths can undergo a long-range dispersal, typically traveling 50–100 km downwind, and as far as 600 km.

Population Dynamics

It appears that periodic irruptions of spruce budworm have long been a feature of certain boreal landscapes. Blais (1965) established a budworm outbreak chronology that extended back to the eighteenth century in an area in Quebec. The evidence for historical outbreaks was based on ring-width measurements of living host trees, which recorded historical reductions of radial growth caused by defoliation. Blais concluded that outbreaks had taken place at rather even intervals, and that infestations had begun in 1704, 1748, 1808, 1834, 1910, and 1947, for an average periodicity of about 35 years (Royama, 1984; Blais, 1985).

During its endemic phase of small population density, spruce budworm is scarce. Only 10 spruce budworm larvae were found in more than 1000 insect collections made in New Brunswick during an endemic phase between 1939 and 1944 (Blais, 1985). This represents a density of about five individuals per host tree (Miller, 1975). At the beginning of an outbreak density increases rapidly, to about 2000 larvae per tree within 4 years. The density then increases to > 20,000 per tree during the epidemic phase of the outbreak. This large density is typically sustained for 6–10 years, after which the outbreak usually collapses.

Outbreaks tend to be synchronous over a large region of susceptible forest (Royama, 1984). However, on a more local scale the amplitude of the oscillation varies, so that there are large differences among stands in the maximum density of budworm. According to Royama (1984), the occurrence and persistence of a population outbreak is governed by several intrinsic, density-dependent factors that affect mortality. Especially important are the abundance of insect parasitoids, disease, and an undefined "complex of unknown factors" (a recent outbreak collapsed for no discernible reason, hence the vague third group). Royama discounted the

Cordwood stacked along a logging road on the Cape Breton highlands. After the decision was made not to spray the spruce budworm-infested forest, a great deal of tree mortality occurred. Much of the dead forest was salvage harvested, using very large skyline-to-skyline clearcuts. The wood in the photo had been stored at roadside for more than 5 years and, because of decomposition, it was then only marginally useful for the production of paper. (Photo courtesy of B. Freedman.)

roles of predation, food shortage, weather, mortality during the dispersal phases, and the spraying of infested stands with insecticide. He also hypothesized that the invasion of a stand by a large number of egg-bearing moths was not important in upsetting the population equilibrium during the endemic phase. He felt that such an invasion only played a role in accelerating an already-occurring increase in abundance.

Forest Damage

The spatial extent of spruce budworm outbreaks appears to have increased in modern times. The outbreak that began in 1910 involved some 10 million ha, while another that began in 1940 involved 25×10^6 ha, and the one that began in 1970 involved 55×10^6 ha (Fig. 8.3). Blais (1983, 1985)

hypothesized that the trend to an increasingly widespread infestation was due to a combination of factors that may have been important in creating large areas of optimal budworm habitat. These are (1) forest management practices such as clear-cutting; (2) fire protection; and (3) the spraying of infested stands with pesticides, which keeps the habitat suitable for budworm and may thereby help to prolong the infestation [note that Royama (1984) has a contrasting view on part 3]. These practices have had the effect of either producing or keeping alive a mature fir-dominated forest, which is the preferred habitat of spruce budworm.

Stands that are particularly vulnerable (i.e., apt to suffer damage) during a spruce budworm outbreak are mature and dominated by balsam fir (*Abies balsamea*) (MacLean, 1980, 1985). Immature balsam fir stands are less vulnerable, followed

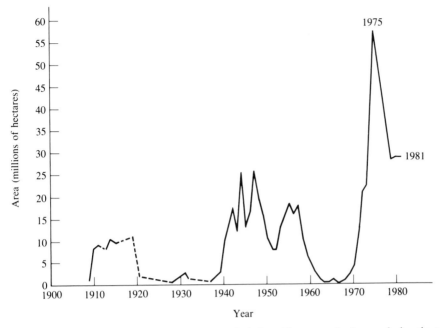

Fig. 8.3. The forest area moderately or severely infested by spruce budworm during the twentieth century. From Kettela (1983).

by mature spruce, and then immature spruce stands. In terms of tree species, balsam fir is most vulnerable, followed by white spruce (*Picea glauca*), red spruce (*P. rubens*), and then black spruce (*P. mariana*). Blum and MacLean (1985) summarized the stand characteristics that are believed to increase susceptibility to spruce budworm: (1) species composition is mainly balsam fir, followed by white spruce; (2) stand age > 60 years; (3) a large basal area of susceptible species; (4) an open stand with spike tops of host species penetrating the canopy; (5) location within an extensive area of susceptible forest; (6) location downwind of an ongoing infestation; (7) elevation <700 m, and south of 50°N latitude; and (8) a relatively dry or wet site condition.

A forest with a vulnerable character is widespread in eastern Canada and to a lesser extent the northeastern United States, comprising about 60.7 × 10⁶ hectares. This is reflected in the distribution of budworm-damaged forest in 1981 (Figure 8.4). Between 1977 and 1981, the average tree mortality caused by spruce budworm in this region was more

than 38×10^6 m³/year, and in 1983 mortality occurred over an area of about 26.5×10^6 ha (Ostaff, 1985).

In balsam fir trees, the first year of severe defoliation causes a decrease in volume growth of about 15–20%, followed by 25–50% in the second year of defoliation, and 75–90% in subsequent years (MacLean, 1985). The growth loss of spruce trees is somewhat less, ranging up to 50–60%. Height growth stops entirely during a severe defoliation; in fact, tree height can decrease because of top-kill. Mortality caused by budworm begins in fir stands that have been continuously defoliated for 4–5 years, and after 7–8 years in spruce stands. The mortality is progressive with time. MacLean (1984) examined mortality of balsam fir in unsprayed, mature stands in Cape Breton. He found that 9% of the trees were dead after 4 years of heavy defoliation, 22% after 5 years, 37% after 6 years, 48% after 8 years, 61% after 9 years, and 75% after 10 years. Spruce budworm-caused mortality usually ends about 12 years after the beginning of the outbreak.

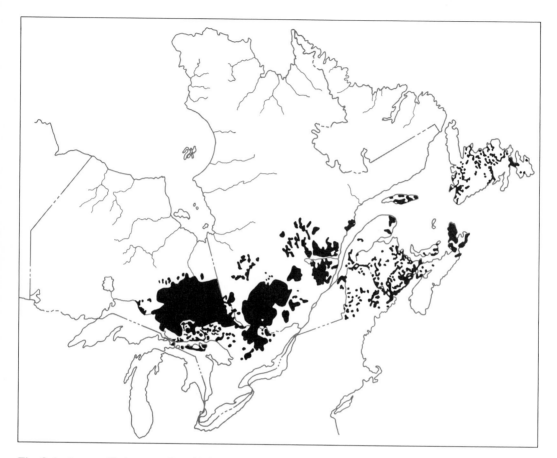

Fig. 8.4. Areas suffering mortality of balsam fir in 1981 as a result of spruce budworm infestation. From Kettela (1983).

The average final mortality was 85% (maximum of 100%) in mature balsam fir stands in eastern Canada, 42% in immature fir stands, 36% in mature spruce stands, and 13% in immature spruce stands (MacLean, 1980). Note that a great deal of mortality can take place after the spruce budworm population has collapsed, as weakened trees succumb to environmental stresses that might otherwise be tolerated. Blais (1981) studied a situation in Quebec where a budworm outbreak had collapsed in 1975. At that time the mortality of balsam fir averaged 44%, but this increased to 91% after four postoutbreak years. Mortality of spruce increased from 17% in 1975, to 52% 4 years later.

In the understory of stands that have been damaged or killed by spruce budworm, there is an advance regeneration of tree species that is generally dominated by the same species that comprised the original stand. For example, stands on Cape Breton Island that had been severely defoliated for 4–6 years had an average stocking of 45,000 individuals of balsam fir/ha and 3250 spruce/ha (MacLean, 1984). Almost all of these small individuals of tree species survived the spruce budworm infestation, and after the death of the overstory they were available to establish the succeeding fir–spruce forest. Therefore, the dynamic budworm–fir system can be viewed as a cyclic succession with a longer-term ecological stability. This cycle has probably recurred on the landscape for thousands of years (Baskerville, 1975a; MacLean, 1984; Blais, 1985). Evidence that supports the hypothesis of long-term stability includes (1) the presence of a large area of relatively pure, even-aged balsam fir forest; (2) the observed successional trajectory after stand mortality caused by spruce budworm; and (3) dendrochronological data that indicate that budworm outbreaks are an ancient and apparently cyclic phenomenon.

In contrast to this long-term ecological stability, spruce budworm causes a severe economic instability, which makes difficult the longer-term exploitation and management of the forest by humans. Industries that harvest the forest are in direct competition with spruce budworm for the same natural resource. Because budworm causes an unstable wood supply, there is severe economic instability (Baskerville, 1975a,b; MacLean, 1984).

As a way of coping with this resource crisis, forests have been sprayed with insecticides in an attempt to limit defoliation and prevent tree mortality. The objective of this management strategy is not to eradicate the spruce budworm, but to lessen the degree of the damage that it causes to the forest resource. In a study of fir-dominated stands in New Brunswick, Clowater and Andrews (1981) found that an average of 67% of the balsam fir volume was dead or severely damaged in unsprayed stands, compared with 38% in sprayed stands. In another study in New Brunswick, MacLean et al. (1984) found that balsam fir mortality in unsprayed spruce-dominated stands averaged 59%, while in sprayed stands it was 39%.

In a more theoretical study, MacLean and Erdle (1984) modeled the effects of spruce budworm on wood supply in New Brunswick. The model forecast that if forest protection were withdrawn as a management practice, the maximum sustainable harvest of the spruce–fir forest would be reduced by 46–64% under a severe outbreak condition, and by 23–36% under a moderate outbreak condition. Projections such as these provide a powerful economic justification to mount an insecticide spray program in a region where the forest is infested with spruce budworm.

Insecticide Use and Residues

The first insecticide that was used against spruce budworm was calcium arsenate. This was first used in 1927–1929 in Nova Scotia, where it was applied aerially at a rate of 10–40 kg/ha (Nigam, 1975; Armstrong, 1985b).

However, the real era of operational spraying of infested forests did not begin until after the Second World War, when DDT began to be used in a large quantity. In 1949, about 108,000 ha of forest was sprayed with DDT in Oregon and Washington; in 1952, 113,000 ha was sprayed in Quebec and New Brunswick, and in 1953, 804,000 ha was sprayed (Armstrong, 1985b). In New Brunswick alone between 1952 and 1968, a total of about 5.75 million kg of DDT was sprayed on infested stands (Pearce, 1975).

New Brunswick has had the largest and most continuous program of forest spraying for protection against spruce budworm, with millions of hectares being treated with various pesticides in most years since 1950 (Table 8.5). Much of this terrain is resprayed in consecutive years (Eidt and Weaver, 1986). For example, in 1982 about 1.7×10^6 ha was sprayed. Of this, 51% was resprayed in 1983; 17% was resprayed in 1984; 14% was resprayed in both 1983 and 1984; and 3.4% was resprayed in 1983, 1984, and 1985 (i.e., during 4 consecutive years).

The typical spray rate and selected toxicity data for the most important insecticides that have been used against spruce budworm are summarized in Table 8.6. Note that among these insecticides DDT is least toxic to the budworm itself, and that it is relatively toxic to salmonid fish. It was largely because of a widespread kill of commercially and recreationally important Atlantic salmon (*Salmo salar*) and brook trout (*Salvelinus fontinalis*) caused by DDT spraying that this chemical was banned for forestry purposes in New Brunswick (see below). The other insecticides listed in Table 8.6 are more

Table 8.5 Insecticide spraying of forests in New Brunswick for the control of spruce budworm[a]

Year	Area Sprayed (10⁶ ha/year)	Year	Area Sprayed (10⁶ ha/year)
1952	0.08	1969	1.3
1953	0.73	1970	1.7
1954	0.45	1971	2.4
1955	0.45	1972	1.9
1956	0.81	1973	1.8
1957	2.1	1974	2.3
1958	1.1	1975	2.7
1959	0.00	1976	4.2
1960	1.1	1977	1.6
1961	0.89	1978	1.5
1962	0.57	1979	1.6
1963	0.28	1980	1.6
1964	0.81	1981	1.9
1965	0.85	1982	1.7
1966	0.81	1983	1.7
1967	0.40	1984	1.3
1968	0.20	1985	0.8
		1986	1.9

[a]DDT was the major insecticide used until 1968. After this time DDT was replaced by phosphamidon, fenitrothion, and matacil. The latter two are currently the most important chemicals being used. Modified from McEwen and Stephenson (1979), Mitchell and Roberts (1984), and Eidt and Weaver (1986).

toxic to budworm than is DDT, and therefore they can be sprayed at a smaller operational rate. In addition, these insecticides are less toxic to fish than is DDT. Toxicity to other organisms varies; the relatively great toxicity of phosphamidon to birds

caused much avian mortality during operational spraying, and this was the major reason why this chemical was no longer used after 1975.

Another important problem with DDT is its long persistence in the environment. As was discussed earlier, the persistence of DDT, coupled with its great solubility in lipids, meant that it could accumulate to a large concentration in organisms, especially in top predators.

Yule (1975) examined an area in New Brunswick that had been aerially treated with DDT between 1956 and 1967, for a cumulative spray equivalent to 4.9 kg DDT/ha. In 1968, the year after the final application of DDT, there was a residue of total DDT in soil and the forest floor that was equivalent to 0.77 kg/ha. Between 1968 and 1973 the half-life of DDT in this compartment was about 10 years. DDT was also persistent in conifer foliage, even if the tree had not been sprayed for several years (Table 8.7). The concentration of DDT was smallest in foliage produced after spraying stopped in 1967, but DDT was nevertheless present in a fairly large concentration in unsprayed tissue. Some of the DDT made its way into herbivores and predators. The residues in some mammal species in New Brunswick could be large, with a concentration of up to 156 ppm in fatty tissue of a bobcat (Table 8.8).

The residues of insecticides that have been used more recently against spruce budworm are much less persistent than DDT. Sundaram and Nott

Table 8.6 Data for acute toxicity of insecticides to fifth-stage spruce budworm larvae and selected vertebrate animals under controlled laboratory conditions[a]

Insecticide[b]	Application Rate (kg a.i./ha)	Contact Toxicity to Budworm (μg/cm²) 72-hr LD₅₀	72-hr LD₉₅	Salmon 48-hr LC₅₀ (ppm)	Mallard LD₅₀ (mg/kg)	Rat Oral LD₅₀ (mg/kg)
DDT (1)	0.3–2.2	1.3	6.6	0.05	2,240	87–500
Phosphamidon (2)	0.3	0.39	0.75	11.0	3.1	15–33
Fenitrothion (3)	0.21	0.31	0.67	1.4	1,190	250–600
Matacil (4)	0.07	0.04	0.11	1.3	23	30
Mexacarbate (5)	0.07	0.04	0.13	13.0	3.0	15–63

[a]Modified from Nigam (1975).
[b]Insecticides: (1) chlorinated hydrocarbon, 1,1,1-trichloro-2,2-bis(p-chlorophenyl) ethane; (2) organophosphate, 2-chloro-N,N-diethyl-3-hydroxycrotonamide, dimethyl phosphate; (3) organophosphate, O,O-dimethyl O-(4-nitro-m-tolyl) phosphorothioate; (4) carbamate, 4-dimethylamino-m-tolyl methylcarbamate; (5) carbamate, 4-dimethylamino-3,5-xylyl methylcarbamate.

Table 8.7   Total DDT residues in foliage of balsam fir and red spruce in an experimental spray plot in New Brunswick[a]

Year When Foliage Was Produced	Concentration of DDT (ppm)					
	Balsam Fir			Red Spruce		
	1967	1969	1973	1967	1969	1973
1960	7.7					
1961	6.7					
1962	5.4			2.1		
1963	7.4			2.5		
1964	5.1			3.4		
1965	5.9	9.1		2.3	3.4	
1966	3.8	7.1		2.3	3.1	
1967	3.7	6.3		2.7	1.0	
1968		4.6			1.4	
1969		1.5	0.78		1.9	0.96
1970			0.76			0.54
1971			0.16			0.02
1972			0.13			0.01
1973			0.01			<0.01

[a]The plot was sprayed annually between 1956 and 1967 with a total of 4.9 kg DDT/ha. Foliage was collected in 1967, 1969, and 1973, and was sorted into year classes prior to its analysis for DDT. Modified from Yule (1975).

(1985) reported the residues of aminocarb after two sprays of 0.07 kg active ingredient (a.i.)/ha each (the usual operational procedure is to treat each stand twice, with about 5–7 days between sprays). Immediately after the first spray there was 2.2 ppm of aminocarb in balsam fir foliage. The residues decreased to 0.73 ppm after 2 days, 0.40 ppm after 5 days, and 0.29 ppm after 7 days. A similar pattern was observed after the second spray. The initial concentration in forest litter and soil averaged

Table 8.8   Average DDT residues in mammals in an area of New Brunswick that was sprayed annually between 1956 and 1967 with a total of 4.9 kg DDT/ha[a]

Species	Tissue	Sample Size	Total DDT (DDD + DDE + DDT) (ppm wet weight)
Herbivores			
Red squirrel (*Tamiasciurus hudsonicus*)	Whole body	2	0.34 (0.20–0.48)
Beaver (*Castor canadensis*)	Fat	1	3.0
	Liver	1	0.08
Snowshoe hare (*Lepus americanus*)	Muscle	5	0.003 (tr–0.008)
White-tailed deer (*Odocoileus virginiana*)	Muscle	13	0.042 (0.014–0.108)
Carnivores			
Raccoon (*Procyon lotor*)	Fat	2	0.53 (0.31–0.75)
Short-tailed weasel (*Mustela erminea*)	Whole body	8	1.8 (0.65–3.0)
Mink (*Mustela vison*)	Whole body	2	3.9 (3.3–4.5)
Bobcat (*Lynx rufus*)	Fat	4	50 (9.3–156)
	Liver	10	6.1 (0.13–41)
	Muscle	2	1.8 (1.1–2.5)

[a]Modified from Buckner and McLeod (1975).

<0.11 ppm and <0.06 ppm, respectively. In another field study in which soil was sprayed directly, Sundaram *et al.* (1985) found that the residues of aminocarb disappeared quickly. The initial concentration in the surface 1 cm of mineral soil was 0.050 ppm. The residue decreased to <0.007 ppm between 1 and 3 days, and to <0.003 ppm after 4 days.

Residues of fenitrothion are similarly short-lived. Immediately following two operational sprays of 0.21 kg a.i./ha, the concentration in balsam fir foliage was as large as 2.9 ppm, while it was up to 1 ppm in the forest floor, and 0.60 ppm in mineral soil (Sundaram and Nott, 1984). The residue in the forest floor and soil is not persistent, and it quickly decreases to <0.01 ppm (Ayer *et al.*, 1984; Sundaram and Nott, 1986). In contrast, the residue in needles of balsam fir appears to decrease to a threshold of about 0.5–1.5 ppm f.w., indicating that foliage serves as a weak sink for fenitrothion (Ayer *et al.*, 1984; Eidt and Mallet, 1986; Eidt and Pearce, 1986; Sundaram and Nott, 1986).

Insecticide spray is often deposited to surface water during an operational spray program. However, the concentration in water disappears fairly quickly, as a result of dilution, downstream transport, and degradation. Kingsbury (1977) applied fenitrothion at 0.42 kg/ha to a small lake in Ontario. The initial postspray concentration was 21 ppb, but this declined to 8 ppb after 6 hr, 2–3 ppb after 12–48 hr, <0.6 ppb after 8 days, and <0.03 ppb after 1 year. Fish accumulated the fenitrothion rapidly, to a concentration of 1.0 ppm in white sucker (*Catastomus commersoni*), 0.76 ppm in fallfish (*Semotilus corporalis*), 0.44 ppm in brown bullhead (*Ictalurus nebulosus*), and 0.34 ppm in smallmouth bass (*Micropterus dolomieu*). However, no deaths or distress behavior were observed in these fish.

There has been some replacement of synthetic organic insecticides in spruce budworm spray programs with biological pesticides. One such insecticide is based on the bacterium *Bacillus thuringiensis* (abbreviated as B.t.), which can be applied as a high-potency, low-volume aerial spray. The use of B.t. against spruce budworm has been highly variable in eastern North America (Armstrong, 1985a;

Cunningham, 1985; Carter and Lavigne, 1986). B.t. is the only insecticide that was used against spruce budworm in Nova Scotia between 1980 and 1984, during which $20.6–31.2 \times 10^3$ ha was treated annually. B.t. has also been relatively important in Ontario (accounting for 17–91% of the $3.2–20.3 \times 10^3$ ha sprayed between 1979 and 1984), Newfoundland (100% of the $5.9–11.8 \times 10^3$ ha sprayed in 1979–1980, but <10% of the $48–240 \times 10^3$ ha treated in 1981–1983), and Maine (11–33% of the $332–491 \times 10^3$ ha treated in 1980 to 1984). In 1987, B.t. was the only insecticide used against spruce budworm in Ontario, Quebec, Maine, and Nova Scotia. The use of B.t. is also increasing in New Brunswick, the Canadian province with the largest insecticide spray program against spruce budworm (<1% of the $1320–1900 \times 10^3$ ha sprayed in 1979–1984, but 12% of the 701×10^3 ha sprayed in 1985).

Effects on Nontarget Arthropods

An important characteristic of B.t. is its relatively small nontarget toxicity (in large part, the toxicity of B.t. is specific to lepidopterans). Eidt (1985) studied the toxicity of B.t. to a variety of aquatic insects. He only found a toxic effect on larvae of the blackfly *Simulium vittatum*, but only at an unrealistically large dose.

In contrast to B.t., fenitrothion and aminocarb are both toxic to a wide spectrum of arthropods, and a broadcast forest spray with these chemicals causes a large nontarget kill of insects and spiders. According to Varty (1975), a typical spray of fenitrothion in early June can kill 2.4–7.5 million arthropods of several hundred species per hectare of fir–spruce forest (more than 90% of the biomass of the kill is typically spruce budworm). In spite of such a large nontarget kill, neither long-term surveys of arthropod abundance in sprayed areas nor surveys of stands a short time after spraying have indicated a measurably reduced abundance of nontarget arthropods in sprayed forest in New Brunswick (Varty, 1975, 1977; Otvos and Raske, 1980). In large part, such observations are due to the great spatial and temporal variations that must be contended with during the quantitative sampling of forest arthropods. Small "signals" can be difficult to detect in such "noisy" systems.

In the view of Varty (1975, 1977), the most important factors that affect the abundance of arthropods in balsam fir forest are (1) the biomass of young foliage of fir, which is of course greatly diminished during and following a severe infestation of spruce budworm, so that heavily defoliated stands have a relatively small abundance of arthropods; (2) predator–prey relationships, which can cause cyclic or irruptive variations in abundance that overwhelm any spray effect; (3) density-independent influences on mortality, particularly inclement weather; and (4) insecticide intervention, which produces 3–5 days of increased mortality, usually followed by a rapid recovery. From his studies of nontarget arthropod effects in sprayed forests of New Brunswick, Varty concluded that "the arthropod community on balsam fir has not been drastically affected by fenitrothion larvicide treatments. Undoubtedly an unchecked budworm outbreak without insecticide usage would have had a far more profound influence."

Effects on Pollination of Plants by Insects

Various species of bee have been negatively affected by the drift of insecticide to off-target, non-forest sites in New Brunswick (Kevan, 1975; Plowright *et al.*, 1978; Wood, 1979). Bees are important commercially, in that they are the principal pollinators of lowbush blueberries (*Vaccinium angustifolium* and *V. myrtilloides*), a regionally important agricultural crop. Data from several studies indicated a smaller abundance of bees and a decreased fruit set of blueberry in some sprayed areas. Subsequently, this damage was largely eliminated by the prohibition of insecticide spraying within 3.2 km of any blueberry field.

A survey of plant fecundity in sprayed conifer forest showed an effect on the reproductive success of insect-pollinated native plants, which was attributed to a small population of pollinators, especially bees (Thaler and Plowright, 1980). For exam-

ple, fruit was produced by 71% of inflorescences of *Aralia nudicaulis* in unsprayed stands, but only 49% in sprayed stands; *Clintonia borealis* was 94% versus 78%; *Cornus canadensis*, 17% versus 9%; and *Maianthemum canadense*, 21% versus 15%. However, this degree of reduced fecundity may not be of much importance to the longer-term population dynamics of these long-lived plants. Vegetative growth is the most important mechanism by which their populations are maintained in a mature forest habitat, and in undisturbed situations their seedlings are rarely observed.

Effects on Aquatic Arthropods

Effects of budworm spray programs on aquatic arthropods have also been monitored, because spray can reach surface water by direct deposition and by drift from applications elsewhere. In streams, the aquatic fauna is exposed to a pulse of insecticide, which rapidly diminishes in concentration because of dilution and downstream transport. After applications of fenitrothion at 0.21–0.28 kg a.i./ha in New Brunswick, there was a maximum concentration in water of 15 ppb (Hall *et al.*, 1975). This decreased to <1 ppb within 1–2 days of spraying.

Several studies have shown an increased drift of killed aquatic insects after a spruce budworm spray. This damage was especially severe with DDT, and in some streams in New Brunswick the density of grazing insects was decreased to such an extent that a mat of filamentous algae covered the streambed (Kingsbury, 1975). With fenitrothion drift has sometimes been large, but there is no long-term reduction in the abundance of aquatic insects. The effect of aminocarb on drift has been trivial (Eidt, 1975, 1977; Hall *et al.*, 1975; Kingsbury, 1975; Holmes, 1979). For example, the data of Table 8.9 show no apparent effect of spraying with fenitrothion or aminocarb on the drift of aquatic and terrestrial insects in streams in Quebec. In another study, Eidt (1975) found that for a few days after spraying with fenitrothion there was a drift of aquatic insects of 95,000 animals/24 hr, compared with a prespray drift of 13,000/24 hr. However, this effect on mortality did not result in a measurable change in the

Table 8.9   Effect of insecticide spraying on the drift of aquatic and terrestrial arthropods in forest streams in Quebec[a]

Treatment	Drift of Aquatic Arthropods		Drift of Terrestrial Arthropods	
	4 Days Prespray	4 Days Postspray	4 Days Prespray	4 Days Postspray
Unsprayed	0.46	0.51	0.27	0.78
0.053 kg a.i./ha Aminocarb (a)	2.61	0.72	0.53	0.39
(b)	1.49	1.58	0.68	2.27
(c)	0.53	0.96	0.82	0.84
(d)	0.20	0.23	2.43	3.85
(e)	0.31	0.46	1.73	1.01
0.21 kg a.i./ha Fenitrothion	0.45	0.43	0.19	0.35

[a]Sampling was done for 4 days before and after the spray application. Calculated from Holmes (1979).

total density of the benthos, or of the most sensitive benthic genera (i.e., the stoneflies *Leuctra* spp. and *Amphinemoura* spp., and the mayfly *Baetis* spp.).

Effects on Birds

The avifauna of a budworm-infested forest is relatively abundant, because several insectivorous birds display a strong numerical response to the abundance of this insect (Kendeigh, 1948; Morris *et al.*, 1958; Crawford *et al.*, 1983). For example, the bay-breasted warbler increased in abundance from 2.5 pairs/10 ha in uninfested stands, to 300 pairs/10 ha during an epidemic; blackburnian warbler increased from 25–50 pairs/10 ha to 100–125 pairs/10 ha; and Tennessee warbler increased from 0 pairs/10 ha to 125 pairs/10 ha (Morris *et al.*, 1958; see also Table 8.11).

During an outbreak of spruce budworm, these and other insectivorous species rely heavily on abundant budworm larvae for food for the rearing of nestlings, and there is a relatively successful rate of reproduction. Crawford *et al.* (1983) calculated that in a Maine forest with an epidemic population of spruce budworm, avian predation consumed about 89,000 larvae and pupae per hectare, compared with 54,000/ha in transitional stands, and 5,600/ha

in stands with an endemic budworm population. The most important avian predators are five species of warbler (bay-breasted, Blackburnian, Cape May, black-throated green, and northern parula) and the white-throated sparrow. However, birds are only quantitatively important in causing a reduction of budworm density in the endemic stands, in which they were estimated to have removed 87% of the budworm population, compared with only 2.4% in epidemic stands.

During the era of DDT spraying, no measurable short-term changes in the abundance of non-game forest birds were documented, even though some examples of avian mortality were identified (Pearce, 1975). The lack of a measurable effect on avian abundance could have been influenced by such factors as (1) movement of birds in and out of the spray area; (2) a floating surplus of nonbreeding individuals that could rapidly replace killed territorial birds; (3) errors inherent in forest bird censuses; and (4) an underestimation of direct mortality because of the difficulty in finding sick or dead birds, due to their low density, scavenger activity, etc. (Stewart and Aldrich, 1951; Pearce *et al.*, 1979; Richmond *et al.*, 1979). There was a reduced abundance of a game species, the woodcock (*Philohela minor*). This species had a relatively large concentration of DDT residues, with an average in breast muscle of 2.1 ppm f.w. in sprayed areas, compared with <0.1 ppm in unsprayed habitat (Dilworth *et al.*, 1974).

The effects of spraying with phosphamidon, fenitrothion, and aminocarb on the abundance of forest birds have generally been fairly small (Buckner and McLeod, 1977; Pearce *et al.*, 1979; Kingsbury and McLeod, 1980, 1981; Pearce and Busby, 1980; Kingsbury *et al.*, 1981; McLeod and Millikin, 1982; Spray *et al.*, 1987). However, phosphamidon was an occasional exception. Because it is relatively toxic to birds (Table 8.6), it has sometimes caused great mortality of some species after operational sprays (McLeod, 1967; Pearce and Peakall, 1977; Pearce *et al.*, 1979). One calculation estimated that as many as 376,000 individuals of a relatively susceptible species, the ruby-crowned kinglet, may have died in New Brunswick in 1975 as a result of exposure to phosphamidon and, to a lesser extent, fenitrothion (Pearce and Peakall, 1977). It is largely because of its effects on avifauna that phosphamidon is no longer used in spruce budworm spray programs.

It should be noted that the margin of toxicological safety for birds exposed to fenitrothion is rather small. Much greater mortality of adult and nestling white-throated sparrows was observed after a double application of this chemical, compared with the usual operational spray rate (Pearce and Busby, 1980; Busby *et al.*, 1983a). Overexposure to pesticides is not an infrequent occurrence, because of (1) the overlap of spray swaths caused by difficulties in the navigation of spray airplanes treating a large area of forest; (2) atypical atmospheric conditions that affect spray deposition; and (3) calibration and mixing errors in the delivery systems. The toxic effects of insecticides on forest birds can also be influenced by (1) spray droplet size and other effects related to the delivery system; (2) the particular pesticide and formulation that is being used; (3) foraging height and other avian behavioral traits that influence exposure; and (4) variations of physiological sensitivity among bird species (Busby *et al.*, 1981, 1982, 1983a,b; Grue *et al.*, 1983).

The effect of a spruce budworm spray with fenitrothion on the abundance of breeding birds in an infested forest in Quebec is described in Table 8.10. When considering these data, it is necessary to compare the relative trends of the experimental sprayed plot with those of the unsprayed reference treatment. The reason for this is that during the prespray census in mid-May not all of the migratory species had returned to this breeding habitat (in particular, Swainson's thrush and most of the warbler species). As a result, the total avian abundance at this time was smaller than the breeding density later in the season. If the data are considered with this relative comparison in mind, it is apparent that there was no clear effect of fenitrothion spraying on the abundance and species richness of the bird community of this fir–spruce forest.

There is a much larger effect on the bird community when the forest is badly damaged by an uncontrolled defoliation by spruce budworm. This is illustrated in Table 8.11, which compares the abundance of birds during and after a period of budworm infestation. Stand A suffered little long-term damage to trees as a result of the defoliation. The decreased density of many bird species reflects their numerical response to the reduction of insect prey after the collapse of the outbreak (e.g., solitary vireo, many warbler species). Stand A suffered intense damage to trees because of the budworm infestation, and death of the balsam fir overstory caused a large change in the plant species composition and physical structure of the habitat. The most important habitat changes were the development of a large density of snags, a dense stratum of angiosperm shrubs, a vigorous regeneration of small balsam fir trees of the understory, and a lush growth of monocotyledonous and dicotyledonous herbs. Therefore, in Stand B some bird species naturally declined in abundance because of the decreased abundance of budworm prey and other habitat changes (e.g., solitary vireo, Tennessee warbler, black-throated green warbler, Blackburnian warbler, and bay-breasted warbler). Other species responded positively to the habitat changes (e.g., least flycatcher, magnolia warbler, and white-throated sparrow).

Table 8.10 Effects of aerial spraying with fenitrothion on the populations of birds in a spruce budworm-infested fir–spruce forest in the Gaspé region of Quebec[a]

Species	Prespray		Postspray 1		Postspray 2	
	Sprayed Area	Reference Area	Sprayed Area	Reference Area	Sprayed Area	Reference Area
Boreal chickadee (*Parus hudsonicus*)	2.5	4.3	1.5	1.5	1.4	1.1
Winter wren (*Troglodytes troglodytes*)	1.0	3.9	2.3	4.6	1.7	3.0
American robin (*Turdus migratorius*)	3.5	3.6	4.8	3.3	5.3	1.8
Swainson's thrush (*Catharus ustulatus*)	0.0	0.0	0.0	0.6	3.7	2.0
Ruby-crowned kinglet (*Regulus calendula*)	8.6	4.5	5.9	3.8	4.4	6.4
Tennessee warbler (*Vermivora peregrina*)	0.0	0.0	0.3	1.5	2.3	5.8
Nashville warbler (*Vermivora ruficapilla*)	0.0	0.0	0.3	0.4	0.3	0.6
Magnolia warbler (*Dendroica magnolia*)	0.0	0.0	0.1	1.5	0.8	3.1
Cape May warbler (*Dendroica tigrina*)	0.0	0.0	2.0	3.1	4.3	4.0
Yellow-rumped warbler (*Dendroica coronata*)	0.5	0.3	7.7	6.3	10.8	8.2
Black-throated green warbler (*Dendroica virens*)	0.0	0.5	0.5	2.3	2.6	3.9
Bay-breasted warbler (*Dendroica castanea*)	0.0	0.0	0.3	0.3	2.5	4.5
Blackpoll warbler (*Dendroica striata*)	0.0	0.0	0.0	0.0	1.1	2.8
Northern waterthrush (*Seiurus noveboracensis*)	0.0	0.0	0.6	0.2	1.3	1.6
Evening grosbeak (*Hesperiphona vespertina*)	1.3	0.4	1.8	2.4	0.8	1.4
Purple finch (*Carpodacus mexicanus*)	1.3	0.7	1.1	1.2	1.1	2.8
Pine grosbeak (*Pinicola enucleator*)	0.6	0.5	1.3	0.9	0.2	0.4
Pine siskin (*Carduelis pinus*)	0.2	0.0	0.9	1.8	1.6	1.5
Dark-eyed junco (*Junco hyemalis*)	9.6	4.1	4.2	1.8	4.5	0.5
White-throated sparrow (*Zonotrichia albicollis*)	8.9	9.6	9.7	10.5	10.4	8.4
Fox sparrow (*Paserella iliaca*)	2.1	3.8	3.3	4.2	2.5	3.1
Total bird abundance	46.1	39.1	52.1	57.7	68.8	75.8
Number of species	23	23	41	35	41	39

[a]Birds were censused for 5 days prior to spraying, for 7 days after the first spray on May 21, 1976, and for 5 days after the second spray on May 30. The total fenitrothion spray was 0.56 kg a.i./ha. Bird density is expressed as number/10 ha. Only prominent species are listed. Modified from Kingsbury and McLeod (1981).

Table 8.11 Abundance of selected bird species in spruce budworm-infested balsam fir stands in New Brunswick[a]

Species	Stand A Outbreak	Stand A Post-outbreak	Stand B Outbreak	Stand B Post-outbreak
Yellow-bellied flycatcher (*Empidonax flaviventris*)	4.6	p	5.1	0.0
Least flycatcher (*Empidonax minimus*)	3.9	2.8	2.1	15.9
Boreal chickadee (*Parus hudsonicus*)	2.0	4.7	2.0	p
Winter wren (*Troglodytes troglodytes*)	5.9	8.2	2.2	3.3
Swainson's thrush (*Catharus ustulatus*)	15.1	11.9	14.0	8.7
Golden-crowned kinglet (*Regulus satrapa*)	8.9	3.0	2.3	0.0
Ruby-crowned kinglet (*Regulus calendula*)	p	10.1	1.1	2.7
Solitary vireo (*Vireo solitarius*)	8.3	p	7.7	0.0
Tennessee warbler (*Vermivora peregrina*)	9.9	0.0	25.7	0.0
Nashville warbler (*Vermivora ruficapilla*)	p	3.0	0.0	p
Magnolia warbler (*Dendroica magnolia*)	22.0	7.9	1.3	14.9
Yellow-rumped warbler (*Dendroica coronata*)	7.1	1.6	5.1	2.1
Black-throated green warbler (*Dendroica virens*)	11.9	4.2	11.8	4.2
Blackburnian warbler (*Dendroica fusca*)	17.1	5.3	19.5	1.5
Bay-breasted warbler (*Dendroica castanea*)	55.8	16.4	78.2	2.0
Blackpoll warbler (*Dendroica striata*)	p	4.9	1.1	5.1
American redstart (*Setophaga ruticilla*)	3.9	2.3	2.1	p
Dark-eyed junco (*Junco hyemalis*)	5.9	3.5	8.4	4.1
White-throated sparrow (*Zonotrichia albicollis*)	9.3	5.0	11.4	19.5
Total density	191.6	94.8	201.1	84.0

[a]Stand A was censused for 8 years up to 1959 during an epidemic, and then for 6 postoutbreak years. No long-term damage was caused to trees, and the average age of balsam fir was >120 years during both time periods. Stand B was censused for 5 epidemic years up to 1959, and then for another 5 postoutbreak years. Intense damage was caused in Stand B; the average age of balsam fir declined from an initial >80 years, to <10 years after the collapse of the outbreak. Bird data are in number/40 ha; p = present. Modified from Gage and Miller (1978).

Effects on Fish

During the era of DDT spraying in New Brunswick and elsewhere, fish kills were documented after spraying, and the population of sport fish was decreased in some areas (Kingsbury, 1975; Logie, 1975). Because of its commercial importance, Atlantic salmon (*Salmo salar*) was relatively intensively studied. Caged salmon in sprayed streams suffered a mortality rate of 63–91% within 3 weeks of spraying with DDT, compared with negligible mortality in unsprayed streams (Kingsbury, 1975). Young salmon were especially sensitive to DDT. After spraying, the abundance of stages younger than 1 year old was decreased by as much as 90%, while 2- and 3-year-old juvenile stages were re-

duced by 70% and 50%, respectively. In the Miramichi River of New Brunswick, the abundance of juvenile salmon (i.e., ≤ 3 years old) decreased from 67/100 m^2 of breeding habitat, to 15/100 m^2 after a spray with 0.5 kg DDT/ha (Logie, 1975). It was predicted that 2 further years of spraying with DDT at this rate would cause an additional decline in abundance, to 3.1/100 m^2 after 2 spray years, and to 1.0/100 m^2 after 3 spray years. Eventually, DDT spraying in the watershed of this river caused a 50–60% decrease in the number of salmon migrating from the sea. After the use of DDT against spruce budworm was suspended, the population of salmon in the Miramichi River recovered in about 3 years.

In contrast, the insecticides that replaced DDT in

the spruce budworm spray programs have caused no or few measurable effects on populations of freshwater fish (Kingsbury, 1975; Gillis, 1980).

Overview of Spruce Budworm

Overall, it appears that the measurable ecological impacts of the modern, post-DDT spray program against spruce budworm are rather short-term and moderate. The persistence of insecticide residues is short-lived, and there is no food-chain accumulation. Although there are effects on nontarget organisms, the damage does not appear to be "unacceptable" from a resource management perspective. For example, although spraying causes a large kill of nontarget insects and spiders, the abundance of these arthropods is not measurably affected. Similarly, although some forest birds are affected by pesticide toxicity under certain conditions, this does not measurably affect their breeding populations.

However, it must be stressed that the spray strategy has certain important drawbacks from a forest management perspective. It is true that spraying infested stands with insecticide will keep the forest "alive and green," and therefore available to supply the needs of the economically important forest industry. However, spraying also maintains the prime habitat of budworm in a susceptible condition, and thus the infestation may be prolonged. Therefore, once spraying is begun, agencies become "locked" into a spray strategy, and must continue the practice in order to maintain an economically viable forest resource. While this may be the best available short-term strategy, it is certainly not preferable on the longer term to have to mount an annual spray program over a large area. [However, note that Royama (1984) has a fundamentally different view. Although spraying causes short-term decreases in abundance of budworm and therefore prevents some damage, according to his theory of budworm dynamics the longer-term infestation will increase or decrease regardless of the nature of spray intervention.]

A possible alternative to spraying could involve some combination of forest management practices that reduce the susceptibility of stands to infesta-

tion. For example, less vulnerable species such as black spruce could be planted. In addition, the landscape could be structured so that there is not a large, continuous tract of mature (> 50 years old) stands. If a smaller-scale mosaic of less vulnerable stands were ideally structured, then the rate of harvesting by the forest industry could equal the rate at which stands mature during succession. In this way, the area vulnerable at any time to infestation by budworm could be reduced. However, nothing is known about the longer-term stability or viability of this potential forest management strategy (Baskerville, 1975b; Anonymous, 1976b; Batzer, 1976; Holling and Walters, 1977; Blum and MacLean, 1985).

An important perspective is gained from the observation that, in the short term, the ecological impacts of spraying are much less than those that take place when severely infested stands are not protected with insecticide. If an infested stand is not sprayed, a large proportion or all of the mature individuals of balsam fir may die, as may many of the spruce trees. For a number of years, the moribund forest of dead, dry, standing timber may present an explosive fire hazard, especially in the early spring before the flush of new angiosperm foliage has grown. In a controlled-burn experiment in Ontario, it was found that in a budworm-killed forest (1) ignition took place very readily; (2) the fire spread almost immediately to the tree crowns; (3) the rate of lateral spread was rapid, at up to 80 m/sec; (4) downwind "branding" was frequent as bark peeled, ignited, and was convectively transported away; and (5) a large fraction of the fuel was consumed by the intense combustion (Stocks, 1985, 1987). The fire hazard was severe for about 10 years after the trees died. Afterward, regeneration of the stand and the decomposition of dead trees reduced the fire hazard.

In addition, budworm-killed forest has a relatively small abundance of arthropods, birds, and mammals for as long as a decade or more after the death of the stand. After this time, a vigorous natural regeneration restores the quality of the habitat and wildlife populations recover. As might be expected, during the initial years of recovery the wildlife is dominated by early successional species. It takes several decades for the forest to recover to the degree that it provides habitat for the late successional species that were present before and during the budworm infestation (see Chapter 9 for a parallel discussion of the effects of forest harvesting by humans). Overall, the changes in the abundance and species composition of wildlife that take place in a budworm-killed forest are much larger than those that are caused by the post-DDT spray programs.

From a forest management perspective, it is not desirable to allow a large tract of mature forest to be devastated by defoliation by spruce budworm. Such an event causes great difficulties of supply of raw materials for the forest industry, and thereby causes great economic disruption. In 1976, the government of Nova Scotia decided to not allow the spraying of insecticides to reduce the impacts of an infestation of spruce budworm on the highlands of Cape Breton Island. As a result, a large area of infested forest died, comprising most of the resource base of the province's largest pulp mill. The total loss of wood was about 21.5 million m^3, equivalent to 10% of the total softwood resource of Nova Scotia, and half that of Cape Breton Island before the budworm infestation (Bailey, 1982). The longer-term economic implications of this event are not yet clear. The pulp mill has managed to keep operating, in part by processing more costly wood obtained elsewhere.

In addition, the industry and provincial government were faced with the fact that there was a large area of moribund and dead forest on Cape Breton Island. If harvested quickly, that wood was potentially useful for some purposes, and tremendous skyline-to-skyline clear-cuts were used to salvage some of the killed timber. Within the 5-year period of the salvage program, about 3.4 million m^3 of wood was recovered from a clear-cut area of about 16,000 ha. The largest contiguous clear-cut is an impressive 6,000 ha (E. Bailey, personal communication). There were important, but essentially undocumented, environmental consequences of the large-scale, intensive salvage harvesting of the dead forest of Cape Breton Island. If the forest had been protected with insecticide during the budworm in-

festation, then these extensive clear-cuts would not have been required, and a mature forest would still blanket the highlands of Cape Breton. However, that forest might still be infested by spruce budworm, and a yearly spray program would have been required to keep it alive. [However, note that according to Royama (1984) the epidemic might have collapsed, even if susceptible stands were kept alive by spray intervention.] Clearly, the forest management decision is difficult: "To spray, or not to spray."

Silvicultural Use of Herbicides

Silvicultural Objectives

The most frequent objectives of the use of herbicides in forestry are (1) to release young conifers from competition with economically undesirable (at least from the forestry perspective) angiosperm species or (2) to prepare a site for the planting of young conifers. By these uses, herbicides can help to temporarily or permanantly reduce the potential dominance of productive forest sites by unwanted "weedy" vegetation, while allowing the desired conifer regeneration to more rapidly attain a greater site dominance (Newton, 1975; Daniel *et al.*, 1979; Newton and Knight, 1981; Wenger, 1984; Malik and Vanden Born, 1986). Other relatively minor uses of herbicides in forestry are for the spraying of vegetation beside forest roads, and for the manipulation of the habitat of wildlife that can injure conifer regeneration by browsing, for example, rabbits and hares. Forestry usage comprises a relatively small proportion of the total use of herbicides in most areas, for example, 3.6% of the quantity of phenoxy herbicides used in Ontario in 1978, 5.1% in the United Kingdom in 1979, and less than 5% in the United States in 1980 (Kilpatrick, 1980; Mullinson, 1981; Stephenson, 1983). If herbicides other than the phenoxys are considered, the relative usage in forestry is even smaller.

In meeting the silvicultural objectives described above, herbicides have important advantages over alternative procedures such as mechanical scarification, the removal of vegetation by hand cutting, and

The vegetation of a 4-year-old clearcut of a coniferous forest in Nova Scotia. The dominant plants in this vigorously regenerating ecosystem are red raspberry (*Rubus strigosus*), pin cherry (*Prunus pensylvanica*), red maple (*Acer rubrum*), hay-scented fern (*Dennstaedtia punctilobula*), and various herbs of the Asteraceae. The regeneration of commercially desired conifers can be inhibited or prevented by this intensely competitive "weedy" vegetation. An increasingly frequent silvicultural prescription for this sort of site is an aerial herbicide spray treatment. (Photo courtesy of B. Freedman.)

prescribed burning. These advantages include (1) economy and safety, especially in comparison with manual treatments; (2) time savings, in that a large area can be treated quickly; (3) fewer retreatments to control the regeneration of weeds; and (4) selectivity, so that damage to conifers can be minimized. Important disadvantages of the use of herbicides for silvicultural purposes include (1) success in meeting the silvicultural objective depends on the appropriate selection of a herbicide, its formulation, and

application rate—if these are not planned and executed properly, then conifers may be damaged, or weeds may not be adequately controlled; (2) the timing of the application is critical to maximizing conifer tolerance while still achieving control of weeds; (3) there can be nontarget, off-site damage because of the drift of herbicide spray; (4) fewer people are employed in vegetation management programs; (5) herbicide application may detract from the ability of weedy plants to serve as a biological "sponge" for mobile nutrients such as nitrate, which might otherwise leach from the disturbed site (see Chapter 9); and (6) there is often a strongly negative social reaction to the broadcast spraying of herbicides (and insecticides) in forestry.

Controversy

The reasons for the social controversy are various, but they include fears of (1) potential epidemiological consequences to humans, caused by the exposure to herbicides; (2) the loss of employment opportunities, which are much greater if an alternative method such as manual treatment is used; (3) a landscape dominated by a monoculture of conifers; and (4) the degradation of wildlife habitat [Daniel *et al.*, 1979; General Accounting Office (GAO), 1981; Newton and Knight, 1981; Freedman, 1982; Wenger, 1984].

In some areas, the opponents of herbicide spraying in forestry have caused the interruption, reduction, and even cessation of this silvicultural practice. There have also been isolated incidents of vandalism of spray equipment. In Nova Scotia there was a case in which a spray apparatus was damaged, causing herbicide to leak onto the ground and then into a stream. In other places, valuable machinery has been broken, and some equipment has been burned. In a few cases, groups of citizens have filed lawsuits and obtained injunctions against the silvicultural use of herbicides. In 1982–1983 a coalition of 15 landowners took Nova Scotia's largest forest corporation to trial, in order to stop a proposed spray program using 2,4,5-T and 2,4-D. That legal action stimulated intense public interest and discussion, and there was a 1-month trial with testi-

An aerial application of herbicide to a regenerating clearcut, in a silvicultural conifer release program in Nova Scotia. The helicopter has to fly at a relatively high altitude of more than 20–30 m because of the uneven terrain, the surrounding forest, and the presence of residual uncut trees on the clearcut. Partly as a result of this, there can be a considerable drift of herbicide, and nearby off-target vegetation can suffer some damage. (Photo courtesy of B. Freedman.)

A 4-year-old clearcut of a coniferous forest in Nova Scotia, 1 year after it was aerially treated with the herbicide glyphosate. The extensively defoliated stump-sprout shrubs are *Acer rubrum*, which was considered to be one of the most important "weeds" on the site. The abundance of all other plants (except conifers) was initially drastically reduced, but after only 1 postspray year, there was a substantial recovery of the weedy vegetation. (Photo courtesy of B. Freedman.)

mony from 49 witnesses, including international environmental and medical experts. In a 182-page decision, the judgement went strongly against the plaintiffs, and the legal conclusion was that 2,4,5-T and 2,4-D could be used safely for forestry purposes in Nova Scotia (Nunn, 1983). However, the silvicultural use of herbicides in Nova Scotia and elsewhere has remained a controversial and emotional issue, with a strong polarization of the positions of the opponents and proponents of the practice.

In the following, we will discuss some issues that are related to the silvicultural use of herbicides, with particular focus on the ecological aspects.

Weeds in Forestry

After the disturbance of a forested site by fire or harvesting, the regeneration is comprised of many plant taxa that compete for space and other resources. In the first decade or so of secondary succession, the young community is dominated by plants other than the conifers that are desired by foresters. Most important in this respect are various perennial herbs and woody angiosperm species. This is illustrated for some 4- to 6-year-old clearcuts of conifer forest in Nova Scotia (Table 8.12). Among the four sites, the economically desired regeneration of spruce and fir comprise only 3–8% of the total plant cover. Much more prominent in the young seral community are ferns, monocots such as

sedges and grasses, dicotyledonous herbs (various Asteraceae were most prominent), raspberries and blackberries (*Rubus* spp.), and woody dicot species, especially birch (*Betula* spp.), red maple (*Acer rubrum*), and pin cherry (*Prunus pensylvanica*). In such a situation, the degree of site domination by the "undesirable" plants can be sufficient to inhibit the establishment and growth of the commercially desired conifers.

The ecological "strategy" of weedy plants is to take advantage of the relatively great abundance of environmental resources that is available for a short time after the death of the forest overstory. After this initial stage of the postdisturbance secondary succession, competition from the regenerating forest canopy eliminates most of the relatively intolerant, ruderal species from the site.

Some of the species that dominated the regenerating cutovers in Nova Scotia (Table 8.12) were also present in the mature, uncut forest. They managed to survive the disturbance and, because they experienced a release from the competitive stress exerted by the previous conifer overstory, they expanded their relative degree of site dominance. Examples of such long-term site occupants are the hay-scented fern (*Dennstaedtia punctilobula*), and the tree *Acer rubrum*, which regenerates after cutting by a prolific stump-sprouting. Other ruderals establish from a persistent seedbank (e.g., *Prunus pensylvanica* and *Rubus* spp.), or by

Table 8.12 Dominant vegetation of four 4- to 6-year-old conifer clear-cuts in Nova Scotia[a]

	Site 1	Site 2	Site 3	Site 4
Bryophytes	7	5	15	3
Lichens	1	<1	1	<1
Pteridophytes	22	29	15	1
Conifers	3	5	8	4
Monocots	9	5	5	26
Dicots, herbs	17	19	14	39
Dicots, woody shrubs	21	16	31	10
Dicots, *Rubus* spp.	20	21	11	17
Total plant cover (absolute)	137%	153%	110%	113%

[a]Data are relative cover (i.e., percent of total cover), average of 48 1-m² quadrats per site.
Modified from Morash and Freedman (1987).

an influx of airborne seed to the disturbed site (e.g., Asteraceae, *Betula* spp., *Populus* spp.).

Many studies have shown that competing vegetation has the capacity to reduce the productivity of tree species, and that the growth of conifers can be increased by a reduction of this stress (see Stewart *et al.*, 1984, for a detailed bibliography of the effects of competition on conifers). This effect can be illustrated by the study of MacLean and Morgan (1983), who examined the vegetation of a site that had been treated with 2,4,5-T and 2,4-D 28 years previously (Table 8.13). Before they were sprayed, all three study plots had a vigorous growth of competing angiosperm vegetation, which formed a closed canopy at a height of about 2 m that overtopped the shorter conifer species. The average density of raspberry (*Rubus* spp.) was 81,000 stems/ha, mountain maple (*Acer spicatum*) was 7000/ha, and other angiosperm shrubs totaled 2500/ha. The silvicultural herbicide treatment released the suppressed conifers from some of the stress of competition with intolerant hardwoods, and 28 years after spraying the volume of balsam fir on the two treatment plots averaged about three times larger that on the unsprayed reference plot, while its diameter averaged 21% greater. Overall, conifers comprised a much larger percentage of the stand basal area in the spray plots. Therefore, the herbicide treatment resulted in a more rapid establishment of a conifer-dominated forest, effectively shortening the length of the harvest rotation. It is important to note that herbicide spraying did not eliminate the weedy angiosperms from the spray plots; after 28 postspray years, hardwoods still averaged 6–13% of the total basal area, compared with 56% on the reference plot. It is also important to realize that because the most prominent of the competing hardwoods are intolerant species, they will eventually be eliminated from even the reference plot by the conifers, which are much more tolerant of the stressful conditions of a closed forest.

Of course, the broadcast spraying of herbicides on regenerating cutovers can have important ecological consequences in addition to the release of economically desirable conifers from competition with weeds. These effects include direct toxicity to nontarget vegetation and other biota, and an indirect influence on animals caused by changes to their habitat. These are discussed below.

Persistence of Herbicide Residues

Generally speaking, the herbicides that are used in forestry have a moderate persistence in the ter-

Table 8.13 Effects of suppressing hardwoods on the plant community composition and growth of balsam fir in New Brunswick[a]

	Herbicide Plot A	Herbicide Plot B	Reference Treatment
Total tree density (stems/ha)	4200	4460	3350
Total tree basal area (m^2/ha)	44.7	36.6	23.3
Total volume, balsam fir (m^3/ha)	191	135	52
Mean diameter, balsam fir (cm)	10.8	9.3	8.3
Species composition (% of basal area)			
Balsam fir (*Abies balsamea*)	85	83	42
Spruce (*Picea* spp.)	9	4	2
Pin cherry (*Prunus pensylvanica*)	2	2	19
White birch (*Betula papyrifera*)	2	3	30
Mountain ash (*Sorbus americana*)	2	8	7

[a]The study plots were three 0.4-ha, 6-year-old clear-cuts initially with 9500 stems/ha of hardwood trees and 81,000 stems/ha of *Rubus* spp. Two plots were treated with a 50 : 50 mixture of 2,4,5,-T and 2,4-D, and the other plot was a reference treatment. This table summarizes characteristics in 1981, 28 years after herbiciding. Modified from MacLean and Morgan (1983).

restrial environment, so that their residues decrease to a small concentration within a few months of application [Norris *et al.*, 1977, 1982; Norris, 1981; Smith, 1982; U.S. Department of Agriculture (USDA), 1984]. Norris *et al.* (1977) monitored the residues of 2,4,5-T after an aerial application in Oregon. The concentration in graminoid plants, which are not killed by this herbicide, decreased from an initial postspray value of 120 µg/g, to 3.4 µg/g after 1 month, 0.58 µg/g after 3 months, 0.14 µg/g after 6 months, 0.12 µg/g after 1 year, and to an undetectable concentration after 2 years. There was a similar pattern of disappearance in conifer foliage. Morash *et al.* (1987) studied the persistence of glyphosate, another herbicide that is commonly used in forestry. The initial postspray residue was relatively small in conifer foliage (0.1 µg/g), but it was larger in the foliage of target species (5.2 µg/g in maple, 13 µg/g in graminoids and raspberry). The residues decreased fairly quickly in all types of foliage. In graminoid foliage, the concentration decreased to 1.4 µg/g after 1 week, to 0.76 µg/g after 2 weeks, and to 0.46 µg/g after 8 weeks.

Changes in Vegetation

The objective of the silvicultural use of herbicides is to change the character of the vegetation on the site. We previously described an example of the longer-term change in woody vegetation (Table 8.13). An example of the short-term effects on the plant community is summarized in Table 8.14 for two young conifer clear-cuts that were sprayed with glyphosate. On both clear-cuts, the vegetation of the reference treatment exhibited a vigorous and progressive development over the 3-year study period. In contrast, on the herbicided plots the abundance of the most important competing plants (woody shrubs and *Rubus* spp.) was decreased for a short time. In effect, the herbicide treatment set back the postcutting secondary succession to an earlier stage of development, while releasing small conifer trees for that period of time. Morash and Freedman (1987) found that no plant species were eliminated within a grid of permanent quadrats located in the herbicided clear-cuts. However, there was of course a large

change in the relative abundance of the various plant species, because of (1) differences in their susceptibility to glyphosate and (2) differences in their rate of postspraying recovery by the growth of surviving plants and by the establishment of new individuals from seed.

It is typical that a mixed community of plants develops on the site after a silvicultural herbicide treatment (MacLean and Morgan, 1983; Cain and Yaussy, 1984; Morrison and Meslow, 1984a,b). In itself, the silvicultural use of herbicides is not intended to produce an absolute softwood monoculture. Rather, this forestry practice is used to enhance the rate of growth of conifers in the regenerating forest, so that the harvest rotation can be shortened.

Effects of Silvicultural Herbicide Use on Wildlife

Toxicity to Animals of Silvicultural Herbicides. All pesticides are toxic substances, and they should be handled with caution in order to avoid an occupational exposure to applicators and to avoid unneccesary environmental contamination. However, compared with insecticides and other pesticides that are targeted to animals, most herbicides have a relatively small toxicity to vertebrate wildlife (Table 8.2). In fact, all of the herbicides that have been frequently used in forestry in North America (i.e., 2,4,5-T, 2,4-D, glyphosate, hexazinone, atrazine, and some other relatively minor chemicals in terms of quantity used) have a toxicity to vertebrates that is less than that of caffeine and nicotine (Table 8.2). This observation does not, of course, mean that the use of herbicides is without toxicological risk. However, it does help to put risks into perspective. Because of the relatively small toxicity of most herbicides to animals, it can generally be concluded that at the rate of application used for typical forestry applications, a direct toxic effect is unlikely (Kenaga, 1975; Morrison and Meslow, 1983; Norris *et al.*, 1983; USDA, 1984).

Because herbicide use in forestry has been such a high-profile environmental issue, the practice has been closely scrutinized with respect to the occupa-

Table 8.14 Changes in vegetation after the aerial application of the herbicide glyphosate in a silvicultural conifer release program in Nova Scotia[a]

Vegetation Category	Prespray Cover	First Year Postspray Cover (% change)	Second Year Postspray Cover (% change)
Site A: (1) herbicided plot			
Monocots	8	8 (0)	14 (+75)
Dicot shrubs	29	12 (−59)	14 (−52)
Rubus spp.	28	9 (−68)	20 (−29)
Dicot herbs	21	24 (+14)	39 (+86)
Other plants[b]	10	12 (+20)	18 (+80)
Total	96	65 (−32)	105 (+9)
Site A: (2) reference plot			
Monocots	5	6 (+20)	8 (+60)
Dicot shrubs	13	28 (+115)	34 (+161)
Rubus spp.	17	20 (+18)	26 (+53)
Dicot herbs	12	21 (+75)	28 (+133)
Other plants[b]	32	41 (+28)	45 (+41)
Total	79	116 (+47)	141 (+78)
Site B: (1) herbicided plot			
Monocots	12	3 (−75)	16 (+33)
Dicot shrubs	38	11 (−71)	14 (−63)
Rubus spp.	12	1 (−92)	9 (−25)
Dicot herbs	11	8 (−27)	17 (+55)
Other plants[b]	26	15 (−42)	21 (−19)
Total	99	38 (−62)	77 (−22)
Site B: (2) reference plot			
Monocots	7	6 (−14)	8 (+14)
Dicot shrubs	19	30 (+58)	43 (+126)
Rubus spp.	2	4 (+100)	10 (+400)
Dicot herbs	8	16 (+100)	20 (+150)
Other plants[b]	37	46 (+24)	59 (+59)
Total	73	102 (+40)	140 (+92)

[a] Spray rates were 1.4 kg a.i./ha. Vegetation data are percent cover grouped by major classes of competing vegetation, average of 40 permanent quadrats per treatment. Change is expressed as postspray values relative to prespray. Site A was a 3-year-old clear-cut dominated by red maple and herbaceous dicots. Site B was a 3-year-old clear-cut dominated by ericaceous shrubs. Modified from Morash and Freedman (1987).

[b] Sum of bryophytes, lichens, pteridophytes, and conifers.

tional and more general epidemiological hazards to humans. The issue of human health hazards is controversial and has not been resolved to everyone's satisfaction. (Because of the nature of the scientific investigation of epidemiological problems, few such controversial issues are ever fully resolved.) This issue will not be dealt with here in detail. However, it is notable that a number of critical reviews of the available scientific information by government and medical task forces in various parts of the world have all concluded that when silvicultural herbicides are used in accordance with the recommended operational guidelines, the practice is safe for both sprayers and people living in the vicinity of a spray area [McQueen *et al.*, 1977; Turner, 1977; Kilpatrick, 1979, 1980; Royal Society of New Zealand (RSNZ), 1980; Beljan *et al.*, 1981; Mullinson, 1981; Walsh *et al.*, 1983; Hatcher and White, 1984; Walstad and Dost, 1984; USDA, 1984]. This appears to be the "mainstream" scientific opinion on this matter. For alternative viewpoints, see Warnock and Lewis (1978), Whiteside (1979), or Hay (1982). Also, see Chapter 11 for a brief discussion of the toxicology of the very poisonous trace contaminant of 2,4,5-T known as TCDD, a dioxin isomer.

Therefore, the direct, toxic effect of herbicides on wildlife is not very important. However, herbicide use causes large changes in the physical structure and species composition of the plant community. We might anticipate that these habitat changes would cause secondary effects on wildlife.

Effects on Birds and Their Habitat. Several studies have investigated the effects of silvicultural herbicide use on the abundance of breeding birds. Morrison and Meslow (1984a) examined clear-cuts in Oregon that had been sprayed with 2,4-D or with a mixture of 2,4-D and 2,4,5-T, along with an un-

sprayed reference treatment (Table 8.15). One year after spraying with 2,4-D, and 4 years after spraying with 2,4,5-T plus 2,4-D, there was no large difference in the abundance, species richness, or diversity of birds between the herbicided and reference treatments. (Note that because of the study design, we do not know how similar these plots were before they were sprayed.) Certainly, any effects that might be attributable to herbiciding are much smaller than the large changes in the avifauna that must have taken place when the original, mature conifer forest was clear-cut (see Chapter 9).

The study design that Morrison and Meslow (1984b) used to examine the effects of silvicultural glyphosate spraying was better than that just described, because there are prespray data that can be used to evaluate the initial similarity of the treatment and reference plots (Table 8.16). In this case, the two plots were similar in the species composition, total abundance, and diversity of birds, but the relative abundance of particular species differed somewhat. In the first postspray year, there was a decrease in the total abundance of birds on both the herbicided and the reference plot, and no large differences in the relative abundance of particular bird

Table 8.15 Effects of herbicide treatment of clear-cuts in western Oregon on breeding birds[a]

Species	1 Year after 2,4-D Treatment		4 Years after 2,4,5-T/2,4-D Treatment	
	Spray	Reference	Spray	Reference
Willow flycatcher (*Empidonax traillii*)	37	34	26	29
American goldfinch (*Carduelis tristis*)	33	48	34	23
Rufous-sided towhee (*Piplio erythrophthalmus*)	46	23	28	45
Dark-eyed junco (*Junco hyemalis*)	21	13	6	19
White-crowned sparrow (*Zonotrichia leucophrys*)	39	72	76	23
Song sparrow (*Melospiza melodia*)	45	60	48	41
Swainson's thrush (*Catharus ustulatus*)	52	64	40	42
MacGillivray's warbler (*Oporornis tolmiei*)	20	29	21	22
Orange-crowned warbler (*Vermivora celata*)	32	30	27	34
Wilson's warbler (*Wilsonia pusilla*)	17	25	12	37
Rufous hummingbird (*Selasphorus rufus*)	48	43	54	72
Total density	416	460	396	410
Species richness	14	14	14	14
Species diversity (H')	2.5	2.4	2.4	2.5

[a]Data are in units of birds/40.5 ha. Modified from Morrison and Meslow (1984a).

Table 8.16 Effect of glyphosate treatment of a clear-cut in western Oregon on the density of breeding birds[a]

Species	Prespray		1st Year Postspray		2nd Year Postspray	
	Spray	Reference	Spray	Reference	Spray	Reference
Rufous hummingbird (*Selasphorus rufus*)	74	37	54	37	63	40
MacGillivray's warbler (*Oporornis tolmiei*)	24	36	10	33	44	37
Wilson's warbler (*Wilsonia pusilla*)	11	25	10	31	13	35
Rufous-sided towhee (*Piplio erythrophthalmus*)	35	16	17	13	75	41
White-crowned sparrow (*Zonotrichia leucophrys*)	86	116	33	62	56	59
Willow flycatcher (*Empidonax traillii*)	31	39	14	10	24	40
Orange-crowned warbler (*Vermivora celata*)	56	20	63	15	34	22
Dark-eyed junco (*Junco hyemalis*)	30	14	14	5	31	12
American goldfinch (*Carduelis tristis*)	42	48	40	21	30	32
Swainson's thrush (*Catharus ustulatus*)	27	63	28	63	83	131
Song sparrow (*Melospiza melodia*)	58	57	32	31	81	102
Bewick's wren (*Thryomanes bewickii*)	4	2	3	8	6	4
American robin (*Turdus migratorius*)	5	7	3	3	6	13
Black-headed grosbeak (*Pheucticus melanocephalus*)	5	11	1	2	10	20
Total density	488	491	322	334	556	588
Species Diversity (H')	2.4	2.3	2.3	2.3	2.4	2.4

[a]Data are in units of birds/40.5 ha. Modified from Morrison and Meslow (1984b).

species. Therefore, any effects attributable to herbiciding appear to be small. In the second postspray year, the total abundance of birds was larger on both plots, but again there was no large change in the relative abundance of various species. The rather small effects on the avifauna that are apparent in Tables 8.15 and 8.16, occurred in spite of large changes in vegetation that were caused by the herbicide treatment. However, these changes in habitat were not sufficiently large to cause a measurable change in the bird community. Other field studies of the effects of silvicultural herbicide use have also reported a fairly small effect on breeding birds (Beaver, 1976; Slagsvold, 1977; Freedman *et al.*, 1988a; Mackinnon and Freedman, 1988). An exception is the study of Savidge (1978) of the use of 2,4,5-T in eastern California. They found that total bird abundance was only 30.9/10 ha on a 6-year-old herbicided plot, compared with 65.0/10 ha on a reference plot, while species richness was 8 versus 14.

It might also be mentioned that the selective injection of systemic herbicides into individual trees has been investigated as a management tool for use in the production of standing dead trees suitable for excavation by woodpeckers and other species (Conner *et al.*, 1983). As is discussed in Chapter 9, snags are a very important habitat feature for many species of wildlife.

Effects on Mammals and Their Habitat. Studies of the effects of silvicultural herbicide use on deer have largely focused on the availability of browse on sprayed clear-cuts, rather than directly on the abundance of the animals (they are very difficult to census). Since angiosperm shrubs are important weeds in forestry, and also a preferred browse of *Odocoileus* deer, it should be anticipated that the quantity of at least some browse species would be reduced on sprayed clear-cuts. Lyon and Mueggler (1966) examined the effects of spraying with 2,4,5-T plus 2,4-D in Idaho, and found a decreased abundance of the shrub *Ceanothus sanguineus*, the most desirable browse species. Savidge (1978) also found a decreased abundance of important browse species (*Ceanothus velutinus* and *Symphoricarpos*

spp.) on plots sprayed with 2,4,5-T in California. There was also a reduced abundance of fecal pellet groups (an index of abundance) of mule deer (*Odocoileus hemionus*). In another study of 2,4-D spraying, Krefting and Hansen (1969) found that the most affected shrubs (*Corylus* spp.) were not desirable as browse for deer. In their study area, the herbicide treatment stimulated the production of desired browse species and of grass. As a result, white-tailed deer (*Odocoileus virginicus*) were attracted to the sprayed area for winter browsing and summer bedding, and pellet group counts were more abundant in the herbicided area for 8 postspray years. Therefore, the effects of herbicide treatment on deer browse appear to be site-specific, and to make a prediction for management purposes knowledge is required of the relative susceptibility of particular browse species to the herbicide that is being used.

Studies have also been made of the effects of silvicultural herbicide use on small mammals. Both Borrecco *et al.* (1979) and Freedman *et al.* (1988a) found a small, sporadic effect. However, Savidge (1978) found that small mammals were about twice as abundant on a sprayed plot, a phenomenon that was attributed to more favorable habitat because of an increased abundance of composites, graminoids, and the dicot shrub *Ribes*.

In an interesting but very artificial study, Thalken and Young (1983) censused small mammals at a site in Florida that had been used to test aerial spray delivery systems for the military herbicide program that was used in the Vietnam War (see Chapter 11). The 172-ha study area had received a massive dose of 2,4,5-T that was equivalent to 426 kg a.i./ha, from numerous spray trials between 1962 and 1970 (this is more than 100 times the average rate of application during a single conifer release treatment in forestry). Because the spray area had been cleared of the regional pine–oak forest, and since monocots are not very sensitive to 2,4,5-T, the vegetation of the spray test area developed into a distinct habitat island dominated by the grasses *Andropogon virginicus*, *Panicum virgatum*, and *P. lanuginosum*. In studies done between 1973 and 1978, a total of 341 species of organism was identified on the intensively sprayed test area, including an abundant population of beach mice (*Peromyscus polionotus*). Studies of the residues of TCDD and histopathological examination of 225 mice indicated a minimal effect on the health and reproduction of this species. The small effects on beach mice, plus the general abundance and richness of biota on what must be the world's most intensively herbicided site, are notable.

Aquatic Effects.   Finally, mention should be made of the apparent effects that occur in aquatic systems as a result of the use of herbicides in forestry. The most frequently used herbicides in silviculture are not very mobile in soil, and thus most aquatic contamination takes place from the direct deposition of spray to the water body. Since direct spraying over water is usually avoided in forestry spray programs, most investigations of operational spraying have not found detectable residues in water in the vicinity of sprayed areas (Matida *et al.*, 1975; Norris *et al.*, 1982, 1983; Morash *et al.*, 1987). Norris *et al.* (1982) studied an experimental situation in Oregon where direct deposition was not avoided, and found that stream-flow discharge contained 0.35% of the aerially applied picloram, and 0.014% of 2,4-D. This contamination took the form of a well-defined, short-lived pulse.

Reviews of aquatic toxicological data for the most frequently used herbicides in forestry indicate that the biological implications of aquatic contamination appear to be small (Norris *et al.*, 1983; USDA, 1984). In a field experiment, Matida *et al.* (1975) applied 2,4,5-T plus 2,4-D at 6.0 kg a.i./ha to a 9.5-ha forested watershed in Japan. The application did not result in a detectable residue in stream water (the detection limit was 0.06 ppm). No statistically significant changes were measured in the abundance of benthic invertebrates, and there was no mortality of fish.

Integrated Pest Management

It could be concluded that the documented ecological impacts of the post-DDT insecticide spray pro-

grams against spruce budworm, and of the silvicultural use of herbicides, are relatively small. Certainly, many politicians and public and private sector resource managers have decided that, at least in the short term, the environmental "costs" of these programs are "acceptable" in view of their important management and economic benefits. Of course, similar decisions have been made for the much larger-scale and more intensive spray programs that are annually mounted in most modern agroecosystems. However, it is arguable whether it is desirable on the longer term to rely on the broadcast spraying of wide-spectrum pesticides to cope with resource management problems. It would be much better if less reliance could be placed on such nonspecific practices.

A preferable approach is integrated pest management (IPM). Within the context of IPM, acceptable pest control is achieved by employing an array of complementary approaches. These can include (1) the use of natural predators, parasites, and other biological controls; (2) the use of pest-resistant varieties of crop species, which can be produced using standard breeding techniques, and also using modern, high-tech genetic engineering techniques (e.g., Fischhoff *et al.*, 1987); (3) the modification of environmental conditions so as to reduce the optimality of the pest habitat; (4) a careful monitoring of pest abundance; and (5) the use of pesticides only when they are required as a necessary component of the integrated pest management strategy (CEQ, 1979; Bottrell and Smith, 1982; Flint and van den Bosch, 1983). If successfully implemented, an IPM program can greatly reduce, but not necessarily eliminate, the reliance on pesticides. For example, the widespread use of an IPM scheme for the control of boll weevil (*Anthonomus grandis*) in Texas cottonfields was largely responsible for a reduction of insecticide use for this purpose from 8.77 million kg in 1964, to 1.05 million kg in 1976 (Bottrell and Smith, 1982).

An important component of IPM is the use of procedures that are as pest-specific as possible, so that nontarget damage can be avoided or reduced. There are precedents for such pest-specific practices in resource management. Some examples are the biological control of certain introduced pests in agriculture (Swan and Papp, 1972):

1. The cottony-cushion scale (*Icyera puchasi*) was a serious threat to the citrus industry in the United States, where it had been introduced from its native Australia. However, virtually total control of this pest was achieved by the introduction in 1888 of the vedalia lady beetle (*Vedalia cardinalis*) and a parasitic fly (*Cryptochetum iceryae*) from Australia.
2. The Klamath weed or common St. John's wort (*Hypericum perforatum*) was introduced from Europe to North America, where it became a serious weed of pastures of the U.S. southwest because of its toxicity to cattle. This pest was successfully controlled by the introduction in 1943 of the herbivorous leaf beetles *Chrysolina hyperici* and *C. gamelata*.
3. The prickly pear cactus (*Opuntia* spp.) was imported from North America to Australia for use as an ornamental plant, but it became a serious weed of rangelands. This pest has almost totally been controlled by the introduction of the moth *Cactoblastis cactorum*, whose larvae feed on the cactus.

There are also several cases of the successful, species-specific biological control of native pests of livestock and crops:

1. The common vampire bat (*Desmodus rotundus*) is a serious pest of livestock in Central and South America. It can be controlled by capturing individual bats and treating them with a topical application of petroleum jelly containing the anticoagulant diphenadione. The pesticide is spread to other bats during social grooming in cave roosts. Other bat species are not affected by the treatment, including two other relatively uncommon and specialized species of vampire bat (*Diphylla acaudata* and *Diaemus youngii*) (Mitchell, 1986).
2. The screw-worm (*Callitroga hominivorax*) is a serious pest of cattle, because of

damage done when its larvae feed on open wounds. In some areas this fly has been successfully managed by the release of a large number of sterile individuals into the wild population. The sterile flies were produced by the gamma-irradiation of populations that were mass reared in the laboratory. Since the female fly only mates once, a copulation with a sterile male does not result in successful reproduction. The technique operates by the swamping of the wild population of the screw-worm with sterile males, so that there are relatively few successful matings, and the abundance of the pest declines to an acceptable level (Baumhover *et al.*, 1955).

Unfortunately, biological control has not been successful in the majority of cases in which it has been attempted, and the technique may not be suitable for all pest problems. This seems to be especially true of forest pest problems, such as spruce budworm and competing weeds. In the case of spruce budworm, there has been active investigation of the potential roles of bacterial, viral, and other pest-specific disease agents, of the use of parasitoids, and of the use of species-specific sex pheromones to disrupt mating. With the exception of the use of a pesticide formulated with the bacterium B.t. (discussed earlier), these biological methods have not yet proved to be sufficiently successful to serve as a viable alternative to the broadcast spraying of synthetic pesticides (Morris, 1982; Hulme *et al.*, 1983; Miller *et al.*, 1983; Sanders *et al.*, 1985). However, this field of research is being actively pursued, and hopefully in the future biological control will prove to be more useful. This will probably take place in conjunction with other methods in a program of integrated pest management, for example, the management of habitat to make it less susceptible to budworm infestation (Schmidt *et al.*, 1983). However, in the short term IPM is not yet a viable alternative to the presently used control practices for spruce budworm. The broadcast spraying of insecticides to control this pest will continue, as will the use of herbicides in silviculture, and the use of pesticides for many other pest management problems.

9

HARVESTING OF FORESTS

9.1

INTRODUCTION

Forests cover about one-third of Earth's land surface. A closed-canopy forest covers about 4–5 billion ha, while about 2 billion ha is relatively open woodland and savannah (Tables 9.1 and 9.2). Geographically, the most heavily forested regions are in North and South America, Europe, and the Soviet Union, each with more than 30% forest cover (Table 9.1). Temperate plus boreal forests cover a comparable area to that of tropical forest, but their production is only about 50% as large, and there is only 61% as much total biomass (Table 9.2).

Humans clear forests for many reasons. Most important are the creation of new agricultural land, and the harvesting of biomass to manufacture lumber and paper and to burn to produce energy. Globally, a huge area of mature, forested ecosystems is impacted each year by these activities. In 1978 about 20 million ha of forest was cleared, and the global yield of wood was about 2.8 billion m³. The 1980 global consumption of wood included 455 million m³ of sawn timber (26% of which was used in North America), 109 million m³ of wood panels such as plywood (38%), 180 million metric tons of paper (39%), and 1450 million m³ of fuel-wood (about 85% of which was consumed in underdeveloped countries) [Food and Agriculture Organization (FAO), 1982; WRI, 1986]. The 1978 global harvest of 20 million ha represented an increase from 16×10^6 ha in 1950, 10×10^6 ha in 1900, and 6×10^6 ha in 1800 (FAO, 1982; Houghton et al., 1983).

In many countries, the forest resource is being severely depleted by an excessively large harvest. For example, between 1981 and 1985 the annual rate of forest clearing in Central America and northern South America ranged from 0.9%/year in Panama, to 4.6%/year in Paraguay. In Nepal it was 3.9%/year, in Nigeria 4.0%/year, in the Ivory Coast 5.9%/year, and in Haiti 3.4%/year. The net global deforestation was about 33% from preindustrial times up to 1954, but between 1980 and 1985 it proceeded at about 1% per year, a rate that, if extrapolated linearly, would predict a forest-area half-life of only 70 years (WRI, 1986).

The global net primary production of forests has been estimated as 48.7 billion metric tons/year, of which a remarkable 28% is appropriated in various ways for human use (Vitousek et al., 1986). Of this appropriated production, 45% is accounted for by the clearing of forest for shifting cultivation in less developed countries, 18% by a more permanent for-

Table 9.1 Global forest resources[a]

Region	Forested Land (10⁶ ha)	Forests as % of Total Land Area	Total Average Wood Volume (m³/ha)	Percent Coniferous	Percent Broadleaf
North America	630	34	93	74	26
Central America	65	22	92	32	68
South America	730	30	172	1	99
Africa	800	6	133	1	99
Europe	170	30	94	67	33
U.S.S.R.	915	35	106	83	17
Asia	530	15	96	16	84
Pacific	190	10	31	23	77
World total	4030	21	110		

[a]Modified from Persson (1974).

est conversion to agricultural land, 16% by the harvest of forest biomass, and 12% by growth in forest plantations of biomass that is destined for use by humans, while 10% was lost during harvests (Vitousek *et al.*, 1986).

The vigorous, forest-based resource industry has a huge economic impact. In 1984, the value of forest products traded internationally was estimated at about $107 billion (U.S. dollars), comprising global exports of $51 × 10⁹, and imports of $56 × 10⁹ (FAO, 1986; WRI, 1986). The most important trading regions were Western Europe ($45 × 10⁹), North America ($28 × 10⁹), Eastern Europe and the U.S.S.R. ($6 × 10⁹), and Asia ($6 × 10⁹) (FAO, 1986).

To supply the global demand for fiber, and to achieve the great economic benefits of forestry, a very large area of mature forest must be harvested or otherwise disturbed each year. This can be illustrated by the case of Canada, a nation whose econo-

Table 9.2 Forest vegetation by generalized ecosystem type[a]

	Area (10⁶ km²)	Total Net Primary Production (10¹⁵ g C/year)[b]	Total Mass of Vegetation (10¹⁵ g C)
Tropical rain forest	17.0	16.8	344
Tropical seasonal forest	7.5	5.4	117
Temperate conifer forest	5.0	2.9	79
Temperate angiosperm forest	7.0	3.8	95
Boreal forest	12.0	4.3.	108
Woodland and shrubland	8.5	2.7	22
Savannah	15.0	6.1	27
All other continental vegetation	77.0	10.8	35
Total continental	149	52.8	827
Total marine	361	24.8	1.74
Total global	510	77.6	828

[a]Modified from Woodwell *et al.* (1978).
[b]C = carbon.

my is strongly dependent on forestry. An average of 785,000 ha was harvested each year between 1975 and 1983, 86% by clear-cutting (Kunke and Brace, 1986). This anthropogenic disturbance of natural forests was followed by an average of 176,000 ha/year of site preparation, 65% of which was mechanical scarification, 23% prescribed burning, and 12% other methods, including the use of herbicides. Between 1980 and 1983, 63% of the regeneration on this site-prepared terrain was achieved by planted conifers, 22% by direct seeding, and 15% by natural seeding and advance regeneration. In addition, between 1975 and 1983 an average of 3.6×10^6 ha/year of forest was treated to control pests, mostly in eastern Canada where insecticides are used in an attempt to control an epidemic of spruce budworm (*Choristoneura fumiferana*) in fir and spruce forests (Chapter 8).

World-wide, an immense area of forest of about 20 million ha is cleared each year. There are many long- and short-term consequences of the environmental changes that are caused by this type of land use. In most cases the impacts are relatively moderate, and they could be considered to be an acceptable ecological "cost" that must be borne in order to harvest the forest as a potentially renewable natural resource. Some of the longer-term effects include a potential decrease in site fertility caused by nutrient losses, the alteration of wildlife habitat, a possibly permanent loss of old-growth forest habitat, and climatic change and disruption of the atmospheric CO_2 cycle (Chapter 2). Shorter-term effects include an increase in erosion, changes in watershed hydrology, inadequate forest regeneration, and temporary changes in wildlife habitat. In some situations, these environmental impacts can be so severe that they should preclude the harvesting of particular stands or larger areas of forest.

In this chapter, some of the ecological impacts of forest harvesting will be described. The focus will be on forestry in north temperate latitudes, because the ecological impacts are relatively well known in this region. However, it is very important to realize that the less well-documented impacts of forest harvesting in other regions of the globe, particularly the tropics, can be severe and are important in terms

of resource degradation, climate change through the enhancement of CO_2 flux to the atmosphere (Chapter 2), changes in local and regional evapotranspiration and other hydrologic effects, and the loss of biodiversity (Chapter 10). The topics that will be covered in this chapter are the implications of nutrient removals during harvesting for subsequent fertility of the site, and the effects of forest harvest on erosion, hydrology, and the habitat and abundance of wildlife. Relevant topics that are dealt with in other chapters are (1) the effects of silvicultural insecticide and herbicide spray programs (Chapter 8) and, as just mentioned, (2) the effects of deforestation on atmospheric CO_2, and (3) the effects of tropical deforestation on global species richness.

9.2
CONSEQUENCES OF FOREST HARVESTING FOR SITE FERTILITY

Background

The harvest of a forest can vary greatly in intensity, from a relatively "soft" selection-tree cut, to an intensive whole-tree (all above-ground biomass) or complete-tree (all above- and below-ground biomass) clear-cut. The intensive harvests increase the short-term yield of biomass from the forest, but they also cause a substantially larger removal of nutrients. Potentially, by decreasing site fertility the nutrient removals can result in a deterioration of the inherent capacity of the site to sustain the production of tree biomass (Kimmins, 1977; Norton and Young, 1976; Hornbeck, 1977; Gordon, 1981; Freedman, 1981; Smith, 1985). In agriculture, the syndrome of site impoverishment caused by intensive cropping is a well-recognized problem. In severe cases depleted land must be abandoned for some or all agricultural purposes. Usually the problem can be more-or-less managed by the application of fertilizers, but often the structural degradation of the soil is too severe to allow this simple ameliorative strategy to be successful. However, in forestry the harvest rotation is much longer than in agriculture (in which it is usually annual), and there are very few data that allow the comparison of the productivity of subsequent forest rotations on the same site. Therefore, the situation with respect to nutrient impoverishment by forest harvesting is less clear than it is in agriculture.

A frequently cited example of a decline of site capability in forestry is for second-rotation plantations of introduced radiata pine (*Pinus radiata*) in Australia and New Zealand (Keeves, 1966; Whyte, 1973; Will, 1985). The most important cause of the reduced productivity of many second rotations of *P. radiata* is believed to be nutrient impoverishment of the site caused by the forest harvest. However, the comparison of subsequent rotations of *P. radiata* stands is complicated by such factors as (1) soil changes other than those that directly involve nutrient supply, such as soil compaction, podsolization, changes in soil organic matter and acidity, etc.; (2) differences between rotations in the density and genetic quality of crop trees; and (3) the nature of weedy competition, and any release treatment that may be used.

Because of the problems associated with the present data base, the syndrome of decreased site fertility caused by forest harvest is best viewed as a potential problem. Even though there is not yet hard evidence that this problem is occurring on a large scale, from ecological principles it might be anticipated that decreased fertility could emerge after a number of forest harvests from the same site. A wide-scale impoverishment of forested sites would, of course, have important implications for the management of forests as a renewable natural resource.

The question of nutrient impoverishment of forest sites will be examined by consideration of the magnitude of nutrient removals during harvest, the size of the soil nutrient pool, the net accretion and depletion of nutrients in the forest, and the ways in which these variables can interact to influence forest productivity.

Conceptual Models

The problem of site impoverishment caused by intensive harvesting is illustrated by the conceptual

models of Figure 9.1. A long-term decrease in the magnitude of the nutrient capital would represent a decreased ability of the site to sustain a large rate of tree productivity. The effect of rotation length is conceptually illustrated in Fig. 1a. A long rotation (100 years) allows a sufficient passage of time for the harvested nutrients to be replenished by inputs from rainfall, weathering, nitrogen fixation, etc. Such a postharvest recovery period has been labeled an "ecological rotation" (Kimmins, 1977), since harvesting a forest on this basis is sustainable— the potentially renewable natural resource is not "mined" when an ecological rotation is practiced. Under the 50-year rotation of Fig. 1a, the harvested nutrients are not completely replenished between successive harvests, and this causes a long-term deterioration of site quality. Such a deterioration can also take place if the intensity of the harvest is increased, even if a 100-year rotation is used (Fig. 1b). In this example, the whole-tree harvest removes twice as much nutrient as the stem-only harvest, and therefore the nutrients are not completely replenished before the next harvest takes place. Figure 1c summarizes the influence of site quality, that is, inherently fertile sites can support relatively intensive harvests over shorter rotations, in comparison with less fertile sites. In general, fertile sites have a relatively rapid rate of nutrient replenishment, because of some combination of (1) a rapid rate of mineralization of nutrients from relatively immobile organic and mineral forms, to water-soluble ions that can be assimilated by plant roots; (2) a high rate of input of nutrients from the atmosphere; (3) a high rate of nitrogen fixation; and (4) a low rate of nutrient loss by leaching to below the tree rooting depth.

Factors Affecting Nutrient Removals by Harvesting

When trees are harvested, there is a substantial removal of nutrients that are fixed in the biomass. The quantity of nutrients that is removed is influenced by several factors, including (1) tree species, (2) age of the stand, (3) fertility of the site, (4) intensity of the harvest, and (5) time of year, especially for

(a) Variation in Rotation Length: Fixed Utilization

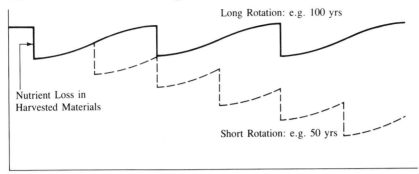

(b) Variation in Utilization: Fixed Rotation Length

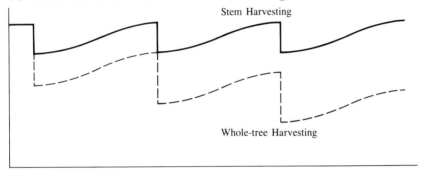

(c) Variation in Rates of Replacement of Nutrient Losses: Fixed Rotation and Utilization

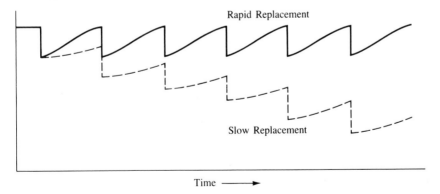

Time ⟶

Fig. 9.1. The influence of rotation length, harvest intensity, and rate of nutrient replenishment on site nutrient capital, and ultimately on the ability of the site to sustain a large rate of productivity. See text for discussion. Modified from Kimmins (1977).

seasonally deciduous species harvested during the dormant season. Some of these effects are described below.

The influence of stand age is illustrated in Fig. 9.2 for a chronosequence of Scotch pine (*Pinus sylvestris*) stands (Ovington, 1959). As expected, the older the stand, the larger its above-ground content of the macronutrients N, P, K, Ca, and Mg. Of course, this reflects the net accumulation of above-ground biomass by the stand, since the nutrients are largely organically bound. We can therefore gener-

alize that the biomass and nutrient removals by a given harvest method are influenced by age of the stand, or by the length of the rotation. This is especially true of young and intermediate-aged stands younger than about 100 years. In older stands, the net above-ground accumulation of biomass and nutrients may be in a state of dynamic balance between the death of old trees, and the positive net production of living individuals (Likens *et al.*, 1977; Bormann and Likens, 1979).

The species of tree that is being harvested also

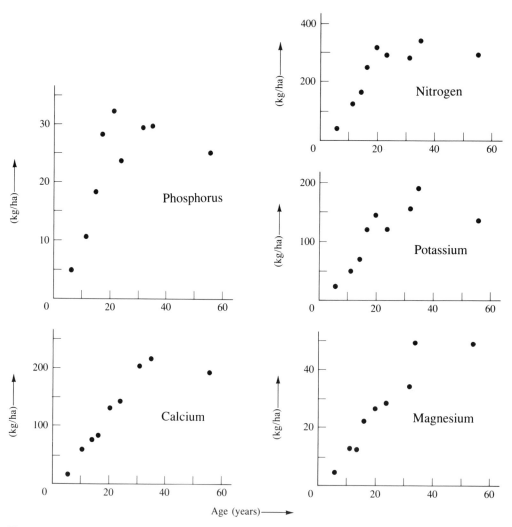

Fig. 9.2. The influence of stand age on the above-ground standing crops of nutrients in *Pinus sylvestris* plantations. Modified from Ovington (1959).

Table 9.3 Effects of tree species on the above-ground quantities of biomass and nutrients in adjacent 40-year-old stands growing on a fine sandy-loam soil in Minnesota[a]

Species	Age	Biomass (MT/ha)	N (kg/ha)	P (kg/ha)	K (kg/ha)	Ca (kg/ha)	Mg (kg/ha)
Pinus resinosa	40	199	155	42	175	291	58
Pinus banksiana	40	141	118	25	97	199	38
Picea glauca	40	141	102	57	229	719	40
Populus tremuloides–P. grandidentata	40	167	199	47	287	848	58

[a]After Alban *et al.* (1978).

influences the yield of biomass and nutrients by a forest harvest. Table 9.3 compares the above-ground quantities of biomass and nutrients in adjacent plantations of red pine (*Pinus resinosa*), jack pine (*Pinus banksiana*), and white spruce (*Picea glauca*), and a natural poplar forest (*Populus tremuloides* and *P. grandidentata*). The variations in the quantities of biomass and nutrients were large, amounting to a ratio of 1.4 for biomass (calculated as the largest/smallest value among the stands), 2.0 for nitrogen, 2.3 for phosphorus, 2.0 for potassium, 4.3 for calcium, and 1.5 for magnesium. Except for P, the most nutrient-demanding trees were the poplars. The poplar stand also had a relatively large net productivity of biomass. However, its nutrient-use efficiency (which is related to the ratio of biomass to nutrient) was smaller than that of the more productive red pine stand.

The influence of site fertility is illustrated by a comparison of the above-ground biomass and nutrients in plantations of sycamore (*Platanus occidentalis*) and white spruce, each growing on two different sites (Table 9.4). The net production and nutrient accumulation by the stands differ greatly between the sites. The inherent fertility of the sites is probably the most important determinant of these differences.

The influence of harvest intensity is illustrated by the relative removals of biomass and nutrients by bole-only and whole-tree clear-cuts of a conifer stand (Table 9.5). The whole-tree clear-cut resulted in a 30% larger harvest of biomass than the stem-only clear-cut. In terms of short-term economics

this can be an advantage of the whole-tree method, especially if the stand is being harvested to produce energy or pulp, in which quality of the biomass is not a very important consideration. However, the increased yield of biomass is largely due to the removal of nutrient-rich tissues such as foliage and small branches. As a result, the removal of nutrients by the whole-tree clear-cut was increased by a substantially larger degree than was biomass: by a factor of 2.0 for nitrogen, 1.9 for phosphorus, 1.7 for potassium, 1.5 for calcium, and 1.8 for magnesium. To use an economic analogy: the relatively small increase in yield of biomass was "purchased" at the "expense" of a much larger increase in the rate of nutrient removal.

Use of Simple Box Models to Evaluate the Consequences of Nutrient Removals

The implications of nutrient removals by a forest harvest can be appreciated by comparing the quantity of harvested nutrients with the amounts in other important compartments and fluxes. Most important are (1) the nutrient quantity in the forest floor and mineral soil within the tree rooting depth; (2) nutrients in residual, nonharvested plants such as shrubs and ground vegetation; and (3) the flux of nutrients incoming from the atmosphere, moving about within the stand as throughfall, stem flow, and litterfall, and leaving the site with drainage, ultimately to streamflow or groundwater.

The conceptually simple, box-model approach

Table 9.4 Effect of site on the above-ground quantities of biomass and nutrients for particular tree species

Species	Site	Age	Biomass (MT/ha)	N (kg/ha)	P (kg/ha)	K (kg/ha)	Ca (kg/ha)	Mg (kg/ha)	Reference
Platanus occidentalis	a)	3	9.2	52	10	10	46	17	Wood *et al.* (1977)
	b)	3	13.7	90	21	53	53	24	
Picea abies	a)	47	263	705	82	226	507	85	Ovington (1962)
	b)	47	140	331	37	161	212	39	

Table 9.5 Effect of harvest intensity on the yield of biomass and nutrients from a *Picea rubens—Abies balsamea* stand in Nova Scotia[a]

Harvest	Component	Biomass (kg d.w./ha)	N (kg/ha)	P (kg/ha)	K (kg/ha)	Ca (kg/ha)	Mg (kg/ha)
Conventional	Merchantable stem	105,200	98.0	16.3	91.7	180.9	17.0
Whole-tree	Merchantable stem	117,700	120.1	18.2	76.2	218.9	20.4
	Tops, branches, foliage	34,800	119.0	17.0	56.4	117.6	16.5
	Total harvest	152,500	239.1	35.2	132.6	336.5	36.9
	Percent increase	29.6%	99.1%	93.4%	74.0%	53.7%	80.9%

[a]In the conventional clear-cut only the stems of the trees were removed. The whole-tree clear-cut included the harvesting of tops, branches, and foliage. Modified from Freedman *et al.* (1981b).

of a comparison of compartments and fluxes is illustrated for a mature maple–birch stand in Table 9.6. The whole-tree biomass (154,600 kg/ha) is 1.4 times as large as the biomass of the merchantable stem (113,000 kg/ha). Therefore, for this stand a whole-tree clear-cut harvest could potentially result in 40% more yield of biomass than a bole-only clear-cut. As in Table 9.5, the whole-tree/merchantable stem ratios for nutrients are considerably larger than for biomass. The whole-tree quantity of nitrogen was 2.2 times as large as that in the stems, while phosphorus was 2.6 times as large, potassium 2.2 times, calcium 1.7 times, and magnesium 1.9 times. In summary, a moderate increase in yield of biomass (40%) would be achieved by a whole-tree harvest, but at the expense of much larger increases in the rates of nutrient removal.

Because the stand described in Table 9.6 was comprised of seasonally deciduous angiosperm trees, the nutrient removal by harvesting would be smaller if the harvesting took place during the leafless dormant period. The midsummer quantity of foliar nutrients in this stand was 63.2 kg N/ha, 6.1 kg P/ha, 29.4 kg K/ha, 19.3 kg Ca/ha, and 5.8 kg Mg/ha (Freedman et al., 1982). These are equivalent to 18% of the whole-tree quantity of N, 16% of P, 16% of K, 4% of Ca, and 13% of Mg. However, some of these nutrients would be resorbed back into perennial tissues or leached from the foliage before leaf drop in the autumn; the resorbed nutrients would be removed by a harvest of the stand during the dormant period. Ryan and Bormann (1982) examined a recently clear-cut and a 55-year-old hardwood stand in New Hampshire, and found that about half of the N, P, and K were retained in foliage after senescence; the rest was resorbed or leached from foliage. Ca and Mg were largely retained in the senescing foliage. Therefore, only about half or less of the growing season quantities of foliar N, P, and K would be removed by a harvest when the trees were leafless.

The largest pools of biomass and nutrients in a typical forest are in three compartments: (1) the trees, (2) the forest floor, and (3) the mineral soil. In aggregate, these essentially comprise the biomass and nutrient "capital" of the site, and therefore comparisons among them are important. In the maple–birch stand (Table 9.6), the whole-tree biomass is equivalent to 45% of the biomass content of the forest floor plus mineral soil, 6.8% of the total nitrogen, 4.1% of the total phosphorus, 1.4% of the total potassium, 50% of the total calcium, and 3.9% of the total magnesium. Only the calcium comparison indicates a short-term cause for concern with respect to an impoverishment of site nutrient capital by harvesting. This reflects the small calcium content of many soils that are derived from granite, gneiss, and other oligotrophic glacial tills (Cann et al., 1965; Boyle and Ek, 1972; Boyle et al., 1973; Weetman and Webber, 1972; Weetman and Algar, 1983; Freedman et al., 1986). In some respects, the apparent "importance" of calcium seems anomalous, since available nitrogen (or much less often phosphorus or potassium) is the nutrient to which temperate forests most frequently respond in fertilization trials (Weetman et al., 1974; Czapowskyj, 1977; Foster and Morrison, 1983).

Another important comparison to make within Table 9.6 is between the size of the potential whole-tree nutrient removal, and the "plant-available" soil nutrient pool. In this comparison, the whole-tree quantities are larger, being equivalent to 4.4 times the size of the water-soluble ammonium–nitrogen plus nitrate–nitrogen content of the forest floor plus mineral soil, 1.0 times the extractable phosphate–phosphorus content, 1.0 times the exchangeable potassium, 1.7 times the exchangeable calcium, and 0.39 times the exchangeable magnesium. Clearly, quite a different conclusion could be reached about the potential effects of a whole-tree harvest of this stand, depending on whether the nutrient removals are compared with the "available" or the "total" soil pools. The total pool, along with the biota, represents the gross nutrient "capital" of the site. However, most of the total nutrients occur in chemical forms that are not available to plants. Availability requires the mineralization of insoluble forms of the nutrient, either by microbial oxidations or by inorganic weathering processes.

It is also important to realize that although the available soil nutrient pool is much smaller in magnitude than the total pool, it is relatively ephemeral

Table 9.6 Standing crops and fluxes of biomass and nutrients in a mature stand of *Acer saccharum–A. rubrum–Betula alleghheniensis* in Nova Scotia[a]

Category	Biomass	N	P	K	Ca	Mg
Standing crops						
Trees, total above ground	154,600	355.4	38.4	183.1	448.1	45.4
	(1.0)	(1.0)	(1.0)	(1.0)	(1.0)	(1.0)
Trees, stem only	113,000	159.4	15.0	84.6	256.4	23.5
	(1.4)	(2.2)	(2.6)	(2.2)	(1.7)	(1.9)
Ground vegetation	218	3.2	0.3	2.8	0.5	0.3
	(719)	(111)	(128)	(65)	(896)	(151)
Forest floor, total	18,700	314	22.2	56.2	50.8	25.8
	(8.4)	(1.1)	(1.7)	(3.3)	(8.8)	(1.8)
Forest floor, available	—	1.0	0.3	4.3	20.8	3.5
		(355)	(128)	(43)	(22)	(13)
Mineral soil, total	324,600	4945	921	12,580	855	1139
	(0.48)	(0.072)	(0.042)	(0.0015)	(0.52)	(0.040)
Mineral soil, available	—	79.8	37.0	181	244	113
		(4.5)	(1.0)	(1.0)	(1.8)	(0.40)
Fluxes						
Litterfall	4,900	33.7	2.5	5.4	34.1	4.8
	(32)	(11)	(15)	(34)	(13)	(9.5)
Incident precipitation	—	3.7	<1.0	2.2	2.7	1.1
		(96)	—	(83)	(166)	(41)
Throughfall	—	0.9	0.3	9.9	3.6	1.2
		(395)	(128)	(18)	(124)	(38)
Stem flow	—	0.009	0.030	1.1	0.26	0.07
		(39,500)	(1280)	(166)	(1723)	(649)
Weathering[b]	—	0	0.6	5.0	18	5.0
			(64)	(37)	(25)	(9.1)
Nitrogen fixation[b]	—	10	—	—	—	—
		(36)				
Streamflow[b]	—	0.32	<0.1	2.7	14.6	9.7
		(1110)	—	(68)	(31)	(4.7)
Net flux[b]	—	+7	+0.4	−0.6	−7	−2
		(51)	(128)	(305)	(64)	(23)

[a] Standing crops are in kg/ha; fluxes are in kg/ha-year; throughfall and stemflow are in kg/ha over the growing season. Data in parentheses are ratios of the quantity in the whole-tree compartment, relative to the quantity in other standing crops and fluxes. Modified from Freedman *et al.* (1986).

[b] These were not measured for this specific site; they were estimated by a review of the literature.

because its turnover time is short. For example, in a northern hardwoods forest in New Hampshire, the turnover time of available nitrogen was estimated as only 1.2 years, and 7 years for calcium (Likens *et al.*, 1977; Bormann and Likens, 1979). As such, it may not be important that an apparent "depletion" of the available pool is calculated in a simple box model budget. In fact, as is described in Section 3 of this chapter, some studies have shown that there is a short-term increase in the quantity of available nutrients (especially nitrate) following clear-cutting and some other site disturbances (Likens *et al.*, 1977; Vitousek *et al.*, 1979; Krause, 1982).

Bulk precipitation is an important source of some nutrients to forests, and over the rotation period it helps to replace some of the nutrients that are removed by a harvest. For the maple–birch stand, 96 years of precipitation input are equivalent to the potential whole-tree removal of nitrogen, 83 years for potassium, 166 years for calcium, and 41 years for magnesium (Table 9.6). Therefore, over a moderately long rotation (i.e., > 50 years), a substantial fraction of the nutrients removed during harvesting could be replaced by precipitation inputs. Other nutrient inputs take place by the weathering of minerals, by nitrogen fixation, and by the dry deposition of gases and particulates.

Of course, inputs of nutrients do not take place in the absence of outputs. The most important output is the leaching of soluble forms of nutrients to below the tree rooting depth, and ultimately to ground or surface water. In the case of the maple–birch site, the potential whole-tree harvest removal is equivalent to the potassium loss in 68 years of stream flow from an undisturbed forested watershed, 31 years for calcium, and 4.7 years for magnesium (Table 9.6). The stream-flow losses of soluble nitrogen and phosphorus are of much smaller magnitude. (However, as is discussed in Section 3, the flux of nitrate in stream flow can be greatly increased after disturbance.)

In order to integrate the inputs and outputs of nutrients, net flux is calculated as the difference betweeen the total inputs and the total outputs. A positive value of net flux (inputs > outouts) indicates that the nutrient in question is aggrading on the site over time, while a negative value (outputs >

inputs) indicates that its quantity is decreasing. According to the estimates of net fluxes for undisturbed forested watersheds in Table 9.6, both nitrogen and phosphorus are aggrading; 51 years of net flux of nitrogen and 128 years of phosphorus would replace the removals by a whole-tree harvest. The other three nutrients are degrading, and in 305 years of net flux of potassium, 64 years of calcium, and 23 years of magnesium, a quantity of these nutrients would be lost from the site that is equivalent to the potential whole-tree harvest removal. A negative net flux of potassium, calcium, and magnesium also contributes to acidification of the site, a process that is exacerbated when an additional quantity of these base cations is removed by harvesting (Oden, 1976; Rosenqvist *et al.*, 1980; see also Chapter 4).

Overall, it appears that a single whole-tree clear-cut of the maple–birch stand might not cause an important depletion of site nutrient capital, since (1) the soil reserve is relatively large and (2) over the rotation period a large quantity of nutrients is regenerated. However, an important exception may be calcium. A large depletion of its site capital could take place, possibly resulting in decreased productivity and increased site acidification.

Use of Simulation Models to Evaluate the Consequences of Nutrient Removals

Several teams of forest researchers have developed computer simulation models of forest productivity, nutrient cycling, and forest floor dynamics that can be parameterized to respond to variations in the intensity of stand harvest (Aber *et al.*, 1978, 1979; Kimmins *et al.*, 1981). These models predict the postharvest successional dynamics of these variables, and they are based on the knowledge or estimation of the sizes of the biomass and/or nutrient compartments and the rates of transfer between compartments. Unlike the simplistic box-model approach described above, the simulation models have the valuable capability of a dynamic response to changes over time in the rates of forest processes.

The simulation models of Aber *et al.* (1978, 1979) incorporated the intensive database of the

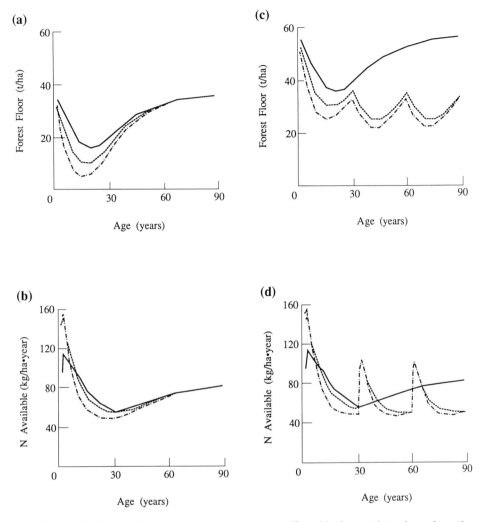

Fig. 9.3. Simulation models of forest floor dynamics, as affected by harvest intensity and rotation length. In all cases, ———, is stem-only clear-cut; ----, whole-tree clear-cut; ·······, complete-tree clear-cut. (a) Effect of harvest intensity on forest floor biomass; (b) effect of harvest intensity on available nitrogen; (c) effect of rotation length on forest floor biomass; (d) effect of rotation length on available nitrogen. Modified from Aber *et al.* (1978).

Hubbard Brook Ecosystem Study in New Hampshire (Likens *et al.*, 1977, Bormann and Likens, 1979). Their simulations included six strategies of forest harvest: three cutting intensities, and three rotation lengths. Figure 9.3a describes the effects on biomass of the forest floor of three harvest intensities (stem-only, whole-tree, and the rarely used complete-tree clear-cutting) over a 90-year rota-tion. The three types of harvest have different effects on the biomass of the forest floor, with the complete-tree treatment having as little as 33% the forest floor weight of the stem-only harvest during early to mid succession. However, by late succession (> 40–50 years) the forest floor biomasses of all three cutting treatments converge. These effects are largely due to a difference in the fraction of the total site biomass that is removed by the three harvests: the whole-tree harvest removes foliage, twigs, and branches that are left behind as slash during the stem-only harvest; the complete-tree harvest also removes below-ground biomass. There are negative ecological implications of a decreased biomass of the forest floor during much of the postharvest secondary succession. This compartment is important for cation exchange, moisture and nutrient retention, and other processes that may affect the overall capability of the site to sustain a high rate of ecosystem productivity.

The annual rate at which nitrogen is made available for plant uptake (Fig. 9.3b) was modelled on the basis of the net effects of (1) mobilization of nitrogen from unavailable forms such as organic N, to available forms such as nitrate and ammonium; plus (2) the bulk input from the atmosphere; minus (3) the biological immobilization of soluble available N as unavailable organic N of decomposing litter and other organic fractions. The results of the nitrogen simulations are more complex than were those for weight of the forest floor. For about the first 10 years, available nitrogen is present in a larger quantity in the whole-tree and complete-tree treatments. This is probably due to some combination of (1) a larger rate of nitrogen immobilization in the relatively organic-rich forest floor of the stem-only treatment, because of decomposition of slash and (2) a greater stimulation of nitrification by the more severe disturbance associated with the relatively intensive methods of harvest. Since there is little aggrading plant biomass to take up the available nitrogen during the early postdisturbance period, it may leach from the site (as nitrate; see Section 3). Therefore, the residual organic matter from the stem-only harvest may play an important role in the conservation of site nitrogen capital, by immobilizing soluble, ionic forms of nitrogen as organic nitrogen, and thereby preventing the leaching of inorganic nitrogen to below the rooting depth.

The effect on forest floor biomass of a decrease of the rotation length to 30 years, while using a whole-tree or complete-tree clear-cut, is described in Fig. 9.3c. Compared with the 90-year stem-only rotation, the weight of the forest floor is greatly reduced by the shorter rotations, by a factor of about one-half. The relative effect of a shorter rotation is much larger than that of harvest intensity. This is likely caused by a generally enhanced condition for litter decomposition, because of the more frequent site disturbance associated with the use of a shorter rotation.

The effect of rotation length on the availability of nitrogen is shown in Fig. 9.3d. The 30-year rotation causes three times as many peaks of nitrogen availability as the 90-year cutting cycle. This results in more frequent pulses of nitrate loss to stream water, and a larger loss overall. The net effect is that during the later stages of the postcutting succession, when net production is potentially large, nitrogen availability is smaller for the more frequent rotation, by about 33% after 60 years, and 41% after 90 years. The model does not predict much of an effect of harvest intensity on this response. The overall implications of these effects on nitrogen availability are that the longer-term net production is likely to be smaller in frequently harvested stands.

In another simulation, Aber et al. (1979) varied harvest intensity and rotation length, and their effects on total net ecosystem production and total harvest yield over a 90-year period (Table 9.7). The total net production was largest for the 90-year rotations, irrespective of harvest intensity. Of course, the yield of these three treatments was affected by harvest intensity. The next largest net production was for a strategy of two 45-year rotations over the 90-year simulation period. These had a net production that averaged about 22% smaller than the 90-year rotation. The 30-year rotation ranked last in terms of net ecosystem production, averaging about 56% smaller than the 90-year rotation. Another important observation is that the relatively intensive harvests removed a larger fraction of the total net production. Most efficient in this respect was the complete-tree, 90-year harvest, which removed 24% of the 90-year net production. Least efficient was the whole-tree 45-year treatment, which removed 13% of net production, and the stem-only 90-year treatment, which removed 14%. If we were to formulate an optimal management strategy based

Table 9.7 Simulations of total net ecosystem production and total harvest yield over a 90-year period: There are three intensities of harvest, and three rotation lengths[a]

Type of Cutting	Length and Number of Rotations	Total Net Production (t/ha per 90 years) and Rank	Total Harvest Yield (t/ha per 90 years) and Rank	Percent of Total Production Harvested
Clear-cutting	90 (1)	1090 (2)	154 (4)	14
Whole tree	90 (1)	1120 (1)	197 (2)	18
Whole tree	45 (2)	853 (4)	108 (6)	13
Whole tree	30 (3)	478 (6)	93 (7)	19
Complete forest	90 (1)	1055 (3)	252 (1)	24
Complete forest	45 (2)	841 (5)	171 (3)	20
Complete forest	30 (3)	476 (7)	150 (5)	32

[a]From Aber *et al.* (1979).

solely on the simulations summarized in Table 9.7, it would be to use the complete-tree treatment on a 90-year rotation. Of course, this assumes that this treatment does not have negative ecological impacts that are not taken into account in the simulation model, for example, on the rate of tree regeneration, effects on the forest floor, effects on nutrients other than nitrogen, etc.

9.3

LEACHING OF NUTRIENTS FROM DISTURBED FOREST WATERSHEDS

Various studies have demonstrated that soluble nutrients can leach from forested watersheds after disturbance by cutting or wildfire (Tamm *et al.*, 1974; Corbett *et al.*, 1978; Hornbeck and Ursic, 1979; McColl and Grigall, 1979; Vitousek *et al.*, 1979; Hornbeck *et al.*, 1987a). This process causes a reduction of site nutrient capital that is incremental to the nutrients that are removed with harvested biomass. Leaching is especially important for nutrients that are relatively mobile in soil, such as nitrate and potassium.

A frequently cited example of this effect of forest disturbance is a study done at Hubbard Brook, New Hampshire. This large-scale experiment involved the clear-felling of all trees on a 15.6-ha watershed,

but without the removal of biomass—the cut trees were left lying on the ground. Subsequent regeneration was prevented for the next three growing seasons by treatment of the watershed with the herbicides Bromacil and 2,4,5-T. Therefore, this experiment examined the effects of a perturbation of biological control via deforestation, on nutrient cycling and other watershed processes. This experiment did not examine the effects of a typical forest management practice.

Over a 10-year postdisturbance period, the deforested watershed had a stream-water loss of 499 kg/ha of NO_3-N, 450 kg/ha of Ca, and 166 kg/ha of K. These were much larger than the losses from an undisturbed reference watershed of 43 kg/ha of NO_3-N, 131 kg/ha of Ca, and 22 kg/ha of K (Bormann *et al.*, 1974; Likens *et al.*, 1978) (see also Table 9.8). The effect on nutrient flux in streamwater was partly due to an average 3-year increase of 31% in the yield of water from the cut watershed, which was caused by the disruption of transpiration. More important, however, was an increase in nutrient concentration in the stream water, by factors of 40 for NO_3, 11 for K, 5.2 for Ca, 5.2 for Al, 3.9 for Mg, and 2.5 for H^+ (Bormann and Likens, 1979). In total, the net increases in the stream-water losses of N, Ca, and Mg from the disturbed watershed were similar in magnitude to the contents of these nutrients in the above-ground biomass of the hardwood forest at Hubbard Brook (i.e., 371 kg/ha of

Table 9.8   Annual net flux (atmospheric inputs minus stream-water exports) of dissolved substances for a deforested and a reference watershed at Hubbard Brook, New Hampshire[a]

Element	Net Flux (kg/ha)	
	Deforested Watershed	Reference Watershed
Ca	−77.7	−9.0
Mg	−15.6	−2.6
K	−30.3	−1.5
Na	−15.4	−6.1
Al	−21.1	−3.0
NH_4-N	+1.6	+2.2
NO_3-N	−114.1	+2.3
SO_4-S	−2.8	−4.1
Cl	−1.7	+1.2
HCO_3-C	−0.1	−0.4
SiO_2-Si	−30.6	−15.9
Total	−307.8	−36.9

[a]Weighted average data over June 1966 to June 1969. Modified from Bormann and Likens (1979).

N, 403 kg/ha of Ca, and 155 kg/ha of K; Whittaker *et al.*, 1979).

As mentioned previously, this deforestation experiment did not involve a typical forest management practice, and the measured impacts were extreme. However, some other watershed studies of the effects of more typical clear-cutting and other management practices have shown a qualitatively similar impact on the leaching of nutrients. Stream-flow losses in the first 2 years after clear-cutting a hardwood watershed in New Hampshire were 95 kg N/ha and 89 kg Ca/ha, compared with 144 kg N/ha and 221 kg Ca/ha that were removed with the harvested biomass (Pierce *et al.*, 1972; Likens *et al.*, 1978). In a wider-scale study in New Hampshire, a comparison was made of nine clear-cut and five reference watersheds (Martin *et al.*, 1986). Averaged over these two treatments, 4 years of stream-water loss of NO_3-N from the clear-cuts averaged 71 kg/ha per 4 years, compared with 14 kg/ha per 4 years for the reference watersheds. Calcium export from these treatment-watersheds averaged 111 kg/ha per 4 years, compared with 50 kg/ha per 4 years, while K exports were 23 kg/ha per 4 years, compared with 8 kg/ha per 4 years. The 10-year stream-water losses from a strip-cut hardwood watershed in New Hampshire were increased by 48% for Ca, 135% for K, and 50% for N, while on a clear-cut watershed they were increased by 29%, 218%, and 128%, respectively (compared with values for an uncut reference watershed of 166 kg Ca/ha per 10 years, 22 kg K/ha per 10 years, and 45 kg N/ha per 10 years; Hornbeck *et al.*, 1975, 1987a). The 3-year loss attributed to clear-cutting a 391-ha watershed in New Brunswick was more moderate, at 19 kg NO_3-N/ha (Krause, 1982). These effects parallel, but are quantitatively much smaller than, those that were observed for the devegetated watershed at Hubbard Brook that was described first. In addition, it should be mentioned that some other studies of the effects of forest management on water quality have shown little or no effect, especially if only a portion of the watershed was cut (G. W. Brown *et al.*, 1973; Verry, 1972; Aubertin and Patric, 1974; Richardson and Lund, 1975; Hetherington, 1976; McColl, 1978; Burger and Pritchett, 1979; Stark, 1980).

A relatively minor effect on stream-water chemistry has also been observed after burning (Smith, 1970; Wagle and Kitchen, 1972; Grier, 1975; Wright, 1976; Tiedemann *et al.*, 1978; Schindler *et al.*, 1980; Neary and Currier, 1982). For example, the first-year export of NO_3-N from a burned 122-ha watershed in South Carolina was 0.67 kg/ha, compared with 0.05 kg/ha for an unburned reference watershed (Neary and Currier, 1982).

The effects of forest disturbance on nutrient loss via stream flow are influenced by such variables as soil and stand type, intensity of the disturbance, speed and vigor of the regeneration, watershed hydrology, and climate. Because operational forest harvests cause an inconsistent loss of nutrients in stream flow, and these are generally fairly small in comparison with the nutrient capital of the site, some researchers have concluded that a decrease in site fertility would not normally be anticipated through this mechanism (Sopper, 1975; McColl and Grigall, 1979). However, if most of the watershed is severely disturbed, if the harvest of biomass is intensive, and if the regeneration is not vigorous, then the leaching of soluble nutrients may be of greater importance.

The loss of nitrate is of most concern, because this is the nutrient that is most often lost in a large quantity, and because available nitrogen is the most frequent limiting nutrient for forest productivity. Nitrate is a relatively mobile ion in soils. For anions, the mobility series increases as $PO_4^{3-} < SO_4^{2-} < NO_3^- \simeq Cl^-$, and for cations the series is $Mg^{2+} \simeq Ca^{2+} < NH_4^+ < Na^+ \simeq K^+$ (Russell, 1973; McColl and Grigall, 1979).

There are several reasons why nitrate and other ions are leached from watersheds after disturbance.

1. After disturbance there is an increase in the rate of decomposition of organic matter, causing the release of soluble forms of nutrients. The rate of decomposition can be increased by a complex of environmental factors related to disturbance, including a warmer surface soil, an influx of organic matter to the forest floor, and an increased availability of inorganic nutrients and moisture, due in part to a decreased uptake by higher plants in the first years after disturbance (Cole and Gessel, 1965; Cole *et al.*, 1975; Likens *et al.*, 1970; Bormann *et al.*, 1974; Bormann and Likens, 1979; Piene, 1974; Harvey *et al.*, 1976, 1980; Jurgensen *et al.*, 1979).
2. Ammonification converts organic N to ammonium, which can be oxidized to nitrate by the bacterial process of nitrification. This process enhances the potential leachability of the fixed nitrogen capital of the site, since nitrate is very mobile in soil. In some situations, the rate of nitrification is greatly increased after disturbance; in other cases it is not, especially if the soil is acidic (Likens *et al.*, 1970; Reinhart, 1973; Bormann and Likens, 1979; Wallace and Freedman, 1986).

From the perspective of many foresters, the vigorous plant regeneration that develops after clear-cutting is considered to be detrimental to silvicultural objectives. This is because the highly competitive situation may inhibit or preclude the successful establishment of a new forest dominated by economically desirable tree species, usually conifers. Indeed, most herbicide and much mechanical site preparation in forestry is aimed towards reducing the degree of site dominance by the herb- and shrub-dominated stage of the secondary forest succession (see Chapter 8). However, the rapid revegetation of a disturbed watershed can reduce the loss of nutrients. Important actors in this initial process of ecological recovery are as follows.

1. Early successional, ruderal species. Important forest ruderals in temperate and boreal North America include blackberries and raspberries (*Rubus* spp., e.g. *R. strigosus*), cherries (*Prunus* spp., e.g. *P. pensylvanica*), alders (*Alnus* spp.), and many herbaceous taxa such as grasses (Poaceae), sedges and bullrushes (Cyperaceae), and taxa of the aster family (Asteraceae, especially *Aster* spp. and *Solidago* spp.).
2. Species that are tolerant of the environmental conditions of the mature forest. After disturbance, the tolerant species may regenerate by the establishment of seedlings, or by the vegetative growth of unkilled individuals. Vegetative regeneration can take place by the stump sprouting of cut trees and shrubs [e.g., birches (*Betula* spp.), maples (*Acer* spp.), and poplars (*Populus* spp.)], or by a competitive release in the case of plants of the ground vegetation.

The various taxa of the vegetation of revegetating clear-cuts can quickly reestablish a large rate of net primary production and nutrient uptake. As such, the vegetation can act as a biological reservoir (or "sponge") for some of the nutrient capital that might otherwise leach from the site. As development of the stand proceeds after disturbance, the early successional species are progressively eliminated or reduced in dominance because of competitive stress exerted by more tolerant species, or in some cases because of senescence. The nutrients that were immobilized in their biomass are then made available for uptake by trees, after recycling by litterfall and decomposition (Marks and Bormann, 1972; Marks, 1974; Bormann and Likens, 1979).

Therefore, because foresters usually perceive a vigorous regeneration of plants other than the desired crop trees to be a negative ecological value, while many ecologists perceive the same vegetation to be desirable, the ecological importance of nutrient conservation by the weedy vegetation of regenerating clear-cuts is controversial.

9.4
SOIL EROSION RESULTING FROM DISTURBANCE

Severe erosion has occurred on many deforested watersheds, particularly those containing steep slopes. Frequently, erosion is triggered by poor forestry practices, including the faulty planning and construction of logging roads, the use of streams as a skid-trail, the clear-cutting of forest immediately adjacent to water bodies, running skid trails down slopes instead of along them, and removing forest from steep slopes that are hypersensitive to soil loss. Severe erosion of soil has many ecological impacts, including (1) the loss of mineral soil substrate, which in severe cases can expose bedrock; (2) the loss of soil nutrient capital; and (3) secondary effects on recipient aquatic systems, including siltation, flooding hazard, and the destruction of fish habitat. Because of the damage that can be caused by erosion after forest harvesting on some sites, the phenomenon has been the focus of much research, and several reviews have been published (Rice et al., 1972; Fredriksen et al., 1975; Anderson et al., 1976; Corbett et al., 1978; Hornbeck and Ursic, 1979; McColl and Grigall, 1979).

The most important factors that characterize erosion from harvested lands are (Rice et al., 1972) (1) most logging activities increase the rate of erosion from forested lands; (2) erosion is spatially variable on harvested sites; (3) initially there is a large rate of sedimentation in streams after disturbance, followed by a rapid reduction to the predisturbance

condition, usually within 2–5 years; (4) landslides and creep are the quantitatively most important erosional processes in mountainous areas; (5) steep slopes are especially vulnerable; and (6) road building is a very important factor in the initiation of erosion, especially the inadequate number or size, or the improper installation, of culverts.

Studies of erosion from harvested watersheds have shown a wide range of effects, with soil loss measured as suspended sediment ranging from much less than 1 to 5 MT ha^{-1} year^{-1} (McColl and Grigall, 1979). Megahan and Kidd (1972) studied a harvested *Pinus ponderosa* watershed in Idaho with an average slope of 70%, and found a 6-year average erosion loss of 4000 kg ha^{-1} year^{-1}, compared with 90 kg ha^{-1} year^{-1} from an uncut reference watershed. Haupt and Kidd (1965) reported a much smaller effect in another *P. ponderosa* watershed in Idaho with a 35–55% slope, where the 5-year postharvest sediment loss averaged 120 kg ha^{-1} year^{-1}, compared with essentially zero for an uncut watershed. In Montana, a watershed with a conifer forest of *Larix occidentalis*, *Pseudotsuga menziesii*, and *Picea engelmannii* on an average slope of 24% was clear-cut, and then the logging slash was burned (DeByle and Packer, 1972; Packer and Williams, 1976). The treatment watershed had a sediment loss of 50 kg/ha in the first postcutting year, 150 kg/ha in the second year, 13–15 kg/ha in years 3 and 4, and about 0 kg/ha 7 years after harvesting (the loss was also essentially zero on an uncut, reference watershed). E. L. Miller et al. (1985) studied a conifer-dominated forest in the Ouachita Mountains of Oklahoma and Arkansas, and reported a 4-year sediment loss of about 345 kg/ha per 4 years from a clear-cut watershed (250 kg/ha in the first post-cutting year), compared with 73 kg/ha per 4 years in a reference watershed. At a second location, they found a 3-year sediment loss of 450 kg/ha per 3 years from a clear-cut watershed (215 kg/ha in the first postcutting year), compared with 150 kg/ha per 3 years for a less intensively harvested selection-cut watershed, and 82 kg/ha per 3 years for a reference watershed. In the relatively extreme devegetation experiment described earlier for a 16-ha hardwood watershed with 12–13% slopes at Hubbard Brook, New Hampshire, the

4-year posttreatment sediment yield averaged 190 kg ha^{-1} year^{-1} (maximum of 380 kg ha^{-1} year^{-1}), compared with 30 kg ha^{-1} year^{-1} for a reference watershed (Bormann *et al.*, 1974; Bormann and Likens, 1979). However, in this case erosion from the disturbed watershed was actually minimized since there was no road building, and no damage was caused to the forest floor or stream banks by the skidding of logs, since wood was not removed from the cut watershed.

Many studies have shown that a large erosional loss from watersheds managed for forestry purposes can be prevented by following certain operational guidelines (Packer, 1967a; Kochenderfer, 1970; Rothewell, 1971; Simmons, 1979; Miller and Sirois, 1986). These include (1) the careful planning of forest roads; (2) the installation of a sufficient number of adequately sized culverts; (3) the avoidance of direct disturbance to stream beds by heavy equipment; (4) leaving a buffer strip of uncut forest along water courses; (5) the use of skidding techniques that minimize the physical disturbance of the forest floor, for example, cable or skyline yarding; (6) allowing or encouraging a vigorous regrowth of vegetation so as to speed the reestablishment of biological control over erosion; and ultimately (7) a decision to leave hypersensitive sites uncut.

Many researchers have identified logging roads that were poorly planned, constructed, or maintained as the primary factor causing erosion from many harvested watersheds (Haupt and Kidd, 1965; Dyrness, 1967; Fredriksen *et al.*, 1975; Corbett *et al.*, 1978; McColl and Grigall, 1979). Dyrness (1967) found that following a severe rainstorm in a 6000-ha watershed in mountainous terrain in Oregon, 72% of 47 massive erosion events were associated with roads, even though roads accounted for only 2% of the terrain. Of 79 cases of erosion caused by forestry that were examined in a study in Maine, 37% were associated with roads, 55% with skidding in or across watercourses, and 8% with yarding [Land Use Regulatory Commission (LURC), 1979]. The highly compacted, mineral substrates of roads and skid trails are susceptible to erosion because they encourage the overland flow of water. Compacted soil of this sort was found to be present on 20% of a harvested area in Nova

Scotia (Henderson, 1978), and 10% of another area in Newfoundland (Case and Donnelly, 1979).

A buffer strip of uncut forest adjacent to streams, rivers, and lakes can reduce erosion after harvesting of the forest. A buffer strip can also reduce or eliminate temperature increases in water, preserve riparian habitat for wildlife, and reduce the esthetic impacts of harvesting (Van Groenwoud, 1977; McColl and Grigall, 1979). While it is generally accepted that buffer strips have these mitigative effects, there is no concensus as to the width of uncut strip that should be used. This is an economically important consideration, since a large area of merchantable timber can be withdrawn from the potential harvest when uncut buffer strips are designated. Trimble and Sartz (1957) recommended a strip width of only 8 m between logging roads and streams on level sites in the U.S. Northeast, and an increase in width of 0.6 m for each 1% increase in slope of the land. However, they recommended that the strip width be twice as large on municipal watersheds, since the requirement for good water quality is greater in the case of drinking water. A wider strip has been recommended for eastern Canada: 15–20 m on each side of watercourses on level terrain, and wider strips on slopes (Van Groenwoud, 1977). The guidelines currently in place in New Brunswick are for a 65-m buffer on each side of watercourses, 100 m on each side of access roads, and the preservation of all deer yards. If rigorously applied, these restrictions would cause a 20% reduction in the annual allowable cut in that province (Reed and Associates, Ltd., 1980). In Nova Scotia, the guideline is for a 30-m-wide buffer strip on each side of watercourses; this standard could result in the withdrawal of about 8% of the potentially harvestable area [Nova Scotia Department of Lands and Forests (NSLF), 1975].

9.5
HYDROLOGIC EFFECTS OF THE DISTURBANCE OF WATERSHEDS

Forest vegetation exerts a powerful influence on watershed hydrology, and this can be important in terms of preventing erosion and downstream flooding, and in reservoir management. This hydrologic effect is caused because transpiration by vegetation evaporates a large quantity of water into the atmosphere. In the absence of this process an equivalent quantity of water would exit the watershed as stream flow or seepage to deep groundwater (actually, this overstates the effect somewhat, because in the absence of a forest canopy the rate of nonbiological evaporation is enhanced). Freedman *et al.* (1985) studied four gauged forested watersheds with shallow soil and impervious bedrock in Nova Scotia. On an annual basis, evapotranspiration ranged from 15% to 29% of the total precipitation input. Runoff via streams or rivers accounted for the remaining 71–85% of the atmospheric input of water.

There is a marked seasonality in the relative contributions of evapotranspiration and streamflow to the hydrologic budget of forested watersheds, especially in temperate latitudes. This effect is illustrated in Fig. 9.4, which shows the average monthly hydrology for a 723-km² forested watershed in Nova Scotia. The total input of water as precipita-

tion averaged 1458 mm/year, of which 18% arrived as snow. On an annual basis, 38% of the input was partitioned into evapotranspiration, and 62% into riverflow. Although there was a slight tendency for November to January to have more precipitation, the seasonal variation in the quantity of atmospheric input was fairly small. In contrast, the seasonalities of evapotranspiration, runoff, and groundwater storage in the watershed were quite marked. Evapotranspiration was relatively large during the growing season of May to October, and as a consequence runoff was relatively small. Runoff was somewhat larger in autumn–early winter when transpiration was small; however, most of the precipitation input during this period served to recharge groundwater storage capacity, which had been depleted by withdrawals by vegetation during the growing season. Runoff was largest during mid to late winter and especially in the spring, when the accumulated snowpack melted over a short period of time and caused a flush of stream and river flow. This region of Nova Scotia experiences a relatively mild, maritime climate so that there are frequent episodes of snowmelt during the winter. In a more continental

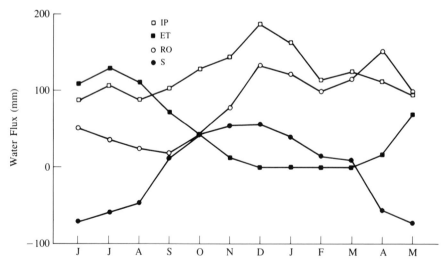

Fig. 9.4. Mean monthly hydrology for the 723-km² watershed of the Mersey River, Nova Scotia. Incident precipitation (IP) and runoff (RO) were measured between 1968 and 1982. Potential evapotranspiration (ET) was calculated using a climate-based model. Watershed storage (S) was calculated as IP − RO − ET. Modified from Ambler (1983) and Freedman *et al.* (1985).

temperate climate, there is an even more pronounced peak of spring runoff, since snowpack accumulates to a greater depth during the relatively cold winter, and it melts over a shorter period of time.

The influence of the forest on watershed hydrology is affected by disturbance caused by cutting or wildfire. Effects include changes in the timing and amount of stream flow, with secondary downstream effects such as flooding and erosion. In addition, on an imperfectly drained site there can be a change in the height of the water table (Hibbert, 1967; Douglass and Swank, 1972; Hornbeck and Ursic, 1979).

The increase in stream flow from a watershed is roughly proportional to the severity of the harvest, that is, in terms of the relative amount of foliar transpiration surface that is removed (Troendle, 1987). The increase in flow can be as large as 40% in the first year after the clear-cutting of an entire watershed. Rothacher (1970) found a stream-water yield increase of 32% in a totally clear-cut Douglas-fir (*Pseudotsuga menziesii*) watershed in Oregon, compared with a 12% increase in another watershed that was clear-cut over 30% of its area. Reinhart *et al.* (1963) compared four cutting methods that differed in the intensity of tree removal from mixed-hardwood watersheds in West Virginia. The most intensive technique was clear-cutting, which caused an increase in water yield of 19% in the first postcutting year. In comparison, a less intensive diameter-limit cut caused a 10% increase, and a selection cut caused only a 2% increase.

Usually, the largest increase in stream flow takes place in the first postcutting year, with a rapid recovery to the precutting condition after 3–5 years because of revegetation and the consequent reattainment of transpirational surface area of foliage. In many temperate regions, the largest increases in stream flow occur during late spring, summer, and early autumn, when transpiration normally exerts its strongest influence on watershed hydrology. Douglass and Swank (1975) monitored stream flow from a clear-cut mixed-hardwood watershed for 6 years. In the first postcutting year there was an increase in yield of 11.4 cm; this decreased to 7.3, 5.0, 3.3, 2.0, and 0.9 cm in subsequent years. Hy-

drologic effects were measured for the previously described 16-ha hardwood watershed at Hubbard Brook that was clear-felled and then herbicided for 3 years. In the first 3 postdisturbance years there were increases in streamwater yield of 40%, 28%, and 26%, respectively (Fig. 9.5). However, in the second full season after cessation of the herbicide treatment of the experimental watershed, the hydrologic effect largely disappeared because of the vigorous regeneration of vegetation (Likens, 1985). A similar effect was observed in a clear-felled and herbicided 60-ha watershed in West Virginia, where 5 years after the cessation of herbiciding the streamwater enhancement was reduced to about 20% of the initial impact, again because of a rapid revegetation of the site (Kochenderfer and Wendel, 1983).

Forest management can also affect the size and timing of the peak stream-water flow from a watershed. These effects can occur both for storm flow and for spring meltwater flow (Hewlett and Helvey, 1970; Leaf and Brink, 1972; Verry, 1972; Hornbeck, 1973; Swanson *et al.*, 1986; Hornbeck *et al.*, 1987a). The storm-flow effect takes place because deforested terrain has a relatively small ability to moderate the speed of the lateral flow of water, and thereby enhance its rate of percolation into the ground and ultimately to stream water. This occurs because the compacted soils that are often present in harvested watersheds encourage overland water flow. The spring meltwater effect is caused when the rate of snow melt is increased by the relatively unshaded condition of a clear-cut. In addition, the accumulated snowpack is sometimes larger in forests than in clear-cuts, because the relatively exposed conditions of the clear-cuts can favor the evaporation of snow by sublimation (Golding and Swanson, 1986)

A change in the nature of the forest community can also affect hydrology. After the conversion of a mixed hardwood forest to a white pine (*Pinus strobus*) forest in the southern Appalachians of the eastern United States, there was a decrease in annual stream flow of about 20% (Douglass and Swank, 1975). The reason for this effect was that the pines had a more extended transpiration season

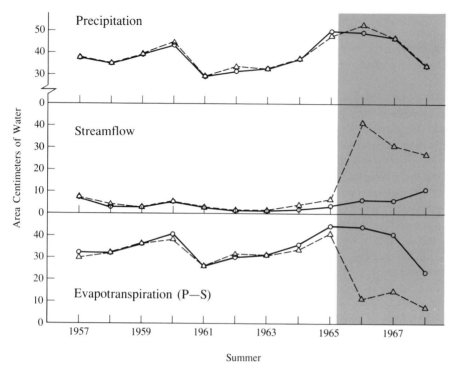

Fig. 9.5. Comparison of summer hydrology (June 1 to Sept. 30) of a devegetated (dashed line) and reference (solid line) watershed at Hubbard Brook, New Hampshire. The shaded area represents the period when the devegetated watershed had all trees felled and herbicides applied for three growing seasons. The preceding period was used for calibration, and reflects annual variations in summer hydrology. From Bormann and Likens (1979).

than did the seasonally deciduous angiosperms that they replaced. In the same region, the conversion of a mixed-hardwood forest to a grassland resulted in a long-term increase in water yield (Douglass and Swank, 1975).

9.6
FOREST HARVESTING AND WILDLIFE

The disturbance of forested terrain by clear-cutting causes a drastic change in the suitability of sites as habitat for some animals. At the same time, new habitat and opportunities are created for early successional, ruderal species. Habitat changes that are

of particular importance to wildlife include (1) a modification of the physical structure of the ecosystem by a change in the spatial distribution of plant biomass; (2) a change in the plant species mixture; and (3) effects on the rates of processes such as primary production, nutrient cycling, and litter decomposition. On the short term, the net effect of harvesting is that the habitat of some wildlife species is improved by forest harvesting, while at the same time the habitat of species that require mature, undisturbed forest is destroyed. In this section, we will discuss the effects of certain forest harvest practices on deer, birds, and sport fish. The effects of silvicultural insecticide and herbicide spray programs on wildlife were dealt with in Chapter 8.

Big-Game Species

Certain species of deer that feed by browsing woody stems require brushy habitat for at least part of the year. These deer can benefit from human activities that create such habitat, for example, the harvesting of forests and the abandonment of poor-quality agricultural land. These land-use practices can create a habitat that is dominated for as long as 15 years by woody shrubs, usually with a rich and lush understory of forbs (i.e., herbaceous dicotyledonous plants) and monocotyledonous species. Such a habitat can favor a large population of certain species of deer in an area where they were rare when the landscape was covered with continuous mature forest.

In North America, the best examples of such deer species are the white-tailed deer (*Odocoileus virginianus*) and the mule deer (*O. hemionus*). In recent decades, both of these species have increased greatly in abundance in many parts of their range because of an increased availability of suitable habitat, caused by regrowth after the disturbance of forest by harvesting, and after the abandonment of poor-quality farmland. In some areas, a decreased population of natural predators has also been important.

For example, white-tailed deer were uncommon in Nova Scotia at the time of its settlement by Europeans. The species was quickly extirpated by hunting, but it was then reestablished by deliberate introductions and by natural immigration (Banfield, 1974; Benson and Dodds, 1977). At present, white-tailed deer are probably more abundant in Nova Scotia than at any other time since deglaciation. It is the province's most important big-game species, and its population supported an all-time high legal harvest of at least 66,000 individuals in 1985, and a hunter success (i.e., number of deer killed per licence) of 0.68 (Patton, 1986).

The most important reason for the modern abundance of white-tailed deer in Nova Scotia is the widespread availability of early successional, shrubby habitat as a result of regeneration after clear-cutting, and to a lesser extent after the abandonment of poor-quality pasture lands. The clear-cuts are generally distributed over the landscape as a mosaic of stands in various stages of secondary succession, within a larger matrix of mature forest. This spatial arrangement optimizes the suitability of the habitat, because it allows for a large amount of habitat edge, a large production of nutritious and palatable browse species in the immature stands, and good cover in mature forest. The latter "yarding" habitat is especially important in northern areas where the winter climate is severe. Many studies have demonstrated the importance to deer of these various habitat features, especially if they are present within a mosaic of stands of different age. This sort of landscape is more favorable to white-tailed deer than one that contains either very small or very large clear-cuts, or an unbroken expanse of mature forest (Krefting, 1962; Harlow and Palmer, 1967; Telfer, 1967; Patton, 1974; Lyon and Basile, 1980; Lyon and Jensen, 1980; McNicol and Timmermann, 1981).

White-tailed deer eat a great variety of woody plants and herbs, with the relative importance of particular species varying between and within regions (Martin *et al.*, 1951; Telfer, 1972; Skinner and Telfer, 1974; Rogers *et al.*, 1981). These preferred plant species may be particularly abundant on cut-over sites, as compared with mature forest. In addition, the biomass of both browse and forbs increases progressively after cutting. Shrubs increase to a maximum after about 8–15 years, followed by a large decline as the maturing tree canopy progressively shades the understory. These temporal patterns are illustrated for 22 hardwood stands of different age (Table 9.9). The younger stands in this chronosequence were created by clear-cutting of the mature forest. Shrub biomass increased from less than 1 MT/ha in 1-year-old stands, to a maximum of 17–19 MT/ha at 8–13 years, and it then decreased to less than 4 MT/ha at 50–75 years. Herbaceous plants exhibited a similar pattern, but their biomass peaked earlier at 2–6 years after cutting.

In addition, the nutritional quality of browse on cut-overs is generally better than in mature forest. Recently sprouted, rapidly growing twigs have larger concentrations of protein, nitrogen, and

Table 9.9 Shrub and herb biomass in stands of different age in a region of maple–birch forest in Nova Scotia[a]

Stand Age (year)	Number of Stands Sampled	Shrub Biomass (MT d.w./ha)		Herb Biomass (MT d.w./ha)	
		Average	Range	Average	Range
1	2	0.45	0.41–0.49	1.1	0.93–1.2
2	2	1.9	1.8–2.1	1.6	1.4–1.7
3–6	4	8.2	6.6–12.2	1.7	1.5–2.1
8–13	5	17.6	16.9–18.7	0.57	0.27–1.0
20	1	10.9	—	0.14	—
30–40	3	4.8	0.8–7.9	0.25	0.10–0.47
50–75	5	2.4	1.1–3.9	0.23	0.7–0.45

[a]Stands 20 years old and younger originated by clear-cutting of the mature forest. Older stands were part of a wildfire-created mosaic. Unpublished data of M. Crowell and B. Freedman.

phosphorus, and they are more succulent and more easily digested than is the older browse of mature forest (Halls and Epps, 1969; Short *et al.*, 1975).

Therefore, on clear-cuts deer food is present in a relatively large quantity, and it is of good nutritional quality. However, food availability is not necessarily the major factor that influences the suitability of habitat to white-tailed deer. Other important factors include (1) the size and shape of the clear-cut, especially as these influence the ratio of edge:area and the distance of lines-of-vision; (2) ongoing disturbance and harassment; (3) barriers to movement; and (4) the previous experience or tradition of the animals.

In general, a very large clear-cut may preclude the use of the central portion of the habitat by deer, as these animals prefer to not be excessively far from a protective forest cover. For optimal white-tailed deer habitat in New Brunswick, Boer (1978) recommended that a clear-cut be no larger than 4 ha, while Euler (1978) recommended less than 2-ha clear-cuts in southern Ontario. Reynolds (1962, 1966, 1969) suggested 9–14 ha as the maximum size of clear-cut that deer would fully utilize in a variety of habitats in the U.S. Southwest.

It is generally felt that the more irregular-shaped the clear-cut, the more favorable is the resulting habitat for deer and other species of wildlife (Halls, 1978; Lyon and Basile, 1980; Lyon and Jensen,

1980). Irregular shapes have a larger ratio of edge to surface area than do round, square, and rectangular shapes, and therefore they create relatively more ecotonal habitat. In addition, an irregular shape makes more of the central clear-cut habitat accessible, if it is assumed that deer will not venture further than some maximum distance from forest cover into a clear-cut. Furthermore, an irregular-shaped clear-cut breaks up lines of vision, and helps to make deer and humans less visible to each other.

Some characteristics of clear-cuts can create barriers to deer movement, and these could restrict the use of otherwise suitable habitat. A barrier could take the form of a tangle of logging slash, the presence of a logging road with frequent traffic, or the presence of relatively deep snow on some clear-cuts because of a lack of canopy interception, especially by a conifer canopy (Pengelly, 1972; Beall, 1976; Lyon, 1976; Lyon and Basile, 1980). Sometimes, the above snowpack effect occurs for only a short time after snowfall events; under some climatic regimes the accumulated snowpack is greater in the forest, because of the greater amount of evaporation of snow from relatively open clear-cuts (Golding and Swanson, 1986). Snow depth is important, since deer mobility is restricted by snow depths greater than about 50 cm (Pengelly, 1972; Newman and Schmidt, 1980). It is important that forest management in an area with severe winters is conducted in a way that continues to provide suitable refuge habitat of mature forest in which the microclimate is less severe, browse is available, and snow depth is not extreme. The availability of such winter yarding habitat may be more important than the quantity and quality of summer habitat in the annual use of a managed area by deer (Boer, 1978; Telfer, 1978; Moore and Boer, 1979). Since particular winter yards are often used for many years by deer from a large surrounding area (Telfer, 1978; Lyon and Jensen, 1980), these traditional and necessary habitat features should be identified and protected from any cutting that could detract from this use.

Finally, it should be mentioned that some types of ongoing disturbance in a forest management area can be detrimental to the use of otherwise suitable habitat by deer. These can include frequent vehicular traffic along roads, noise from a harvesting operation, and excessive hunting pressure (Fletcher and Busnel, 1978; Halls, 1978; Lyon and Jensen, 1980).

Other important North American deer species that can benefit from certain forest harvesting practices include moose (*Alces alces*) and elk (*Cervus*

elaphus). Like the *Odocoileus* deer, moose are primarily browsers, although they also feed on aquatic and terrestrial herbs during the summer. Elk primarily graze on graminoids and herbaceous dicots during the growing season, but they browse in the winter when herbs are unavailable (Martin *et al.*, 1951; Peek, 1974; Crete and Bédard, 1975; Joyal, 1976; Telfer, 1978). Since the availability of browse and many herbs can be increased by forest harvesting, an integrated program of forestry and wildlife management could potentially improve the habitat for these species (Dodds, 1974; Krefting, 1974; Parker and Morton, 1977; Telfer, 1978; McNicol and Timmermann, 1981). In general, moose and elk are somewhat less favored by forest harvesting than are white-tailed deer and mule deer.

Crouch (1985) examined the effects of clearcutting a subalpine conifer forest in Colorado on the abundance of mule deer and elk (Table 9.10). He examined three habitat types: (1) four uncut, reference blocks of mature forest on mesic–xeric sites, (2) four mesic–xeric clear-cuts dominated by shrubs and herbs, and (3) one mesic–hydric clear-cut dominated by herbs. After clear-cutting of the

Table 9.10 Deer density on uncut and clear-cut habitats of a subalpine forest in central Colorado[a]

Species	Index of Abundance					
	Before Logging	Years after Logging (1978–1982)				
		1	2	3	4	5
Elk (*Cervus elaphus*)						
Uncut forest	57	82	44	49	69	44
Mesic–xeric clear-cuts	44	7	0	37	69	94
Mesic–hydric clear-cut	74	0	271	173	222	304
Mule Deer (*Odocoileus hemionus*)						
Uncut forest	32	44	37	44	86	136
Mesic–xeric clear-cuts	44	12	37	99	148	235
Mesic–hydric clear-cut	74	25	99	173	222	272

[a]The uncut habitat (*n* = 4) was a mature forest of subalpine fir (*Abies lasiocarpa*), lodgepole pine (*Pinus contorta*), and Engelmann spruce (*Picea engelmannii*) growing on mesic–xeric sites. The mesic–xeric clearcuts (*n* = 4) were initially similar to the uncut habitat, but after 5 postharvest years they were dominated by the shrubs *Vaccinium* spp., *Pachystima myrsinites*, and *Rosa* spp., and by a vigorous growth of graminoids and herbaceous dicots. The mesic–wet clearcut (*N* = 1) was initially a *Picea engelmannii* stand, but after 5 postharvest years it was dominated by a lush growth of graminoids and herbaceous dicots. Deer and elk density were indexed as the mean number of fecal groups per hectare. Modified from Crouch (1985).

mesic–xeric stands, the abundance of mule deer decreased for 1 year, and then progressively increased until by the fifth postcutting year it averaged 5.3 times larger than the precutting level. On the mesic–hydric site abundance was also greater after five postcutting years, by a factor of about 3.7. Interpretation of these trends is complicated by the observation that mule deer were also more abundant in the reference habitat in the fourth and fifth postcutting years, by a factor of 4.3 in year 5. Although it cannot be unequivocally concluded that the abundance of mule deer was increased by clearcutting, they certainly were not affected negatively within the period of this study. The effects on elk were more straightforward. After an initial decline in abundance for 1–2 years in the clearcut habitats, their abundance increased by a factor greater than 2, while on the reference plots abundance was more stable.

In another study, Lyon and Jensen (1980) examined the abundance of deer in 87 clear-cuts of varying size and age, and in adjacent uncut coniferous forests in Montana. Abundance was indexed as the number of fecal groups per hectare. Averaged across all sites, they found an elk density of 77 groups/ha on clear-cuts and 123/ha in forests. The density of white-tailed deer plus mule deer was 81/ha on clear-cuts and 113/ha in mature forest. Although deer use of clear-cuts averaged somewhat less in this study, certain types of clear-cuts were preferred over forest. These clear-cuts were generally characterized by good cover and growth of forage, ease of movement through logging slash, and they were not too close to an active logging road or other ongoing disturbance.

Another important big-game deer species is the woodland caribou (*Rangifer tarandus*; known as the reindeer in Eurasia). For the most part, this species requires an extensive habitat of mature coniferous forest, particularly during winter when the so-called "reindeer moss" lichens (especially *Cladina alpestris*, *C. mitis*, *C. rangiferina*, and *C. uncialis*) comprise the bulk of their diet (Scotter, 1967). In the very short term, cutting the forest can increase the availability of highly palatable arboreal lichens,

because these can be abundant in recently deposited logging slash (Klein, 1974; Eriksson, 1976). However, in the longer term, logging greatly decreases the abundance of reindeer mosses and other lichens that are important winter foods (Klein, 1974; Eriksson, 1976). As a result, woodland caribou would not be expected to benefit in the longer-term from forest harvesting and its associated activities, unlike white-tailed deer, mule deer, moose, and elk (McNicol and Timmermann, 1981; Eriksson, 1976).

A 2-year-old clearcut of a hardwood forest in Nova Scotia. Typical breeding birds in this habitat include white-throated sparrow (*Zonotrichia albicollis*), dark-eyed junco (*Junco hyemalis*), song sparrow (*Melospiza melodia*), and common snipe (*Capella gallinogo*). (Photo courtesy of B. Freedman.)

An 8-year-old clearcut of a hardwood forest in Nova Scotia. Typical breeding birds in this habitat include alder flycatcher (*Empidonax alnorum*), chestnut-sided warbler (*Dendroica pensylvanica*), common yellowthroat (*Geothlypis trichas*), and white-throated sparrow (*Zonotrichia albicollis*). At this stage, an upper canopy is beginning to form and some species more typical of mature forest are beginning to breed, including red-eyed vireo (*Vireo olivaceus*), American redstart (*Setophaga ruticilla*), and veery (*Catharus fuscescens*). (Photo courtesy of B. Freedman.)

Breeding Birds

Many species of bird use forests, or plant communities that are part of a postcutting forest succession, as habitat for breeding, migrating, or wintering.

Some of these birds are game species. Notable North American examples include ruffed and spruce grouse (*Bonasa umbellus* and *Canachites canadensis*), wild turkey (*Meleagris gallopavo*), and common bobwhite (*Colinus virginianus*). These birds are all favored by a habitat mosaic that comprises both mature forest and younger brushy stands, with much edge between these types. The ruffed grouse is the most important upland game bird in North America, with about 6 million harvested annually (the total harvest of all other grouse and ptarmigan species is about 2.4×10^6 per year; Johnsgard,

1983). Ruffed grouse live in a wide variety of habitats, but they prefer a landscape that is dominated by hardwood forest with some conifers mixed in, especially when there is a major component of poplars (particularly *Populus tremuloides*) and birches (especially *Betula papyrifera*) (Edminster, 1947; Gullion, 1967; Johnsgard, 1983). Ruffed grouse primarily feed on the foliage, young twigs, catkins, and buds of woody plants, but they also eat seasonally abundant fruit (Brown, 1946; Martin *et al.*, 1951). Gullion (1969, 1986) found that clear-cut aspen stands in Minnesota became suitable for ruffed grouse after 4–12 years of regeneration. The stands were then used as breeding habitat for 10–15 years. The older, mature stands were most important as wintering habitat. Morgan and Freedman (1986) studied a post–clear-cutting chronose-

A 3-year-old shelterwood cut in a hardwood forest in Nova Scotia. During the cut, about 40% of the tree basal area was removed, leaving the "best" trees to promote their growth into high-value sawlogs and to promote a good advance regeneration of desired tree species. This silvicultural treatment results in a habitat with a vertically complex structure. At this stage, there is a diverse and vigorous ground vegetation, a thick shrub layer largely comprised of groups of stump-sprouts of cut trees, and a sparse tree canopy. The bird community contains species typical of both clearcut and mature forests. (Photo courtesy of B. Freedman.)

quence in Nova Scotia, and found ruffed grouse in clear-cuts 5 years and older. To optimize habitat for ruffed grouse in the U.S. Northeast, Gullion (1977, 1986) recommended a checkerboard mosaic of different-aged stands less than about 10 ha in area, and with adjacent blocks differing in age by 10–15 years. Such an arrangement would provide cover at all seasons, and abundant, accessible food in the form of browse and seasonal berries.

There is a much greater variety of birds of forested landscapes that are not hunted for sport. These are sometimes considered together as "nongame" species, but they can be economically important as predators of injurious insects and small mammals (Bruns, 1960; Martin, 1960; Crawford *et al.*, 1983), and in recreation. These benefits are considerable, but they are difficult to quantify in terms of dollars. It has been estimated that expenditures for the enjoyment (mainly bird-watching and winter feeding) of nongame birds in North America could exceed several billions of dollars annually; sales of bird seed alone account for more than $200 million per year (Payne and DeGraaf, 1975; DeGraaf and Payne, 1975; George *et al.*, 1981). In 1980, an estimated 83 million people in the United States were engaged in some degree of "nonconsumptive" wildlife use, compared with 47 million who were involved in hunting (WRI, 1987).

The effects of forestry on nongame birds can be considered from two directions: (1) the effects on individual species and (2) the effects on the bird community in terms of its overall density, species richness, diversity, etc. Both bird species and the bird community are thought to be strongly influenced by the physical structure and plant species composition of their habitat.

The hypothesis that vegetation structure has an important influence on the size and species composition of bird communities has attracted a great deal of attention from ecologists. Pitelka (1941) suggested that the life forms of the dominant plants of the vegetation were more important to birds than were the actual plant species. Many studies have clearly shown a strong relationship between avian community characteristics and vegetation structure. These include studies of the relationship between spatial complexity and bird species diversity (MacArthur, 1964; MacArthur and MacArthur, 1961; MacArthur *et al.*, 1966; Karr and Roth, 1971; Roth, 1976), and between habitat structure and bird community composition (Johnson and Odum, 1956; James, 1971; Shugart *et al.*, 1978; Collins *et al.*, 1982; Morgan and Freedman, 1986). Notable variables within the theme of habitat structure include vertical structure, horizontal structure, total vegeta-

A 3-year-old progressive strip cut of a hardwood forest in Nova Scotia. The clearcut section is about 40 m wide and 300 m long, and there are intervening similar-sized, uncut strips between the cut strips. After about 5–8 years, the uncut strips will also be harvested. An area composed of a mosaic of cut and uncut strips has a bird community that contains species typical of both clearcut and mature forests, but these segregate among the strips. (Photo courtesy of B. Freedman.)

tion cover, and the distinctness of ecotones at habitat discontinuities. In the seminal papers by MacArthur (1964), MacArthur and MacArthur (1961), and MacArthur *et al.* (1966), vertical structure was divided into three functional strata: the ground vegetation, shrubs, and trees. Horizontal structure involves the spatial complexity of patches of distinct habitat within a larger landscape mosaic. Considerations include the shape of patches with a secondary effect on the ratio of edge to area, and also patch size since small and isolated habitat islands will not sustain bird species that have a large territory (Moore and Hooper, 1975; Formann *et al.*, 1976; Galli *et al.*, 1976; Crawford and Titterington, 1979; Noon *et al.*, 1979; Robbins, 1979; Temple *et al.*, 1979). Another important feature of the habitat is the species composition of the vegetation. In temperate latitudes the degree of mixture of coniferous and angiosperm tree species is particularly impor-

tant (Karr, 1968; Balda, 1975; Crawford and Titterington, 1979). The presence of snags (erect but dead trees) suitable for nesting and other purposes can also be a critical habitat feature for many bird species (Conner *et al.*, 1975; Conner, 1978; Evans and Conner, 1979; Maser *et al.*, 1979; Miller and Miller, 1980; Scott *et al.*, 1980). Finally, the occurrence of an epidemic of a defoliating insect, such as spruce budworm (*Choristoneura fumiferana*), can dramatically increase the abundance of certain insectivorous birds (Bruns, 1960; Martin, 1960; Gage and Miller, 1978; Welsh and Fillman, 1980; Crawford *et al.*, 1983) (see Chapter 8).

These observations and hypotheses have suggested to ecologists that it should be possible to determine whether a given site is suitable for a particular bird species or suite of species by examining the structural and compositional character of the vegetation of the habitat. Amateur ornithologists have long known of this principle in a qualitiative sense. They call it "bird watching by habitat"—particular species and assemblages of birds are expected to be present in certain ecological situations.

It follows that it should be possible to manage a breeding bird community or particular species by manipulation of the habitat (Verner, 1975; Noon *et al.*, 1979). For example, prescribed burning has been used in Michigan to create homogeneous jack pine (*Pinus banksiana*) stands, so as to manage the habitat of the rare and endangered Kirtland's warbler (*Dendroica kirtlandii*). The entire breeding population of this species comprises only about 200 pairs, all of them nesting in central Michigan. The optimal habitat of Kirtland's warbler is even-aged, 1.5–6.0 m tall, 7 to 20-year-old stands of jack pine. The availability of this habitat is maintained and enhanced by the deliberate burning of older stands (Mayfield, 1960; Buech, 1980). The manipulation of habitat structure and stabilization of its plant species composition, coupled with an intensive effort to reduce the impact of nest parasitism by the brown-headed cowbird (*Molothrus ater*), have allowed the maintenance of the small breeding population of Kirtland's warbler (Mayfield, 1961, 1977).

The effects of forest harvesting on bird species are conceptually simple, since each species has its own particular habitat requirements, although these may vary in different parts of its range. On a particular site, habitat that is suitable to bird species of mature forest is modified or destroyed by forest harvesting, while opportunities for some early successional species are created by the same disturbance.

These effects can be illustrated by comparison of the breeding birds of mature hardwood stands and adjacent clear-cuts in Nova Scotia (Table 9.11). The uncut forest in this study was mixed maple–birch, and its bird community was dominated by ovenbird, least flycatcher, red-eyed vireo, black-throated green warbler, and hermit thrush. The total abundance of breeding birds in the forest averaged 663 pairs/km², while species richness averaged 12, with a range of 9–16 among the three census plots. Although the three 3 to 5-year-old clear-cuts had a

very different habitat structure and plant species composition (they were very shrubby, and had a species-rich and dense ground vegetation dominated by a diverse array of graminoids and forbs), their bird density averaged only slightly different from that of the forest at 588 pairs/km², while species richness averaged 8 (7–10). However, the bird species that were present were almost completely different on the clear-cuts, which were dominated by chestnut-sided warbler, common yellowthroat, white-throated sparrow, and dark-eyed junco.

Freedman *et al.* (1981a) also examined stands that were harvested in a way that created a habitat that was intermediate in structure to that of young clear-cuts and mature forest. These harvest methods were (1) shelter-wood cuts, in which about 50% of the tree basal area was removed, with the other trees left so as to encourage regeneration and to produce high-quality sawlogs, and (2) progressive strip cuts, in which about one-half of the area was clear-cut in long (300 m) and narrow (30 m) strips, with intervening uncut strips left in order to encourage the regeneration of tolerant tree species. These intermediate habitats were occupied by a mixture of bird species that are typical of either clear-cuts or mature stands, while bird density and species richness were little affected.

In another study in the same area of Nova Scotia, Morgan and Freedman (1986) examined bird populations and their habitat in a chronosequence of 23 hardwood stands. All stands 20 years old and younger originated after clear-cutting, while older stands were part of a fire-caused landscape mosaic. There was little variation in overall bird community variables among stands of different age (Fig. 9.6a and b), in spite of great differences in habitat. The range of total avian density of clear-cuts 3–10 years old fell within the range of variation of density for the older, mature stands (Fig. 9.6a). Only the 1- and 2-year-old clear-cuts had a bird population that was smaller than that of the uncut stands. In addition, a distinct suite of bird species was present relatively early in the post–clear-cutting succession; these were later replaced by another group as the forest matured (Fig. 9.6c and 6). The transition between these two bird communities took place in stands 12–

Table 9.11 Breeding birds of three mature hardwood forest plots and three adjacent clear-cuts
3–5 years old[a]

Species	Clear-Cut Plots			Mature Forest		
	A	B	C	A	B	C
Common snipe (*Capella gallinago*)	0	0	0	10	15	0
Ruby-throated hummingbird (*Archilochus colubris*)	0	0	0	25	30	15
Least flycatcher (*Empidonax minimus*)	290	120	0	0	0	0
Hermit thrush (*Catharus guttatus*)	60	40	30	0	0	0
Veery (*Catharus fuscescens*)	50	10	0	25	0	0
Solitary vireo (*Vireo solitarius*)	60	30	0	0	0	0
Red-eyed vireo (*Vireo olivaceous*)	80	50	30	0	0	0
Black-and-white warbler (*Mniotilta varia*)	15	50	40	0	0	0
Northern parula warbler (*Parula americana*)	15	30	40	0	0	0
Black-throated green warbler (*Dendroica virens*)	50	30	30	0	0	0
Chestnut-sided warbler (*Dendroica pensylvanica*)	0	0	0	100	40	190
Ovenbird (*Seiurus aurocapillus*)	150	120	200	0	0	0
Mourning warbler (*Oporornis philadelphia*)	0	0	0	0	0	90
Common yellowthroat (*Geothlypis trichas*)	0	0	0	25	300	130
American redstart (*Setophaga ruticilla*)	15	80	100	0	0	0
Rose-breasted grosbeak (*Pheucticus ludovcianus*)	15	10	0	0	0	0
Dark-eyed junco (*Junco hyemalis*)	15	20	15	50	70	30
White-throated sparrow (*Zonotrichia albicollis*)	0	20	0	90	190	100
Song sparrow (*Melospiza melodia*)	0	0	0	90	70	0
Total density (pairs/km²)	815	660	515	435	745	585
Total number of species	12	16	9	10	8	7

[a]The mature forest had a closed canopy dominated by maple and birch (*Acer saccharum, A. rubrum, Betula papyrifera, B. allegheniensis*). Regeneration on the clear-cuts was vigorous, with a shrub stratum dominated by maple and birch stump sprouts, raisinbush (*Viburnum cassinoides*), and pin cherry (*Prunus pensylvanica*), and a dense ground vegetation of raspberry (mainly *Rubus strigosus*), graminoids, and dicotyledonous herbs. Some uncommon bird species are not included. Modified from Freedman *et al.* (1981a).

20 years old. During this period hardwood stump sprouts and saplings had thinned to relatively few stems with a canopy at 8–12 m. At the same time the shrub and ground vegetation strata were declining in abundance because of shade and other competitive stresses exerted by the developing tree canopy. These habitat changes allowed the progressive invasion of the stand by bird species that are characteristic of mature forest. The distinctness of the "early" and "late" bird communities was shown by a cluster analysis of the data matrix of bird species abundance versus stand age. This multivariate procedure strongly separated the bird communities of stands aged 1–12 years old from those of stands 20–74 years old. Clusters of stands within these two age groups were much weaker, indicating that their bird communities were similar.

A similar pattern of change in the breeding bird community has been reported in postlogging chronosequences of conifer forest. Welsh and Fillman (1980) described the changes in breeding birds among nine clear-cuts that ranged in age from 3 to 24 years, plus a reference black spruce (*Picea mariana*) forest in northern Ontario (Table 9.12). The greatest abundance of breeding birds was present in moderate-aged cutovers. Stands aged 11–24 years had from 1020 to 1970 pairs/km², considerably more than were present in the mature, uncut forest (561 pairs/km²). The smallest avian abundance was in the 3-year-old cutover with about 200 pairs/km², but by 5 postcutting years the abundance of birds had recovered to 690 pairs/km².

The stands studied by Welsh and Fillman had not been completely clear-cut—they all had occasional uncut islands of trees, or scattered uncut stems. As a result, some of the birds that were dominant in the

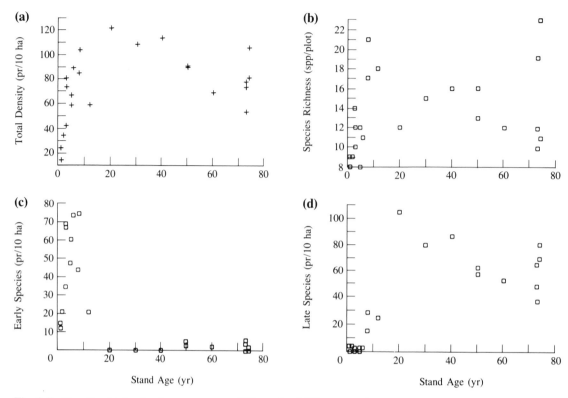

Fig. 9.6. Breeding birds of a chronosequence of 23 stands of different age in a region of hardwood forest in Nova Scotia. Note the replication of some stand ages. All stands younger than 20 years originated from clear-cutting of the mature forest. Stands 20 years and older were part of a fire-created landscape mosaic. (a) Total density, or the sum of density of all species; (b) species richness, or the number of species breeding on each plot; (c) sum of densities of species that were prominent in young stands (i.e., alder flycatcher, *Empidonax alnorum*; chestnut-sided warbler, *Dendroica pensylvanica*; common yellowthroat, *Geothlypis trichas*; dark-eyed junco, *Junco hyemalis*; white-throated sparrow, *Zonotrichia albicollis*; and song sparrow, *Melospiza melodia*; (d) sum of densities of species that were prominent in older stands (i.e., least flycatcher, *Empidonax minimus*; hermit thrush, *Catharus guttatus*; veery, *Catharus fuscescens*; red-eyed vireo, *Vireo olivaceous*; black-throated green warbler, *Dendroica virens*; American redstart, *Setophaga ruticilla*; and ovenbird, *Seiurus aurocapillus*). Modified from Morgan and Freedman (1986).

mature forest were also present on relatively young clear-cuts. Obviously, these species did not require a continuous, closed forest as their habitat. A good example of such a species is the Tennessee warbler, which was present in all stands except the 3-year-old cutover. It was the most prominent species in the uncut stand, where it comprised 18% of all breeding birds. However, the Tennessee warbler was most abundant in cutovers aged 11–24 years old, where its density ranged from 217 to 678

pairs/km². The most prominent species in the 3-year-old cutover were Lincoln's sparrow, savannah sparrow, and song sparrow. Only the latter two species were restricted to this young habitat—Lincoln's sparrows were present on almost all stands in the chronosequence. Overall, the effects on breeding birds of the clear-cutting of this lowland black spruce forest were not great. If anything, the overall avian abundance was greater in the vigorously regenerating, moderately aged cutovers.

Table 9.12 Breeding birds in a post-clear-cutting chronosequence in a lowland black spruce (*Picea mariana*) forest in northern Ontario[a]

Species	Uncut Forest	24-Year Cutover	19-Year Cutover	17-Year Cutover	13-Year Cutover	11-Year Cutover	9-Year Cutover	6-Year Cutover	5-Year Cutover	3-Year Cutover
Tennessee warbler (*Vermivora peregrina*)	100	489	678	283	622	217	78	106	94	
Yellow-rumped warbler (*Dendroica coronata*)	67	78	28	11	28	11		p		
Spruce grouse (*Canachites canadensis*)	50				11		p			
Least flycatcher (*Empidonax minimus*)	50	61	39	44	22					
Nashville warbler (*Vermivora ruficapilla*)	50	78	50	67	67	50	67	6	11	
Cape Map warbler (*Dendroica tigrina*)	50	117	22	67						
Ruby-crowned kinglet (*Regulus calendula*)	34		22	11	11					
Alder flycatcher (*Empidonax alnorum*)	25	6	200	39	206	133	83	56	67	
White-throated sparrow (*Zonotrichia albicollis*)	25	128	256	267	211	183	228	78	111	14
Lincoln's sparrow (*Melospiza lincolnii*)	25		133	28	133	50	100	67	78	61
Yellow-bellied flycatcher (*Empidonax flaviventris*)	17	11		11	6					
Cedar waxwing (*Bombycilla cedrorum*)	17									
Magnolia warbler (*Dendroica magnolia*)	17	167	139	83	67	106	39	56	22	
Blackburnian warbler (*Dendroica fusca*)	17	33								
Common yellowthroat (*Geothlypis trichas*)	17		244	6	222	133	106	111	150	11
Swainson's thrush (*Catharus ustulatus*)		67	67	17	50	22				
Hermit thrush (*Catharus guttatus*)		61	22	28						

(continued)

Table 9.12 (*Continued*)

Species	Uncut Forest	24-Year Cutover	19-Year Cutover	17-Year Cutover	13-Year Cutover	11-Year Cutover	9-Year Cutover	6-Year Cutover	5-Year Cutover	3-Year Cutover
Bay-breasted warbler (*Dendroica castanea*)		61								
Black-and-white warbler (*Mniotilta varia*)		50					11			
Red-eyed vireo (*Vireo olivaceus*)		44					11	p		
Northern waterthrush (*Seiurus noveboracensis*)		39	44		50					
Yellow-bellied sapsucker (*Sphyrapicus varius*)		28					11	11		
Dark-eyed junco (*Junco hyemalis*)		28								
American robin (*Turdus migratorius*)		17						6		8
Solitary vireo (*Vireo solitarius*)		11								
American redstart (*Setophaga ruticilla*)		11								3
Purple finch (*Carpodacus purpureus*)		11	17	33	17	17				
Wilson's warbler (*Wilsonia pusilla*)			11	22	44	83	39	11	50	11
Winter wren (*Troglodytes troglodytes*)				11				11		
Swamp sparrow (*Melospiza georgiana*)				6			11	11	56	6
American bittern (*Botaurus lentiginosus*)					11					

	1	2	3	4	5	6	7	8	9	10
Boreal chickadee (*Parus hudsonicus*)					11					
Veery (*Catharus fuscescens*)					6					
Mourning warbler (*Oporornis philadelphia*)	6	11		33						
Chestnut-sided warbler (*Dendroica pensylvanica*)		44	72	17						
Ruffed grouse (*Bonasa umbellus*)				11						
Chipping sparrow (*Spizella passerina*)				6						
Merlin (*Falco columbarius*)				p						
Song sparrow (*Melospiza melodia*)	25		6							
LeConte's sparrow (*Ammospiza leconteii*)		11								
Killdeer (*Charadrius vociferus*)	6	p								
Savannah sparrow (*Passerculus sandwichensis*)	33									
Common snipe (*Capella gallinago*)	6									
Spotted sandpiper (*Actitis macularia*)	3									
Red-winged blackbird (*Agelaius phoeniceus*)	3									
Total number of species	14	12	17	18	13	17	18	16	21	15
Total density	196	694	619	851	1022	1778	1034	1972	1518	561

[a]Modified from Welsh and Fillman (1980). Data are in pairs/km^2; p, present.

As mentioned previously, a habitat feature that influences the presence and success of certain bird species is the density of dead trees in the forest, either as standing snags or as logs lying on the forest floor. These habitat features are used by birds for a number of purposes, such as nesting in excavated and natural cavities, feeding substrate, and perches for hunting, resting, feeding, and singing. Scott *et al.* (1980) estimated that 30–45% of the breeding bird species of typical conifer and aspen forest of the northwestern United States are cavity nesters, which therefore require snags as a necessary habitat feature. These species include primary excavators such as woodpeckers, secondary cavity users, and species that use natural cavities. Clearly, snag management can be an important variable that influences the impacts of forest harvesting on birds.

The maintenance of the habitat of cavity-nesting birds has become an important consideration in the multiple-use management of many forests of the United States, particularly in the Pacific Northwest, where as many as six species of woodpeckers can co-occur in relatively large populations (Davis *et al.*, 1983; Bull *et al.*, 1986). Thomas *et al.* (1979) noted that woodpeckers of the Blue Mountains of eastern Oregon require the following snag densities as a minimal component of their habitat: pileated woodpecker (*Dryocopus pileatus*), 0.32 snags/ha; common flicker (*Colaptes auratus*), 0.93 snags/ha; black-backed three-toed woodpecker (*Picoides arcticus*), 1.5 snags/ha; and hairy woodpecker (*Picoides villosus*), 4.5 snags/ha. For the same area, Bull *et al.* (1980) recommended that at least 4 snags/ha be left on a harvested site, so as to ensure the maintenance of at least 70% of the potential woodpecker population.

If a forested landscape is to be managed for both fiber and birds, a number of practices should be followed (after Evans, 1978; Freedman, 1982): (1) a proportion of the management area should be reserved for mature forest, in blocks that are as large as possible, and interconnected by corridors of mature habitat whenever possible; (2) cut portions should generally be arranged to produce a mosaic of habitats of different age abutting each other; (3) dead, dying, and girdled cull trees should be left

standing whenever possible, to provide habitat for birds that require snags; (4) less intensive harvest techniques such as thins, shelter-wood cuts, and strip cuts will produce a less severely modified habitat, and in some cases these methods could be used with relatively little detriment to avian species composition or abundance; and (5) if rare or endangered bird species are present in the management area, a specific management plan should be developed to preserve or even enhance their habitat.

Except for item 5, the rationale has previously been discussed for the use of these various practices in a multiple-use management system that involves both fiber, and birds and other wildlife. Point 5 can be illustrated by reference to the case of the spotted owl (*Strix occidentalis*) of the western United States and extreme southwestern Canada. This nonmigratory raptor has an obligate requirement for large tracts of old-growth conifer forest—that is, each breeding pair requires more than about 600 ha of forest older than 140–170 years (Lee, 1985; Dawson *et al.*, 1987). Because forest of this type is of great commercial value and its area has been reduced and fragmented by logging, the abundance of spotted owls has been critically decreased in many areas (Harris, 1984; Gutierrez and Carey, 1985). [Note that although old-growth forests are a very valuable natural resource because of their considerable standing crop of large-dimension timber, such stands are rarely managed by foresters as a renewable natural ecosystem. Rather, old-growth resources are usually "mined" by harvesting, which converts them to ecosystems of younger character, which are only allowed to successively redevelop into middle-aged forest before they are harvested. The reason for this management strategy is that old-growth forest has little or no positive net production (at the stand level, the production by living trees is approximately balanced by the death of other individuals by disease, senescence, or accident). Hence, if the primary management objective is to optimize the production of tree biomass on the landscape, then it is much better to harvest the middle-aged stand soon after its productivity suffers a large decrease, i.e., before it becomes a "decadent" old-growth forest.]

In recognition of the threat posed to the spotted owl by the logging of old-growth forests of Washington, Oregon, and California, specific management plans that incorporate the conservation of the habitat of this species have been formulated for many areas. In National Forests in this region, stands that are suitable for spotted owls have been surveyed for the presence of breeding pairs of this species, and in the great majority of cases occupied habitat is not harvested as long as it remains suitable for this owl. The longer-term objective of this management practice is to conserve sufficient old-growth forest to ensure a viable breeding population of the spotted owl in the western United States, even if some economic detriment is suffered by the forest industry because of the withdrawal of potentially harvestable, high-value timber (Gutierez and Carey, 1985).

The esthetic and economic benefits of multiple-use management practices can be further enhanced by providing bird-watching opportunities, for example, by identifying and publicizing unique or diverse birding areas, by providing facilities where birds are easily visible to people, and by developing interpretation programs to promote an understanding and appreciation of birds. Such management practices would, of course, also enhance the value of wildlife other than birds, such as game wildlife, small mammals, and vegetation. If implemented, such a scheme could have sociological, educational, and economic benefits that would increase the value of the overall management area, and of the wildlife resource itself.

Freshwater Biota

Forestry practices can detrimentally affect populations of fresh-water biota in three major ways: (1) by causing siltation of habitat; (2) by causing an increase in water temperature by the removal of stream-side shading vegetation; and (3) by the blockage of stream channels with logging slash and other debris. In some cases, damage can also be caused by chemical and fuel spills, and as a result of pesticide spraying.

Compared with reference watersheds, Welch *et al.* (1977) found that clear-cut watersheds in New Brunswick had 17% fewer brook trout (*Salvelinus fontinalis*) and 26% fewer benthic invertebrates, but 200% more sculpin (*Cottus cognatus*), a nongame species of fish. Murphy and Hall (1983) did not find consistent differences in the abundance of salmonid fish and insects between streams running through logged and uncut conifer forest in Oregon. In fact, stream sections in clear-cuts 5–17 years old tended to have a larger biomass, density, and species richness, compared with sections in old-growth forest. In another case, Haefner and Wallace (1981) reported a doubling of the abundance of aquatic insects in an Appalachian stream running through habitat that 10 years previously had been converted from a mature hardwood forest to a grassland, by clear-cutting, herbiciding, and fertilization.

Erosion resulting from poorly controlled forestry operations can cause serious problems of turbidity and siltation of fresh-water habitat (Packer, 1967b; Burns, 1972; Corbett *et al.*, 1978; Murphy *et al.*, 1981). Stream-water turbidity caused by fine inorganic particulates is rarely directly lethal to fish, since they can tolerate a large concentration of suspended solids, especially if the particulates are not abrasive. Cordone and Kelley (1961) found no detectable behavioral response by salmonids to a short-term bioassay concentration of turbidity of less than 20,000 ppm, and no lethal effect in short exposures unless the concentration exceeded 175,000 ppm. In contrast, Phillips (1970) found that a longer-term bioassay exposure of only 200–300 ppm was lethal to salmonids. Even these latter concentrations of turbidity are considerably larger than are likely to occur for a protracted period in streams that are affected by erosion caused by forestry. However, indirect effects of turbidity could be of importance to fish. By reducing light transmission, turbidity could decrease primary productivity and make feeding difficult for sport fish, most of which are visual predators (Hesser *et al.*, 1975; Corbett *et al.*, 1978).

Coarse particles that have been mobilized by erosion can be of greater importance, since they can settle out of the water and cover critical breeding habitat of fish (Beschta and Jackson, 1979). Salmonid and many other types of fish require a clean gravel stream bed in which to spawn in the autumn. Salmonid eggs hatch in early spring in the gravel interstices, where the alevins remain until mid to late spring, when they emerge to the stream water as fry. The gravel bed must be well flushed in order to maintain a high oxygen concentration and to remove toxic metabolites. If the interstices are filled with eroded sediment, the circulation of water is impeded, with a deleterious effect on these early life-history stages. Even if the eggs and alevins do manage to survive, their emergence from the gravel can be prevented by silt deposits (Hall and Lantz, 1969; Corbett et al., 1978). Clean gravel is also required to provide young fish with cover from predators and from high water velocity, and to provide an overwintering site.

Therefore, erosion caused by forestry operations can have a severe effect on the habitat of fresh-water fish, particularly for gravel-spawners such as salmonids. As noted previously in this chapter, these effects of erosion can be greatly reduced if care is taken during the construction of roads and culverts, in the skidding of logs, and by leaving a buffer strip of uncut forest beside water courses.

If it is excessive, the accumulation of slash and other debris can degrade fresh-water habitat, mainly by creating obstructions to the movement of fish. However, in a smaller quantity this debris can be beneficial by providing habitat structure and cover (Narver, 1970; Slaney et al., 1977a,b). The use of uncut buffer strips beside water courses and the directional felling of trees away from water can almost eliminate debris inputs into streams (Slaney et al., 1977a).

Many studies have shown that the temperature of a water course can increase after the removal of stream-side vegetation by forest harvesting. This effect is caused by direct heating of the water after the removal of the shading vegetation (Brown, 1969, 1970). Studies in West Virginia showed that the stream-water temperature averaged 4.4°C higher in the first year after clear-cutting of the surrounding forest, compared with uncut, reference parts of the watercourse. There was a maximum

summer temperature of 26°C on the cut portion. Water temperature returned to the reference level after 4 years of regeneration (Kochenderfer and Aubertin, 1975). Another study in West Virginia found that the maximum daily stream-water temperature increased to 26°C after clear-cutting and herbiciding, compared with a reference value of 18°C (Corbett et al., 1978). In North Carolina, the maximum stream-water temperature reached 29°C after clear-cutting, which was as much as 7°C higher than the reference value. The temperature returned to the reference level in the fourth postcutting year (Swift and Messer, 1971). After a streamside clear-cut in Oregon, there was a water temperature increase of as much as 16°C. The annual maximum temperature increased to 29°C in the first postcutting year, compared with a reference value of 14°C (Brown and Krygier, 1970).

Holtby (1988) measured changes in water temperature in a stream draining a watershed in British Columbia that had 41% of its area clear-cut. The increase in average water temperature ranged from 0.7°C in December to 3.2°C in August. In this case, the warmed water resulted in an earlier postwinter emergence of fry of coho salmon (*Oncorhynchus kisutch*), causing an extended growing season of about 6 weeks, a consequently larger size and increased overwinter survival of fingerlings, and then a 47% increase in the abundance of smolt-sized fish. However, the warmed water also caused an earlier seaward migration of smolts, which probably caused a decreased rate of survival to adulthood. Overall, it was estimated that the net effect of logging was to cause a 9% increase in the abundance of adult salmon.

In general, a large increase or fluctuation of water temperature in a forest stream is not desirable, since salmonid fishes such as trout and salmon are sensitive to high temperatures. This is particularly true during the late summer, when the water of even undisturbed rivers and streams can attain quite high temperatures. For example, Huntsman (1942) reported lethal high temperatures in the Moser and St. Mary's Rivers in Nova Scotia during the droughtly summer of 1939. He attributed the death of both adult and immature (parr) Atlantic salmon (*Salmo salar*) and brook trout to the high temperatures that occurred on sunny days during a period of low flow.

The optimum range of water temperature for adult Atlantic salmon and brook trout is about 15–20°C. These salmonids begin to show signs of stress at temperatures greater than about 20–21°C (Fisher and Elson, 1950; Stroud, 1967; Swift and Messer, 1971; Hokenson *et al.*, 1973). For adult and juvenile life stages of trout and salmon the upper lethal limit is about 25°C (Fry *et al.*, 1946; Alabaster, 1967; Hynes, 1971), although Huntsman (1942) re-

ported that Atlantic salmon parr could survive up to 33°C, and grilse up to 29°C. However, salmonid eggs have an upper lethal temperature of only about 14°C (Hynes, 1971). Other fish species can tolerate higher temperatures: up to 36°C by white suckers (*Catostomus commersoni*) and 38°C by carp (*Cyprinus carpio*) (Huntsman, 1942; Hynes, 1971). Water temperatures as high as these have been reported after the removal of vegetation from forest stream sides in many areas. However, this impact can easily be avoided by leaving a buffer of uncut, shading forest beside watercourses.

10

LOSS OF SPECIES
RICHNESS

INTRODUCTION

The extinction of a species represents an irrevocable and regrettable loss of a portion of the biological richness of the earth. Extinction can be a natural process, being caused by random catastrophic events, by biological interactions such as competition, disease, and predation, by chronic physical stresses, and by frequent disturbance. However, with the recent ascendance of humans as the dominant large animal on earth and the perpetrator of global environmental changes, there has been a dramatic increase in the rate of extinction of species. The recent wave of anthropogenic extinctions has included such well-known cases as the dodo, the passenger pigeon, the great auk, and others. There are also many high-profile species that humans have brought to the brink of extinction, including the plains buffalo, the whooping crane, the ivory-billed woodpecker, and various species of marine mammal. Most of these instances were caused by an unregulated and insatiable overexploitation of species that could not sustain a high rate of mortality, often coupled with an intense disturbance of their habitat. However, in the future, anthropogenic hab-

itat destruction will be an increasingly dominant cause of extinction.

Over and above the tragic and well known cases of the extinction or endangerment of large vertebrate species, the earth's biota is facing, and is already experiencing, an even more tragic loss of it's species richness. In large part, this loss is being caused by the conversion of large areas of tropical ecosystems, particularly moist forests, to agricultural or otherwise ecologically degraded habitats. A large fraction of the species richness of tropical biomes is comprised of endemic taxa with local distributions. Therefore, the conversion of large areas of natural tropical forest to habitats that are unsuitable for the continued presence of these specialized taxa inevitably causes the extinction of most of the locally endemic biota. Remarkably, the species richness of tropical forests is so large, particularly in insects, that most of it has not been described taxonomically. We are therefore faced with the prospect of a mass extinction of perhaps millions of species, before they have even been recognized by science. For esthetic reasons, this is an unfortunate and reprehensible way to proceed. This is also a strategically foolish way for humans to play the game of global dominance, because these unique and irrevocable organisms could disappear before their importance as components of ecosystems has been discovered, and before they have been examined for their possible utility in medicine or agriculture.

In this chapter, the problem of the anthropogenic impoverishment of the earth's biota will be examined.

10.2
SPECIES RICHNESS
OF THE BIOSPHERE

About 1.7 million organisms have been identified and designated with a binomial name. About 6% of the identified species live in boreal or polar latitudes, 59% in the temperate zones, and the remaining 35% in the tropics (Table 10.1). However, there is a very incomplete knowledge of the global richness of species, especially in tropical latitudes. If an estimate is made of the number of undescribed tropical taxa, then the fraction of the global species richness that lives in the tropics increases to at least 86% (Table 10.1).

Invertebrates comprise the greatest proportion of described species, with insects making up the bulk of that total, and beetles (Coleoptera) comprising most of the insects (Table 10.2; see also Wilson, 1988). Furthermore, it is believed that there is a tremendous richness of undescribed insects in the tropics, possibly as many as another 30 million species (Erwin, 1982, 1988). This remarkable conclusion has emerged from recent experiments in which tropical forest canopies were fogged with an insecticidal mist, and the "rain" of dead arthropods was collected using ground-level sampling trays. This innovative sampling procedure indicated that (1) a large fraction of the insect species richness of moist tropical forest is undescribed; (2) most insect taxa are confined to a single type of forest or even particular plant species, that are themselves of restricted distribution; and (3) most tropical forest insect species have a very limited dispersal ability (Erwin, 1982, 1983a,b, 1988). For example, Erwin (1983a,b) fogged four types of tropical rain forest in Amazonian Brazil. Beetles comprised the largest fraction of the total number of previously or newly named species, and 58–78% of these taxa were restricted to only one forest type, and were probably narrowly endemic (Table 10.3). The field work associated with these canopy fogging experiments is expensive and logistically difficult, and it takes years to sort the samples and name the many new species. As a result, very few of these measurements have been made. However, the emerging conclusion from this descriptive work is that there is a remarkable number of undescribed species of insects and other invertebrates in moist tropical forests.

Although it is often ignored in conservation programs that tend to focus on large vertebrates or on particularly noteworthy plants, the species richness of insects and other invertebrates is ecologically important in its own right. According to Janzen (1987), these animals "are more than just decora-

Table 10.1   The estimated number of species in three major climatic zones[a]

Zone	Number of Identified Species ($\times 10^6$)	Estimated Total Number of Species	
		Assume 5×10^6	Assume 10×10^6
Boreal	0.1	0.1	0.1
Temperate	1.0	1.2	1.3
Tropical	0.6	3.7	8.6
Total	1.7	5.0	10.0

[a]Modified from World Resources Institute (WRI) (1986).

Table 10.2 Estimated number of species in various classes of organisms[a]

	Number of Identified Species	Estimated Total Number of Species
Nonvascular plants	150,000	200,000
Vascular plants	250,000	280,000
Invertebrates	1,300,000	4,400,000[b]
Fishes	21,000	23,000
Amphibians	3125	3500
Reptiles	5115	6000
Birds	8715	9000
Mammals	4170	4300
Total	1,742,000	4,926,000

[a]Modified from World Resources Institute (WRI) (1986).
[b]This figure is conservative. Some recent estimates (see text) suggest that there may be as many as 30 million species of insects in tropical forests alone.

tions on the plants; rather they are the building blocks and glue for much of the habitat." These conclusions were based on the facts that (1) insects are the primary food for most small vertebrate carnivores in tropical forest; (2) insects are important predators of seeds and they thereby influence the plant species composition of the forest; (3) insects are important pollinators, often in an obligate species-specific relationship with particular plant species; and (4) other observations that indicate that insects have a strong influence on the structure and functioning of tropical ecosystems.

The species richness and endemism of other tropical forest biota are better known than for arthropods. For example, a plot of only 0.1 ha in a moist tropical Ecuadorian forest had 365 species of vascular plants (Gentry, 1986). Consider some other examples of the species richness of woody plants in tropical rainforest: (1) 98 woody species with diameter at breast height (DBH) larger than 20 cm in 1.5 ha of moist forest in Sarawak (Richards, 1952); (2) 90 species > 20 cm DBH in 0.8 ha of forest in Papua New Guinea (Paijmans, 1970); (3) 44–61 species > 20 cm DBH (total of 112 species) among five 1-ha plots of forest on Barro Colorado Island, Panama (Thorington et al., 1982); (4) more than 300 species of woody plants on a 50-ha forest plot on Barro Colorado Island (Hubbell and Foster, 1983); and (5) 283 tree species in a 1-ha plot in Amazonian Peru, with 63% of species represented by only 1 individual, and another 15% by only 2 (Gentry, 1988).

These are much more species-rich than the fewer than 12–15 tree species in a typical temperate forest, and 30–35 species in the Great Smokies of the United States, one of the richest temperate forests in the world (Leigh, 1982).

There have been few systematic studies of all of

Table 10.3 Distribution of species of adult Coleoptera among four forest types at Manaus, Brazil[a]

Forest Type	Number of Restricted Species	Expressed as % of the Total Number of Species in Each Forest Type
Black-water forest	179	64%
White-water forest	129	58%
Mixed-water forest	325	71%
Terra firma forest	266	78%
Total	899	

[a]The samples were obtained by intensively sampling the "rain" of dead insects after fogging the forest canopy with insecticide. Out of the total of 1080 species that were observed among 24,350 individuals in the four forests, 83% were restricted to only one forest type. Modified from Erwin (1983a).

the biota of particular tropical ecosystems. In one case, a savannah-like dry tropical forest in Costa Rica was studied for several years (Janzen, 1987). It was estimated that a particular 108-km^2 reserve had about 700 plant species, 400 vertebrate species, and a remarkable 13,000 species of insect, including 3140 species of moths and butterflies.

10.3 RATIONALIZATION FOR THE PRESERVATION OF SPECIES

In essence, there are three classes of reasons why the extinction of a species, or a more general loss of species richness, is regrettable:

1. One class of reasons is essentially philosophical, and it revolves around the esthetics of extinction (Ehrlich and Ehrlich, 1981). The central questions are (1) whether humans have the "right" to act as the exterminator of unique and irrevocable species of wild biota and (2) whether the human existence is somehow impoverished by the tragedy of anthropogenic extinction. These are deeply philosophical issues, and they are not scientifically resolvable. However, it is certain that few people would applaud the extinction of a unique species of organism.
2. A second class of reasons is more utilitarian. Humans are not isolated from the biosphere. We take advantage of other organisms in myriad ways for sustenance, shelter, and other purposes, including the functions that they may play in regulating or carrying out ecological processes. If species become extinct, then their unique biochemical, ecological, and other properties are no longer available for actual or potential exploitation by humans.
3. The third class of reasons is essentially ecological, and involves the possibly essential roles of species in maintaining the stability and integrity of ecosystems, and their roles

in nutrient cycling, productivity, trophic dynamics, and in other important aspects of ecological structure and function. In only a very few cases do we have sufficient knowledge to evaluate the ecological "importance" of particular species. It appears likely that an extraordinary number of species could disappear in the current wave of anthropogenic extinctions before they have been studied in this respect.

There are many cases where research on previously unexploited species of plants and animals has revealed the existence of products of utility to humans as food, medicinals, or for other purposes. These are comprehensively described in several recent books and monographs (NAS, 1975b, 1979b; Ehrlich and Ehrlich, 1981; Myers, 1983; Bolandrin *et al.*, 1985; Farnsworth, 1988; Plotkin, 1988), and will not be dealt with here in detail. To illustrate this important point, we will refer to only one example: the rosy periwinkle (*Catharanthus roseus*), an angiosperm that is native to the island of Madagascar (Myers, 1983; OECD, 1985).

During the course of an extensive screening of a diverse array of wild plants for their possible anticancer properties, an extract of the rosy periwinkle was found to counteract the reproduction of cancer cells. Subsequent research identified the active ingredients as several secondary alkaloids that are present in the foliage of the rosy periwinkle, probably to deter herbivores. These natural alkaloids are now used to prepare the important anticancer drugs known as vincristine and vinblastine, which counteract the reproduction of cancer cells by interfering with their mitotic division. Chemotherapy based on these chemicals has been notably successful in the treatment of several human malignancies, most notably a cancer of the lymph system known as Hodgkin's disease, and childhood leukemia. About 530 metric tons of plant material is required to produce 1 kg of vincristine, which in the early 1980s had a value of about $200,000/kg, and annual sales of $100 million. So far the anticancer alkaloids of the rosy periwinkle cannot be economically synthesized by chemists, and thus all of the chemothera-

A powerful justification for the preservation of species richness is that there are a great many uses of plants and animals that have not yet been discovered by scientists. If species become extinct before their utility is discovered, then humans are irrevocably deprived of their potentially beneficial function. Illustrated here is the rosy periwinkle (*Catharantus roseus*), an angiosperm plant with a very restricted natural distribution on the island of Madagascar. An extensive screening of plants for potential anti-cancer properties identified the possible utility of this rare species, which subsequently proved to be highly efficacious for the treatment of several types of human cancers. This valuable plant is now cultivated in abundance for the production of its chemotherapeutic chemicals. However, because it is endangered in its natural habitat, the rosy periwinkle could easily have become extinct prior to the discovery of its utility to humans. Most of the natural habitat of Madagascar has been converted to agricultural uses or disturbed in other ways, and this degradation is continuing apace. (Photo courtesy of Steele, World Wildlife Fund.)

peutic material must be extracted from plants grown for that purpose. Therefore, a plant whose existence was until very recently only known to a few botanists has proven to be of great benefit to humans. This once-obscure plant prevents many deaths from a previously incurable disease, and is the basis of a multi-million-dollar economy.

Undoubtedly, there is a tremendous undiscovered wealth of other biological products that are of potential use to humans. Many of these natural products are present in tropical species that have not yet been "discovered" by taxonomists.

10.4
EXTINCTION AS A NATURAL PROCESS

Most of the species that have ever lived on Earth are now extinct, having disappeared "naturally" for some reason or other. Most likely, they could not successfully cope with changes that took place in their inorganic or biotic environment (e.g., a change in climate, or in the nature and intensity of predation or competition). Alternatively, the ex-

Virtually all species and most populations of large terrestrial carnivores are endangered in the wild. Illustrated here is a jaguar (*Panthera onca*), whose formerly extensive range included the southwestern United States and most of the northern half of South America. Jaguars have suffered precipitous decreases in abundance, largely because of a loss of habitat, persecution as predators of livestock, and hunting for their valuable pelage. This individual was photographed in a reserve that was specifically created to preserve the jaguar habitat in Belize. (Photo courtesy of World Wildlife Fund.)

tinct species could have succumbed to some unpredictable, catastrophic event (Fisher *et al.*, 1969; Raup, 1984, 1986a,b; Vermeij, 1986).

From the geological record it is well established that species and larger phylogenetic assemblages of organisms have appeared and disappeared over time (Clark and Stern, 1968; Dobzhansky *et al.*, 1977). For example, entire divisions of plants have appeared, radiated, and then disappeared from the fossil record (e.g., the seed ferns Pteridospermales, the cycad-like Cycadeoidea, and the woody Cordaites). There are also extinct orders of invertebrates, including the gastropod mollusc orders Ammonoidea and Belemnoidea, and the arthropod orders Eurypterida and Trilobita. Of the 12 classically recognized orders within the class Reptilia, only three are extant. Clearly, the fossil record is replete with evi-

dence of extinctions, including several within the hominid lineage (Walker, 1984).

The rate of extinction has not been uniform over geological time. Long periods characterized by a fairly uniform rate of loss of taxa have apparently been punctuated by about nine catastrophic episodes of mass extinction (Eldredge and Gould, 1972; Gould and Eldredge, 1980; Knoll, 1984; Rampino and Stothers, 1984; Stanley, 1984; Raup, 1988). Probably the most intense event of mass extinction took place at the end of the Permian period some 250 million years ago, when a remarkable 96% of marine species are estimated to have become extinct. Another prominent example is the apparently synchronous extinctions of many vertebrate animals that took place over a geologically short period of time about 65 million years ago at

The giant panda (*Ailuropoda melanoleuca*) is perhaps the most popularly recognized example of a rare and endangered species. Its native range is restricted to a region of mountainous forest in the interior of China, but the extent of its habitat has decreased because of agricultural conversion, logging, and other disturbances, and the species has also been hunted. The government of China is now committed to the preservation of the giant panda through the establishment of protected reserves and, with international cooperation, it is engaging in an intensive program of research on the ecology of this animal, which will hopefully allow for effective management in the future. (Photo courtesy of MacKinnon, World Wildlife Fund.)

the end of the Cretaceous era. The most notable extinctions were of the reptilian orders Dinosauria and Pterosauria, but many plants and invertebrates also became extinct at about the same time. In total, perhaps one-half of the global fauna became extinct during this biotic spasm (Raup, 1986a,b, 1988). A popular (but controversial) hypothesis for the cause

of this catastrophic event of mass extinction is that a 10-km-wide meteorite impacted the earth. This spewed a great quantity of fine dust into the atmosphere, which secondarily caused a climatic deterioration that many large animals could not tolerate (Alvarez *et al.*, 1980, 1984).

It is known that there is a relatively high probability of both natural and anthropogenic extinction in small and isolated populations, such as those that exist on oceanic islands (Diamond, 1984; Richman *et al.*, 1988). In general, the smaller and more isolated the island, the higher is the rate of extinction (Fig. 10.1). This is a primary reason for the area-dependent impoverishment of the biota of oceanic islands, and of habitat islands within larger land masses (Fig. 10.2).

There have been very few direct observations of natural extinction. Some researchers have observed the extinction (usually short-term) of local populations of wide-ranging species on oceanic islands and on continental habitat "islands" (Diamond, 1984). These events were usually caused by some catastrophic disturbance, usually a brief period of inclement weather. For example, Ehrlich *et al.* (1972; see also Ehrlich, 1983) had been studying an isolated, subalpine population of the butterfly *Glaucopsyche lygdamus* in Colorado for several years. Early in the summer of one year, a late snowstorm killed the flower primordia of the lupine *Lupinus amplus*, the only food plant of the larvae of this butterfly. This event caused the local extinction of the population of this butterfly, although because of its great vagility it later recolonized the site.

Extinction has also been documented after the physical isolation of populations of species in relatively small "island" habitats. Local extinctions of birds were documented after Barro Colorado Island was isolated from continuous forest, by the creation of Lake Gatun in 1914 during construction of the Panama Canal (Karr, 1982; Terborgh and Winter, 1980). Initially, there were at least 218 resident bird species on Barro Colorado Island. By 1981, at least 56 species or 26% of the resident avifauna were known to have become extinct. Of the extinct species, 26 are typical of young, disturbed habitats, which disappeared over time as the forest on the

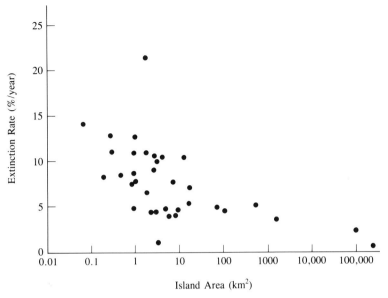

Fig. 10.1. The risk of extinction of species of breeding landbirds of northern European islands, as a function of island area. The vertical scale is the average percentage of the bird species richness that goes extinct from one year to the next. From Diamond (1984).

island developed successionally. In addition, eight aquatic birds disappeared because of a reduction in the abundance of their prey after the introduction of a piscivorous fish. The other 22 known extinctions were of bird species that occur in mature forest (the actual number may have been as large as 50–60 species, since the birds of mature forest were incompletely known in 1921). The reasons for the extinctions of the forest species are unclear, especially in view of the fact that the area of forest on Barro Colorado Island actually increased between 1921 and 1981 because of succession. The forest species presumably became extinct because of such factors as (1) a change in habitat quality; (2) an inadequate area of suitable habitat to support a breeding population over the longer term; and (3) some sporadic extinction event, possibly caused by an unpredictable episode of inclement weather; coupled with (4) a limited vagility, or an inability to disperse across water to recolonize the island after a local extinction.

10.5
ANTHROPOGENIC LOSSES OF SPECIES RICHNESS

Many species have been brought to, or beyond, the brink of extinction by the direct and indirect consequences of human activities. The most important anthropogenic influences that have caused the extinction or endangerment of organisms are (1) overhunting, (2) the effects of introduced predators, competitors, and diseases, and (3) habitat destruction.

Because of these anthropogenic influences, both the rate of extinction and the number of taxa that are threatened with extirpation have increased greatly in the last few centuries. This effect is best documented for vertebrates (as noted previously, most invertebrate species, particularly insects, have not yet been described). Over the last 200 years, a global total of perhaps 100 species of mammals, 160 birds, and many other taxa have become ex-

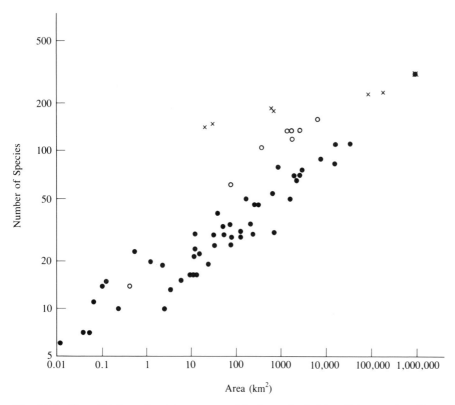

Fig. 10.2. Plot of bird species richness versus island area in the New Guinea region. The oceanic islands occur variously around New Guinea. The land-bridge islands were isolated from the New Guinea land mass by a rise in sea level at the end of the Pleistocene. The mainland census areas were habitat islands. x, New Guinea mainland; ○, land-bridge islands; ●, oceanic islands. From Diamond (1984).

tinct, all because of some anthropogenic influence (Fitter, 1968; Wood, 1972; Soule, 1983).

In addition, a distressingly large number of plant and animal taxa are facing an imminent prospect of extinction. For example, consider the conservation status of the major orders of resident terrestrial vertebrates of Latin America, as summarized in Table 10.4. Note that the percentage of the total fauna that is threatened or endangered appears to be larger for mammals and birds than for reptiles and amphibians. However, this apparent trend does not represent the reality of the situation—it only reflects the relatively impoverished state of knowledge of the conservation status of the reptiles and amphibians of Latin America. In fact, because of incomplete

information for all of the classes of animals in Table 10.4, the data for percent threatened or endangered should be considered as the bare minima. The actual state of affairs is cause for more pessimism, and it is progressively deteriorating as ever more of the native habitat is disturbed or converted to agricultural ecosystems.

The most current and comprehensive global catalogues of threatened organisms are the Red Data Books of the International Union for the Conservation of Nature and Natural Resources (IUCN) (e.g., King, 1981; Groombridge, 1982; Thornback and Jenkins, 1982; Wells et al., 1983). These catalogues are most accurate and complete for relatively large and conspicuous species, particularly those

The American ginseng (*Panax quinquifolium*) is a formerly abundant species of the understory of rich angiosperm forests of eastern North America. Similarly, the true ginseng (*P. ginseng*) is now very rare in its native forest habitat of eastern Asia. Both species have been badly overharvested throughout their natural range, because of a lucrative and virtually inexhaustible demand in China and Korea for the dried roots of this plant, which are believed to have many medicinal properties, including those of an aphrodisiac. Both ginseng species are now abundant in cultivation, but they remain very rare in the wild. (Photo courtesy of Charron, World Wildlife Fund.)

that live in relatively well-researched temperate and higher latitude ecosystems. In contrast, the information base is woefully incomplete for the multitude of relatively small and poorly known taxa that live in tropical ecosystems. Many of these endangered tropical species, particularly insects, will disappear before they have been described by taxonomists. This will take place as their geographically restricted and unexplored tropical habitat is cleared for agriculture or some other purpose.

In the following, a few cases of losses of species richness that have been caused by some anthropogenic influence will be examined. For much more detailed treatments of cases of anthropogenic extinction and endangerment, see Ziswiler (1967), McClung (1969), Myers (1979), Ehrlich and Ehrlich (1981), or Wilson (1988).

Unsustainable Exploitation as the Cause of a Loss of Species Richness

The most prominent and readily appreciated anthropogenic cause of the extinction of organisms, which largely applies to relatively large and conspicuous species, is exploitation by hunting or harvesting at a grossly unsustainable rate. The insatiable and unsustainable "mining" of a potentially renewable natural resource has caused the extinction of many species, as is described below.

Pleistocene Extinctions. At about the end of the Pleistocene Epoch, there were great waves of apparently synchronous extinctions of large animal species in various places. These mass extinction events took place at different times, but they all

Table 10.4 Status of resident terrestrial vertebrates in 1985 in selected countries in Latin America[a]

Region or Country	Mammals		Birds		Reptiles		Amphibians		Total	
	Total Number	%	Total Number	%	Total Number	%	Total Number	%	Total Number	%
Latin America + Caribbean	1234	11	3812	11	2420	5	1833	1	9299	7
Mexico + Central America	559	8	1318	12	930	4	505	1	3312	7
Tropical Andean Region	602	14	2518	10	823	6	818	0	4761	8
Guiana Region	330	12	1354	11	286	10	229	0	2199	10
Tropical Atlantic Region	419	19	1598	13	475	8	491	0	2983	11
Southern South America	284	17	1007	12	266	8	156	0	1713	11
Caribbean Islands	141	13	639	16	434	12	138	1	1352	13
Mexico	439	7	958	13	702	5	273	0	2372	8
Brazil	394	20	1561	13	462	8	479	0	2896	11
Peru	359	17	1640	10	291	8	233	0	2523	10
Ecuador	280	18	1449	11	344	8	349	0	2422	10

[a]The designation of species as threatened or endangered (i.e., % in table) was based on their listing as such by one or more of the U.S. Department of Interior, the Committee on Trade of Endangered Species (CITES), or the International Union for Conservation of Nature and Natural Resources (IUCN). According to the IUCN, a species is endangered if it is in imminent danger of extinction; vulnerable implies that the species will soon be endangered if nothing is done to protect it. Modified from World Resources Institute (WRI) (1986).

The peregrine falcon (*Falco peregrinus*) became extinct or endangered over much of its natural range in North America and Europe because of various toxic effects caused by chlorinated hydrocarbons, especially the insecticide DDT and the industrial chemicals, polychlorinated biphenyls (PCBs). Peregrine falcons are beginning to experience a recovery of abundance, because of a great decrease in the use of these chemicals, coupled with a vigorous program of captive breeding and release of young birds (see Chapter 8, pp. 196–198). (Photo courtesy of Oliphant, World Wildlife Fund.)

apparently coincided with the colonization of land masses that were previously unknown to humans. The compeling implication is that overhunting caused or contributed to the extinction of many species of large animals, because they could not adapt to the sudden onslaught of anthropogenic predation and habitat changes (Martin, 1967, 1984; Diamond, 1982). In North America, the wave of extinctions took place around 11,000 years ago. Within a short period of time, there were extinctions of at least 27 genera comprising 56 species of large mammals (i.e., larger than 44 kg), two genera and 21 species of smaller mammals, and several large birds. Some of the notable losses include: 10 species of the horse *Equus*, the giant ground sloth *Gryptotherium listai*, four species within the camel family Camelidae, two species of *Bison*, a native species of cow in the genus *Bos*, the saiga antelope *Saiga tatarica*, four species within the elephant family Proboscidea, including the relatively widespread *Mammut americanum* and *Mammuthus primigenius*, the sabertooth tiger *Smilodon fatalis*, the American lion *Panthera leo atrox*, and the 25-kg raptorial scavenger *Terratornis merriami*. Many of these extinct large animals were found as part of a fossil fauna in the Rancho la Brea tar pits in southern California. Today, the only extant representative of the large animal fauna of those tar pits is the critically endangered California condor (*Gymnogyps californianus*).

Similar mass extinctions coincided with the first arrival of aboriginal hunters elsewhere. In Australia and New Guinea there was a wave of extinctions about 50,000 years ago that involved many species of large marsupial, large flightless birds, and tortoises. In New Zealand an extinction wave took place less than 1000 years ago. This swept away 30 large bird species, including the 3-m-tall giant moa *Dinornis maximus*, 26 other species of moa, large species of lizard and frog, and fur seals (*Arctocephalus forsteri*) from North Island. The arrival of humans to Madagascar about 1500 years ago coincided with the extinction of 14 species of large and giant lemurs (another 10 species of lemur are still extant on this island), about 6–12 species of flightless elephant birds (Aepyornithidae), two giant tortoises, and various other large animals. Other events of mass extinction that were caused by overhunting by newly colonizing, aboriginal humans took place in Hawaii, New Caledonia, Fiji, the West Indies, and South America.

More Recent Anthropogenic Extinctions on Islands.   In general, plants and animals that live on islands have been relatively susceptible to anthropogenic extinction (Fisher *et al.*, 1969; Ehrlich

Conservation issues are not usually straightforward; they can be greatly complicated by political, economic, and cultural considerations that are related to the historical patterns of use of natural resources. The bowhead whale (*Balaena mysticetus*) of the Western Arctic is one of the world's most endangered species of cetacean, numbering perhaps 5000 individuals or 5–10% of its preexploitation abundance. In spite of its precarious status, the bowhead whale still sustains a small aboriginal hunt that results in the death of about 20–40 individuals per year. Its habitat may also be undergoing degradation by chemical and acoustic pollution associated with exploration for hydrocarbons in the offshore of the North Slope of Alaska and in the Beaufort Sea. Although Alaskan Inuit have been engaged in bowhead hunting since prehistoric times and the practice is an important component of their cultural identity, there has been controversy over the longer-term sustainability of the continued exploitation of this rare species. This 1980 photo depicts a 9-m-long bowhead whale being hauled onto shorefast ice near Barrow, Alaska, prior to its butchering. (Photo courtesy of T. F. Albert.)

and Ehrlich, 1981; Soule, 1983; Gentry, 1986; Richman *et al.*, 1988). In many cases, these organisms had not been subject to intense predation during their recent evolutionary history, and hence they were poorly adapted to coping with a sudden onslaught of mortality caused by human predation. For example, the bird fauna of unexplored oceanic islands often had a large frequency of species that were flightless, relatively large, and fearless of potential predators. Furthermore, species on islands often did not co-occur with closely competing organisms, so that they were easily displaced by introduced species that filled a "generalist" niche. In addition, the colonization of oceanic islands by people,

and particularly Europeans, usually resulted in a great degradation of the native habitat, because of the clearing of endemic vegetation for agriculture, urban, and tourism development, and because of habitat degradation caused by introduced plants and mammalian herbivores and predators (discussed later).

Because of these factors, the biotas of many remote islands have been subject to especially large rates of anthropogenic extinction. This can best be illustrated with extinctions of birds. Of 167 bird taxa (including 95 full species) considered to have become extinct worldwide since 1600, only nine were continental in their distribution (Fisher *et al.*, 1969; King, 1981). Four of the continental species

The natural range of the orangutan (*Pongo pygmaeus*) is forests of the Indonesian islands of Sumatra and Borneo. The abundance of this species has decreased to a level of endangerment because of hunting, the capture of individuals for the pet trade and for export to zoos and circuses, and the conversion of its habitat to agricultural purposes. The photo illustrates an orangutan rehabilitation center on Sumatra, which was established to facilitate the return of individuals to the wild after they had been kept in captivity and had lost many of their essential survival skills. (Photo courtesy of the World Wildlife Fund.)

disappeared from Asia (1) the pink-headed duck, *Rhodonessa caryophyllacea*, in 1944; (2) the Himalayan mountain quail, *Ophrysia superciliosa*, in 1868; (3) Jerdon's courser, *Cursorius bitorquatus*, in 1900; and (4) the forest spotted owlet, *Athene blewitti*, about 1872. The other five species disappeared from North America: (5) the Labrador duck, *Camptorhynchus labradorium*, in 1875; (6) Cooper's sandpiper, *Pisobia cooperi*, in 1833; (7) the passenger pigeon, *Ectopistes migratorius*, in 1914; (8) the Carolina parakeet, *Conuropsis carolinensis*, in 1914; and (9) Townsend's finch, *Spiza townsendii*, in 1833. The other 158 bird taxa that

have become extinct since 1600 only lived on islands.

The syndrome of an extinction-prone island biota can be illustrated by the case of the Hawaiian Islands, an ancient archipelago of volcanic projections in the Pacific Ocean that is 1600 km from the nearest island group and 4000 km from the nearest major land mass (Simon, 1987). At the time of first contact with Polynesian hunters, there were at least 68 endemic taxa of birds on the Hawaiian Islands, out of a total richness of land birds of 86 species. Of the initial 68 endemics, 24 are now extinct and 29 are perilously close to the edge of survival (Ehrlich

The whooping crane (*Grus americana*) is the tallest and one of the rarest birds in North America. This species breeds in the boreal muskeg habitat in Wood Buffalo National Park in northern Canada, migrates through the Great Plains, and winters in the coastal saltmarsh habitat of the Gulf of Mexico near Aransas, Texas. Largely because of habitat loss and overhunting during its migration and in its wintering habitat, the whooping crane decreased in abundance to only 15 individuals in 1941. However, vigorous conservation efforts by agencies of the United States and Canadian governments have encouraged an increase in abundance to more than 160 individuals in 1988. This photograph taken in the Northwest Territories shows an immature bird flying between its parents. (Photo courtesy of the Canadian Wildlife Service/World Wildlife Fund.)

and Ehrlich, 1981; King, 1981; Vitousek, 1988). Especially hard hit has been an endemic family of birds, the Hawaiian honeycreepers (Drepanididae). Thirteen taxa of this endemic family are believed to be extinct, and 12 are endangered (King, 1981; Freed *et al.*, 1987). Since 1780 more than 50 alien bird species have been introduced by humans to the Hawaiian islands (Vitousek, 1988), but this gain can hardly be considered to compensate for the loss and endangerment of the specifically evolved endemic birds.

Similarly, the native flora of Hawaii is estimated to have been comprised of 1765–2000 taxa of angiosperm plants, of which a remarkable 94–98% were endemic to the islands. (In addition, humans are estimated to have introduced another 4600 taxa

of vascular plants to the Hawaiian islands, of which about 700 can survive in the wild.) In the last several hundred years, more than 100 native plant taxa have become extinct, and the survival of at least an additional 500 taxa is threatened or endangered (Ayensu, 1978; Fay, 1978; Vitousek, 1988). The extreme rarity and endemism of the Hawaiian flora can be illustrated by the six species of the endemic tree genus *Hibiscadelphus*, which has an aggregate total of only 14 known individuals in the wild, and one species that is represented by only a single wild individual (Gentry, 1986). The most important causes of extinction of Hawaiian biota have been habitat conversion to agricultural and urban landscapes, and the introduction of alien biological agents, such as predators, competitors, destructive

An aerial view of a pristine rainforest covering an expansive lowland near Tortugeuro, Costa Rica. Mature tropical forest of this type contains a tremendous richness of plant and animal species, most of which are of relatively local distribution. By far, the greatest fraction of the species richness of the biosphere occurs in tropical forests. Moreover, a very large amount of carbon is stored as organic carbon of plant biomass. If a mature forest of this or any other type is converted to an agricultural ecosystem with a much smaller standing crop of organic carbon, then the difference in carbon content is made up by a large emission of carbon dioxide to the atmosphere, with possible secondary climatic changes taking place through the so-called "greenhouse effect" (see Chapter 2, Section 2.5). (Photo courtesy of E. Greene.)

herbivores such as feral goats (*Capra hircus*), and diseases to which the native biota were not adapted (Simon, 1987; Vitousek, 1988).

There are many other examples that describe how the unique and fragile biota of remote oceanic islands has become impoverished because of the direct and indirect depredations of humans. Consider the island of Madagascar, which has been continuously isolated from the African continent for at least 20 million years, and has developed a unique biota. Madagascar has a native flora of about 10,000 taxa, of which 90% are endemic, including 95 endemic genera [World Wildlife Fund (WWF), 1984]. Within the fauna, perhaps the most interest-ing group is the lemurs (Lemuridae), an endemic family of primitive primates. We previously noted that the 14 largest species of lemur disappeared dur-ing a wave of extinctions about 1500 years ago, after the first aboriginal colonists arrived on Madagascar. In addition, the 15 extant lemur spe-cies are all decreasing in abundance and are in dan-ger of extinction, as are the members of two other primate families (the Indriidae with eight taxa, and the Daubentoniidae with one species—the aye-aye, *Daubentonia madagascariensis*, one of the world's rarest mammals) (Fitter, 1968; Fisher *et al.*, 1969; Mittermeier, 1988). The extreme endangerment of the native biota of Madagascar is further evident

A very large area of previously forested landscape in the tropics has been converted to various types of agricultural agroecosystems. Some notable effects are a great loss of species richness and a large flux of carbon dioxide to the atmosphere. This grazing ecosystem in Costa Rica was created after the clearing of a continuous expanse of dry tropical forest. (Photo courtesy of E. Greene.)

from the fact that in the early 1980s only about 5% of that island's primary vegetation was still intact and essentially undisturbed (Myers, 1988).

Another example illustrates the sensitivity of many island biotas to mortality caused by generalist predators. The boid snake *Boiga irregularis* was accidentally introduced in the late 1940s to the island of Guam in the Pacific Ocean (Pimm, 1987; Savidge, 1987). Although the biota and natural habitats of Guam have been impacted by many factors, including warfare, residential and agricultural development, pesticides, and introduced predators other than *Boiga*, it appears that predation by this snake may have been a key factor in the recent decline of the island's avifauna. Proir to the introduction of *Boiga*, the native avifauna of Guam comprised 18 species, most of which were abundant. By the mid-1980s seven species were extinct and another four were critically endangered, including two endemics of Guam. The pattern of avian de-

cline closely correlated with the pattern and timing of range expansion of *Boiga*, and careful observations of the snake's natural history showed it to be a very effective predator of birds and their nests. Therefore, it appears that the severe decline of the extinction-prone avifauna of Guam has been caused by the impacts of many anthropogenic stresses, including the accidental introduction of *Boiga*. Sensitive island faunas elsewhere have been similarly impacted by introduced generalist predators such as mongooses (Viverridae), domestic cats (*Felis catus*), domestic dogs (*Canis familiaris*), and pigs (*Sus scrofa*), as well as generalist herbivores such as goats (*Capra hircus*) and sheep (*Ovis aries*).

Some Species Made Extinct or Endangered by Overharvesting. Hunting at an unsustainable rate has been the most prominent cause of the anthropogenic extinction or endangerment of large species that have value as a market commodity. If

An expanse of cleared, slashed, and burned tropical forest in Indonesia, in the process of conversion to an agricultural use. Virtually none of the species that comprised the previously tremendous richness of species of the primary forest will be able to survive in the new agricultural landscape. (Photo courtesy of Lloyd, World Wildlife Fund).

the rate of exploitation is greater than the rate of recruitment, then a precipitous decline in abundance can take place in a remarkably short period of time. This cause-and-effect relationship can hold for even extremely abundant species. This phenomenon will be illustrated by reference to the dodo, the passenger pigeon, the great auk, and a few other notable cases.

(a) The Dodo. The first well-documented extinction of an animal that was caused by overhunting by humans occurred in 1681. This involved the dodo (*Raphus cucullatus*), a large, turkey-sized, flightless member of the pigeon order Columbiformes (McClung, 1969; Campbell and Lack, 1985). The extinction of the dodo has been immortalized in our everyday language by the phrase "dead as a dodo," which connotes an irrevocable loss, and by the general use of the term "dodo" to

describe an old-fashioned or stupid person. This etymology recognizes the unfortunate dodo's apparent inability to adapt to or even recognize the threat posed by humans.

The dodo lived on Mauritius, a small island east of Madagascar in the Indian Ocean. Mauritius was discovered by the Portuguese in 1507, and was colonized by the Dutch in 1598. The colonists hunted the dodo as game, gathered its eggs, cleared its habitat for agriculture, started devastating ground fires, and released feral cats, pigs, and monkeys that predated dodo eggs in their ground-level nests. As a result, the dodo quickly became extinct. We do not even have a complete skeleton of the species. There had been a stuffed bird at Oxford University, but it was damaged by a fire in 1755, and only the head and foot were saved. The only other member of the family Raphidae was the dodo-like Rodrigues solitaire (*Pezophaps solitaria*), but in 1715–1720 it disappeared from its habitat on several islands near to Mauritius, for similar reasons as the dodo.

Interestingly, the dodo may have coevolved with the endemic tambalacoque tree (*Calvaria major*), in a mutualism in which the bird benefited from a food of abundant and nutritious seeds, while the plant benefited from the fact that passage through the crop and gut of a dodo scarified its seeds and allowed them to germinate (Temple, 1977, 1979). The tambalacoque tree is now perilously rare in the wild, with only about ten surviving mature individuals, each older than 300 years. No young individuals are known in the wild, even though the mature trees produce fertile fruit. It has been suggested that the need for dodoiferous scarification is obligate, so that recruitment of the plant was impossible following the demise of the dodo. Temple (1977) found that seeds of tambalacoque germinated well after they had passed through the gizzard and gut of a turkey, and he speculated that he may have grown the first seedlings of this species since the disappearance of the dodo. Human intervention may save the day for this tree, by the use of artificial scarification followed by the outplanting of seedlings (Owadally, 1979).

(b) The Great Auk. The first well-described anthropogenic extinction of an animal whose range was at least partly North American was that of a seabird, the great auk (*Pinguinus impennis*) (McClung, 1969; Nettleship and Evans, 1986). In memoriam to this unfortunate species, the journal of the American Ornithological Union has been named "The Auk." The great auk was commonly known to mariners as the original "pennegoin," although it is a member of the Alcidae, and not related to the superficially similar southern hemisphere penguin family, Spheniscidae.

Originally, the great auk had an amphi-Atlantic distribution. It bred on a few north Atlantic islands off eastern Newfoundland, in the Gulf of St. Lawrence, around Iceland, and north of Scotland. Because the great auk bred on only a few islands, it was vulnerable to extirpation by uncontrolled hunting. Although the great auk was initially abundant

on its breeding islands, it was easy to catch or club because it was flightless, and therefore it could be killed in large numbers.

For many centuries, the great auk had been exploited by aboriginal Newfoundlanders and European fishermen as a source of fresh meat, eggs, and oil. However, when its feathers became a valuable commodity for the stuffing of mattresses in the mid-1700s, a systematic and relentless exploitation began that quickly caused the extinction of the species. A harvesting and processing operation was described in 1785 at one of the largest breeding colonies of the great auk and other alcids, on Funk Island off eastern Newfoundland (from Nettleship and Evans, 1986): "it has been customary of late years, for several crews of men to live all summer on that island, for the sole purpose of killing birds for the sake of their feathers, the destruction of which they have made is incredible. If a stop is not soon put to that practice, the whole breed will be diminished to almost nothing, particularly the penguins." The slaughter of great auks and other seabirds on Funk Island was so great in the late 1700s that much of the soil that presently occurs there has been formed from their carcasses (Olson *et al.*, 1979; Kirkham and Montevecchi, 1982). About 85 years after the extirpation of the great auk from Funk Island, the common puffin (*Fratercula arctica*) began to breed there. This seabird requires soil that is sufficiently deep for the excavation of a nesting burrow. On Funk Island, the auk-derived soil was apparently suitable, and today puffins can sometimes be observed carrying bones of the extinct great auk out of their excavations.

The great auk became extinct on Funk Island in the early 1800s. The last known individuals of this species were killed on June 4, 1844, on the Island of Eldey Rock, by three Icelanders who were searching for specimens for a bird "collector" in Reykjavik. They killed the only two adult birds that they saw, and smashed the only egg they found, because it had been cracked and therefore was not a good specimen. They were reputed to have said: "What do you mean the great auk's extinct? We just killed two of them!"

(c) The Passenger Pigeon. One of the few birds to have become extinct from a continental range was the passenger pigeon (*Ectopistes migratorius*) (Ziswiler, 1967; McClung, 1969; Blockstein and Tordoff, 1985). Prior to the beginning of its demise some 300 years ago, the passenger pigeon may have been the world's most abundant landbird, possibly numbering some 3–5 billion individuals, and comprising one-fourth of the total population of birds of North America. The passenger pigeon migrated in tremendous flocks that were sufficiently dense to obscure the sun, and that could take many hours to pass. In 1810, the American ornithologist Alexander Wilson guesstimated 2 billion birds in a single flock that was 0.6 km wide and 144 km long. The range of the passenger pigeon was the northeastern United States and southeastern Canada, and its habitat was mature, mast-producing forests containing oak, beech, and chestnut (*Quercus* spp., *Fagus grandifolia*, and *Castanea dentata*).

The overwhelming abundance of the passenger pigeon, coupled with its tastiness and certain biological characteristics that made it easy to kill in large numbers (i.e., its propensity to migrate and breed in very large and dense groups), made this species an attractive target for commercial hunters who sold the birds in urban markets. The passenger pigeon was killed in incredible numbers by clubbing, shooting, netting, and incapacitation with smoke. The size of some of the annual kills is staggering to the imagination—for example, an estimated 1 billion birds were taken in 1869 in Michigan alone. In the face of this exploitation coupled with destruction of its breeding habitat, the passenger pigeon suffered a precipitous decline in abundance. The last observed nesting attempt was in 1894. Martha, the last known individual, died a lonely death in the Cincinnati Zoo in 1914. In only a few decades the world's most abundant species of bird had been rendered extinct.

In spite of the terrific kill of passenger pigeons, it is possible that a large death rate caused by overhunting was not the primary cause of its extinction (Blockstein and Tordoff, 1985). Instead, it has been suggested that the annual interference with reproduction at its colonial breeding sites caused a succession of nesting failures, so that the rate of recruitment was much too small to make up for the intense anthropogenic predation. In addition, the passenger pigeon was deprived of much of the original area of its habitat by the clearing of mature, mast-producing forest for agricultural purposes.

(d) The Eskimo Curlew. The eskimo curlew (*Numenius borealis*) is a large North American sandpiper. As late as one century ago, the eskimo curlew was extremely abundant. This fact, coupled with its large size, fine taste, tameness, and gregarious nature, encouraged its relentless exploitation by market hunters during its migrations through the prairies of Canada and the United States, and on the pampas and coasts of South America, where the eskimo curlew wintered. Because of the uncontrolled hunting, this species became very rare by the end of the nineteenth century. The last observation of a nest of the eskimo curlew was in 1866, and the last "collection" of a specimen was in Labrador in 1922 (McClung, 1969). However, it seems that the eskimo curlew is not extinct, although it certainly perches on the edge of the precipice of existence. There have been a number of reliable sightings of individuals and small flocks of this species, mostly in migratory habitat in Texas and elsewhere, but also in its breeding habitat in the Canadian low arctic (McClung, 1969; Blankenship and King, 1984; Gollop *et al.*, 1986).

(e) Some Overharvested Marine Mammals. Marine mammals have suffered the consequences of overharvesting in many parts of the world. As will be discussed later in this chapter, some previously heavily exploited and badly depleted species have made an excellent recovery. However, several species of marine mammals have become extinct because of overexploitation, including the following.

1. The Steller's sea cow (*Hydrodamalis stelleri*) of the Bering Sea, which was discovered by explorers employed by the Russian Czar in 1741, and was apparently

extinct by 1768, after only 26 years of hunting (McClung, 1969; Fitter, 1968).

2. The Caribbean monk seal (*Monachus tropicalis*), formerly abundant in the Carribean Sea and Gulf of Mexico, where it was "discovered," and eaten, on Christopher Columbus's second voyage to the New World in 1494. This species was decimated by an eighteenth century seal "fishery," and then exterminated by hunting by fishermen and loss of habitat (Thornback and Jenkins, 1982).

Many other species of marine mammals have been endangered by human activity, especially hunting. Some of the most notable of these are listed here (after Thornback and Jenkins, 1982).

1. The Guadalupe fur seal (*Arctocephalus townsendi*), which was historically abundant on the west coast of Mexico, possibly numbering as many as 200,000 individuals. This seal was hunted relentlessly for its valuable pelage, and in the 1920s it was believed to be extinct. However, a breeding colony was discovered in 1965 off Baja California, and the species may currently number 1300–1500 individuals.

2. The Juan Fernandez fur seal (*Arctocephalus philippii*) was historically abundant off southern Chile. It was intensively harvested for its pelt; between 1797 and 1804 more than 3 million individuals were killed. This seal was believed to have been extinct until 1965, when it was rediscovered. It presently numbers at least 2400 individuals.

3. The Mediterranean monk seal (*Monachus monachus*) ranges over most of the Mediterranean Sea and the Black and Adriatic Seas, while the Hawaiian monk seal (*M. schauinslandi*) lives off the Hawaiian Islands. Neither species was ever very abundant, and both suffered a decrease in population because of hunting, deterioration of their habitat, and pollution in the case of the Mediterranean monk seal. Both

species currently number in the low thousands.

4. All four extant species of the order Sirenia are considered to be endangered or vulnerable (Steller's sea cow was also in this order). The dugong (*Dugong dugon*) has a paleotropical distribution, and it has suffered a large population decrease over virtually its entire range. The manatees (genus *Trichechus*, including *T. manatus* of the Caribbean Sea and coastal Florida) are also considered to be vulnerable to extinction, because of declines in their population that have been caused by hunting, habitat loss, and pollution.

Certain of the great whales are other examples of marine mammals that have been greatly reduced from their preexploitation abundance. These include (after Northridge, 1984) the following.

1. The right whale (*Eubalaena glacialis*), which ranges over all temperate waters of the northern and southern hemispheres. The current estimated world population is about 2000 individuals. The only estimate of the preexploitation abundance is for the North Pacific, where the right whale is thought to have numbered 10,000 individuals, compared with 200–500 at present.

2. The bowhead whale (*Balaena mysticetus*) of Arctic waters, currently numbering about 5000 individuals, or 5–10% of its estimated original abundance.

3. The blue whale (*Balaenoptera musculus*), of virtual worldwide distribution, and currently numbering 10,000–20,000 individuals, compared with an estimated initial abundance of 250,000.

Extinction of Competitors of Humans

Some cases of anthropogenic extinction have involved species that were relentlessly extirpated because humans percieved them as an economically important competitor for some common resource, usually an agricultural product or big game. These can be considered to be special cases of extinction caused by overexploitation.

Many species of large predator have been persecuted because they were considered to be an important competitor (in a very few cases, they were also predators of humans). Examples of such animals that are now extinct or endangered over much of their range include the wolf (*Canis lupus*) and other members of the genus *Canis*, the brown and grizzly bears (*Ursus arctos*), and most of the large cats within the Felidae, such as the cougar (*Felis concolor*) in North America.

Often, pest control measures for predators have involved the use of poisoned bait and carcasses, a practice that can cause a large incidental kill of non-target scavengers. Indiscriminate poisoning (along with hunting, and excessive collecting of eggs and adults by ornithologists) has been an important cause of the decline and possibly imminent extinction of the California condor (*Gymnogyps californianus*) (King, 1981).

In at least two cases, species that were viewed as important competitors were extirpated largely by pest control efforts:

1. The original range of the Carolina parakeet (*Conuropsis carolinensis*) was the southeastern United States (Ziswiler, 1967; McClung, 1969). This brightly colored frugivore was fairly common, living mostly in mature bottomland and swamp forest, where it foraged and roosted communally. Although they were occasionally killed for their colorful feathers, the Carolina parakeet was not a valuable natural commodity. This bird was extirpated because it was considered to be an agricultural pest, due to the damage that flocks would cause while feeding in fruit orchards. For this reason, the Carolina parakeet incurred the enmity of humans, and it was relentlessly persecuted. Unfortunately, it was an easy mark for extirpation because of its conspicuous-

ness, communal nesting and feeding, and the propensity of birds to aggregate around wounded colleagues, so that an entire flock could be wiped out by hunters. The last definite record of a flock of Carolina parakeets was in Florida in 1904, and the last known individual of the species died in the Cincinnati zoo in 1914, at the same place and in the same year that Martha, the last passenger pigeon expired.

2. The Tasmanian wolf or thylacine (*Thylacinus cynocephalus*) became extinct in Australia about 4000–5000 years ago, probably because of its inability to compete with anthropogenically introduced dingos (*Canis dingo*). However, dingos were absent on the island of Tasmania, so the thylacine survived there and was the largest extant carnivore when that island was colonized by Europeans (Fisher *et al.*, 1969; Thornback and Jenkins, 1982). Unfortunately, the thylacine developed a taste for the sheep and chickens that were raised on Tasmanian ranches, and it was therefore percieved to be an important agricultural pest. A bounty was established for dead thylacines, and it was relentlessly persecuted by stockholders. It became extinct over most of its range in the early 1900s. The last definite record of a wild animal was made in 1933, and this individual was captured and kept in a zoo, where it died in 1936.

Habitat Destruction as a Cause of a Loss of Species Richness

A number of species have become extinct or are endangered largely because of the conversion of their native habitat to an agricultural purpose, or because of other anthropogenic disturbances that rendered the habitat unsuitable for their survival. This phenomenon will first be illustrated with several notable examples of species that have suffered from this effect. The broader implications for global species richness of the anthropogenic destruction of tropical forest habitat will then be considered.

The American Ivory-billed Woodpecker. The ivory-billed woodpecker (*Campephilus principalis principalis*) was probably never abundant in its North American range, which comprised most of the southeastern United States (McClung, 1969; King, 1981). The habitat of this large woodpecker was mature, bottomland angiosperm forest and cypress (*Taxodium distichum*) swamp, most of which was heavily logged for sawlogs or cleared for agricultural purposes by the early 1900s. There have been no published reports of this species since the early 1960s, and it may now be extinct in North America. Similarly, the closely related Cuban ivory-billed woodpecker (*Campephilus principalis bairdii*) is critically endangered, as is the congeneric imperial woodpecker (*C. imperialis*) of Mexico.

The Dusky and Cape Sable Seaside Sparrows. The dusky seaside sparrow (*Ammodramus maritimus nigrescens*) was a locally distributed subspecies of the seaside sparrow that lived in *Spartina bakerii*-dominated salt marsh habitat on the east coast of Florida (King, 1981). In early systematic treatments this bird was considered to be a distinct species, *Ammospiza nigrescens*, but recent treatment has lumped it and other related taxa within a seaside sparrow "superspecies" [American Ornithologists' Union (AOU), 1983]. The dusky seaside sparrow became functionally extinct in the early 1980s (several male birds still lived in captivity at that time, but there were no known female individuals), and the last known individual died in 1987. The causes of its extinction were habitat loss caused by drainage and construction activity, and possibly the toxic effects of the spraying of insecticides to control salt marsh mosquitoes in its habitat, much of which was close to areas that are used for tourism or residential land use. The closely related Cape Sable seaside sparrow (*Ammodramus maritimus mirabilis*, formerly considered to be a separate species, *Ammospiza mirabilis*) of south Florida has also suffered much habitat loss, because of the invasion of its native saltmarsh and fresh-water wet prairie habitat by introduced shrubs and trees, habitat deterioration caused by saltwater intrusion and wildfire, and the drainage of its habitat for construction and other purposes. The Cape Sable subspecies as a whole is endangered, and several of its former populations are extinct.

The Black-footed Ferret. The black-footed ferret (*Mustela nigripes*) was first "discovered" in the southwestern prairie of North America in 1851 (McClung, 1969; Thornback and Jenkins, 1982). This mustelid predator may never have been abundant during historical times. It has become endangered because of habitat changes caused by the widespread conversion of the short-grass and midgrass prairie to an agricultural ecosystem, and because of the great reduction in abundance of its principal food, the prairie dog (*Cynomys ludovicianus*). The prairie dog has declined because of a loss of habitat, and because it was considered to be a competitor for cattle on rangelands, it was relentlessly persecuted by poisoning.

The Furbish Lousewort. The Furbish lousewort (*Pedicularis furbishiae*) is endemic to a 230-km stretch of the valley of the St. John River in Maine and New Brunswick (Lucas and Synge, 1978; Day, 1983). The last known observations of this rare plant had been made in the early 1940s, and it was considered to be extinct until a botanist "rediscovered" it in Maine in 1976. At that time, its known habitat was threatened with obliteration by inundation caused by a proposed hydroelectric development on the St. John River. This project was controversial for a number of reasons in addition to the lousewort (it would have dammed one of the largest remaining free-flowing rivers in the northeastern United States), and its development appears to have been put on a long-term hold.

Tropical Deforestation and Global Species Richness. As was described previously, the great bulk of the global biological diversity is comprised of millions of as yet undescribed taxa of tropical insects. Because of the extreme endemism of

In 1979, the World Wildlife Fund began an innovative research program called the Biological Dynamics of Forest Fragments Project, in which forest "islands" of various size are created during the operational conversion of continuous Amazonian rainforest to pastureland. The purpose of the research is to study the longer-term ecological dynamics and stability of isolated forest fragments, with a view toward determining the optimal size and shape that reserves of tropical forest should be in order to provide suitable habitat for various species. This airphoto shows newly created forest islands of 1 ha and 10 ha. (Photo courtesy of R. Bierregaard.)

many of these arthropods and of other tropical biota, there is a likelihood of the extinction of many taxa as a result of the clearing of tropical forest, and its conversion to other types of habitat.

In the mid-1980s the extent of modern global deforestation was about 11%, while the loss of closed forest in temperate and boreal latitudes was about 19% (WRI, 1986). These were considerably larger than the 4% loss of closed tropical plus subtropical forest up to that time (in the mid-1980s, the area of these forests was about 12.5×10^6 km², or 96% of their preagricultural area of 13.0×10^6 km²).

However, it appears that the amount of deforestation in the moist tropics is increasing rapidly, in contrast to the situation at higher latitudes where the total forest cover is relatively stable. This situation is illustrated for recent decades in the Americas in Table 10.5. Between the mid-1960s and the mid-1980s there was little change (-2%) in the total forest cover of North America. However, in Central America forest cover decreased by 17%, and in South America it decreased by 7% (but by a larger percentage in equatorial countries of South America). The global rate of clearing of closed tropical rainforest in the mid-1980s was about $6-8 \times 10^6$ ha/year, equivalent to 6–8% of that biome per year (Grainger, 1983; Melillo *et al.*, 1985). Of course, some of the cleared forest would regenerate by a secondary succession, which would ultimately produce another mature forested ecosystem. However, little is known about the rate and biological character of succession in the tropical rainforest biome, or how long it would take to restore a fully diverse ecosystem after disturbance (Lugo, 1988). Nevertheless, if a 6–8% rate of clearance is accept-

Table 10.5 The change in forested area between 1964 and 1985 in selected countries in North and South America[a]

	Total Land Area (10³ km²)	Forested Area in 1964–1966 (10³ km²)	Forested Area in 1981–1983 (10³ km²)	Percent Change (%)
North America	18,348	6056	5965	−2
Canada	9221	3135	3227	+3
United States	9127	2921	2738	−6
Central America	2617	831	691	−17
Mexico	1923	558	481	−14
Nicaragua	119	62	42	−31
Cuba	115	16	20	+25
Honduras	112	53	39	−26
Guatemala	108	51	38	−26
Panama	76	46	41	−11
Costa Rica	51	30	17	−43
Others	114	15	14	−7
South America	17,437	9880	9231	−7
Brazil	8457	6004	5666	−6
Argentina	2737	602	602	0
Peru	1280	742	704	−5
Bolivia	1084	596	564	−5
Colombia	1039	655	520	−21
Venezuela	882	397	344	−13
Chile	749	157	157	0
Paraguay	397	214	206	−4
Ecuador	277	172	144	−16
Guyana	197	181	164	−10
Uruguay	174	5	5	0
Suriname	161	155	155	0

[a]Modified from World Resources Institute (WRI) (1986).

ed and simplistically projected in a linear fashion into the future, an alarming biome half-life of only 9–12 years is calculated!

It is readily apparent that the modern rate of disturbance and conversion of tropical forest will have important consequences for global species richness. Because of a widespread awareness and concern about this important ecological problem, there has recently been much research and other activity in the conservation of tropical forests. As of 1985, several thousand sites, comprising more than 160 million ha, have received some sort of "protection" in low-latitude countries (Table 10.6). Of course, the operational effectiveness of the protected status varies greatly, depending on factors that influence the real commitment of governments to conserva-

tion, including (1) political stability, (2) political priorities, and (3) the finances that are available to mount an effective warden system to control the poaching of animals and lumber, and to prevent other disturbances. Nevertheless, an encouraging level of conservation activity is beginning to take place in some areas, and there appears to be a real, emerging commitment to the conservation of threatened ecosystems in the tropics.

Within Central America, Panama and Costa Rica are the nations with the most progressive policies with respect to the conservation of natural ecosystems. In 1985, 8.7% of the area of Panama had been given park or reserve status, as had 8.1% of Costa Rica, where the ultimate plan is to conserve 10% of the landscape as natural forest (Hartshorn,

Table 10.6 Protected areas in 1985 in selected low-latitude biomes[a]

Biome	Region	Number of Protected Areas	Total Area Protected (ha × 10^6)
Tropical humid forest	Afrotropical	44	8.91
	Indomalayan	122	5.09
	Australian	53	7.78
	Neotropical	61	17.28
	Total	280	39.06
Tropical dry forest and woodland	Afrotropical	240	48.67
	Indomalayan	238	10.42
	Australian	10	0.93
	Neotropical	93	5.50
	Total	581	65.52
Evergreen sclerophyllous forest	Nearctic	6	0.05
	Palaearctic	122	3.37
	Afrotropical	41	1.62
	Australian	301	6.92
	Neotropical	5	0.04
	Total	475	12.00
Subtropical and temperate rainforest and woodland	Nearctic	18	8.13
	Palaearctic	48	1.74
	Australian	26	0.90
	Antarctic	145	2.78
	Neotropical	38	8.85
	Total	275	22.40
Tropical grassland and savannah	Australian	18	2.04
	Neotropical	12	7.01
	Total	30	9.05
Mixed island systems	Palaearctic	9	0.05
	Afrotropical	4	0.02
	Indomalayan	177	10.43
	Oceanian	51	4.11
	Neotropical	26	1.19

[a]Modified from World Resources Institute (WRI) (1986).

1983; WRI, 1986). This sort of vigorous conservation activity is badly needed in Costa Rica, since the current rate of deforestation is about 4% per year (it is about 0.9%/year in Panama; WRI, 1986). For perspective, we could note that the relative area of protected land in Panama and Costa Rica in 1985 compares very favorably with those of Canada (2.5%) and the United States (7.1%), in spite of great differences in wealth and industrialization among these countries. In other countries in Central America, conservation efforts have been badly disrupted by civil war and other problems of political instability (Hartshorn, 1983). For example, the percentage of territory that had received protected status in 1985 was 0.0% in Haiti, 0.1% in Nicaragua, 0.5% in Guatemala, and 3.8% in Honduras (WRI, 1986).

The Amazonian basin comprises the world's largest expanse of tropical rainforest. It has been broadly estimated that Amazonia could contain as much as 10% of the biotic diversity of the earth (Lovejoy, 1985). At present, this rich tropical region is largely composed of mature rain forest that has been little influenced by modern agricul-

A 1-year-old, 100-ha reserve of tropical forest created as part of the Biological Dynamics of Forest Fragments Project of the World Wildlife Fund. This fragment is connected to a continuous rainforest by a corridor, and it has a fairly heterogenous shape in order to increase its ratio of edge : area over that of a square or a circle. (Photo courtesy of R. Bierregaard.)

ture, lumbering, and other anthropogenic disturbances. However, there is an accelerating exploitation of the Amazonian region, including (1) the agro-industrial conversion of great expanses of rainforest to establish cattle ranches and (2) the smaller-scale clearing of patches of forest by poor farmers who have migrated from other parts of Brazil in search of agricultural land. Between 1975 and 1986 the population in Amazonian Brazil increased 10-fold to more than 1 million persons, and in 1985 the rate of forest clearing was about 17,000 km²/year (Myers, 1988). It was officially estimated that up to 1979, only about 1% of the closed Amazonian forest had been "deeply altered" by modern anthropogenic disturbance (de Gusmao-Camara, 1983). However the actual extent of disturbance is not well quantified. Lovejoy (1985) estimated that deforestation in Amazonian Brazil in the early 1980s was as great as 10%. In any event, while the government of Brazil is committed to a vigorous program of "development" of its Amazonian terri-

tory, it also appears to be committed to a program of conservation. There are official plans to keep 1.8 × 10⁶ km² of the Amazonian rain forest of Brazil in a natural state, and up to 1982 the government had established about 87,000 km² of parks and reserves (de Gusmao-Camara, 1983).

An important aspect of the conservation strategy that will be employed in the Amazon basin is the selection of reserves using criteria that depend strongly on the existence of centers of endemism and great species richness. It has been theorized that such areas with a high frequency of endemism represent the likely locations of forest "island" refugia during the Pleistocene and earlier glacial eras (Haffer, 1969, 1982; Prance, 1982; but see Endler, 1982, for a contrasting view). At those times, the regional climate may have been relatively dry, so that the tropical rainforest was contracted into relatively small nuclei surrounded by savannah. Because these forest refugia were periodically isolated during glacial epochs, they favored an enhanced

The edge of a forest fragment, just after clearing of the adjoining forest to create pasture for cattle. At first, the ecotone is very sharp, but with time, a dense growth of secondary vegetation develops into a 10- to 25-m-wide windbreak that softens the transition between the forest and the pasture. (Photo courtesy of R. Bierregaard.)

rate of speciation of the tropical forest biota. On this basis, the putative refugia might be reasonable locations for the siting of large ecological reserves. If the reserves are sufficiently large in area, they could preserve much of the Amazonian species richness. Unfortunately, the "boundaries" of these forest refugia are not accurately known, and their determination (if they exist at all) could be a major stumbling block in the utility of such a system for the designation of ecological reserves (Endler, 1982).

The size of ecological reserves is another important consideration with respect to the long-term stability and conservation of species richness in the tropics (and elsewhere). As might be expected from

predictions of the theory of island biogeography (MacArthur and Wilson, 1967; Diamond, 1972; Diamond and May, 1976), newly created reserves that comprise tropical forest "islands" could lose some of their species richness (a process called "relaxation") if they are too small to accomodate the following (Terborgh, 1974; Soule, 1986):

1. A viable breeding population of species that have a very large home range. Note that within this factor, consideration must be made not only of space, but also of time, since habitats change successionally.
2. The relatively large rate of extinction that occurs in small islands. This is due to the deleterious effects of genetic drift, and to the tendency of small, isolated populations to be extirpated by random or sporadic catastrophic events.

We earlier described the relaxation of bird species richness on Barro Colorado Island after its isolation by construction of the Panama Canal (Terborgh and Winter, 1980; Karr, 1982). In another long-term study that began in 1979, a series of forest fragments of various size are being created experimentally, in order to investigate the importance of size and amount of habitat edge to the stability of Amazonian forest reserves (Lovejoy et al., 1984, 1986). Isolated forest remnants are being created from continuous rainforest in the vicinity of Manaus, Brazil, by controlling the operational clearing of forest for the production of pasture for the grazing of cattle. When the experiment is fully established, there will be eight reserves of 1 ha, nine of 10 ha, five of 100 ha, two of 1000 ha, and one of 10,000 ha. Some of the preliminary, short-term observations from this experiment are as follows.

1. After isolation, there are large microclimatic effects, caused by hot, dry winds that blow into the reserves from the surrounding clearing. These effects are especially pronounced close to the edge of the reserve, and they are important in causing an increase in the rate of tree mortality after the stand is isolated. With time, a dense growth

of regenerating secondary vegetation produces a 10- to 25-m-wide windbreak, which effectively shades and protects the interior of the reserve.

2. Initially, isolation causes a short-term increase in the abundance and species richness of forest birds, due to an influx of individuals that have been displaced from the surrounding clearcut forest. This effect only occurs in 1-ha and 10-ha reserves, but it is quickly followed by a relaxation of the bird species richness of these small stands. There is a "negative edge effect" on bird species richness, with only about 28 species present within 10 m of the edge of a reserve, 47 species at 50 m, and 50 species further into the stand. The negative edge effect translates into a small species richness for relatively small reserves, since these have a relatively large ratio of edge:area compared with larger reserves.

3. Primates are strongly affected by the area of the reserve. Most species decline in reserves that are ≤ 10 ha in size. Only one species, the red howler (*Alouatta seniculus*), was able to persist for the first 4 years after the isolation of a 10-ha reserve. In total, within a few years of isolation a 1-ha reserve could only support three species of nonflying mammal, while a 10-ha stand had five species, and intact forest had 20 species.

4. There is a large, positive edge effect for some insects, notably for large species of butterfly. This is due to the use of the vigorous secondary vegetation by many light-loving butterfly species, some of which penetrate 200–300 m into the interior of the reserve. However, other insects decline in reserves that are ≤ 10 ha in area. These include Euglossine bees, an important group of pollinators, and army ants (*Echiton burchelli*) and their associated avifauna, which follow army ants in order to feed on displaced arthropods.

10.6
BACK FROM THE BRINK— A FEW CONSERVATION SUCCESS STORIES

A number of species that had been close to extirpation because of anthropogenic disturbance have recovered in abundance after the initiation of conservation measures. In some cases the recovery has been sufficiently vigorous that the species is no longer in imminent danger of extinction. A few examples of these successes of conservation are described below.

Some Notable Examples

The Northern Fur Seal and Some Other Pinnipeds

The range of the northern fur seal (*Callorhinus ursinus*) is the north Pacific Ocean (Walker *et al.*, 1968; Northridge, 1984). This seal was relentlessly exploited for its pelage, to the degree that it was reduced in abundance from several million individuals to about 130,000 in 1920, and it was believed that further exploitation would place it in danger of extinction. As a result, an international treaty was signed in 1911 that strictly regulated the harvest of these animals, by a large reduction in the total kill, by a ban of the pelagic catch, and by only allowing fur seals to be taken by land-based hunters. The population of northern fur seals responded vigorously to these conservation measures, so that there are now about 1.8 million individuals. This presently abundant species is again being exploited for pelts, hides, and oil, and it also suffers considerable mortality from entanglement with pelagic fishing nets. However, the harvest is closely regulated so that it will hopefully be sustainable on the long term.

The northern fur seal is the best documented example of a seal that was excessively exploited but then rebounded in abundance after the cessation or strict regulation of hunting. Other notable examples of seals that had been badly depleted by unregulated exploitation up to about the first decade or so of the present century, and that then recovered after conservation measures were initiated, include (after Walker *et al.*, 1968; Northridge, 1984) (1) the southern fur seal (*Arctocephalus australis*) of sub-Antarctic waters off South America; (2) the harp seal (*Phoca groenlandica*) of the North Atlantic and Arctic; (3) the gray seal (*Halichoerus gryptus*) of Atlantic temperate waters; and (4) the northern elephant seal (*Mirounga angustirostris*) of the Pacific coast of Mexico and southern California.

The Whaling Industry

The best documented recovery of a whale species after the cessation of unregulated exploitation is that of the grey whale (*Eschrichtius robustus*) of the Pacific coast of North America (Northridge, 1984). Its population is now about 11,000 individuals, which is slightly less than its estimated abundance prior to exploitation (15,000), but substantially recovered from the 1880s to 1930s when there were only about 1000–2000 grey whales. However, another very small population of grey whales in the western Pacific is still threatened, and an Atlantic population became extinct several centuries ago.

Other large baleen whales remain greatly depleted in abundance as a result of their commercial exploitation. Some prominent examples include (after Northridge, 1984) (1) the blue whale (*Balaenoptera musculus*), with a prewhaling abundance of about 250,000, but currently numbering only 10,000–20,000; (2) the fin whale (*B. physalus*), initially 700,000, but now 163,000; (3) the humpback whale (*Megaptera novaeangliae*), initially >100,000, but now 10,000; and (4) the sperm whale (*Physeter macrocephalus*), initially 2 million, but now fewer than 1 million. Although the abundance of these species of large whales (and of other marine mammals) is notoriously difficult to measure, it appears that they may be starting to recover in response to several decades of complete or partial protection, which has resulted from economic and political pressures in favor of more "scientific" management and conservation of these ani-

mals. This rebound of abundance may intensify in the near future due to the almost complete elimination of the legal hunting of large whales in the late 1980s.

However, it is possible that the equilibrium abundance of some of the recovered stocks of large baleen whales, particularly in the Antarctic, could be smaller than their population prior to commercial exploitation (Laws, 1977; Bengston and Laws, 1985). The reason for this is that there has been a great increase in the abundance of other much smaller marine vertebrates with which the baleen whales compete for a pelagic food of the crustacean macrozooplankter known as krill (*Euphausia superba*). The most notable krill-eating species whose abundance appears to have increased in response to decreased competition from large whales are the crabeater seal (*Lobodon carcinophagus*), which now numbers about 15 million individuals, several other less numerous species of seal, and several species of seabirds, especially penguins (Spheniscidae) and possibly petrels and shearwaters (Procellariidae).

Unfortunately, some other badly depleted species of large whale have shown little sign of a recovery of abundance, despite the almost complete elimination of hunting for at least 50 years (Northridge, 1984). These include (1) the right whale (*Eubalena glacialis*), with an estimated world population of only 2,000 individuals, and (2) the bowhead whale (*Balaena mysticetus*), now numbering about 5000 individuals, and still exploited by a legal aboriginal hunt in northern coastal Alaska. Both of these species presently number only 5–10% of their abundance prior to commercial exploitation.

The American Bison

Prior to its intensive commercial exploitation, the American bison or buffalo (*Bison bison*) was North America's most populous large mammal, with an estimated abundance of 60 million individuals (after McClung, 1969; Grzimek, 1972; Wood, 1972). An appreciation of the remarkable abundance of the plains bison can be gained from the size of some of its migratory herds, one of which was described as being 40 km × 80 km in size, another as 320 km deep, and another as advancing over a 160-km-wide front! The density of animals in these migrating groups was variously described as being "as thick as they could graze," but it probably ranged from about 25 to 38 animals/ha.

At the time of settlement of North America by Europeans, the bison ranged over most of the continent. However, the bison was exploited relentlessly for its meat and hide, and it suffered a precipitous decline in abundance. The eastern subspecies (*B. b. pennsylvanicus*) originally ranged over much of the eastern United States (witness the name of Buffalo, the largest city in western New York State). However, because of overhunting and the destruction of its habitat for agricultural purposes, the eastern bison declined and by 1832 it was probably exterminated east of the Mississippi River.

The plains bison (*B. b. bison*) was by far the most populous subspecies, ranging throughout the prairies of North America. Its near extermination by commercial hunting in the nineteenth century may actually have been encouraged by the United States government of the time, because (1) the plains bison disrupted the emerging prairie agricultural system by its mass migrations and (2) extermination of the bison was viewed as a means of disruption of the ecosystem of the Plains Indians, in order to ease their displacement in favor of European agricultural colonists.

There are many remarkable descriptions of the wanton destruction of the plains bison. During the construction of the transcontinental Union Pacific Railroad (beginning in 1869), bison were killed in large numbers to feed railroad workers. The most famous hunter was William F. Cody or "Buffalo Bill," who reportedly killed 4280 bison in 18 months, and 250 in a single day. After the railroad was built, sporting excursions were organized in which a train would be stopped near a buffalo herd, and passengers would leisurely shoot animals from the coach. Later, the tongues (a delicacy) would be cut from some of the dead animals, but in the main the carcasses were left where they had fallen. The railroad also allowed for a well-organized commer-

cial hunt of the plains bison, since it made possible the rapid shipment of carcasses to urban markets. It was estimated that between 1871 and 1875, market hunters killed about 2.5 million bison per year. Furthermore, by this time the Plains Indians had acquired rifles, and they also initiated a massive hunt of bison.

Obviously, this sort of unregulated butchering of a wild population of large animals was unsustainable, and the plains bison underwent a precipitous decline in abundance. In 1889 it was estimated that there were fewer than 1000 bison left in the United States, including a protected herd of 200 individuals in Yellowstone National Park. The number surviving in Canada was undetermined, but it was also believed to be alarmingly small. Because of concerns over the imminent extinction of the plains bison, a number of closely guarded preserves were established, and captive breeding programs were initiated. As a result, the number of plains bison has increased, to the point where they now number more than 50,000 individuals. Because their original habitat is largely gone, the plains bison will never again achieve an abundance similar to its pre-exploitation level. However, its future as a viable species appears to be secure, because (1) it breeds easily in captivity and (2) it appears that a viable breeding population can be maintained in relatively small but managed reserves.

The wood bison (*B. b. athabascae*) is a third subspecies of the American bison. This animal was also hunted intensely, and it declined to a small abundance. After the demise of the wood bison seemed imminent, its single known wild population was preserved in and around a large reserve in Wood Buffalo National Park in northern Canada (Reynolds and Hawley, 1987). However, the genetic integrity of this population has suffered because of intergradation with introduced plains bison. Fortunately, in 1960 in a remote area of Wood Buffalo National Park, a previously unknown population of wood bison was discovered that had not yet suffered from interbreeding with the plains subspecies. Some of these animals were used to establish an isolated wood bison population northwest of Great Slave Lake. Unfortunately, in recent years introduced diseases such as tuberculosis and brucellosis, along with predation by wolves (*Canis lupus*) and humans, have taken a toll of the wild-ranging bison populations of northwestern Canada, and there is concern over the long-term viability of their remote, protected populations.

Overall, however, it appears that vigorous conservation efforts have preserved the American bison. It will survive, but in relatively small wild populations, and in captivity.

Other Recoveries from Depleted Abundance

It will be useful to briefly consider a few other examples (all taken from North America) of species that have exhibited a substantial recovery of abundance after the cessation of overexploitation (after McClung, 1969).

1. The sea otter (*Enhydra lutris*) was the object of intense exploitation for its dense and lustrous fur, and as a result it was greatly reduced in abundance. In fact, it had been thought to be extinct until the 1930s, when small populations were rediscovered along the west coast of California and the Aleutian Islands. The sea otter is now exhibiting a substantial rebound of abundance, and it is being reintroduced to some areas from which it had been extirpated.

2. The pronghorn antelope (*Antilocapra americana*) of the western plains was badly overhunted by the late nineteenth century, and had been diminished to an estimated 20,000 individuals in its range north of Mexico. However, because of strong conservation efforts it now numbers more than 0.5 million individuals, and it sustains a sport hunt over most of its range.

3. The trumpeter swan (*Olor buccinator*) is the world's largest member of the Anatidae. It was greatly diminished by harvesting for its meat and skin. However, because of the cessation of a legal hunt and efforts at habitat conservation, it has re-

covered to an abundance of more than 5000 individuals.

4. The wild turkey (*Meleagris galloparvo*) was greatly diminished in abundance by hunting and habitat loss (a domestic variety is, of course, abundant in captivity). However, in many places the wild turkey has recovered substantially, because of conservation measures and its introduction to areas from which it had been extirpated. Many populations of this large game bird now sustain a sport hunt.

5. The wood duck (*Aix sponsa*) was greatly diminished in abundance by hunting for its beautiful feathers and for meat, and by deterioration of its habitat caused by lumbering and the drainage of wetlands. However, the wood duck has recovered substantially in abundance, in part because of a widespread program of provision of nest boxes for this cavity-nesting species. The same nest-box program benefits another relatively uncommon duck, the hooded merganser (*Lophodytes cucullatus*). An unrelated terrestrial nest-box program has been important in the recent increase of abundance of eastern and western bluebirds (*Siala sialis* and *S. mexicana*).

6. The fur trade that stimulated much of the exploration of northern and western North America was largely based on the valuable pelage of the American beaver (*Castor canadensis*). Because of its overexploitation, this species declined to a low level of abundance, and it was extirpated from much of its range. However, because of conservation measures and a decreased demand for its fur, the beaver has recovered to its original abundance over most of its range where the habitat remained suitable.

7. Finally, the whooping crane (*Grus americana*) may be a case of an incipient conservation success story. The whooping crane was probably uncommon prior to the time of the maximum anthropogenic impact on its abundance in the 1890s, with a likely

original population of only 1300 individuals (King, 1981). However, because of hunting and the deterioration of its winter habitat, the whooping crane declined further in abundance, to as few as 15 individuals in 1941. Since then, strong conservation measures for the wild population and their habitat, coupled with a captive-breeding program, have increased the number of whooping cranes to more than 150 individuals in the mid-1980s. As a result, there is guarded optimism for the survival of this species.

Conservation Activities in General

The conservation of wild species of plants and animals is regarded by almost all human societies as a laudable and important objective. As a result, there are (1) a great deal of government activity in this field, (2) much research and training of scientists at universities and other educational institutions, and (3) many nongovernment organizations that are active at the local, regional, national, or international levels. All of these contribute to such important conservation activities as (1) the identification, acquisition, and management of sites where rare and endangered organisms live; (2) increasing the awareness of the general public about important conservation issues; (3) research into the biology and autecology of rare and endangered species, and of the ecosystems of which they are a part; and (4) raising or providing funds for all of the above.

The intensity of these conservation activities is greatest in relatively industrialized countries, but awareness and activity are also beginning to emerge in less developed countries of lower latitudes. (Of course, a conservation ethic has always been an integral component of a number of religions that developed in tropical countries, for example, Buddhism and Hinduism. In spite of this prevailing attitude in some countries, wildlife and natural habitats have suffered very badly because of the conversion of natural ecosystems to agricultural ones, and for other reasons.) The worldwide and progressive development of conservation activities can itself be regarded as an emerging "success story."

It is far beyond the purpose of this chapter to describe the many agencies and activities that are of importance to the conservation of the earth's biota. It will be sufficient to briefly describe a joint program of three of the most important international agencies whose mandate centers on the conservation of the world's biota and ecosystems, namely, the International Union for Conservation of Nature and Natural Resources (IUCN), the World Wildlife Fund (WWF), and the United Nations Environment Program (UNEP). Jointly, these organizations have developed a framework for a global program called "The World Conservation Strategy" (IUCN, 1980). The objectives of this program (after WRI, 1986) are (1) to maintain essential ecological processes and life support systems on earth; (2) to preserve biotic diversity; and (3) to ensure the sustainable development of the earth's natural resources. Therefore, the World Conservation Strategy links biological conservation with the sustainable development of global resources.

A critical component of the World Conservation Strategy is the priorization of conservation activities by participating countries, within the framework of a national conservation strategy. The national strategy is developed through the use of expertise in government, academia, public conservation organizations, and other sectors. A high priority is placed on the designation of ecological reserves for the protection of the natural flora and fauna, and on the conservation of endangered or threatened species.

It is too early to appraise the success of the World Conservation Strategy, since the program only began in the late 1970s. However, the existence of this comprehensive international effort is very encouraging, as is the active participation in it of more than 40 countries representing all stages of socioeconomic development. If the use of the earth's resources for the benefit of humans is take place on a sustainable basis, including adequate conservation of rare and endangered species and ecosystems, then a program like the World Conservation Strategy will certainly be required.

11

ECOLOGICAL
EFFECTS OF
WARFARE

11.1
INTRODUCTION

The destructiveness of warfare to humans and their civilization is well appreciated. Warfare can cause a tremendous loss of human life. For example, the First World War of 1914–1918 caused an estimated 20 million fatalities, the Second World War of 1939–1945 ~38 million, the Korean War of 1950–1953 ~2.9 million, and the Second Indochina War of 1961–1975 ~2.4 million (Sivard, 1986, 1987). These are the most prominent of the twentieth century wars, which up to the mid 1980s had resulted in the death of about 84 million people (Sivard, 1986)

Military activities also appropriate a remarkable amount of economic activity (Sivard, 1982, 1986, 1987). In 1987, the global military expenditures were equivalent to $930 billion (current U.S. dollars), or about 6% of all economic activity on Earth. Total arms sales in 1985 were $250 billion, including $55 billion in international arms trade. The remaining military economic activity was comprised of expenditures for salaries, training, infrastructure, logistic support, research, and other activities.

The global military spending of $825 billion in 1986 represents a tremendous increase from the 1960 expenditure of $345 billion, and the 1970 expenditure of $523 billion (all in constant U.S. 1984 dollars).

Military activities also appropriate a large human resource (Sivard, 1982, 1987). In 1986 there were about 26 million people in the armed forces worldwide. Another 75 million people were in military reserves, in paramilitary forces, or were civilians producing weapons or providing other direct services to the military.

In spite of the great destruction that has been caused by warfare, and the overwhelming importance of the military economic sector, there is remarkably little hard information on the ecological damage that is caused by warfare. Nevertheless, in this chapter some of the ecological impacts of warfare will be investigated. Topics that will be covered include the effects of the explosion of conventional munitions, the use of poisonous gases, the military use of herbicides, and the direct and indirect consequences of nuclear warfare. As will be seen, the potential ecological implications of a large-scale nuclear exchange are particularly horrific. In fact, predictions have been made of a collapse of biosphere processes, caused by the potential climatic consequences of a nuclear war.

11.2
DESTRUCTION BY
CONVENTIONAL WARFARE

A tremendous quantity of explosive munitions is used in warfare. In the Second World War, an estimated 21.1×10^9 kg of explosives was expended by the combatants, equivalent to about 25.3×10^{15} J of explosive energy (36% of this was used by the United States, 42% by the Germans, and the rest by other combatants) (Westing, 1985b). In the Korean War, the total expenditure of munitions was about 2.9×10^9 kg, of which 90% was used by U.S. forces. In the Second Indochina War the total expenditure was about 14.3×10^9 kg, more than 95%

by the U.S. and South Vietnamese forces. The explosion of these great quantities of munitions caused direct, though poorly documented, damage to ecosystems during these conflicts.

World Wars

In terms of a loss of human life, the two greatest wars waged in human history were the First and Second World Wars. Both of these widespread conflicts caused a terrible mortality, destroyed cities, and disrupted civilization in many other ways. Among the important impacts of these wars was damage caused to ecosystems. These ecological effects have not been subject to detailed scientific documentation, and therefore they are poorly quantified. Moreover, the ecological impacts are not generally perceived as notable impacts of these wars—the human tragedy is most strongly emphasized.

"The way up to Passchendaele Ridge, and some of the German defences," in France in 1917. Long-term static confrontations between large armies typically caused this sort of churned-up morass of mud over most of the Western Front on the poorly drained coastal lowlands of western Belgium and France. The water-filled depressions are old explosion craters of artillery shells. In this case, the terrain was initially agricultural fields. (Photo courtesy of *The Times*, 1919.)

"German trenches in a wood wrecked by artillery," probably in Belgium in 1918. The lowlands of the Western Front were largely agricultural, but the small remnant woodlands were strategically important because they provided cover for troops and equipment, fuelwood, and lumber for the construction of fortifications and other structures. In the foreground is a latticework made of local forest materials, intended to strengthen the walls of the trenches in order to keep them from collapsing inward. In the background is a woodland devastated by repeated artillery barrages. (Photo courtesy of *The Times*, 1918.)

To a limited extent, we can gain an appreciation of the ecological impacts of these human conflicts from literary images of the time, which allow an appreciation of the intensity of the devastation caused to forests and other ecosystems. This is especially true of literature that emerged from the Western Front of the First World War. That particular conflict involved virtually static confrontations between huge, well-equipped armies. The most massive battles took place in the coastal plain lowlands of Belgium and France. For as long as 4 years, armies fought back and forth over a well-defended terrain that was webbed with elaborate trenchworks. In these circumstances territorial gains were very difficult, in spite of the wastage of many men and much material during offensives. The intense and long-lasting confrontations devastated the agricultural terrain and woodlands of the battlefields.

These effects were especially great in the flat and poorly drained lowlands of the Flanders region of Belgium. The city and vicinity of Ypres was devastated: "In this landscape nothing existed but a measureless bog of military rubble, shattered houses, and tree stumps. It was pitted with shell craters containing fetid water. Overhead hung low clouds of smoke and fog. The very ground was soured by poison gas. . . ." (Wolff, 1958).

The soils of Flanders were largely developed from a deep deposit of clay. Such soils were readily churned into a sticky, gelatinous morass by intense bombardments and the movements of men and machinery: "Because of the impervious clay, the rain cannot escape and tends to stagnate over large areas . . . the low-lying, clayey soil, torn by shells and sodden with rain, turned to a succession of vast, muddy pools . . . the ground remains perpetually

"Gun crew . . . firing 37 mm gun during an advance," somewhere in France in 1918. Another view of a woodland devastated by artillery barrages. (Photo courtesy of *The Times,* 1919.)

saturated . . . gluey, intolerable mud . . . liquid mud . . . molasses-like topsoil" (excerpts from military dispatches and other literature of the time; from Wolff, 1958).

The churned-up land was largely devegetated by warfare. However, some weeds managed to flourish briefly during the summer, most notably the red-flowered poppy (*Papaver rhoeas*):

> I've seen them, too, those blossoms red,
>   Show 'gainst the trench lines' screen,
> A crimson stream that waved and spread
>   Thro' all the brown and green:
> I've seen them dyed a deeper hue
>   Than ever nature gave,
> Shell-torn from slopes on which they grew,
>   To cover many a grave.
>
> (W. C. Galbraith)

Other prominent weeds of the battlefields of Flanders were the coltsfoot (*Tussilago farfara*), lesser celandine (*Ranunculus fricaria*), cornflowers (*Centauria cyanus*), and various graminoids (Gladstone, 1919).

The literature from battlefields of the Western Front also speaks of the devastation of forests. From military dispatches and other descriptions (Wolff, 1958): "The scene in No Man's Land . . . was indeed a chilling one. . . . Woods were empty fields masked by what seemed to be a few short poles stuck in the ground. . . . Gaunt, blackened remnants of trees drip in the one-time forests. . . . Houthulst Forest, shelled day and night throughout six hundred acres of broken tree stumps, wreckage, and swamps—the acme of hideousness, a Calvary of misery." Similar images can be exerpted from poetry of the Western Front: "Seared earth, stark, splintered trunks, proclaim the Bois-Etoile once was fair" (Bois-Etoile, by E.M. Hewitt); "Ruins of trees whose woeful arms vainly invoke the sombre sky—stripped, twisted boughs and tortured boles" (Ruins, by G.H. Clarke). A final passage describes the dismemberment of a forest during a bombard-ment: "When a copse was caught in a fury of shells the trees flew uprooted through the air like a handful of feathers; in a flash the area became, as in a magi-cians trick, as barren as the expanse around it" (Wolff, 1958).

Forests in the battle zones of World War I were pretty much devastated by the fighting, and by the harvesting of wood for fuel and lumber. The de-forestation of the Western Front and elsewhere was described in a series of essays in the journal *American Forestry* that were well illustrated with graphic photographs of the damage and the stark landscapes (Dana, 1914; Risdale, 1916, 1919a–d; Buttrick, 1917; Graves, 1918).

In addition, beyond the battle zones forests were harvested at a frantic rate to supply the war effort. In total, Belgium lost most of her forested area, while France lost about 10%. In the British Isles, where there was no fighting, the rate of forest harvest was increased by a factor of at least 20, and about one-half of the forest area was harvested during the period of the war. The most important demand on the British forests was to supply timber to be used as pit props in underground coal mines, because there was a tremendous demand for this strategic fuel from industry and the navy (Risdale, 1919a–d). A similarly large increase in the harvesting of the for-ests of Europe took place during the Second World War.

A number of botanists described the flora and ecological succession that took place in and around the city of London after bombing during the Second World War (Salisbury, 1943; Castell, 1944, 1954; Wrighton, 1947, 1848, 1949, 1950, 1952; Jones, 1957). Salisbury (1943) noted that sites that had been bombed and also burned were initially domi-nated by the fireweed *Epilobium angustifolium* and the bryophytes *Ceratodon purpureus*, *Funaria hygrometrica*, and *Marchantia polymorpha*. Un-burnt bombed areas were more species-rich and were dominated a by various urban weeds, includ-ing members of the aster family (especially *Eriger-on canadensis*, *Galinsoga parviflora*, *Senecio vul-garis*, *Sonchus oleraceus*, *Taraxacum officinale*,

and *Tussilago farfara*), grasses (especially *Agrostis alba*, *Lolium perenne*, and *Poa annua*), clovers (*Trifolium pratense* and *T. repens*), chickweed (*Stellaria media*), and plantain (*Plantago major*). Eventually, the London bomb sites supported a flo-ra of more than about 350 species (Jones, 1957). Depending on particular site characteristics (espe-cially the availability of a soil substrate, as opposed to concrete or brick rubble), the vegetation devel-oped into a shrub-dominated community, or into an open grassland-herb community (Jones, 1957).

Wildlife has also been affected by warfare, but again there is little hard documentation. Gladstone (1919) discussed some of the direct effects on birds during the First World War. He described mortality of seabirds (alcids, gulls, and sea ducks) caused by contamination with oil from torpedoed wrecks. There was also concussion injury during gunnery practice, because gulls would be attracted to the target sites by the large number of dead fish that could be scavenged from the surface of the sea.

Remarkably, Gladstone (1919) described bird-life on land as "almost normal" within a short dis-tance of the front line trenches of the Western Front and elsewhere. Only in the bombardment zone was there a large decrease in species richness. Some bird species were remarkably resilient to the disturbance caused by warfare on the Western Front, and they persisted in the battle zone, but undoubtedly at a relatively small density. A notable example is the house sparrow (*Passer domesticus*), which was "plentiful and unconcerned" in "half-felled or-chards and ruined houses." Other prominent land-birds of the front, including "No Man's Land," were the kestrel (*Falco tinnunculus*), woodlark (*Lul-lula arborea*), skylark (*Alauda arvensis*), reed war-bler (*Acrocephalus scirpaceus*), black-cap (*Sylvia atricapilla*), willow warbler (*Phylloscopus troch-ilus*), robin (*Erithacus rubecula*), nightingale (*Lus-cinia megarhynchos*), blackbird (*Turdus merula*), song thrush (*Turdus philomelos*), great tit (*Parus major*), whitethroat (*Sylvia communis*), and others. Many anecdotes described how individuals of these species nested in the midst of seeming devastation, and how they sang and otherwise went about their business in the midst of bombardments and assaults. Some of these observations found their way into the poetry of the time:

In Flanders fields the poppies blow
Between the crosses, row on row,
    That mark our place; and in the sky
    The larks, still bravely singing, fly
Scarce heard amid the guns below.
                    (In Flanders Fields; J. McCrae)

Several species of wildlife have been brought close to or beyond the brink of extinction as a direct or indirect consequence of warfare. The last semi-wild Pere David's deer (*Elaphurus diavidianus*) were killed by foreign troops during the Boxer Rebellion in China of 1898–1900 (the species still survives in captivity) (Westing, 1980). The European bison or wisent (*Bison bonasus bonasus*) was taken close to extinction during the First World War by hunting to provide food for troops. Its population then recovered somewhat in response to intensive conservation and management, but during the Second World War it was again decimated. However, there are now more than 1000 animals in a managed 12,000-ha reserve in Poland and elsewhere in eastern Europe (Grzimek, 1972; Westing, 1980). In addition, a number of commercially valuable big-game species are presently endangered in Africa as a direct consequence of (1) warfare and rebellion, (2) the associated political instability and inadequate administrative structure and enforcement of game laws, and (3) the widespread availability of automatic weapons to poachers. The financial incentives for poaching are especially great in the case of endangered rhinos (*Ceratotherium simum* and *Diceros bicornis*) because of the great value of their horns as dagger-sheaths and as an aphrodisiac, and for African elephants (*Loxodontia africana*) because of their ivory tusks (Curry-Lindahl, 1971, 1972; Fitter, 1974; Westing, 1980).

As a result of the conflict in the Pacific Ocean during the Second World War, the avifauna of a number of remote islands was badly damaged by warfare. On the small island of Iwo Jima, only three species of landbird and one shorebird were observed 1 month after a devastating bombardment and invasion by United States forces, and most of the individuals that were "collected" for examination had healing or recently healed injuries (Baker, 1946). On the island of Guadalcanal, "the destruction of the jungle by the fighting drove out the forest-loving species, and those frequenting the open country made their appearance" (Donaghho, 1950). In the initial stage of postwar succession on Guadalcanal, the forest was open and "completely devastated," and there was essentially no bird life. As plants began to develop on the disturbed site, the first bird to occur was the little starling (*Aplonis cantaroides*), followed by the common myna (*Acridotheres tristis*), the whiskered tree swift (*Hemiprocne mystacea*), and the dollar bird (*Eurystomus orientalis*). When a dense underbrush developed, the bird community was dominated by the chestnut-bellied monarch (*Monarcha castaneiventris*) and the broad-billed flycatcher (*Myiagra ferrocyanea*). As the site developed into a open forest, and then a dense rainforest, a characteristic and diverse avifauna reappeared (Donaghho, 1950).

On some of the smaller and relatively isolated Pacific islands, warfare and its associated disturbances caused a loss or extreme endangerment of endemic bird taxa that are susceptible to anthropogenic extinction (see Chapter 10). On the twin islands of Midway, there were extinctions of the Laysan rail (*Porzanula palmeri*) and the Laysan finch (*Telespiza contrans*) (Fisher and Baldwin, 1946). On the island of Guam the Guam rail (*Rallus ownstoni*), the Guam broadbill (*Myiagra freycineti*), and the Marianas mallard (*Anas oustaleti*) were greatly diminished in abundance (Baker, 1946; Savidge, 1984). Wake Island lost its endemic species of rail (*Rallus wakensis*) (Westing, 1980).

In a few cases, the abundance of certain wild animals increased as a result of warfare. Usually this happened because of decreased exploitation, as in the case of various North Atlantic fisheries during the Second World War (Clark, 1947; Westing, 1980), and various fur-bearing mammals in northern and eastern Europe during that same conflict (e.g., polar bear *Ursus arctos*, red fox *Vulpes vulpes*, wolf *Canis lupus*, and wolverine *Gulo gulo*; Westing, 1980). In addition, it was reported that during the Second Indochina War, tigers (*Panthera tigris*) increased in abundance in certain areas because of the ready availability of corpses as food (Westing, 1980).

Second Indochina War

Although there are few data on the magnitude of the ecological disturbance that was caused by this conflict, we can gain some insight from knowledge of the tremendous amounts of munitions that were expended. During the Second Indochina War of 1961–1975, the total quantity of munitions used by the United States forces alone was more than 14.3 million metric tons, equivalent to about twice the amount used by the United States during World War II (Martin and Hiebert, 1985; Westing, 1985b). About half of the explosive tonnage was delivered from the air, half by artillery, and less than 1% from offshore ships (Martin and Hiebert, 1985). In terms of the number of explosive devices, the United States dropped about 20 million aerial bombs of various size, fired 230 million artillery shells, and used more than 100 million grenades, plus additional millions of rockets and mortar shells (Martin and Hiebert, 1985; Westing, 1976, 1984a, 1985a,b).

This vast outpouring of munitions caused great damage to the landscape of Indochina. One effect was the creation of explosion craters (Orians and Pfeiffer, 1970; Westing, 1976, 1982). In 1967 and 1968, an estimated 2.5 million craters were formed by the explosion of 225- and 340-kg bombs dropped in a saturation pattern by B-52 bombers flying at > 10,000 m. Each sortie produced a bombed-out area of about 65 ha per plane. The total area of terrain affected in this way was about 8.1 million ha or 11% of the landscape of Indochina, including 4.5 million ha or 26% of the area of South Viet Nam. Westing (1976) quoted the following impression of a military observer: "The landscape was torn as if by an angry giant. The bombs uprooted trees and scattered them in crazy angles over the ground. The tangled jungle undergrowth was swept aside around the bomb craters." Each crater typically had a diameter of about 15 m and a depth of 12 m. They usually filled with fresh water and produced habitat for aquatic biota, including mosquitoes and other insects. In agricultural areas, some of the water-filled bomb craters were subsequently developed for aquaculture (Westing, 1982).

In addition, bomb explosions often started forest fires. It was estimated that more than 40% of the area of South Viet Nam's pine plantations was burned during the war, with most fires being ignited by exploding bombs (Orians and Pfeiffer, 1970).

The Legacy of Unexploded Munitions

An important secondary consequence of the delivery of such great quantities of explosives is the presence of a large number of unexploded devices, which cause a lingering hazard on the landscape. The legacy of explosive remnants following World War II is still considerable (Westing, 1985a). In Poland, more than 88 million explosive items have been discovered and removed since 1945, and this cleansing still proceeds at a rate of about 200,000 items per year. A similar problem exists in all of the battle theaters of that war. In the Second Indochina War an estimated 10% of U.S. munitions did not explode, causing an explosive legacy of about 2 million bombs, 23 million artillery shells, and tens of millions of other high-explosive items (Martin and Hiebert, 1985; Westing, 1984a, 1985a).

11.3
CHEMICAL WEAPONS IN WARFARE

Antipersonnel Agents

Chemical warfare began on a large scale during World War I, when some 100 million kg of lethal antipersonnel agents were used. These were mainly the devastating lung agents chlorine, phosgene, trichloromethyl chloroformate, and chloropicrin, and the dermal agent bis(2-chloroethyl) sulfide, or mustard gas. These caused an estimated 1.3 million casualties, including 10^5 deaths (Westing, 1977). The "harassing agent" CS or *o*-chlorobenzolmalononitrile was used by the U.S. military during the Second Indochina War. A total of 9 million kg of

CS was sprayed aerially at 1–10 kg/ha over more than 1 million ha of South Viet Nam. A sprayed site was rendered uninhabitable for 15–45 days (Westing, 1977). There has been no scientific documentation of the ecological damage caused by these various chemical agents of warfare. However, effects on sprayed areas must have been severe, and they would have included the death of a large number of wildlife (Westing, 1977).

Herbicides in Viet Nam

Better information is available about the use by the U.S. military in Viet Nam of broadcast-sprayed herbicides to deprive their enemy of food production and forest cover. This was a large-scale program, with more than 1.4 million ha being sprayed at least once (about one-seventh the land area of South Viet Nam), including more than 100,000 ha of cropland (Boffey, 1971; Westing, 1971, 1976, 1984b,c; Turner, 1977). The program began in

1961, peaked in 1967, and was stopped in 1971 (Table 11.1).

The most frequently used herbicide formulation was a 50 : 50 mixture of 2,4,5-T plus 2,4-D that was known as Agent Orange, but picloram and cacodylic acid were also sprayed (Orians and Pfeiffer, 1970; Boffey, 1971). In total, more than 25 million kg of 2,4-D, 21 million kg of 2,4,5-T, and 1.5 million kg of picloram were sprayed in this military program (Westing, 1971; 1984b,c). The rate of application was relatively large, averaging about 12 kg 2,4,5-T plus 13 kg 2,4-D per hectare, or about 10 times the application rate for conventional silvicultural purposes (NAS, 1974; Turner, 1977). About 86% of the spray missions were targeted against forest, and the remainder against cropland (Westing, 1984b,c).

As was the military intention, great ecological damage was caused by this defoliation program. The effects could be so severe that opponents of the practice labeled it "ecocide," that is, the intentional

"A gas and flame attack seen from the air," somewhere in the coastal lowlands of Belgium or France in 1918. An estimated 100 million kg of lethal antipersonnel chemicals were used during the first World War, and these caused at least 1.3 million casualties. These blanket ground-level fumigations with toxic chemicals also caused severe but undocumented ecological damage. (Photo courtesy of F. Freidel, 1964.)

Table 11.1 The use of herbicides by the U.S. military during the Second Indochina War[a]

Year	Area Sprayed (ha $\times 10^3$)	Quantity of 2,4,5-T + 2,4-D (10^4 kg acid equivalent)
1961	<10	<10
1962	<10	<10
1963	10	10
1964	30	40
1965	31	45
1966	167	595
1967	416	1214
1968	267	845
1969	409	1245
1970	75	214
1971	0	0
Total	1425	4228

[a]After Westing (1971) and Turner (1977).

use of antienvironmental actions, carried out for years over a large area, as an important tactical component of the military strategy (Westing, 1976). Although the ecological effects of herbicide spraying in Viet Nam were not documented in detail, an impression can be gained from several cursory surveys that were made by ecologists. These are described below.

The most extensively sprayed type of vegetation was forest, an ecosystem that covered more than 10 million ha in South Viet Nam, comprising about 60% of the land surface (Westing, 1971; 1976). Mangrove forest is especially sensitive to herbiciding. About 110,000 ha of coastal mangrove was sprayed at least once, or about 36% of the area of that type of forest (NAS, 1974). Virtually all of the diverse assemblage of halophytic plant species in the mangrove forest (including the dominant tree *Rhizophora apiculata*) is extremely sensitive to phenoxy herbicides. As a result, spraying devastated the mangrove ecosystem, and created a large area of poorly vegetated or unvegetated coastal barrens (Orians and Pfeiffer, 1970; Westing, 1971, 1976, 1984b,c; NAS, 1974; Hiep, 1984; Snedaker, 1984). The regeneration that followed was usually slow and sporadic. By the early to mid 1980s, revegetation of the sprayed mangrove areas was essentially complete, but it usually did not involve the commercially-preferred mangrove tree *Rhizophora apiculata*. More frequently the forest was dominated by less desirable mangrove trees such as *Avicennia alba*, or undesirable species such as the palm *Phoenix paludosa*, the mangrove *Ceriops decandra*, the shrub *Acanthus ebracteatus*, the fern *Acrosticum aureum*, and grasses (Westing, 1982; Hong, 1984a,b, 1987). There has been some reforestation, in which the favored mangrove tree *Rhizophora apiculata* was planted over about 10^5 ha. However, the success rate of these plantations has only been about 50% in some places (Hong, 1984b, 1987). Overall, the damage caused to the coastal mangrove forest by herbiciding rendered it the ecosystem type that was most severely impacted by the Second Indochina War (Snedaker, 1984; Westing, 1984b,c).

Effects were also severe in the much more species-rich inland forests, including rain forest with a total area of 10.5 million ha. Mature forest of this type has many angiosperm species, especially of the families Dipterocarpaceae (tree species of *Dipterocarpus*, *Hopea*, *Anisoptera*, and *Shorea*) and Leguminoseae (*Dalbergia*, *Pterocarpus*, *Sindora*, and *Pahudia*). The tree height is up to 40 m or taller, and diameter at breast height is up to 2 m

(NAS, 1974; Galston and Richards, 1984). A single aerial application of herbicide initially killed about 10% of the trees of the upper canopy layer. The total defoliation was much more extensive than this, but defoliated trees were not neccessarily killed. Subsequent resprays (about 34% of the sprayed land was treated more than once) killed additional canopy trees, plus understory trees, shrubs, and ground vegetation (Westing, 1971, 1984b,c; Galston and Richards, 1984). The total loss of merchantable timber caused by the military use of herbicides in South Viet Nam was about 47 million m³ (Westing, 1971).

Some plant species are relatively tolerant of phenoxy herbicides, and of course these were released from competivive stress when the upper tree canopy was killed. Examples of relatively herbicide-tolerant tree species are *Irvingia malayana* and *Parinari annamense*. In addition, some important understory plants survived, and these plus invading ruderal species dominated many stands in the postherbiciding secondary succession. Particularly important in this respect were several species of bamboo, especially *Bambusa* spp., *Thrysostachys* spp., and *Oxytenanthera* spp., and the aggressive grasses *Imperata cylindrica* and *Pennisetum polystachyon*. Pioneer woody dicots included *Adina sessilifolia*, *Randia tomentosa*, *Colona auriculata*, and *Sindera cochinchinensis*. The dense stands of these and many other early successional plants made natural tree regeneration a slow process (Westing, 1971; Ashton, 1984; Galston and Richards, 1984). Some reforestation to high-yielding species such as *Acacia auriculaeformis* is being used to supplement the natural regeneration of herbicided forests (Ashton, 1984).

The secondary effects on wildlife of the large-scale herbiciding of forest habitat in South Viet Nam have not been well documented. In the herbicided mangrove ecosystem, there were anecdotal accounts of a great decrease in abundance of most birds, mammals, reptiles, and other wildlife, and a reduced catch in the nearshore fishery, of which the mangrove ecosystem is an integral component (Hong, 1984a,b, 1987; Quy, 1984). A study of an inland valley that had been changed by herbiciding

from a continuous upland tropical forest to an 80% cover of grassland found only 24 bird and 5 mammal species, compared with 145–170 bird and 30–55 mammal species in two unsprayed reference valleys (Quy *et al.*, 1984). Another study of the conversion of a diverse and luxuriant tropical forest to a degraded *Imperata cylindica*–low shrub savannah found a decreased abundance of some mammal species, but an increase in others (Huynh *et al.*, 1984). Species described as common prior to spraying but rare subsequently were wild boar (*Sus scrofa*), wild water buffalo (*Bubalus bubalus*), Sumatran goat (*Capricornis sumatraensis*), tiger (*Panthera tigris*), and various species of deer (family Cervidae). Many other initially less common mammals also decreased in abundance. The relatively few species that increased markedly in abundance after the herbicide-caused habitat change included less desirable species such as the fawn-colored mouse (*Mus cervicolor*), Sladen's rat (*Rattus sladeni*), and rats in general (*Rattus* spp.).

In addition to these effects on wildlife, which may have been largely caused by herbicide-caused habitat changes, there were many anecdotal reports of illness in domestic animals after military herbicide spraying (Westing, 1980). Large mammals frequently became ill, including domestic water buffalo (*Bubalus bubalus*), zebus (*Bos indicus*), and adult pigs (*Sus scrofa*). More intense illnesses and occasional mortality were reported in smaller domestic animals such as young pigs, chickens (*Gallus gallus*), and domestic ducks (*Anas* sp.).

Another important aspect of the use of phenoxy herbicides in Viet Nam was the contamination of the 2,4,5-T formulation by the very toxic dioxin isomer known as TCDD (2,3,7,8-tetrachlorodibenzo-*p*-dioxin). TCDD is an incidental by-product of the manufacturing process of 2,4,5-T. Using post-Viet Nam manufacturing technology, the contamination by TCDD in 2,4,5-T solutions can be kept to a concentration that is well below the maximum level of <0.1 ppm set by the U.S. Environmental Protection Agency (Norris, 1981). However, the 2,4,5-T used in Viet Nam was more grossly contaminated with TCDD. A concentration as large as 45 ppm occurred, and the weighted-average concentration was about 2.0 ppm (Kilpatrick, 1980; Beljan *et al.*, 1981). A total of perhaps 110–170 kg of TCDD was sprayed with herbicides in Viet Nam (Westing, 1982).

TCDD is well known as being extremely toxic, and it can cause birth defects and miscarriage in laboratory mammals (Kilpatrick, 1980; Norris, 1981; Walsh *et al.*, 1983). However, as is often the case, the direct effects on humans are less well understood. The most useful data on humans have come from observations of persons who were (1) occupationally exposed to TCDD during the manufacture of 2,4,5-T or (2) exposed to TCDD after a chemical explosion in a trichlorophenol factory in Seveso, Italy, in 1976. The latter accident caused the death of livestock and other animals within 2–3 days of the accident. Remarkably, it was not until 2.5 weeks had passed that about 700 people were evacuated from a heavily contaminated residential area close to the factory. The exposure to TCDD at Seveso caused 187 cases of a skin condition known as chloracne, which resembles adolescent acne but is more persistent and severe. However, there was apparently no statistically significant increase in the rate of embryotoxic or teratogenic damages in children born of women from the contaminated area (that is, compared to the average worldwide rate; unfortunately, there were no suitable control data for the study area in Italy) (Kilpatrick, 1980).

Because the 2,4,5-T formulation that was sprayed in Viet Nam was badly contaminated with TCDD, there has been much concern and debate over the possible effects on people exposed to these chemicals. Although there have been claims of epidemiological effects in exposed populations, the studies were not of great scientific rigor, and thus there is controversy (Kilpatrick, 1980; Westing, 1984c; White *et al.*, 1984). The statistically most valid epidemiological studies were done on U.S. military veterans who had served in Viet Nam. These studies did not find clear evidence of an increased incidence of health problems in Air Force personnel involved with the spray program (Lathrop *et al.*, 1984), or an increased probability that veterans would father children with a birth defect (Walsh *et al.*, 1983; Erickson *et al.*, 1984). How-

ever, observations such as these do not rule out the possibility of a human health effect caused by exposure to 2,4,5-T, 2,4-D, or formulations of these contaminated by TCDD. In particular, no rigorous study has been made of women exposed to TCDD in Viet Nam. The likelihood of a teratogenic or embryotoxic effect might be greater in the case of a maternal exposure to TCDD, compared with a paternal exposure (White *et al.*, 1984). Nevertheless, the apparent concensus from the scientifically rigorous epidemiological studies that do exist suggests that large effects have not occurred in relatively heavily exposed human populations, and this is encouraging (White *et al.*, 1984). It also seems likely that the effects of TCDD added little to the direct ecological effects of the herbicides that were sprayed in Viet Nam.

11.4
EFFECTS OF NUCLEAR WARFARE

Most people have at least an inkling of the tremendous destructive capacity of nuclear weapons. An all-out or even a limited exchange of strategic nuclear weapons would cause tremendous damage to humans and their civilization, and to natural ecosystems as well. Westing (1987) described nuclear warfare as "the ultimate insult to nature."

Nuclear Arsenals

The world's nuclear arsenal is overwhelmingly, and perhaps uncomprehendingly, large (Table 11.2). The explosive yield of all nuclear weapons is about $11–20 \times 10^3$ megatons (Mt) of TNT-equivalent. This is more than 1000 times larger than the total 11-Mt yield of all conventional explosives used in the Second World War (6.0 Mt), the Korean War (0.8 Mt), and the Second Indochina War (4.1 Mt) (Westing, 1985b). If the mid-1980s nuclear weapons arsenal is standardized on a per capita basis, there is the equivalent of 3–4 tonnes of TNT per person on Earth (Grover and White, 1985). Note that the average explosive yield of a strategic weap-

Table 11.2   The estimated numbers and explosive yield of the world's nuclear arsenals[a]

Country	Strategic Weapons		Tactical Weapons		Totals	
	Warheads (10^3)	Yield (10^3 Mt)	Warheads (10^3)	Yield (10^3 Mt)	Warheads (10^3)	Yield (10^3 Mt)
United States	9–11	3–4	16–22	1–4	25–33	4–8
Soviet Union	6–7.5	5–8	5–8	2–3	11–15.5	7–11
United Kingdom					0.2–1	0.2–1
China					0.3	0.2–0.4
France					0.2	0.1
Total					37–50	11–20

[a]Strategic weapons are delivered over thousands of kilometers; tactical weapons are designed for battlefield or theater use. One megaton (Mt) of explosive yield is equivalent to the energy contained in about 1 million tonnes of TNT, and it equals 4.2×10^{15} J of energy. Modified from Grover and White (1985).

on (about 0.60 Mt of TNT equivalent; these are delivered over thousands of kilometers) is considerably larger than that of the more numerous tactical warheads (0.20 Mt; used locally in a battlefield) (calculated from Table 11.2). In the mid 1980s there were more than 4300 strategic delivery vehicles (mainly missiles, but also long-range aircraft), carrying an average of about six warheads each, and with an aggregate yield of more than 11,000 Mt (Barnaby, 1982; Sivard, 1986).

Hiroshima and Nagasaki

There is little direct information that we can use to describe or predict the possible environmental consequences of the explosion of a large number of nuclear weapons during a human conflict. There have, of course, been two cases of the use of nuclear weapons in warfare. Although there were no studies of their purely ecological effects, we can gain some insight by considering other environmental damages that were caused by these explosions. Both of these devices were used by the United States against Japan at the end of the Second World War. The bomb dropped on August 6, 1945, on the city of Hiroshima had an explosive yield of about 0.015 Mt of TNT, and the one dropped on Nagasaki a few days later had a yield of 0.021 Mt (Barnaby and Rotblat, 1982; Pittock *et al.*, 1985). These bombs

were relatively "small" in comparison to the typical yield of today's strategic warheads, which average about 0.60 Mt, and range up to 6 Mt (Grover and White, 1985).

Each explosion was an air burst at about 500–580 m. The Hiroshima bomb killed an estimated 140,000 people or about 40% of that city's population, while the Nagasaki device killed 74,000 or 26% of the population (Barnaby and Rotblat, 1982). The most important cause of death and physical destruction was the combined effects of blast and thermal energy. An estimated 50% of the explosive energy was expended as blast, which induced a shock wave that traveled at the speed of sound (about 11 km in 30 sec). Blast pressure caused severe structural damage to buildings located as far as 2–3 km from the blast epicenter. Thermal energy accounted for about one-third of the energy of the explosion, and created a fireball that was sufficiently intense to vaporize people near the blast epicenter, could ignite wood at about 2 km, and could cause human skin burns as far as 4 km away. The ensuing fires created an area of "burnout" of 13 km² at Hiroshima and 6.7 km² at Nagasaki. The combined effect of blast and thermal energy was to destroy about two-thirds of the 76,000 buildings in Hiroshima, and one-fourth of the 51,000 buildings in Nagasaki. Ionizing radiation comprised about 15% of the explosive yield of the bombs. About one-third of the ionizing radiation was emitted within 1 min of the explosion, and the rest more gradually by the decay of radioactive fallout material. (After both explosions, there was a "black rain" that was induced by an upward-moving convective air mass caused by the tremendous fires. The rain had been blackened by soot, and was highly radioactive.) Among other effects, the ionizing energy caused radiation sickness in many survivors of the initial explosion. Symptoms included nausea, vomiting, diarrhea, hair loss, fever, weakness, septicemia, and bleeding from the bowels, gums, nose, and genitals. The presence of latent epidemiological damage in the exposed population that was not affected by short-term symptoms is controversial. However, there have been claims of increased rates of incidence of eye diseases, blood disorders, malignant tumors, psychoneurological disturbances, and leukemia, compared with the rates of these in nonexposed populations of Japanese (Barnaby and Rotblat, 1982; Pittock *et al.*, 1985).

Nuclear Test Explosions

There is a limited amount of ecological information from observations made after above-ground nuclear test explosions conducted by the United States in the 1950s. These detonations caused severe damage to ecosystems, which decreased more or less geometrically with increasing distance from the blast epicenter. After the explosions, succession restored ecosystems in the damaged areas. For example, a test area in the Mohave Desert of Nevada was subjected to at least 89 relatively small (most were equivalent to about 10 kt of TNT; the largest was 67 kt) above-ground detonations (after Shields and Wells, 1962; Shields *et al.*, 1963). Initially, the explosions cleared a central area of 73–204 ha of all life. There was severe vegetation damage over an additional 400–1375 ha, but no damage beyond an area of 3255 ha. Beginning in the first postblast year, the denuded and disturbed areas were progresively invaded by pioneer species of plants. Annual species such as the tumbleweed *Salsola kali*, along with *Mentzelia albicaulis*, *Erodium cicutarium*, and other ruderals, invaded rapidly. With

time, these were replaced by longer-lived species. Overall, the pattern and rate of succession were similar to that expected in that desert region following a severe disturbance.

The United States also conducted aboveground nuclear tests on islands in the south Pacific. Palumbo (1962) studied tagged plants of seven species on Belle Island in the Eniwetok archipelago, located 4.3 km from a 1952 blast site on Elugelab Island. He observed an initial plant damage after the detonation, but recovery was rapid and essentially complete within only 6 months. Elugelab Island was essentially obliterated, as it was transformed into a water-filler crater (Westing, 1980). In another study, Fosberg (1959) reported on the effects of radioactive fallout from a large aboveground test on Bikini Island in 1954 on other islands in the Marshall group. On the most heavily exposed island, 13 out of the total flora of 15 species showed conspicuous pathological or other abnormal symptoms that were attributed to radiation damage. Three species were especially sensitive: *Suriana maritima*, *Cordia subcordata*, and *Pisonia grandis*. Two widespread species were very tolerant: *Scaevola sericea* and *Tournefortia argentea*. As expected, the damage was less severe on islands located relatively far from the blast epicenter.

Experimental Exposure to Ionizing Radiation

Several experimental field studies provide another important source of information that can be used to predict the ecological effects of ionizing radiation from nuclear explosions or other sources. One of the best-known projects involved the experimental treatment of an oak–pine forest on Long Island with gamma radiation from an elevated, 9500-Ci, cesium-137 point source (Woodwell and Whittaker, 1968; Woodwell, 1970, 1982). Since the dose of ionizing radiation decreased geometrically with increasing distance from the source (with the complication of shielding by tree trunks, at a rate of about 5%/cm of wood), there was a distinct zonation of vegetation damage. After 6 months of chronic irradiation, an essentially sterilized zone

Aerial view of damage to an oak–pine forest on Long Island, New York, caused by chronic exposure to ionizing gamma radiation for 20 hours per day from a 9500-curie source of cesium-137 (located in a metal pipe visible in the center of the damaged area; the source can be lowered by remote control into a shielded container below the ground, so that the irradiated area can be entered to take measurements). The intensity of exposure decreases more or less geometrically with distance from the point source of irradiation, with dosage ranging to as much as several thousand roentgens per day within a few meters of the source. The photo was taken in 1962, about 6 months after initiation of the experiment. A clear concentric pattern of death and damage to trees is visible in the photograph. This experiment is designed to run for many years, with an aim at elucidating the longer-term patterns of ecological damage caused by chronic exposure to ionizing radiation, as well as the more specific genetic and physiological effects of this stress on organisms. (Photo courtesy of Brookhaven National Laboratory.)

was created at a radiation dose of > 345 Roengten per day (R/day). In addition, five well-defined vegetation zones were created:

1. There was a central devastated zone at $>$ 200 R/day. Vascular plants were not present, but at least some bryophytes and li-

chens survived up to 1000 R/day. A notable resistant species was the British soldier lichen, *Cladonia cristatella*.

2. At <200 R/day and > 150 R/day, there was a sedge zone with almost continuous cover of *Carex pensylvanica*.

3. At <150 R/day and > 40 R/day, there was

Aerial view of the irradiated oak–pine forest on Long Island, photographed in 1976 after 15 years of experimental exposure to gamma radiation. Over this time period, there was a progressive ecological deterioration of sites close to the point source of radiation stress. The photograph shows a clear concentric pattern of a decreasing intensity of damage with increasing distance from the source. At this time, the rate of ecological change had slowed markedly, as the ecosystems had more or less adjusted to the intensities of radiation stress that occur at various distances from the source. (Photo courtesy of Brookhaven National Laboratory.)

a shrub zone dominated by heaths (*Vaccinium angustifolium*, *V. heterophyllum*, and *Gaylusaccia baccata*) and a short-statured species of oak (*Quercus ilicifolia*).

4. At <40 R/day and > 16 R/day, there was a zone dominated by tree-sized oaks (*Quercus alba* and *Q. coccinea*).

5. At <2 R/day, there was an oak–pine (*Pinus rigida*) zone with some acute damage, but little mortality.

The overall effect of ionizing radiation on ecosystem structure was to cause a systematic dissection of the vertical layers of the original oak–pine forest, similar in pattern to that described for forests impacted by gaseous air pollutants in Chapter 2. The initially relatively heterogenous vertical structure of the mature forest was reduced to a simple, almost two-dimensional bryophyte–lichen community in the heavily irradiated inner zone. In part, this pattern reflects differences in species sensitivity, as there was a progressive replacement of taxa along the irradiation gradient, based on their relative susceptibility to the ionizing stress. However, the physical layering of vegetation targets in the highly structured forest also determined which species were initially impacted by the gamma radiation, and which were shielded. This also influenced the pattern of layer-by-layer "peeling" of the forest. Of course, the irradiated zones also had a smaller net productivity, live biomass, and species richness and diversity, compared with reference vegetation. Other field studies of the ecological effects of ionizing radiation have come to qualitatively similar conclusions about differential species sensitivity and effects on ecosystem structure and function (McCormick and Platt, 1962; Platt, 1965; Monk, 1966; Sparrow *et al.*, 1970; Dugle and Mayoh, 1984).

The irradiation of various ecosystem types has revealed differences in their sensitivity to gamma radiation. On average, coniferous forest is relatively sensitive, followed by temperate mixed forest, tropical rain forest, shrub vegetation, grassland, moss and lichen communities, and pure lichen communities (Table 11.3). The range of sensitivity of various taxonomic groups of organisms is summarized in Fig. 11.1. Vertebrates and vascular plants are generally more sensitive than are less complex organisms, and unicellular organisms plus molluscs are least sensitive. Therefore, a strongly irradiated situation would be expected to have a structurally simple ecosystem, dominated by lesser life forms.

Possible Consequences of a Large-Scale Nuclear Exchange

Several approaches have been used to predict the probable human and environmental consequences of a large-scale nuclear exchange. A relatively direct approach has attempted to budget the likely magnitude of the loss of human life, the destruction of property, and the devastation of ecosystems that would result from the blast, thermal, ionizing, and other relatively short-term effects of the actual detonations. Another approach that was first attempted in the early 1980s uses complex simulation models to attempt to predict the global climatic consequences of large-scale detonations of nuclear weapons. Following from the climatic predictions of the simulations, the likely consequences for the biosphere were predicted by panels of expert scientists.

The short-term predictions of the direct effects of a nuclear holocaust are, of course, for terrible consequences. However, there has been a general concensus that life (including human civilization) would "go on" in some form, and that ecological recovery was conceivable. In contrast, the computer-simulations of climatic change and its subsequent ecological consequences have generally predicted a

Table 11.3   The sensitivity of various plant communities to ionizing radiation[a]

Community Type	Dose (10^3 rads[b])		
	Low Damage	Moderate Damage	High Damage
Coniferous forest	0.1–1	1–2	>2
Mixed forest	1–5	5–10	10–60
Tropical rain forest	4–10	10–40	>40
Shrub community	1–5	5–20	>20
Grassland community	8–10	10–100	>100
Moss and lichen community	10–50	50–500	>500
Lichen community	60–100	100–200	>200

[a]From Harwell and Hutchinson (1985), using data of Whicker and Schultz (1982).
[b]Dose: rads = Roengten absorbed dose, a measurement of the dose of ionizing radiation, in terms of energy absorbed. Rads vary with the type of radiation and the physical medium through which it passes.

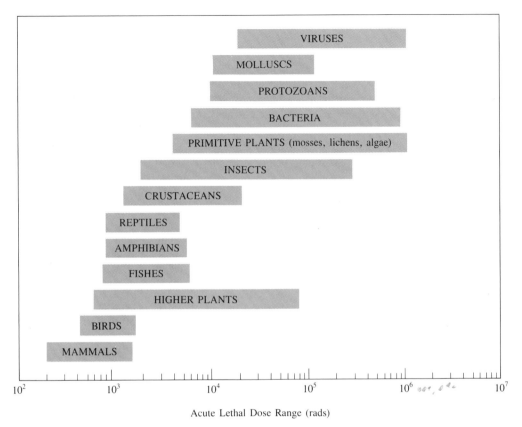

Fig. 11.1. Lethal dose ranges for various groups of organisms. From Harwell and Hutchinson (1985), using data of Whicker and Schultz (1982).

devastation of the biosphere. In the following, the direct, short-term effects of a nuclear exchange will be dealt with first. After this, some of the potential climatic effects of a wide-scale nuclear war and their secondary ecological consequences will be discussed.

Short-Term Effects

Estimates of the size of the global nuclear arsenal were given in Table 11.2. In a large-scale nuclear exchange not all of this capacity would actually be delivered, since many nuclear devices would be destroyed by preemptive strikes, and some capacity would be held back for strategic reasons. Scenarios of the likely scale of the realized yield of the nuclear

explosions vary greatly, but for representative purposes we can cite an estimate of 5000–6000 Mt, almost all of which would be targeted in the Northern Hemisphere (Grover and White, 1985; Harwell and Grover, 1985).

The predicted short-term effects on people are for the death of 20% of the world's population. In the United States, as much as 75% of the population could die from the immediate effect of the explosions, and there would be many additional injuries (Harwell and Grover, 1985). The principal cause of death would be thermal radiation, but the effects of blast, fire, and ionizing radiation would also be important (Harwell and Grover, 1985). This overwhelming loss of human life, in conjunction with the physical devastation and incapacitation of social systems, would destroy civilization as it is now known.

The estimated distributions of energy release from an air burst of strategic nuclear devices of various size are summarized in Table 11.4. In all cases, about 50–54% of the energy is dissipated as a shock wave, 35–38% as thermal radiation, and 6–7% as ionizing radiation. Table 11.5 summarizes estimates of the likely causes of damage to forested terrain as a result of tropospheric or surface detonations of nuclear devices of the size considered in Table 11.4. Close to the blast epicenter, the initial damage to trees would be caused by the power of the blast wave, and to a lesser extent by ionizing radiation. However, a much larger area of forest would be consumed by fires ignited by the intense thermal radiation, and this conflagration would account for most of the short-term damage to vegetation. The most important effects on vertebrates would be caused by thermal radiation, and the area impacted in this way would extend 30–34% beyond the area of ignition of vegetation by thermal radiation. Therefore, a single detonation of any of these variously sized nuclear devices would result in a large-scale devastation of forests and other ecosystems. The simultaneous detonation of nuclear weapons at other sites would cause a similar effect. The aggregate ecological damage would be enormous. However, it would largely be limited to the northern hemisphere, because most strategic targets are located there.

Climatic Effects

There have been attempts to model the global climatic consequences of a large-scale nuclear exchange. The very complex computer models of atmospheric physical properties and circulation that have been used in these studies were developed to simulate the potential climatic consequences of an increasing concentration of atmospheric CO_2 (Chapter 2). In the secondary use of these models to predict climatic effects of nuclear war, key parameters were modified to reflect the expected atmospheric changes that would be caused by a nuclear holocaust. Especially important in this respect are the injections of small carbonaceous smoke particulates and inorganic dust into the upper atmosphere (Crutzen and Birks, 1982; Covey, 1985; NRC, 1985a; Pittock et al., 1985; Stephens and Birks, 1985). One commonly used (but of course, not necessarily correct) scenario is for a 6500-Mt exchange (after NRC, 1985a; Stephens and Birks, 1985), which would cause the explosive injection of 330–825 million metric tons (Tg) of particles into the atmosphere, of which 35% would enter the stratosphere and therefore have a long residence time. These largely inorganic particulates would be accompanied by some 180–300 Tg of sooty smoke from fires, much of which would also end up in the stratosphere. The inorganic dust would increase the albedo of the atmosphere, and thereby decrease the amount of sunlight that penetrates to ground level.

Table 11.4 Distribution of energy released by a nuclear bomb exploded in the troposphere[a]

Energy Form	yield: bomb size:	Energy Released for Bomb Size		
		(10^{12} J) 18 kt	(10^{15} J) 0.91 Mt	(10^{15} J) 9.1 Mt
Blast (shock)		41.9	2.28	22.8
Thermal radiation		29.3	1.59	15.9
Nuclear radiation (first minute)		4.2	0.10	1.0
Nuclear radiation (residual)		8.4	0.21	2.1
Total		83.7	4.19	41.9

[a]From Westing (1977). Note that the data refer to typical air-bursts and that the kT bomb size is a fission bomb, while the other two are half fission and half fusion.

Table 11.5 Damage to biota caused by a nuclear bomb exploded in the troposphere or on the surface[a]

Type of Damage	Area Suffering the Given Type of Damage (ha)					
	Tropospheric Air Burst			Surface Detonation		
	18-kt Bomb	0.91-Mt Bomb	9.1-Mt Bomb	18-kt Bomb	0.91-Mt Bomb	9.1-Mt Bomb
Craterization by blast wave	0	0	0	1	12	57
90% Of trees blown down by blast wave	565	14,100	82,000	362	9040	52,500
Trees killed by nuclear radiation	129	648	1250	148	12,800	63,800
All vegetation killed by nuclear radiation	18	312	759	43	2830	12,100
Vegetation ignited by thermal radiation	1170	33,300	183,000	749	21,300	117,000
Vertebrates killed by blast wave	43	591	2740	24	332	1540
Vertebrates killed by nuclear radiation	318	1080	1840	674	36,400	177,000
Vertebrates killed by thermal radiation	1570	42,000	235,000	1000	26,900	150,000

[a]Modified from Westing (1977).

However, probably more important would be the effect of the sooty smoke, which would absorb incoming solar radiation at a high altitude in the atmosphere, and then reradiate much of the absorbed energy back to space as long-wave infrared radiation. In addition, a high-altitude layer of warm air would create a strong atmospheric stability, and thus retard the rates of the processes by which particles are removed from the atmosphere. The two mechanisms of reflection by dust and absorption and reradiation by sooty smoke could reduce the amount of energy received at the earth's surface by more than 90%.

It is impossible to accurately predict the climatic consequences of effects such as these. This is largely because of uncertainty over the persistence of the particulates at various altitudes in the atmosphere, and because there are various scenarios concerning the scale of the potential exchange of nuclear weapons, the resulting quantity and particle size distribution of dust and smoke, and other important complications. A number of studies using different atmospheric models and nuclear exchange scenarios have concluded that the climatic effects could be severe, and that persistent subfreezing temperatures could occur over much of the northern hemisphere and to a lesser extent in the southern hemisphere (Thompson et al., 1984; Covey, 1985; Pittock et al., 1985). This phenomenon of potentially widespread climatic deterioration has been labeled "nuclear winter." Some other studies have predicted a less consequential cooling effect, by assuming smaller emissions of sooty dusts and more moist smoke, which would be relatively dense and would therefore not penetrate so effectively to the stratosphere (Cess, 1985; Small and Bush, 1985; Penner et al., 1986; Schneider and Thompson, 1988). This smaller climatic effect could be considered to be a "nuclear autumn."

Other atmospheric effects of a nuclear holocaust would be the generation of a large quantity of gaseous pollutants. An estimate for the 6500-Mt nuclear exchange scenario is for the production of 32.5 Tg of gaseous NO from the oxidation of atmospheric N_2 by the heat and pressure of explosions, plus 4.4 Tg of NO produced by fires (NRC, 1985a; Stephens and Birks, 1985). In addition, 225 Tg of CO would be produced by the combustion of some 4500 Tg of organic fuel, and there would be large emissions of SO_2 and other oxides of sulfur, hydrocarbons, asbestos fibers, metals, chlorinated hydrocarbons, and other toxic substances. Some of these pollutants (especially NO_x) would consume stratospheric ozone, resulting in an increased penetration of ionizing solar ultraviolet radiation to the earth's surface, with secondary biological effects (see Chapter 2). For the simulation of a 6500-Mt exchange, it was estimated that there would be a

17% decrease in the concentration of stratospheric O_3 (NRC, 1985a; Stephens and Birks, 1985). On the other hand, the concentration of O_3 in the troposphere could increase due to photochemical reactions involving the large concentrations of NO_x and hydrocarbons (Crutzen and Birks, 1982; see Chapter 2).

Several studies have attempted to assess the likely ecological consequences of a nuclear winter (Harwell and Hutchinson, 1985; Grover and Harwell, 1985; Grime, 1986; Westing, 1987). Important damages to natural and agricultural vegetation would be caused by chilling and freezing injury. This damage would be especially severe in vegetation at low latitudes, since tropical and subtropical plants are not adapted to cold temperatures. For example, during the growing season a typical woody tropical plant cannot tolerate a temperature of 0–4°C for 24 hr without suffering lethal damage, while cultivated tropical plants cannot tolerate 3–7°C (Larcher and Bauer, 1981). A somewhat lower temperature can be tolerated by tropical plants for a shorter period of time, especially if they are exposed during a dormant season (e.g., as low as −3 to −5°C for tropical trees, −4 to −6°C for tropical grasses). Plants of higher latitudes can easily tolerate much colder temperatures than the above when they are in a winter-hardened condition. However, damage can even be caused to arctic plants if they are exposed to a prolonged cold temperature during their growing season (Larcher and Bauer, 1981). Therefore, if the predictions of a prolonged period of low, possibly subfreezing temperatures during a nuclear winter are correct, there would be devastating effects on terrestrial vegetation. These effects would be especially great if the nuclear winter occurred during the growing season, when plants of higher latitudes are not hardened to tolerate cold temperatures. The results would include a catastrophic collapse of agricultural production, and the devastation of most natural ecosystems. The effects on aquatic plants would likely be smaller, since large water bodies provide thermal buffering.

Vegetation would also be affected by the low rates of insolation that are predicted to occur in the aftermath of a nuclear holocaust. The immediate effect would be a decreased rate of photosynthesis, and a loss of net production. However, it is notable that the ability of some plants to survive under a condition of darkness is enhanced by low temperatures (Hutchinson, 1967).

Damage to vegetation would also be caused by (1) an increased surface flux of solar ultraviolet radiation, because of the deterioration of stratospheric ozone; (2) a large dose of ionizing radiation from longer-lived thermonuclear fission products; and (3) effects of phytotoxic air pollutants such as O_3, SO_2, and NO_x.

These various effects of the nuclear aftermath would of course add to the great ecological destruction that was caused by blast, thermal radiation, fire, and ionizing radiation within a short time of the nuclear exchange.

Plants that could survive the initial and aftermath conditions of a nuclear holocaust would probably have a combination of some of the following characteristics (after Grime, 1986): (1) a persistent seedbank that would allow regeneration even if all adult plants were killed; (2) a ruderal strategy, characterized by a large reproductive output, great dispersal ability, and a general ability to do well under disturbed, but not excessively stressful, conditions; (3) be adapted to stressful habitats of small potential net productivity, since plants of such sites are often relatively tolerant of low temperature, low insolation, and other stresses; and (4) be components of an ecosystem that is naturally resilient to the effects of severe periodic or cyclic disturbance, since such plants are generally capable of regeneration after an event of mass mortality. Plants with these various characteristics would be relatively well adapted to survive or regenerate after the nuclear winter, and they would be prominent in the ensuing ecological succession.

Animal life would also be devastatingly affected by the environmental conditions occurring during the nuclear winter. This would largely be due to an intolerance of prolonged low temperatures, the effects of ultraviolet and ionizing radiation, and shortages of food and other critical habitat features as a result of damage to vegetation. Terrestrial animals (including humans) would probably be more greatly affected than aquatic animals.

In summary, the predicted environmental consequences of an all-out nuclear warfare would be horrific. Because of the many possible war scenarios and the incomplete knowledge of atmospheric and ecological processes, it is impossible to predict the effects with a great degree of accuracy. However, the general consensus of several independent modeling experiments is for a very bleak postholocaust future for both humans and the biosphere. Hopefully, the predictions will never be tested in a real-life experiment.

12

EFFECTS OF STRESS ON ECOSYSTEM STRUCTURE AND FUNCTION

12.1
INTRODUCTION

The intensity of ecological stress varies in space and time. When this variable stress impacts a specific ecosystem there is a unique effect. Nevertheless, in spite of particular circumstances, there are commonly observed patterns of ecological response to stress. This phenomenon allows a measure of predictability that transcends the circumstances of unique situations. Some of these frequently occurring patterns will be briefly examined in this chapter.

12.2
ECOLOGICAL EFFECTS OF LONG-TERM CHRONIC STRESS

A particular environment may be inimical to most organisms if severe stress is present for a long period of time. As a result, there can be very little or no ecological development. Examples of such situations include:

1. Sites where severe climatic stress precludes the survival of biota, such as certain terrain in the Arctic, the Antarctic, and at high altitude.
2. Sites where natural or anthropogenic pollution has caused severe toxic stress, such as the most polluted sites at the Smoking Hills (Chapter 2), surface exposures of minerals with a high concentration of toxic elements (Chapter 3), and the immediate vicinity of a smelter that is emitting toxic elements and gases (Chapters 2 and 3).
3. Sites that are disturbed too frequently to allow organisms to establish and persist, such as the ice scour zone of polar oceans, and severely trampled paths.

More frequently, ecosystems are subject to a lesser intensity of long-term stress, which impoverishes, but does not preclude ecological development. Such situations are often characterized by a convergent, or qualitatively similar, pattern of ecosystem structure and function. The High Arctic tundra is an example of a chronically stressed environment. Ecological development is typically restricted by (1) a short and cool growing season and (2) a limited availability of moisture and nutrients, the latter being due to oligotrophic soil and also to climatic suppression of nutrient cycling and litter decomposition (Bliss *et al.*, 1973; Chapin and Shaver, 1985; Henry *et al.*, 1986a). Because of these chronic environmental limitations there is much unvegetated and sparsely vegetated terrain in the High Arctic. Advanced ecological development is limited to a few relatively moist and warm situations called "oases," which comprise only 1–2% of the land area (Bliss, 1977; Freedman *et al.*, 1983; Henry *et al.*, 1986b). The plants that occur most generally in the Arctic are archetypal "stress tolerators" (*sensu* Grime, 1979), and the tundra ecosystem reflects their morphological and ecophysiological characteristics. These stress-tolerant plants are typically long-lived, and they have a very decumbent stature, small annual production, little allocation of biomass below the ground, persistent living and dead foliage, a small requirement for and efficient recycling of nutrients, and reliance on vegetative propogation to persist on the site (Bliss, 1971; Savile, 1972; Maessen *et al.*, 1983; Nams and Freedman, 1987a,b). Field experiments have shown that the characteristics of stress-tolerant arctic species are conservative, that is, they do not exhibit a very plastic phenotypic response to environmental amelioration through fertilization or the use of field greenhouses (Henry *et al.*, 1986a). In contrast, the relatively productive plants that grow in less stressful tundra wet meadows produce much more biomass and flower profusely under ameliorated conditions (Henry *et al.*, 1986a). Because of the severe climate and small primary production of the High Arctic tundra, there is a low species richness and a miniscule standing crop of resident animals. In contrast, the avian biota is largely comprised of about 10–20 migratory species that take advantage of the brief seasonal abundance of arthropods and plant production (Freedman and Svoboda, 1982; Henry *et al.*, 1986b).

In environments where the climatic limitation to ecosystem development is less severe than in the Arctic, stress caused by pollution can cause the impacted ecosystem to exhibit a structure and function that superficially resembles that of the tundra. For example, sites that are strongly influenced by serpentine minerals are stressful to plants because of the toxic effects of nickel and chromium, coupled with nutrient deficiency (Chapter 3). In an otherwise forested landscape, a serpentine site can be a distinct ecological island with a vegetation dominated by low-growing shrubs and herbaceous plants that are tolerant of the stressful edaphic conditions. The serpentine vegetation is not only structurally distinct—it also has a species composition that differs from that of the surrounding terrain. In many cases the serpentine taxa are endemics that are physiologically tolerant of the toxic stress of these sites, and that cannot survive elsewhere because they are poor competitors in less stressful environments.

Under the long-term influences of anthropogenic pollution and other stresses, ecosystem structure is also simplified. This phenomenon causes a gradient of ecosystem structure that is distributed concentrically around a point source of toxic stress. Consider the structure of the terrestrial vegetation along a transect from a large smelter that emits toxic elements and/or gases, for example, near Sudbury or Gusum (Chapters 2 and 3). In the "reference" terrain, the landscape is dominated by a mature forest with a species composition and physical structure that is typical for the geographical region. In contrast, the immediate vicinity of the source of pollution may be characterized by virtually complete ecological degradation, because so few organisms can tolerate the inimical toxic stress. As distance from the point source increases, there is an approximately geometric decrease in the intensity of the toxic stress. As a result, there is a progessive pattern of invasion and/or survival of species, depending on their physiological tolerance of the toxic environmental conditions. As a rule, species that are tolerant of the toxic stress are not present or are uncommon in the uncontaminated reference habitat. Often, the most tolerant plants are a few species of lichens and bryophytes, plus tolerant species or ecotypes of monocots and herbaceous dicots. Together, these comprise a relatively sparse, low-growing ecosystem close to the point source. At a greater distance shrubs may dominate, and still further away particular species of tree may form an open forest. Finally, at a distance where the toxic stress is too small to exert an important ecological effect, the reference forest type is present. It is important to note that these vegetation types are not discrete; they represent identifiable "stages" that are present along a gradient of continuous environmental and vegetational change. The syndrome of spatial decline of the vertical strata of the terrestrial vegetation along a transect from a smelter was described as a "peeling" or "layered vegetation effect" by Gorham and Gordon (1963). A qualitatively similar pattern of ecosystem decline is observed along a transect from a point source of stress by ionizing radiation, as was described for an oak-pine forest impacted by gamma radiation by Woodwell and co-workers (Chapter 11).

In addition to these structural characteristics, severely stressed ecosystems have relatively small rates of such processes as productivity, nutrient cycling, and litter decomposition. As was the case for ecosystem structure and species composition, these functional properties vary rather predictably along a gradient of anthropogenic pollution stress near a large point source, such as the smelters at Sudbury and Gusum.

12.3
ECOLOGICAL EFFECTS OF THE ADVENT OF STRESS

When an ecosystem is suddenly subjected to pollution or some other stress, it may merely serve as a "sink" and absorb the stress without undergoing a measureable change, or it may display a number of disruptions of its structure and function (Bormann, 1982). In a qualitative sense, the latter effects are observed generally regardless of the type of ecosystem or the nature of the stress. Depending on the

intensity of the stress, the impacted ecosystem may become unbalanced energetically, nutrient cycles may become less conservative and "leaky," there may be changes in the species composition and dominant ecological "strategy" of the biota, and the system generally becomes less stable and less complex (Auerbach, 1981; Odum, 1981, 1985; Bormann, 1982; Smith, 1981, 1984; Rapport *et al.*, 1985; Schindler, 1988).

Many of the ecological consequences that might be expected to occur after the initiation of stress are summarized in Table 12.1. Note that not all of these effects are likely to be manifest in any particular, newly stressed situation. The intensity of the stress is important in this respect, as is the period of time during which the ecosystem is exposed to the stress. For example, in many fresh-water systems stress will cause large changes in species composition (e.g., of the phytoplankton) before changes take place in ecological function (e.g., in primary production or respiration) (Schindler, 1988). In such a case the toxic threshold for sensitive taxa occurs at a lower intensity of stress than does the toxic threshold for function, which may be carried out at a community level irrespective of the species composition. Therefore, changes in species composition and demographics would be expected to be relatively sensitive indicators of the initial ecosystem response to the advent of a low or moderate intensity of stress. However, in many terrestrial ecosystems the reverse may be true (Schindler, 1988). In a forest dominated by long-lived plants, a moderate intensity of stress will typically affect functional attributes such as photosynthesis, respiration, and nutrient cycling well before there is mortality of individual plants or elimination of species.

Many such effects are revealed by consideration of the ecological impacts of the various types of stress that have been examined in this book. For example, (1) toxic stress caused by ozone is eliminating the relatively sensitive genotypes of ponderosa pine from montane forests in southern California, and they are being replaced by more tolerant genotypes and by other species of conifer (Chapter 2); (2) the cutting of all trees and subsequent herbiciding of a forested watershed at Hub-

Table 12.1   The trends that may be expected in ecosystems upon the advent of stress[a]

Energetics
1. Community respiration increases.
2. Production/respiration becomes unbalanced (i.e., P/R becomes greater than or less than 1).
3. P/B and R/B (i.e., maintenance to biomass structure) ratios increase.
4. Importance of auxiliary energy increases.
5. Exported or unused primary production increases.

Nutrient cycling
6. Nutrient turnover increases.
7. Horizontal transport increases and vertical cycling of nutrients decreases.
8. Nutrient loss increases (system becomes more "leaky").

Community structure
9. Proportion of r-strategists increases.
10. Size of organisms decreases.
11. Life spans of organisms or parts (leaves, for example) decrease.
12. Food chains shorten because of reduced energy flow at higher trophic levels and/or greater sensitivity of predators to stress.
13. Species diversity decreases and dominance increases; if original diversity is low, the reverse may occur; at the ecosystem level, redundancy of parallel processes declines.

General system-level trends
14. Ecosystem becomes more open (i.e., inputs and outputs become more important as internal cycling is reduced).
15. Successional trends reverse (succession reverts to earlier stages).
16. Efficiency of resource use decreases.
17. Parasitism and other negative interactions increase, and mutualism and other positive interactions decrease.
18. Functional properties (such as community metabolism) may be more robust (homeostatic-resistant to stressors) than are species composition and other structural properties. In systems dominated by long-lived perennial plants (e.g., forests), the reverse may be true.

[a]Modified from Odum (1985).

bard Brook caused a large loss of nutrients in stream flow (Chapter 9); (3) the clear-cutting of a forest and/or the silvicultural herbiciding of a regenerating clear-cut allows a species-rich assemblage of ruderal species to temporarily dominate the site (Chapters 8 and 9); (4) the stressing of certain species of forest trees by an undetermined complex of predisposing agents causes a stand-level decline, often secondarily accompanied by infestations of pathogenic fungi and destructive insects that are normally resisted by healthier trees (Chapter 5); and (5) the acidification of water bodies by atmospheric deposition causes large changes in species composition, and sometimes an oligotrophication of the ecosystem (Chapter 4).

It is useful to know that stressed ecosystems exhibit a broad syndrome of qualitatively similar ecological changes. Such knowledge allows a measure of predictability of the likely ecological consequences of the advent of an anthropogenic stress

such as pollution in a previously unimpacted area. Of course, the ability to accurately predict specific ecological changes is immeasurably improved if there is also knowledge of (1) the physiological tolerance of the specific biota to the new stress and (2) the role of the sensitive biota in the functioning of the impacted ecosystem. For this reason it is important to have an understanding of the effects of similar anthropogenic stresses in other field situations. Insight can also be gained from laboratory experiments that examine the ecophysiological effects of particular stresses on key species of the ecosystem.

Consider a situation where ecologists are asked to participate in an interdisciplinary study of the potential environmental impacts of a new smelter whose proposed location is in forested terrain. Budget calculations by engineers can predict the anticipated emissions of toxic metals and gases from the stack. Subsequent modeling studies by atmospheric and other scientists can predict the likely frequency of ground-level pollution events and the concentrations and depositions of toxic substances at various distances from the point source, as well as their progessive accumulation (if any) in soil, the forest floor, and vegetation. Ecologists can then use (1) these physical scenarios, coupled with (2) knowledge of the toxicity thresholds of key biota from laboratory studies and (3) the effects of similar toxic stresses on forest vegetation near previously studied operating smelters, to predict (4) the likely ecological impacts of the proposed smelter.

The process just described in which predictions are made of the ecological consequences of the advent of a new anthropogenic stress (e.g., the smelter) is a component of the interdisciplinary activity known as Environmental Impact Assessment (EIA) [Canter, 1977; ESSA, 1982; Canadian Environmental Assessment Research Council/National Research Council (CEARC/NRC), 1986]. Sometimes the impacts predicted by the EIA are then measured in the field after the stress is initiated (e.g., after the smelter is built; unfortunately, and remarkably, this important and seemingly fundamental follow-up step is often not required by regulating agencies, and therefore it may not be carried out; Beanlands and Duinker, 1983). The nature of these follow-up studies is largely guided by the predictions of the EIA, but their structure cannot be too rigid, as "surprises" often occur, that is, unpredicted ecological impacts that are unprecedented because they may be particular to the ecosystems and stresses that are being investigated (Loucks, 1985). These surprises have to be investigated by adaptive changes in the monitoring program.

Clearly, the accumulation of knowledge about the effects of particular stresses in various ecological and laboratory situations allows ecologists to better predict, and then measure, the likely impacts of new anthropogenic developments that will disturb previously unimpacted ecosystems.

12.4

THE ALLEVIATION OF STRESS

An ecosystem that is damaged by pollution or by some other disturbance will recover after alleviation of the stress. The postdisturbance successions that follow deglaciation (Chapter 4), wildfire, clearcutting (Chapter 9), herbicide spraying (Chapters 8 and 11), and other disturbances are well known and easily appreciated phenomena. In general, during the relatively early and dynamic phases of succession there are (1) a progressive accumulation of biomass and nutrients in the aggrading vegetation; (2) a change in the species composition of the biota and in the prominence of particular taxa; and (3) a tightening of the nutrient cycle so that there is a smaller loss of nutrient capital and more internal storage and recycling; along with (4) many other changes in ecosystem structure and function (Likens *et al.*, 1977; Bormann and Likens, 1979; Odum, 1983).

Human intervention can ameliorate the conditions of a stressed environment in order to encourage recovery, or to otherwise manage the situation so that a more "acceptable" ecosystem develops. For example, an obvious way of alleviating pollution stress is to reduce or eliminate the emission of the toxic substances. Strategies of pollution abatement have (1) allowed a vigorous recovery of damaged ecosystems close to the Sudbury smelters (Chapter 2); (2) greatly reduced the ecotoxicological impacts of the insecticide DDT on birds and other wildlife (Chapter 8); (3) reduced the amount of crude oil and refined hydrocarbon products entering the oceans and other environments (Chapter 6); and (4) alleviated or reversed the eutrophication of many inland waters (Chapter 7).

In many other situations of anthropogenic stress of ecosystems, society could potentially adopt laudable, but economically and politically difficult, management strategies to alleviate the stress. For example, (1) a vigorous abatement of SO_2 and NO_x emissions would be an excellent way of reducing the ecological impacts of the deposition of acidifying substances from the atmosphere (Chapter 4); (2) the elimination of the use of chlorofluorocarbons would alleviate their effects on stratospheric ozone (Chapter 2); (3) vigorous conservation of tropical rainforest and other threatened ecosystems would prevent the extinction of much of the earth's biota (Chapter 10); (4) decreased atmospheric emissions of CO_2 from the combustion of fossil fuels and deforestation would help prevent global climate change through the greenhouse effect (Chapter 2); and (5) the manufacture of all nuclear weapons into ploughshares would avoid the direct and indirect effects of a thermonuclear holocaust (Chapter 11).

However, in many cases decision makers believe that the abatement of certain anthropogenic stresses is impossible for political, economic, or other reasons. In such a situation the resulting (or potential) ecological degradation is considered (by the decision makers) to be an "acceptable" cost of some desirable human endeavor that unfortunately causes ecological degradation. In other cases, a secondary management strategy can be used to treat the ecological "symptoms" of certain anthropogenic stresses, while still allowing their "cause" to continue. Usually this strategy is adopted on the short term in order to cope with an unacceptable state of environmental degradation, while continued research and political lobbying might hopefully lead to a more sustainable, longer-term solution to the environmental problem. Examples of this sort of "coping" strategy include (1) the periodic liming of acidified surface waters in order to allow the survival of fish until a program of pollution abatement is implemented (Chapter 4); (2) the spraying of spruce budworm-infested forest with insecticides in order to keep the economic resource alive, while working toward a more acceptable system of integrated pest management for use in the future (Chapter 8); and (3) the use of prolonged and tedious international negotiations in order to control or reduce the inventories of conventional, chemical, and nuclear weapons, until a more comprehensive state of disarmament might be implemented (Chapter 11).

In the long term, humans can only prosper on earth if the strategy of their society includes limitations to the abundance of our species, and the management of ecological resources in a way that is sustainable and not destructive in the long term. It is critical that we understand the ways in which environmental degradation is caused by the direct and indirect effects of particular management or exploitative practices, or secondarily by other human activities. The coupling of such knowledge with an appropriate strategy to eliminate or reverse these impacts may be a key to the successful realization of a sustainable prosperity for our species, on the only place in the universe that is known to sustain life.

GLOSSARY

This glossary defines a number of terms and phrases that are specific to ecology, forestry, wildlife biology, environmental sciences, and other disciplines relevant to the ecological effects of pollution and other stresses. However, the names of specific chemicals and organisms are not defined here. The primary intent of this section is to make the contents of this book relatively easily accessible to readers who may not have had an introductory course or other experience in ecology. References consulted during the preparation of parts of this glossary were The Institute of Ecology (TIE) (1973), Abercrombie *et al.* (1978), Bonnor (1978), Hausenbuiller (1978), Newton and Knight (1981), Tver (1981), Lincoln *et al.* (1982), United Nations Environment Program (UNEP) (1982), and Miller (1985).

aboriginal The original people inhabiting a region prior to European contact.

above-ground All biomass or nutrient capital that occurs above the ground surface.

acaricide A pesticide used against spiders and mites.

acclimation A phenotypically plastic response to exposure to some environmental stress, characterized by an enhanced degree of physiological tolerance of that stress.

acidic deposition A shorter but somewhat inaccurate version of the phrase, "the deposition of acidifying substances from the atmosphere."

accretion An increase in the quantity of biomass or nutrients in an ecosystem.

accumulator An organism that sequesters nutrients or toxic substances from the environment to an unusually large concentration.

accuracy The degree to which a measurement of the value of an environmental variable reflects the true value of the variable. Compare with precision.

acid rain See acidic precipitation.

acidic precipitation (1) Rain, snow, or fog water having a pH < 5.65. (2) The deposition of acidifying substances from the atmosphere during a precipitation event.

acidification An increase over time in the content of acidity in a system, accompanied by a decrease in the acid-neutralizing capacity of that system.

acidifying substance A substance that causes acidification. The substance may have an acidic character and therefore act directly, or it may be nonacidic but generate acidity during a chemical transformation (as in the nitrification of ammonium to nitrate).

acidity The ability of a solution to neutralize an input of hydroxide ion. Usually measured as the concentration of hydrogen ion, in logarithmic pH units (see also pH). Strictly speaking, an acidic solution has a pH < 7.0.

acid mine drainage Surface water or ground water that has been acidified by the oxidation of pyrite and other reduced-sulfur minerals that occur in coal mines and coal mine wastes.

acidophilous Refers to organisms that only occur in very acidic habitats.

acid shock A short-term event of great acidity. The phenomenon regularly occurs in fresh-water systems that receive a large pulse of relatively acidic water when the snowpack melts in the spring.

acid sulfate soil An edaphic condition that occurs when a shallow-water marine, estuarine, or saltmarsh system is drained for agriculture. When the previously anaerobic soil rich in pyrites and other reduced-sulfur species is exposed to the atmosphere, oxidation of the sulfides by chemoautotrophic bacteria generates a great deal of acidity.

acute Refers to a relatively large or severe effect caused by a relatively short-term exposure to a toxic environmental agent. Compare with chronic.

acute injury See acute toxicity.

acute toxicity A poisonous effect produced by a single short-term exposure to some toxic agent, and causing measurable biochemical or anatomical damage, or death of the organism.

advance regeneration Young and/or small trees that are established naturally in a stand, and that comprise or contribute to the next forest that regenerates after the overstory trees are harvested.

aerial spray A pesticide application that is delivered to the spray site by a fixed-wing airplane or helicopter.

aerobic Refers to an environment in which oxygen is present.

aerosol A suspension of fine solid or liquid particulates within an atmospheric or aquatic medium. Examples include haze, smoke, and fog.

afforestation The conversion of a nonforested ecosystem to a forest by the planting of trees.

aggrading See accretion.

agroecosystem An ecosystem that is intensively managed to optimize the productivity of agricultural plants or animals.

agronomic Pertaining to an agricultural context.

air burst A nuclear explosion occurring in the atmosphere close to the ground surface, generally at about 500–1000 m.

air mass A portion of the atmosphere with relatively homogeneous conditions of temperature, pressure, and chemistry, and that is moving as a cohesive unit.

airshed A more-or-less definable portion of the atmosphere that overlies a particular area of landscape.

albedo The ratio of reflected to incident electromagnetic radiation for some surface.

alevin A very young salmonid fish, recently hatched and still having a yolk-sac.

algal bloom See bloom.

alkaline Refers to a nonacidic solution with pH > 7.0.

alkalinity The amount of alkali in a solution. In fresh water, alkalinity is mainly comprised of bicarbonates, carbonates, and hydroxides, and it is generally measured by titration with acid to a fixed end point.

allochthonous Refers to an inlux of nutrients or some other material to an ecosystem, but coming from some other place.

ambient Refers to the environmental conditions that affect a body or system, but that are not affected by it.

ammonia volatilization The flux of ammonia to the atmosphere from sewage or after fertilization with ammonium nitrate, urea, or some other nitrogen-containing chemical.

ammonification The conversion of organic-nitrogen to ammonium by soil or aquatic microbes.

amphi-Atlantic A biogeographic distribution that encompasses both western Europe and eastern North America.

anaerobic Refers to an environment in which oxygen is not present, or it only occurs in a very small concentration.

angiosperm A flowering plant in the subdivision Spermatophyta, which is distinguishable from the Gymnospermae by having ovules borne within a specialized closed structure, the ovary, which develops into a fruit.

anion An ion with a negative charge.

annual An organism that completes its life cycle in 1 year.

annual allowable cut (AAC) The amount of wood volume or biomass that can be removed each year from a forested area that is being managed so as to allow a continuous cropping of tree biomass. Strictly speaking, the AAC should not exceed the annual volume increment of the management area.

anoxic See anaerobic.

anthropogenic Occurring because of, or influenced by, the activities of people.

anticoagulant A substance that interferes with the normal clotting of blood that has been exposed to the atmosphere.

arboreal lichen A lichen that occurs on an aerial tree substrate.

Arctic (adj. arctic) Any area north of the Arctic circle, or 66° 32′ N. Sometimes defined as all terrain north of the boreal forest.

atmospheric deposition The influx of a substance from the atmosphere, both with precipitation and in its absence. See also acidic deposition, dry deposition, and wet deposition.

atomic density The ratio of atomic mass to atomic volume.

autecology The ecology of individuals, or of populations of a particular species. Compare with synecology.

autotroph An organism that can synthesize its biochemical constituents using inorganic precursors and an external energy source. Photoautotrophs make use of sunlight through the process of photosynthesis, while chemoautotrophs harness some of the energy content of inorganic chemicals through the process of chemosynthesis.

available concentration   That portion of the total concentration of a nutrient or toxic substance in soil or water that is assimilable by organisms. Availability is related to the degree of solubility in the prevailing aqueous solution of the compartment, and it is measured by various indices. The simplest index used in studies of soil is the water-soluble concentration. See also total concentration.

avifauna   All of the bird taxa in a specified area.

background concentration   The concentration of a substance in a particular environment that has not been measurably influenced by anthropogenic sources.

baleen whale   A whale that filter feeds on planktonic crustacea.

barrens   A generic term used to describe a situation where there is a degraded ecosystem of much simpler structure and function than could potentially occur considering the climatic and edaphic characteristics of the site. Usually refers to a nonforested vegetation occurring in a landscape that potentially supports forest.

basal area   The cross-sectional area of the stem of a tree, or of a stand of trees.

base saturation   The absolute or relative degree of occupation of the total soil cation exchange capacity (CEC) by calcium, magnesium, potassium, and sodium ions. The remaining saturation of CEC is usually comprised of hydrogen and aluminum ions. See also cation exchange capacity.

basic cation (base cation, base)   Refers to Ca, Mg, K, and Na ions in soil or fresh water.

bedrock   The solid rock that underlies the soil or other surficial deposit covering a site.

below-ground   All biomass occurring below the soil surface.

benthos   The biota living on or in the surface sediment of a fresh-water or marine ecosystem.

bilge washings   The hydrocarbon-contaminated water that has been used to rinse the petroleum or fuel-holding compartments of a tanker, and that is then discharged to the environment. See also load-on-top.

bimodal   A distribution of observations that is characterized by two distinct peaks of frequency.

binomial   The proper name of a species, comprised of two latinized words. The first word identifies the genus, and the second the species. Sometimes a third latinized word identifies a distinct subspecies.

bioavailable   See available concentration.

bioaccumulation   The biological sequestering of a substance at a larger concentration than that at which it occurs in the environment.

bioassay   A quantitative estimation of the intensity or concentration of a biologically active environmental factor, measured via some biological response under standardized conditions.

biochemical oxygen demand (BOD)   A test that measures the amount of oxygen consumed by microorganisms in an aqueous solution in a standardized, closed incubation. If all oxygen consumption can be attributed to the oxidation of organic matter, then BOD is effectively biological oxygen demand. However, if oxygen-consuming inorganic chemicals are also present, then biochemical oxygen demand is measured.

bioconcentration   See bioaccumulation.

biodegradation   The process by which biological metabolism, usually microbial, transforms certain toxic chemicals or sewage into other, relatively innocuous chemicals.

biogenic   The emission of toxic or other substances by organisms.

biogeochemical prospecting   The chemical analysis of biota or inorganic matrices such as soil, sediment, or water to discover surface or near-surface occurrences of commercially valuable minerals. May also be supplemented by information about the distribution of indicator plants whose occurrence is specific to the environmental conditions associated with the mineral.

biogeochemistry   The study of the quantity and cycling of chemicals in ecosystems, landscapes, or the earth as a whole.

biogeography   The study of the large-scale distributions of taxa of plants and animals, and the factors that influence those distributions.

biological control   The control of a pest or pathogen by a biological method, such as the use of a tolerant crop genotype, the introduction of a predator, parasite, or disease, etc.

biomagnification   See bioaccumulation.

biomass   The quantity of living and/or dead organic matter, usually standardized per unit area in a terrestrial ecosystem, or per unit volume in an aquatic ecosystem.

biome   A geographically extensive ecosystem, which is usually characterized by its dominant vegetation, for example, coniferous boreal forest, tropical rain forest, or arctic graminoid tundra.

biosphere The thin layer surrounding the earth in which organisms occur, and where biological processes take place.

biota The assemblage of organisms in some defined site or geographical area.

bird-watching The non-professional observation of birds.

birth defect The occurrence of a biochemical disorder or anatomical defect in a newly born animal.

black rain A sooty, radioactive rain that is induced by the convective uplift of an air mass that is induced by the thermal energy of an above-ground nuclear explosion.

blast The kinetic energy of a moving shock wave of air that is caused by an explosion.

bloom A burst of productivity of phytoplankton, resulting in a large standing crop of algal biomass and of chlorophyll, and a restricted transparency of the water column.

blowout An uncontrolled emission of natural gas, petroleum, or water from an oil well.

bog An oligotrophic, acidic wetland that receives all of its input of nutrients by atmospheric deposition, and that has a plant community dominated by *Sphagnum* moss.

bole The stem or trunk of a tree.

boreal forest The conifer forest that occurs in the low Arctic in regions with a long and cold winter and a short growing season, and that gives way to tundra at more northern latitudes.

boundary layer A thin layer of fluid or air that is adjacent to a solid surface, in which there is no turbulent transport of the medium, and through which dissolved substances must move by diffusion.

box model A nondynamic assessment of the quantity of material or energy in various compartments of an ecosystem, and of the fluxes between compartments.

broadcast spraying The treatment of an area so as to achieve an even coverage of pesticide. This method of application is not pest-specific, and it exposes many nontarget organisms to the pesticide.

broad-spectrum Refers to a pesticide or other toxic agent that affects a wide variety of biota.

browse The young twigs, foliage, and reproductive tissues of woody plants that are eaten by mammalian herbivores such as rabbits and deer.

brown water Water that is colored with a brownish hue by the presence of dissolved organic matter. Some-

times termed a dystrophic system. Compare with clear water.

bryophyte A moss or liverwort.

budget An analysis of the quantity of material or energy in the compartments of a system, of the exchanges among compartments, and of the total inputs to and outputs from the system.

buffer A chemical compound that has the capacity to absorb or exchange hydrogen or hydroxide ions, and that allows a system to assimilate a limited amount of these ions without changing appreciably in pH.

buffering capacity The quantitative ability of a solution to absorb hydrogen or hydroxide ions without undergoing a pronounced change in pH.

buffer strip An uncut border of forest that is left beside water bodies or roads during a forest harvest.

bulk collector A collector that is continuously open to the atmosphere, and that samples both wet and gravitational depositions of substances.

bulk deposition The deposition of substances from the atmosphere with precipitation, and the gravitational settling of particulates, as measured with a bulk collector.

bulk precipitation (bulk-collected precipitation) See bulk deposition.

bunker C fuel oil A relatively heavy liquid hydrocarbon fraction that is used as a fuel for ships and oil-fired power plants.

cable logging A system used to remove logs from a harvested forest. Usually, a cable is attached to a tall tree or spar, and logs are winched to that central place and then loaded onto a truck.

calcareous A situation rich in calcium carbonate, with a relatively high pH, and often having a distinctive flora.

cancer A malignant growth or tumor caused by an uncontrolled and undifferentiated division of cells.

canopy The predominant layer of foliage of an ecosystem.

canopy closure A point in succession where almost all of the sky is obscured by foliage when viewed upward from the ground surface.

capital The quantity of biomass or nutrient on a site.

carnivore An animal that eats other animals.

carrying capacity The maximum abundance of a species that can occur in an area, beyond which degradation of the habitat will occur.

case study Consideration of the pertinent aspects of a particularly exemplary situation.

catalyst A substance that decreases the activation energy of a chemical reaction and that thereby increases its rate, without being consumed in itself. See also enzyme.

cation An ion with a positive charge.

cation exchange capacity (CEC) The total amount of exchangeable cations that a given quantity of soil can absorb. See also base saturation.

cavity nester A bird that builds its nest in a hollow of a tree.

census plot A designated area or volume of habitat in which the abundance of animals is determined.

chapparal Vegetation dominated by evergreen angiosperm shrubs, and found in an area with a mediterranean climate with a hot and dry growing season and a mild and wet winter.

chelation The sequestering of a metal by an inorganic chemical that contains a closed ring of atoms, one of which must be a metal.

chemoautotroph See autotroph.

chemosynthesis See autotroph.

chlorofluorocarbon A carbon-based compound containing chlorine and/or fluorine atoms.

chlorophyll A green pigment found in almost all algae and higher plants, and that is responsible for the light capture that drives photosynthesis.

chlorosis A symptom of stress in a vascular plant, characterized by a general or mottled yellow or light green coloration of the foliage, when the healthy condition would be a dark green colour.

chronic injury See chronic toxicity.

chronic toxicity A poisonous effect produced by a long period of exposure to a moderate level of some toxic agent, and causing measurable biochemical or anatomical damage, but not death of the organism. Compare with acute toxicity.

chronosequence A time series of stands or soils of different age that originated after a similar type of disturbance.

circumneutral Having a pH of about 7.

clear-cut A forest harvest in which all merchantable trees have been removed. Usually the trees are delimbed, and branches and foliage are left on the site. See also whole-tree and complete-tree clear-cut.

clear water Water that is not colored with a brownish hue by dissolved organic matter. Compare with brown water.

climate The typical, long-term prevailing conditions of temperature, precipitation, insolation, and cloud cover of an area. Compare with weather.

climax The more-or-less stable plant and animal community that culminates succession under a given set of conditions of climate, site, and biota.

cluster analysis A multivariate data analysis that groups cases on the basis of the common occurrence of statistically correlated attributes.

clutch The eggs laid by a particular bird, or by a number of birds in a particular nest.

codistillation A process in which dissolved substances evaporate with water at a small concentration. Because of this the codistillant occurs in a trace concentration in the condensed vapor.

coevolution The intrinsically linked evolution of two or more species, caused by some close ecological relationship such as predation, herbivory, pollination, symbiosis, etc.

cohort An even-aged group of individuals of the same generation and age.

community An assemblage of interacting plants and animals on a shared site.

compaction See soil compaction.

compartment A designated unit of an ecosystem that contains a quantity of material and energy.

competition A stress-causing interaction between organisms of the same or different taxa, caused by the need for a common resource that occurs in insufficient supply relative to the biological demand.

competitive release A situation where an organism or taxon is relieved of competitive stress, and which therefore becomes more dominant in its habitat.

complete-tree clear-cut Removal of the above-ground and below-ground tree biomass from a harvested site.

complex A chemical compound in which atoms are attached to a central metal atom by coordinate bonds. See also chelation.

concentration The amount of a substance per unit volume or per unit weight of the matrix.

conceptual model A nonquantitative assessment of the dynamics of the most important compartments and fluxes of material or energy in a system, or of changes in population.

congeneric Two or more species of the same genus.

conifer A gymnosperm plant.

conservation The protection, preservation, and careful management of a natural resource.

conservation of electrochemical neutrality In an aqueous solution, the number of cation equivalents equals the number of anion equivalents, so that the solution does not have a net electrical charge.

contact herbicide A herbicide that causes injury at the point of entry to the plant, which is usually the foliage.

contamination   The occurrence of a relatively large concentration of some toxic substance, compared with the normal ambient condition.

content   The total amount of a substance in an organism, compartment, or ecosystem. Sometimes standardized per unit area, or per unit volume in aquatic studies.

continental   Refers to a climate that is not moderated by proximity to a large body of water, and that is characterized by relatively cold winters and hot summers.

control   (1) An experiment in which a variable is held constant, for the purpose of statistical comparison with a parallel experiment in which that variable is altered. (2) A standard of comparison used in the statistical analysis of a scientific experiment.

conventional munitions   Weapons in which the explosive mechanism does not involve a nuclear reaction.

correlation   A statistical relationship that shows the degree of quantitative association between two or more variables.

correlation coefficient   A statistic that quantitatively expresses the degree of correlation between two or more variables.

cover   A measure or estimate of the relative or absolute amount of ground surface that is obscured by plant foliage.

crop   Harvested plant or animal biomass.

crown closure   See canopy closure.

crude oil   See petroleum.

cull trees   Trees that are removed during a silvicultural thin or shelter-wood cut. See also thin and shelter-wood cut.

cultural eutrophication   Eutrophication caused by an anthropogenic nutrient input. See also eutrophication.

culvert   An enclosed water drainage channel that crosses beneath a road, railway, etc.

cuticle   A waxy, superficial layer that covers the foliage of a vascular plant. The cuticle is continuous except for microscopic stomata, through which gases and water vapor pass.

cutover   A tract of land from which trees have been harvested.

cycle   A closed and continuous transfer of material or energy among the compartments of a system. Except at the level of the biosphere, few ecological cycles are truly closed.

cyclic succession   A succession that occurs repeatedly on the landscape, as a result of a disturbance that occurs at intervals.

deciduous   A plant in which all leaves dehisce before a dormant season caused by cold or drought. More properly termed "seasonally deciduous," since the foliage of many so-called evergreen plants is also deciduous, but not all at once.

decline   See forest decline.

decomposition   The heterotrophic microbial oxidation of organic matter.

decrement   A decrease in biomass or productivity.

deer yard   A winter habitat of deer, which is often coniferous and has a relatively shallow accumulation of snow.

defoliation   The removal of plant foliage by mechanical means, acute or chronic injury, or herbivory.

deforestation   The removal of forest from a site.

deglaciation   The retreat of glaciers from the landscape by the mass wasting of ice.

degradation   (1) The process by which a toxic chemical is converted metabolically or abiotically to a less toxic or nontoxic metabolite. See also detoxification. (2) A reduction in the quality of a natural resource.

demography   The science of population statistics.

dendrochronology   The use of patterns of tree ring width to determine historical changes in climate, defoliation, pollution, or some other environmental factor.

denitrification   The microbial oxidation of ammonium to nitrite, and then to nitrate.

density   The number of organisms per unit area, or per unit volume in aquatic studies.

density-independent   A population change that is not influenced by the density of organisms. A density-independent effect is often caused by a catastrophic advent of stress or disturbance.

depauperate   An ecosystem or community with a small richness of species.

deposition   The rate of influx of material per unit area and time. Frequently pertains to influx from the atmosphere.

deposition velocity   A constant that allows the calculation of the rate of deposition of an atmospheric chemical constituent of known concentration, to a vegetation or other surface with particular physical and chemical characteristics.

desertification   A climatic change that involves a decreased precipitation input, causing the diminution or destruction of the biological productivity of the landscape, and leading ultimately to desert-like conditions.

desulfurization   The removal of some or all of the

sulfur-containing fraction of coal or oil, to reduce the emission of sulfur dioxide during combustion.

detergent A surfactant chemical used as a cleaning agent because it facilitiates the formation of an oil-in-water emulsion.

detoxification Reduction of the toxic quality of a chemical by its microbial or inorganic transformation to another, less toxic chemical. See also degradation.

detritivore A heterotrophic microbe or animal that feeds on dead biomass.

devegetated A site that has been severely degraded by the removal of all or most of its biomass by disturbance or toxic stress.

diameter-limit cut A forest harvest in which all trees larger than a commercially determined diameter are removed.

dicotyledonous (dicot) Refers to an angiosperm plant of the Dicotyledonae, characterized by two seed leaves or cotyledons, among other characteristics. The other class of angiosperms is the Monocotyledonae, with a single cotyledon.

dieback A progressive dying from the extremity of any part of a plant. See also forest decline and forest dieback.

dinitrogen fixation See nitrogen fixation.

dioxin A class of chlorinated hydrocarbon, including the very toxic isomer TCDD (2,3,7,8-tetrachlorodibenzo-*p*-dioxin).

direct seeding A method of regeneration of a clear-cut that involves the natural input of seed from nearby uncut forest, or the artificial application of tree seed or seed-bearing cones.

disclimax An unusual climax ecosystem that occurs because of anthropogenic or natural stress.

disease Any impairment of the normal physiological function of a plant or animal, caused by inorganic or pathogenic stress.

disjunct A biogeographic term that refers to a spatially discontinuous distribution.

dispersal The act of scattering of individuals or propagules.

dispersant A chemical agent that encourages the formation of an oil-in-water emulsion of spilled hydrocarbon, so as to reduce the volume of pollution on a shore or on the water surface. See also detergent.

disturbance An episodic environmental influence, usually physical, that causes a measurable ecological change.

diurnal Happening daily, or during the day.

diversity An ecological concept that incorporates both the number of species in a particular sampling area, and the evenness with which individuals are distributed among the various species (the latter is sometimes expressed as the probability of encountering an individual of a particular species).

dominance The degree of ecological influence of a particular species within a community. In plant ecology dominance is usually indexed as the relative biomass of particular species, or as the relative basal area of particular tree species.

dose The quantity of toxic chemical or radiation impinging on or absorbed by an organism, often standardized per unit weight or surface area.

drift (1) The off-site occurrence or deposition of a pesticide spray aerosol. (2) The rate of flux of invertebrate biomass or individuals on the surface of a flowing water body such as a stream.

drought A prolonged period of insufficient water availability to sustain the vegetation characteristic of an area.

dry deposition The influx of gaseous or particulate nutrients or toxic substances from the atmosphere that occurs in the interval between precipitation events.

dry weight (d.w.) The weight of a substance after water has been removed.

duff A layer of the forest floor that is largely comprised of partly decomposed leaf litter, which is no longer recognizable to species, but still contains obvious leaf fragments. See also forest floor.

dwarf shrub A decumbent woody plant whose height is genetically limited to less than about 0.5 m. The usual habitat is tundra or some other highly stressed environment.

dynamic model A model that is capable of responding to temporal changes in key variables.

dynamic response A rapid change in abundance or some ecosystem function, caused by the advent of stress. See also numerical response.

ecocide An intentional antienvironmental action carried out over a large area.

ecological impact A measurable effect on some ecosystem characteristic that is caused by a change in an environmental factor.

ecological release See competitive release.

ecological reserve A tract of land that has been set aside for conservation purposes, often because it is the habitat of rare or endangered species, or because it is representative of a rare or threatened type of ecosystem.

ecological rotation A harvest cycle that is sustainable

on the longer term without causing a degradation of the forest resource or of site quality.

ecologist   A scientist who studies ecology.

ecology   The study of the relationships between and among organisms and their environment.

ecosystem   A generic term for a system that includes a community of organisms and their interactions with the environment.

ecosystem function   Ecological processes such as production, predation, nutrient transformation, the influx and efflux of energy and material, etc.

ecosystem structure   Ecosystem characteristics such as the vertical and horizontal distribution of plant biomass, the quantity of biomass and nutrients in various trophic levels, the size distribution of organisms, species richness and diversity, etc.

ecotone   A zone of abrupt transition between distinct ecosystems or habitats.

ecotoxicology   The study of the effects of toxic stress on ecosystem structure and function. See also toxicology.

ecotype   A local population of a wide-ranging species that is genetically adapted to coping with particular environmental stresses, such as the edaphic toxicity of a site with a large concentration of toxic elements.

edaphic   Pertaining to the soil environment.

edge   See ecotone.

edge effect   A change in the abundance and species richness of wildlife at a habitat discontinuity.

electrochemical   A chemical effect that produces a heterogenous distribution of electric charge.

electromagnetic energy   The energy of photons, having properties of both particles and waves. Sometimes abbreviated as "light" energy.

electrostatic precipitator   A pollution-control device that removes particulates from industrial waste gases by conferring on them an electric charge, and then collecting them at an electrode. The typical collection efficiency is about 99% of the particulate mass.

eluviation   The process by which soluble materials are transported downward through the soil in conjunction with percolating rainwater.

embryotoxic   Pertaining to a substance that causes injury or death of a developing embryo.

emission   The release of a substance to the environment.

empirical   Derived from experiment or observation, rather than from theory.

emulsion   A colloidal suspension in which both phases are liquids, as in an oil-in-water emulsion.

endangered   In imminent risk of extinction, unless there is a vigorous effort to preserve the species and its habitat.

endemic   A distinct race or species that originated locally, and that has a geographically restricted distribution.

endemic phase   A stage of low population density of an irruptive organism.

energy budget   An analysis of the inputs and outputs of energy for a system, and of the internal transformations and storage within the system.

enrichment   Enhancement of the rate of supply of nutrients to a system, causing an increase in productivity.

entrainment   The process by which a small concentration of material is transported within a large moving matrix, as when soil particles are picked up by and carried within an energetic air mass.

environment   The complex of all biotic and abiotic influences on an organism or group of organisms.

environmental degradation   A decrease in the quality of the environment, from an anthropic perspective.

environmental factor   Any biotic or abiotic influence on an organism or group of organisms.

environmental impact assessment (EIA)   An interdisciplinary process by which the environmental consequences of proposed actions and various alternatives are presented and considered. May involve assessments of ecological, sociological, anthropological, geological, and other environmental impacts.

enzyme   A proteinaceous biological catalyst, whose activity is determined by its three-dimensional structure.

epidemic   The widespread occurrence of a pathogenic disease.

epidemiology   The study of the occurrence, transmission, and control of epidemic disease.

epilimnion   The relatively warm surface water of a thermally stratified lake. See also stratified.

epiphyte   A nonparasitic plant that grows on another plant.

episode   An incident of some ecological phenomenon.

epizootic   A disease epidemic affecting a large number of animals.

equilibrium   A condition in which the magnitudes or rates of all influencing forces, chemical processes, or demographic variables cancel each other out, creating a stable and unchanging state.

epicenter   The site immediately proximate to a nuclear explosion, or the site of origin of earthquake tremors.

equivalent   Abbreviation for mole-equivalent, and calculated as the molecular or atomic weight multiplied times the number of charges on the ion. Equivalent units are necessary for a charge-balance calculation. See also conservation of electrochemical neutrality.

erosion   The mass wasting or wearing away of rock or surficial deposits by the action of water or wind.

estuary   The widening channel of a river where it meets the sea, characterized by spatial and temporal variations in salinity of the water.

etiology   The cause or causes of a diease.

euphotic zone   The upper portion of a water column, where light intensity is sufficient to allow net photosynthesis to occur.

eutrophic   The characteristic of being very productive, as a result of a large rate of nutrient loading. Usually refers to an aquatic system. See also mesotrophic and oligotrophic.

eutrophication   The process by which an aquatic ecosystem increases in productivity as a result of an increase in the rate of nutrient input.

evapotranspiration   The evaporation of water from a landscape, including evaporation from inorganic surfaces, and transpirational water loss from foliage.

even-aged   Refers to a population of organisms in which all individuals are approximately the same age.

evolution   The gradual change over time in the genetically based characteristics of a population of organisms.

exchangeable   In soil science, refers to an ion that can take the place of another ion at an ion exchange site on organic matter or a clay particulate. Most important in determining whether exchange will occur are ionic density, the relative number of charges on the ions (related to the ion exchange series), and the relative concentrations of the ions.

exothermic   A chemical reaction that has a positive net production of thermal kinetic energy.

exploitation   The harvest of a natural resource for sustenance or for a commercial purpose

explosive yield   The quantity of energy that is produced by the explosion of a bomb.

experiment   A scientific test that is designed to provide evidence in support of a null hypothesis.

exponential   Refers to a logarithmic relationship, or a mathematical relationship involving a number raised to the power of an exponent. See also logarithmic.

extant   A taxon that survives at the present time.

extinction   The event whereby there are no surviving individuals of a species or other taxon.

extinction coefficient   An index of the transparency of water.

extractable   In soil science, refers to the quantity of a chemical that is solubilized by a particular type of extracting solution, such as a mildly acidic solution.

fallout   Radioactive particulates that settle gravitationally from the atmosphere, or that are removed by precipitation.

fauna   The assemblage of animal taxa in some defined site or geographical area.

fecal pellet group   A cluster of animal excrement. In some studies of animals that are difficult to census directly, the density of fecal pellet groups may be surveyed as an index of abundance.

fecundity   The capacity to produce offspring. Measured as the total number of propagules potentially produced during the lifetime of an individual or population. Compare with fertility.

fermentation   An anaerobic biochemical process by which organic molecules are split into simpler substances, as when yeasts metabolize sucrose and produce ethanol as a by-product.

fertility   (1) The inherent capacity of a site to sustain the production of biomass. (2) The actual number of viable propagules produced during the lifetime of an individual or population. Compare with fecundity.

fertilization   A management practice that involves the enhancement of nutrient supply in order to stimulate a larger rate of productivity.

fertilization trial   A field experiment that is used to evaluate the response of an ecosystem to a particular rate of fertilization with a particular type of nutrient or combination of nutrients.

fire protection   The active suppression of fire in a stand, by preventing the buildup of a large biomass of fuel through the use of prescribed burns, and by the quenching of fires that accidentally ignite.

fixed   Refers to nutrients or toxic substances that are incorporated into organic matter, as in organic nitrogen. See also organically bound.

flora   The assemblage of plant taxa in some defined site or geographical area.

flue gas   Waste gas from an industrial process. The flue gas may be vented directly to the atmosphere, or it may have some or virtually all of the harmful pollutants removed by pollution-control devices, as in flue gas desulfurization to remove sulfur dioxide, or the use of an electrostatic precipitator or physical filter to remove particulates.

flux   The rate of movement of a quantity of material or energy, usually standardized per unit of area and time.

fog   (1) A meteorological condition characterized by a ground-level aerosol of water droplets and very limited visibility. (2) A type of pesticide application in which the above-ground atmosphere of the sprayed ecosystem is saturated with a fine pesticidal aerosol.

food chain   A linear sequence of organisms that are linked by trophic interactions, as in grass–cow–people.

food web   A complex assemblage of organisms that are interlinked by trophic interactions.

forb   An herbaceous plant that is not a grass (Poaceae), sedge (Cyperaceae), or rush (Juncaceae).

forest   A stand that is dominated structurally by tree-sized plants.

forest decline   A syndrome of stand-level dieback or loss of vigor of one or all tree species, caused by an unknown environmental agent or combination of agents. See also forest dieback.

forest dieback   A disease of trees of unknown etiology that is characterized by the progressive death of branches and perhaps the tree, sometimes occurring as a stand-level dieback. See also forest decline.

forester   A person trained in the inventory, harvesting, and management of forested land.

forest floor   The organic layer that overlies the mineral soil in a forest. The forest floor is comprised of three layers: the litter (top), duff (middle), and humus (bottom). See also litter, duff, and humus.

forest management area   A tract of land that is being managed so as to provide a supply of tree biomass for some commercial purpose.

forestry   The science of the planting and caring of trees, and of the management of forests.

formulation   A pesticidal spray, comprised of the active pesticidal ingredient, a solvent, and miscellaneous other chemicals that enhance the effectiveness of the active ingredient for a particular use.

fossil fuel   A mined fuel that is primarily comprised of hydrocarbons, such as coal, petroleum, and natural gas.

frequency   The number of occurrences of a particular observation within a larger population of observations, usually standardized as a percentage.

fresh water   Nonsaline water.

frond   The leaf of a fern or palm, or the thallus of a seaweed.

frugivore   An animal that eats fruit.

frustule   The siliceous outer cell wall of a diatom.

fry   The young of various species of fish.

fumigation   An air pollution incident characterized by a large concentration of toxic gas or particulates.

fungicide   A pesticide used against fungi.

gamma radiation   Very high-energy electromagnetic radiation with a frequency greater that about 3×10^{19} Hz. See also radioactivity.

game   Hunted wildlife.

generation   (1) The average span of time of the life cycle of a species. (2) All of the individuals produced within one contemporaneous life cycle.

genetic   Refers to information that is contained within the base series of the DNA of chromosomes.

genetic drift   A random change in the collective genetic information that occurs in a small and isolated population.

genetic intergradation   A mixing of the genetic information of two populations or distinct taxa as a result of interbreeding.

genotype   The genetic constitution of an organism.

geochemistry   The chemistry of the crust of the earth.

geothermal   Relates to heat that is present in the interior of the earth.

girdle   To cause a lethal injury to a tree, usually by cutting several centimeters into the stem around the circumference of the trunk, and thereby completely severing the vascular tissue.

glacial drift   See glacial till.

glacial till   A variable and unconsolidated mineral debris that remains after the meltback of a glacier.

glaciation   The process by which the landscape becomes covered with glacial ice.

gradient   A continuous change with distance in the magnitude of a variable or variables.

graminoid   A generic term for a plant with a grass-like form, such as grasses (Poaceae), sedges (Cyperaceae), and rushes (Juncaceae).

greenhouse effect   A climatic effect that occurs in a greenhouse, in which solar electromagnetic radiation passes freely through the encasing glass, after which much of it is absorbed within the greenhouse and transformed into thermal kinetic energy. This causes the reradiation of much of the absorbed energy in the form of longer-wave infrared electromagnetic radiation, but much of this is trapped within the greenhouse because it is absorbed by atmospheric moisture and carbon dioxide, and because it cannot pass through the glass, and there is a resulting warming effect. It is believed that carbon dioxide in the earth's atmosphere may act somewhat like the glass and atmosphere of a greenhouse, so that its increasing concentration could cause a global warming.

gross primary production (GPP) The amount of carbon fixed by photosynthesis, including that which is used up in respiration. See also production.

ground spray A pesticide spray that is applied by a hand-held or vehicle-mounted apparatus.

ground vegetation Vegetation with a height less than about 0.5 meter, occurring within a stand of much taller stature, such as a forest.

groundwater Water occurring below the soil surface, and that is held in the soil itself, or in an aquifer comprised of fractures and pores in the bedrock.

groundwater storage The quantity or volume of water that occurs as groundwater.

guard cell One of a pair of cells that controls the aperture width of a stoma.

gymnosperm A flowering plant in the subdivision Spermatophyta, distinguishable from the Angiospermae by having ovules borne naked on the surface of the megasporophylls, and often aggregated into cones.

gyre A circular or spiral oceanic current.

habitat edge See ecotone.

habitat island A spatially isolated habitat that occurs within a much larger matrix of a different kind of habitat, as where a forest remnant is surrounded by agricultural fields.

habitat structure The horizontal and vertical distribution of biomass and species within a habitat. See also ecosystem structure.

half-life The time required for the disappearance of one-half of an initial quantity of material, energy, or radioactivity.

halophyte A plant with a tolerance of, and requirement for, a saline environment.

hardening A physiological process that occurs in plants in the autumn and early winter, and that results in the progressive development of a tolerance to low temperature.

hardwood An angiosperm woody plant.

harvest The removal of biomass for some anthropic purpose.

harvest rotation The length of time between successive harvests.

haze A condition of reduced atmospheric visibility caused by a large concentration of fine inorganic particulates or condensed water vapor.

headwater The uppermost surface water in a watershed, which does not receive a hydrologic input from surface water higher in altitude.

heat island The mass of air over a large city, which averages slightly warmer than the surrounding, ambient condition.

herbaceous Refers to a nonwoody perennial plant in which the above-ground biomass dies back each year.

herbarium A collection of pressed, dried, and identified plant specimens, arranged in a systematic order.

herbivore An animal that feeds on plants.

heritable A characteristic (of an organism) that has a genetic basis, and that can be passed along to offspring.

heterotroph An organism that requires a source of organic matter as food. Compare with autotroph.

hibernaculum The place where a hibernating animal spends the winter.

hidden injury Plant damage, such as decreased productivity, that occurs after exposure to an intensity of toxic stress that is not sufficient to cause acute injury.

histopathology Tissue damage that is observable at a microscopic level.

host An organism that is suceptible to infection by a particular pathogen or to infestation by a pest.

humus The deepest layer of the forest floor, which lies just above the mineral soil, and in which the humified organic matter no longer contains easily identified leaf fragments. See also forest floor.

hybrid An offspring that arises from the interbreeding of two dissimilar taxa. See also genetic intergradation.

hydric Refers to wet site conditions, which are usually manifest by at least a brief period during the growing season in which the water table occurs above the soil surface.

hydrocarbon An organic molecule that only contains atoms of hydrogen and carbon.

hydroelectric Refers to electricity that is generated by harnessing the kinetic energy of falling water.

hydrology The study of the distribution and movement of water in the atmosphere and crust of the earth.

hydrolysis A chemical reaction in which a substrate reacts with water to produce a product.

hydrophobicity Refers to the characteristic of insolubility in an aqueous medium.

hygroscopic Refers to the tendency of a substance to absorb water from the atmosphere.

hyperaccumulator A species that accumulates a nutrient or toxic chemical to a very large concentration. See also accumulator.

hypertrophic   Refers to an aquatic system with a very large nutrient supply and a very large rate of productivity.

hypolimnion   The deeper, relatively cool water of a thermally stratified waterbody. See also stratified.

hypothesis   A suggested explanation for a phenomenon or assemblage of observations, which can be experimentally tested in various ways.

impaction   A particulate removal mechanism that involves the physical interaction with a solid surface by particulates suspended in water or air.

imperfectly drained   Refers to a site in which the water table occurs in a near-surface soil horizon for at least part of the growing season.

incident precipitation   The rate of influx of water, and the influx and concentration of nutrients via precipitation, prior to hydrologic and chemical alteration by interaction with the biota and inorganic surfaces of an ecosystem.

indicator plant   An identifiable plant taxon that is specific to particular site characteristics, and that indicates such characteristics by its presence. See also biogeochemical prospecting.

infestation   The occurrence of a very large population of an irruptive pest.

inflorescence   The part of an angiosperm that consists of the flowering structures.

infrared   Refers to electromagnetic radiation with a wavelength longer than that of visible red or about 0.7 μm, but less than that of radio waves.

insecticide   A pesticide that is used against insects.

insectivorous   An organism that feeds on insects.

integrated pest management (IPM)   A relatively complex pest management scheme that may incorporate the use of resistant host genotypes, cultural practices that reduce susceptibility, biological control, and the use of pesticides when necessary.

intensive harvesting   Refers to the removal of a large fraction of the biomass of a stand, as in clear-cutting.

interbreeding   See genetic intergradation and hybrid.

intertidal   Refers to a marine shore zone occurring between the high water mark and the low water mark.

intolerant   (1) Refers to a species or genotype that is susceptible to the effects of stress. (2) Refers to a species that cannot occur in the understory of a forest because of the stress associated with the restricted availability of light, water, and nutrients.

intoxication   Refers to any case of acute injury caused by a toxic agent.

inversion   Refers to an atmospheric condition in which air temperature increases with increasing altitude, instead of the usual decrease of temperature with increasing altitude. The presence of a stable temperature inversion above a lower air mass leads to stable atmospheric conditions, which can be accompanied by large concentrations of air pollutants if emissions continue during the inversion event.

invertebrate   Any animal that does not have a backbone.

ion   An electrically charged atom or molecule.

ion exchange   The process by which ionic species are exchanged between an aqueous solution and an insoluble exchange surface such as a resin, organic matter, or certain clay minerals.

ionizing radiation   Electromagnetic energy or corpuscular radiation (beta and gamma radiation) that is sufficiently energetic to cause ionization. See also radioactivity.

irrigation   The practice of supplying semiarid land with water in order to promote the growth of agricultural crops.

irruption   A sporadic or rare occurrence of a very great abundance of a species.

island biogeography   A biogeographic theory that predicts a relatively large rate of extinction and a small species richness on small and isolated oceanic islands.

juvenile   A sexually immature individual.

lamination   Refers to the occurrence of distinctly layered sediment, as in a meromictic lake. See also meromictic.

landscape   An extensive area of terrain.

large animal   An animal that weighs more than 44 kg (100 lb).

lateral stem bud   A bud that occurs below the shoot apex. Lateral stem buds generally remain dormant unless the terminal bud is killed or injured, after which dormancy breaks and new shoots emerge from the lateral buds. See also stump sprouting.

latex   A whitish fluid found in diverse plants, that variously contains proteins, starch, alkaloids, and other biochemicals.

LD$_{50}$ The dose of a toxic agent that is required to kill one-half of the organisms in a bioassay.

leaching   The process by which dissolved substances are removed by a percolating water solution.

leaf area index   An estimation of the collective surface area of all foliage in a stand, standardized per square meter of the ground surface.

life form   A structural classification of plants that is based on the characteristic structural features, including the position with respect to the ground surface of the vegetatively perennating tissue.

limiting factor   A metabolically essential environmental factor that is present in least supply relative to biological demand, and which thereby restricts the rate of productivity.

limnology   The study of fresh-water bodies.

line-of-vision   An unobstructed view.

litter   (1) The surface layer of the forest floor, in which leaf detritus is identifiable to species. See also forest floor. (2) Recently dead plant biomass.

litterfall   The flux of dead biomass to the surface of an ecosystem.

littoral   Refers to the relatively shallow, nearshore zone of an aquatic system.

load-on-top (LOT)   A simple technique for reducing the operational discharge of petroleum from oceanic tankers. In LOT, the bilge washings are retained for some time in holding tanks, allowing the oil and water to separate. The relatively clean water is discharged to the sea, and the residual oil is combined with the next load. See also bilge washings.

logarithmic   Refers to a mathematical relationship in which the distance or difference between points is proportional to the logarithms of the numbers, as in $Y = \log X$.

long-distance transport   See long-range transport.

long-range transport   The transport of pollutants within a moving air mass for a distance greater than 100 km.

long-term   In ecology, this refers to a period of time that is longer than several life spans of an organism, or of the dominant organism in an ecosystem.

lysimeter   An instrument that is used to sample soil water in order to determine the concentration or rate of leaching of dissolved substances.

macroalga   An alga with a thallus that is visible without magnification, as in seaweeds.

macronutrient   A nutrient that is required by biota in a relatively large quantity, including C, H, O, N, P, K, Ca, and Mg.

macroscopic   Visible without the aid of magnification.

macrophyte   A macroscopic aquatic plant.

macrozooplankton   A large species of zooplankton, such as krill (Euphasiacea) or shrimp (Decapoda).

management   The care and tending of a renewable natural resource.

mangrove   A coastal marine forest at low latitude, dominated by tree species in the families Rhizophoraceae and/or Avicenniaceae.

manual spraying   The application of a pesticide using a hand-held spray apparatus.

manual treatment   The suppression of weeds using a hand-held cutting tool or apparatus.

maritime   Refers to the moderating climate effect that occurs near a large body of water, particularly an ocean.

mass extinction   The extinction of an unusually large number of taxa in a geologically short period of time.

mass mortality   An event in which a large fraction of the total number of individuals of a population or community dies.

mature   An organism that is old enough to be capable of sexual reproduction, or a stand in which the dominant individuals are sexually mature.

membrane   A very thin, pliable, and sometimes fibrous tissue that covers, lines, or connects biological tissue or cells. See also plasma membrane.

merchantable stem   The part of the bole of a tree that is of commercial value. The diameter limit for merchantability varies depending on tree species, geographic area, and intended use.

merchantable timber   The quantity of tree bole biomass that is valuable for the manufacture of sawn timber.

meristem   A plant tissue that is reponsible for growth, by rapid cell division and differentiation into organ tissue.

meromictic   Refers to a lake in which the water column is stratified permanently or for at least several years, causing anoxia in the hypolimnion, an absence of benthic invertebrates, and laminated sediment.

mesic   Refers to moderate site conditions, particularly with respect to water availability.

mesotrophic   Refers to an aquatic condition of moderate nutrient supply and productivity.

metalliferous   Refers to a mineral, soil, or organism with a large concentration of a metal or metals.

microbe   A microscopic organism.

microclimate   The climatic conditions that immediately affect an organism or a small group of organisms.

microcosm   A miniaturized representation of an ecosystem, usually used for some experimental purpose.

microfauna The assemblage of microscopic animal taxa in some defined site or geographical area.

microflora The assemblage of microscopic plant taxa in some defined site or geographical area.

micronutrient A nutrient that is required by biota in a relatively small quantity, including metals such as Fe, Cu, Zn, etc.

microscopic Refers to an object that is only visible with the aid of magnification.

migration A periodic, long-distance movement undertaken by animals.

mill (1) An industrial facility where raw ore is crushed, followed by the separation of a valuable metal-containing concentrate from the waste tailings, which are disposed of in a tailings dump. (2) An industrial facility where harvested logs are either manufactured into sawn timber, or pulped and processed into paper.

mineral soil The inorganic matrix that serves as the substrate of a terrestrial ecosystem. If the ecosystem has developed a distinct organic surface layer such as a forest floor, then the mineral soil is considered to begin immediately below it.

mineralization (1) An occurrence of minerals containing a large concentration of toxic elements at the soil surface. See also metalliferous. (2) Refers to the conversion of a nutrient or toxic chemical from an organically bound form to a water-soluble inorganic form, as a result of either inorganic or biological chemical reactions.

minerotrophic Refers to a wetland or surface-water body that receives part of its nutrient supply as dissolved substances in water entering from a part of the watershed that is higher in altitude. Compare with ombrotrophic.

miscarriage The spontaneous expulsion of a fetus from the womb.

mixed-hardwood forest A forest comprised of a mixture of two or more angiosperm tree species.

model A conceptual, quantitative, or mathematical representation of an ecological or environmental process. See also box model, conceptual model, dynamic model, and simulation model.

monoculture The cultivation of an agronomic or tree species in the absence or virtual absence of other species.

monospecific Refers to an ecosystem that is comprised of or very strongly dominated by a single species.

morbidity The rate of incidence of a particular disease in a particular place or area.

mortality The rate of death in a particular place or area.

mosaic Refers to a landscape that is characterized by a complex spatial arrangement of distinct habitat types.

mousse A water-in-oil emulsion that is formed by surface water turbulence after a marine petroleum spill.

multiple use The use of a natural resource for more than one purpose, as where a forested watershed is harvested to produce wood for commercial purposes, to provide and regulate the supply of water, for hunting and fishing, and for nonconsumptive recreational purposes such as camping and bird-watching.

multivariate Refers to a system in which there is simultaneous variation in more than one independent variable.

municipal watershed A watershed that is used as a source of water for a centralized system of drinking water treatment and distribution, usually for an urban area.

mutation A sudden change in the genetic information encoded in the DNA of an individual.

mutualism A relationship between two or more organisms that is mutually beneficial.

mycorrhizal fungus A fungus that lives in a symbiotic association with the roots of a vascular plant. The heterotrophic fungus benefits from a source of fixed carbon, and the host plant benefits from a markedly enhanced supply of inorganic nutrients, especially phosphorus.

natural A situation that is not measurably influenced by humans.

natural resource Any naturally occurring resource that can be used by people. Renewable natural resources can potentially by exploited indefinitely if they are not degraded by use that is too intensive. Nonrenewable natural resources are of finite supply, and can only be mined.

natural source A nonanthropogenic emission of toxic or nontoxic substances.

necrosis The death of plant tissue caused by a toxic stress.

nematicide A pesticide used against nematodes.

neoplasia The occurrence of a new, undifferentiated tissue growth or tumor.

neotropics A biogeographical zone consisting of tropical regions within Central and South America and the West Indies.

nest parasitism An avian breeding system in which the parasitic species lays an egg or eggs in the nest of an unsuspecting host, which then incubates the egg and raises the parasitic young.

net flux The difference between the total incoming

quantity of a material or energy, and the total outgoing quantity. Net flux can be positive, negative, or zero.

net primary production (NPP) The difference between gross primary production and respiration, which if positive leads to an accumulation of organic matter by an individual or species. See also production.

neurological Pertaining to the nervous system.

neutralization The process by which an acid interacts with a base to form a neutral salt.

niche The role that an organism or taxon plays in its natural ecosystem, including its activities, resource use, and interactions with other organisms.

nitrification The chemoautotrophic process by which certain bacteria oxidize ammonium to nitrite, and then to nitrate.

nitrogen fixation (dinitrogen fixation) The conversion of atmospheric dinitrogen to ammonia or an oxide of nitrogen, occurring inorganically at high temperature and/or pressure, and biologically via the action of the microbial enzyme nitrogenase.

nongame wildlife Wildlife species that are not hunted, although they may be of commercial importance for other reasons.

nongovernment organization (NGO) A public sector organization without direct links to government.

nonrenewable natural resource See natural resource.

nonselective Referring to a nonspecific pesticide that is toxic to and kills many organisms in addition to the intended pest target.

nontarget Organisms other than the intended pest target of a control action, such as pesticide spraying.

nuclear aftermath The sociological and environmental conditions that would occur after a large-scale exchange of nuclear weapons.

nuclear holocaust A large-scale exchange of nuclear weapons, causing a tremendous loss of human life, destruction of the built environment, climatic change, and other great ecological and sociological effects.

nuclear winter A widespread climatic cooling caused by the probable effects of nuclear warfare on atmospheric conditions, which would reduce the amount of sunlight penetrating to the earth's surface.

numerical response A large increase or decrease in the abundance of a population of organisms, that occurs in response to a change in environmental conditions. Usually refers to a change in the abundance of predators in response to a change in the abundance of prey.

nutrient Chemicals that are required for life. See also macronutrient and micronutrient.

nutrient capital The total quantity of a nutrient that is present in all site compartments, and occurs in the living plus dead biomass and the mineral soil.

nutrient cycling A generic term to describe the influxes and effluxes of nutrients from some designated system, as well as all of the internal fluxes, compartments, and transformations.

nutrient loading The influx of nutrients to a system.

nutrient-use efficiency The amount of nutrient taken up or required by a plant or crop, per unit of biomass that is produced. See also water-use efficiency.

occupational Pertaining to the human working environment.

oil See petroleum.

oil seep A natural emission of liquid petroleum.

oligotrophic Refers to a condition of a restricted supply of nutrients and small productivity. Usually refers to an aquatic system. See also eutrophic and mesotrophic.

ombrotrophic Refers to a situation where there is no input of nutrients from ground water or surface water, so that all nutrient supply arrives from the atmosphere with wet and dry deposition.

operational Refers to conditions occurring during an actual commercial procedure, as opposed to a simulation experiment or some other model system.

opportunistic Refers to an organism that takes advantage of an ephemeral condition of great resource availability, especially soon after disturbance. See also weed.

organic matter Carbon-containing molecules that make up or are derived from the tissue of organisms.

orographic effect The cooling of a moving air mass as it is forced to rise in altitude when it encounters a mountain range, often causing its load of water vapor to condense, and leading to a relatively large rate of precipitation on the windward side.

open-top chamber An experimental apparatus used to simulate the effects on plants of a fumigation by an air pollutant. The chamber is open to the atmosphere on the top, and incursions of ambient air are prevented by maintaining a positive air pressure within the enclosure.

order of magnitude A factor of 10.

organic anion An organic molecule with an overall negative ionic charge.

organically bound Refers to a nutrient or toxic chemical that is structurally incorporated into, or electrochemically bound to, organic matter. See also fixed.

organometallic complex See organically bound. Refers specifically to metals.

outbreak The occurrence of a great abundance of an irruptive pest or pathogen.

outfall A site where there is a large point-loading of domestic or industrial waste material or heat to an aquatic system.

overexploitation The unsustainable exploitation of a potentially renewable natural resource. See also natural resource.

overharvesting The overexploitation of plants or animals. See also overexploitation.

overhunting The overexploitation of animals that are hunted. See also overexploitation.

overland water flow The horizontal flow of water above the soil, forest floor, or bedrock. Can be encouraged by compaction of soil.

overstory The highest level of foliage cover in a vertically complex stand or ecosystem.

oxidation A chemical reaction in which an oxidant causes another substance to undergo a decrease in its number of electrons. In ecological systems, the most frequent oxidant is molecular oxygen.

oxidizing smog An air pollution episode characterized by the occurrence of large concentrations of oxidants such as ozone and peroxyacetyl nitrate. Sometimes called Los Angeles-type smog.

oxygen tension The concentration of molecular oxygen in water.

palaeotropics A biogeographical zone consisting of tropical regions within Europe, Africa, and Asia.

paludification The process by which upland bogs form.

palynology The study of the historical occurrence of local plant communities, as inferred from the record of fossil pollen preserved in lake sediment.

pandemic A very widespread disease epidemic.

parameter One or more constants that determine the form of a mathematical equation. In the linear equation $Y = aX + b$, a and b are parameters, and Y and X are variables.

parameterize To adjust the parameters of a mathematical equation or computer program in order to simulate a particular situation or condition.

parasite An organism that derives benefit from its relationship with a host organism, which is either unaffected or suffers detriment.

parasitoid A parasitic wasp (Hymenoptera).

parent material The original mineral substrate from which a soil has derived. See also soil.

parr A salmon up to 2 years of age.

particulate A small particle. If sufficiently small, it can be suspended in an atmospheric or aquatic medium. If larger, it will settle gravitationally.

pathogen A disease-causing organism.

pelagic Refers to an open-water system.

perennating tissue The meristematic tissue from which vegetative plant growth occurs.

perennial A long-lived organism. Usually refers to plants.

periphyton Algae that occur on the inorganic and macrophytic surfaces of an aquatic ecosystem.

permanent plot A sampling area that can be exactly relocated, and that can be resurveyed over time.

persistence The length of time that a toxic agent occurs in a compartment or environment. Depending on the toxic agent, persistence can be affected by chemical breakdown, radioactive decay, or mass transport processes such as volatilization and erosion of toxin-containing particles.

perturbation The stressing or disturbance of a system.

pest An organism that is considered to be undesirable from the anthropic perspective.

pesticide A chemical or a microbial pathogen that is toxic to pests.

petroleum A naturally occurring liquid comprised primarily of a mixture of hydrocarbon species, which is mined and refined for many uses in the energy and chemical manufacturing sectors. Also known as crude oil.

pH The negative logarithm to the base 10 of the aqueous concentration of hydrogen ion in units of moles per liter. An acidic solution has pH less than 7, while an alkaline solution has pH greater than 7. Note that a one-unit difference in pH implies a 10-fold difference in the concentration of hydrogen ion.

phosphatase enzyme An enzyme that produces phosphate by the hydrolysis of organically bound phosphorus. Phosphatase enzymes can be extracellular or intracellular.

photochemical Refers to a chemical reaction in which the rate is enhanced by particular wavelengths of electromagnetic radiation.

photodissociation A chemical reaction in which the rate of breakdown of a substrate into simpler products is enhanced by particular wavelengths of electromagnetic radiation.

photosynthesis A set of coordinated biochemical reactions that occur in plants, in which visible electromagnetic radiation absorbed by chlorophyll and other pigments is used to synthesize organic compounds from carbon dioxide and water, with the release of molecular oxygen.

phylogenetic   Refers to the putative evolutionary relationships within and between species or higher taxa.

phytoplankton   Microscopic photoautotrophs that occur suspended in the water column.

phytotoxicity   Acute injury to a plant caused by some toxic agent.

piscivorous   An animal that feeds on fish.

planktivorous   An animal that feeds on plankton. Usually refers to zooplankton.

plankton   Plants and animals that are suspended in the water column.

plantation   A tract of land on which commercially desirable trees have been planted and tended.

plant-available   See available concentration.

plasma membrane   A very thin membrane formed of a protein–lipid bilayer that surrounds all cells, and that is responsible for the integrity of the cellular contents and the restricted access of extracellular materials. Physical disruption of the plasma membrane quickly causes death of the cell.

plutonic   Refers to igneous rocks that have derived from intrusive magma that has cooled and solidified beneath the earth's surface.

podsolization   A soil-forming process characterized by acidification of the upper A horizon, the downward leaching of basic cations, metals, and humic substances from the A horizon, and their deposition in the B horizon. Podsolization is most prominent in cool and wet climates, and under coniferous trees, oaks, and heaths.

point source   A situation where a large quantity of pollutants is emitted from a single site, such as a smokestack, a volcano, or a sewage outfall.

pollution   The occurrence of substances or energy in a larger quantity than the environment can assimilate without suffering degradation from the anthropic perspective.

polynya   Open water surrounded by sea ice in the Arctic.

population   (1) The abundance of a species in a specified area. (2) An interbreeding group of individuals of the same species.

population dynamics   The study of the temporal changes in population size.

prairie   A treeless plain dominated by grasses and forbs that occurs in a moderately dry temperate climate.

precipitation   The deposition of rain, snow, or fog from the atmosphere.

precision   The degree of repeatability of a measurement or observation. Compare with accuracy.

predation   A biological interaction in which a predator kills and eats its prey. Usually refers to a carnivore eating another animal.

premature abscission   The shedding of foliage at an earlier time than normal. Premature abscission is often induced by toxic stress or partial defoliation.

premature senescence   The senescence of tissue or an organism before the passing of the usual life span. Premature senescence is usually caused by toxic stress. See also senescence.

prescribed burning   The controlled burning of an ecosystem as a management practice to encourage a desirable type of regeneration, to decrease the abundance of a pathogen, or to prevent a large buildup of fuel that could lead to a more catastrophic wildfire.

primary   Refers to a factor that is first in importance in initiating an ecological phenomenon or a sequence of ecological events.

primary production   Production by autotrophs. See also production.

principle components analysis (PCA)   A statistical technique of ordering samples along multivariate axes.

production   The quantity of organic matter that is produced by biological activity per unit area or volume. Gross production (GP) refers to all production, without accounting for respiratory losses (R). Net production (NP) is GP − R. Primary production is production by autotrophs. Secondary production refers to herbivores, and tertiary production refers to carnivores.

productivity   Production standardized per unit of time and area.

prominence   The relative importance of an individual or species in an ecosystem, often indexed by relative biomass, relative cover, or relative basal area.

protection   Refers to a management practice in which a forest stand is sprayed with insecticide to reduce the abundance of an injurious insect, or where there is an active quenching of fire.

pyrite   A metal sulfide mineral, most frequently an iron sulfide.

qualitative   Distinctions that are not based on measurement. Compare with quantitative.

quantitative   Distinctions that are based on measurement, and that can be represented by numerical values. See also qualitative.

radial growth The growth in diameter of a woody plant.

radiation sickness An illness caused by overexposure of an organism to X-rays or radioactive material.

radiatively active gas A gas that absorbs electromagnetic radiation, converts it to kinetic themal energy and thereby assumes a higher temperature, and then reradiates some of the absorbed energy as longer-wave electromagnetic radiation. Usually refers to gases that absorb long-wave electromagnetic radiation and that can contribute to the greenhouse effect, especially carbon dioxide. See also greenhouse effect.

radioactivity The property of a radioactive material that is characterized by the emission of alpha, beta, or gamma radiation as a result of nuclear transformation. An alpha particle consists of two neutrons and two protons, or a helium nucleus. A beta particle is a high-speed electron. Gamma radiation is very high-energy electromagnetic radiation with a frequency greater than about 3×10^{19} Hz.

radiotracer A radioactive isotope of an ecologically important element or molecule, that is used in a very small but easily measured concentration to study processes important in physiology and biogeochemical cycling.

rainforest A mature forest with large biomass occurring in the temperate or tropical zone, in a site with a large rate of precipitation.

rainout The removal of particulate air pollutants from the atmosphere by their serving as condensation nuclei for the formation of raindrops, or of gases by their dissolution in raindrops within clouds, followed by the removal of the raindrops from the atmosphere as precipitation. See also washout.

rangeland An extensive track of open land on which livestock such as cattle graze.

raptor A bird of prey.

rare A species or other taxon that can be of localized or wide distribution, but that is very infrequently encountered.

reclamation A practice in which toxic or otherwise degraded land or water is managed in order to restore a more acceptable level of ecological development. May involve liming to reduce acidity, fertilizing to treat nutrient deficiency, and other ameliorative techniques.

recovery The natural or anthropogenic restoration of a previous, more acceptable ecological condition.

recruitment The rate at which new individuals enter a population of organisms.

reducing smog An air pollution episode characterized by the occurrence of a large concentration of sulfur dioxide and smoke (i.e., particulate aerosol). Sometimes called London-type smog.

reduction A chemical reaction in which there is a net gain of electrons by a substance.

reference A monitored ecological situation that does not undergo experimental manipulation, and that is used for the purpose of comparison with an experimental treatment. Reference comparisons are used where the experimental design precludes the statistical use of a control treatment. Compare with control.

refinery An industrial facility in which a crude raw material is processed into purer products, as in a metal refinery or a petroleum refinery.

refoliation The process by which a defoliated plant produces new leaves.

refugium A geographic area or region that has remained unaltered by climatic change or glaciation, and which thereby serves as a haven for biota.

regeneration Refers to the regrowth of vegetation on a disturbed site, either by the incursion of sexual propagules or by vegetative growth of unkilled individuals.

release A silvicultural term that refers to the response of conifer vegetation to the suppression of weeds by herbiciding or manual release.

relaxation The progressive loss of species richness in a newly isolated habitat island.

renewable natural resource A natural resource that potentially can be harvested indefinitely, unless the rate of exploitation is sufficiently intense to reduce the capability for regeneration.

replicate The repetition of an experimental procedure or measurement in order to measure experimental error or variation. Replication is required in order to test for statistically significant differences between treatments using commonly used procedures such as t-tests or analysis of variance.

reproduction A sexual or asexual process by which organisms produce new and discrete individuals that are similar to the parent.

reproductive failure A toxic or otherwise stressful situation that can be tolerated by adult organisms, but in which reproduction is not possible, so that there is no recruitment of young individuals.

reproductive potential The number of offspring that a mature adult could potentially produce if there was no mortality of the young. See also fecundity.

reproductive success The number of offspring that a mature individual successfully produces.

reradiation The process by which a body absorbs electromagnetic radiation, converts it to thermal kinetic energy and thereby assumes a higher temperature, and then reradiates some of the absorbed energy as longer-wave electromagnetic radiation. See also radiatively active gas.

reserve   See ecological reserve.

residence time   The length of time that a quantity of substance remains in an environmental compartment. Residence time is influenced by chemical degradation and transformation, and by mass transport processes such as erosion and leaching that can remove the substance from the compartment.

resident   A nonmigratory individual or species.

residue   The quantity of a pesticide or other pollutant that remains in a particular environmental compartment at various times after application.

resilience   Refers to the ability of a system to recover from disturbance.

resistance   Refers to the ability of an organism or population to tolerate a toxic stress. If there is genetic variance for resistance to a toxic stress, then evolution will occur when the stress is encountered, and the trait will rapidly become more frequent in the population.

resorption   The mobilization of a nutrient from a tissue that is about to senesce and become litter, and its translocation to a perennial tissue for reuse or storage.

resource development   The socioeconomic process in which a natural resource is identified, quantified, exploited, and, if the resource is potentially renewable, managed.

respiration   A biochemical process that occurs in all organisms, in which complex organic molecules are broken down by various enzymatic reactions in order to derive energy for metabolism. The ultimate products of respiration are carbon dioxide, water, and other simple inorganic molecules.

revegetation   The regrowth of vegetation on a previously devegetated site.

richness   See species richness.

right-of-way   The corridor of terrain through which a road, power line, or railroad passes.

riparian habitat   Habitat occurring beside flowing water such as a river or stream.

roast bed   A primitive and intensely polluting smelting technique in which a ground-level pile of wood and sulfide metal ore is ignited and allowed to oxidize for several months, after which the metal concentrate is collected and taken away for refining.

rocky intertidal   A hard-rock habitat zone that occurs between the high-water and the low-water levels of the seashore.

rodenticide   A pesticide that is used against rodents.

Roengten (R)   A unit of exposure to ionizing radiation, measured as the ability of the radiation to induce an electrical charge.

rooting depth   The soil depth within which the roots of plants are encountered.

rotation   Refers to the time period between harvests of a biological resource. The rotation is usually one year in agriculture, but 50–100 years in forestry.

r-strategist   An organism that produces a very large number of offspring during its life span.

ruderal   Refers to a plant that occurs on anthropogenically disturbed sites. See also weed.

salinization   The process by which irrigation degrades semiarid land by causing a buildup of salt in surface soil, as a result of the accumulation of soluble inorganic chemicals that remain behind when water evaporates.

salt marsh   A brackish wetland that is periodically inundated by oceanic water, and that is dominated by graminoid vegetation.

saltwater intrusion   A phenomenon that occurs near the ocean, when the withdrawal or fresh groundwater causes a landward movement of saline water, resulting in a degradation of the aquifer.

saprophyte   An organism that feeds on dead organic matter.

savannah   A tropical or subtropical grassland with scattered trees or shrubs.

sawlog   A log that is sufficiently large to be sawn into timber.

scarification   (1) A mechanical disruption of the organic surface of a site to facilitate the establishment of tree seedlings, by the dragging of heavy chains or barrels, ploughing, etc. (2) The mechanical or chemical abrasion of a hard seedcoat in order to stimulate or allow germination.

scenario   A proposed sequence of events.

sclerophyll   A woody plant with small, leathery, evergreen leaves that is frequent in the vegetation of certain dry habitats.

secchi depth   An index of the transparency of a water column, measured as the depth at which a round, black-and-white disc is no longer visible from the surface.

secondary   An indirect effect or relationship.

sector   A subdivision of an economy.

seedbank   The density of viable seeds that is present in the surface organic layer and soil of an ecosystem.

seedling   A recently germinated plant.

seismic   Relating to earth tremors caused by earthquakes, volcanoes, or underground explosions.

selection cut   A forest harvest in which only trees of a desired species and size class are removed.

selection-tree cutting   See selection cut.

seleniferous   Refers to a soil rich in selenium.

semidesert   A region where precipitation is barely sufficient to support the growth of prairie or savannah.

senescence   The process of growing old, or by which there is a general loss of vigor.

sequester   The removal of a substance from the general environment, and into a relatively immobile compartment or chemical complex. See also complex and fixed.

sere   A successional series of communities that occurs as a result of a particular combination of types of disturbance, vegetation, and site conditions.

serpentine   Refers to a group of hydrated magnesium silicate-based minerals, which when they occur at the surface cause the development of a depauperate vegetation because of metal toxicity and deficiency and imbalance of nutrients.

sewage   Wastewater containing human and animal fecal wastes and industrial effluent.

sewage sludge   A dry or semisolid, organic-rich material produced by the anaerobic digestion of sewage, and often applied to agricultural or forested land as a soil conditioner.

shade-tolerant   Refers to a plant that can survive in the stressful conditions of the understory of a closed forest, where there is a limited availability of light, water, and nutrients. See also tolerant.

Shannon–Weiner diversity   An index of species diversity that incorporates both species richness, and the relative abundance of species. See also species diversity and species richness.

shelter-wood cut   A partial harvest of a forest, in which selected large trees are left on the site to favor particular species in the regeneration, and to stimulate the growth of the uncut trees so as to produce high-quality sawlogs at the time of the next cut.

shifting cultivation   A tropical agricultural system in which one to several hectares of forest is cleared and its biomass removed or burned, followed by use of the site for the growth of a mixed agricultural crop for several years, until declining fertility and a vigorous development of weeds require abandonment and the clearing of a new tract of forest.

shortgrass prairie   Prairie in a semidesert region, which is dominated by graminoids less than about 50 cm in height.

short-term   In ecology this is a relative term that de-

pends on the life span of the organisms that are being considered, but it generally refers to a time period of less than one generation.

significant   See statistically significant.

siltation   The filling in of a shallow water body or wetland as a result of the deposition of the sediment load of a river or stream.

silviculture   The branch of forestry that is concerned with the cultivation of trees.

simulation experiment   The use of a dynamic computer model of an ecological or environmental phenomenon or process, to simulate the effects of a change in one or more parameters.

simulation model   A dynamic computer model of an ecological or environmental phenomenon or process.

sink   An environmental or biological repository for a substance, characterized by an output that is nil or very small relative to the rate of input.

site   The particular piece of terrain where a community occurs.

site capability   In forestry, a term that is related to the inherent capacity of a site to support the growth of trees. Sometimes called site productivity or site quality. See also fertility.

site impoverishment   A reduction in site capability.

site preparation.   The treatment of a harvested site prior to the planting of tree seedlings. Site preparation can involve prescribed burning, the crushing or raking of slash, herbiciding, or scarification.

site productivity   See site capability.

site quality   See site capability.

site class   An index of site capability, measured as the number of years that it takes an unsuppressed tree to reach a given height, often about 15 m.

skid trail   The heavily disturbed path by which harvested logs are dragged from the forest.

slash   The debris of branches, foliage, and tree-tops that is left behind during a forest harvest in which the trees were delimbed on the site and dragged away as logs.

slick   A patch of oil floating on water.

slurry   A dense but flowable suspension of solids in water.

smelter   An industrial facility where a metal-rich concentrate is produced by heating and oxidizing an ore or some other metal-containing material. A primary smelter treats raw, mined ore. A secondary smelter treats discarded metal-containing wastes that are being recycled.

smog   An amalgam of the words "smoke" and "fog" that refers to an atmospheric condition of very poor visibility and a large concentration of air pollutants. See also oxidizing smog and reducing smog.

snag   An erect but dead tree.

snowpack   The depth or volume of snow that has accumulated on the landscape.

softwood   A gymnosperm tree.

soil   The variously deep layer of inorganic material that lies on the earth's surface.

soil compaction   A process in which the pore space of the soil is reduced because of compression caused by the heavy weight of machinery or livestock, or because of a loss of tilth caused by the oxidation of soil organic matter.

soil horizon   A visually and chemically distinct horizontal stratum of soil that develop over time as a result of biological and climatic processes. The A horizon lies beneath the surface organic layer, and is characterized by acidification and the leaching out of metal cations and dissolved organic matter. The B horizon is a zone of deposition of material leached from the A horizon. The C horizon is undifferentiated parent material.

soil profile   Refers to the arrangement and characteristics of soil horizons. See also soil horizon.

soil sterilant   A pesticide or steam that is used to fumigate soil in order to kill all organisms.

solar constant   The input of electromagnetic radiation measured at the outer surface of the atmosphere at the earth's average distance from the sun, and having a value of 2.00 cal/cm^2-min.

solar radiation   Electromagnetic radiation that has been emitted by the sun.

solubilization   The process by which materials are made water-soluble, as in the use of detergent to solubilize an oil into water.

somatic   Relating to the body of an organism.

source category   A distinct emissions sector that is used during the quantification of the emission of pollutants for a large geographic area.

spatial   Referring to space.

species   Populations of organisms that actually or potentially interbreed, and produce fertile hybrids. Species are named with a latinized binomial.

species composition   The species present in a defined area. See also species richness.

species diversity   See diversity.

species richness   The number of species present in a defined area, irrespective of their relative abundance.

spongy mesophyll   An internal tissue of a leaf that occurs between the cuticle layers.

sporeling   A recently germinated spore of a fungus, fern, or other simple plant.

spray swath   The width of the area directly impacted by the spray deposited by a ground or aerial application of pesticide.

spring meltwater flush   The hydrologic peak of water flow that occurs when an accumulated snowpack melts during a brief period of time in the spring.

stability   The tendency of a system to persist relatively unchanged over time.

stand   An aggregation of forest or other vegetation in a defined area.

standard   A required or recommended intensity of pollution that should not be exceeded except under extraordinary circumstances.

standing crop   The quantity of biomass per unit area or volume of an ecosystem.

statistically significant   Refers to a level of probability that two or more sampled observations come from different populations, so that in a statistical sense they can be considered to be different. In most ecological studies, the cutoff level is a likelihood of 95% that such a conclusion is correct (i.e., $p = 0.05$). Often abbreviated as "significant."

steady state   A flow-through situation in which the influx of material or energy equals the efflux, so that the net quantity occurring in the compartment does not change.

stem flow   Precipitation water that runs down the stem of trees to the ground surface.

stocking   The density of trees in a stand.

stomata (sing. stoma)   Microscopic pores in the leaf cuticle that occur in a large density, and through which carbon dioxide, oxygen, and water vapor diffuse.

storm flow   The peak of water flow that occurs after a large precipitation event.

strategic nuclear weapon   A nuclear weapon that is delivered over a distance of thousands of kilometers. Compare with tactical weapon.

strategy   The syndrome of adaptive physiological, anatomical, and behavioral traits that characterizes an organism or species.

stratified   A water body that develops a layered water column during the growing season, with relatively dense and cool water in the deeper hypolimnion, and warmer, less dense water in the upper epilimnion. These layers are separated by a zone of rapid change in temperature called the thermocline.

stratosphere   The upper atmospheric layer that caps the troposphere, above an altitude of 8–17 km depending on season and latitude. Within the stratosphere, air temperature varies little with altitude, and there are few convective air currents.

stream-water yield   See water yield.

stress   Physical, chemical, and biological constraints that limit the potential productivity of the biota. Any environmental influence that causes measurable ecological detriment.

stress tolerator   A plant that is well suited anatomically and physiologically to coping with climatic, toxic, or other stress.

strip cut   Refers to a silvicultural system in which there is a series of long and narrow clear-cuts with alternating uncut strips of forest left in between. Several years after the first strip cuts were made, sufficient tree regeneration should have established on them as a result of seeding-in from the uncut strips. At that time, the uncut strips are harvested as well.

structure   See ecosystem structure.

stump sprouting   The production of a large number of vegetative sprouts from the base or roots of a cut stump of certain angiosperm tree species. These shoots progressively self-thin, so that eventually there are only one to three mature, tree-sized stems left.

subalpine   Refers to a coniferous forest zone that occurs beneath the higher elevation alpine tundra of mountainous terrain.

sublittoral   Refers to an intermediate water depth that occurs between the nearshore shallow littoral zone and the deepwater zone.

subspecies   A genetically and anatomically distinct subspecific taxon that is not reproductively isolated from other such subspecies. See also binomial.

subtropical   Refers to a region lying between the tropical and temperate latitudes.

succession   A process that occurs subsequent to disturbance, and that involves the progressive replacement of biotic communities with others over time. In the absence of further disturbance this process culminates in a stable climax ecosystem that is determined by climate, soil, and the nature of the participating biota. Primary succession occurs on a bare substrate that has not previously been modified by organisms. Secondary succession follows a less intensive disturbance, and it occurs on substrates that have been modified biologically.

successional trajectory   The likely sequence of plant and animal communities that is predicted to occur on a site at various times after a particular type of disturbance.

suite   A group of species that tend to co-occur.

sulfuric acid plant   An industrial facility that reduces the emission of sulfur dioxide to the atmosphere by using the gas to manufacture sulfuric acid, which can be sold as a commercial commodity.

surface water   Fresh water occurring free at the surface, as in lakes, ponds, rivers, streams, etc.

surfactant   A substance that serves as a wetting agent, by reducing the surface tension of a liquid such as water, and thereby allowing it to foam or penetrate the small pores of certain solids such as cloth. A detergent is an example.

susceptible   Apt to be infected by a particular pathogen, or infested by a particular pest.

suspended sediment   Particulate matter occurring in the water column.

sustainable development   Exploitation of a renewable natural resource in such a way that does not degrade its capacity to regenerate. See also renewable natural resource.

swamp   A forested wetland.

symbiosis   An obligate mutualism, in which two dependent organisms occur in a mutually beneficial relationship. See also mutualism.

syndrome   A set of conditions that is indicative of a particular disease or disorder.

synecology   The ecology of communities and ecosystems. Compare with autecology.

synergism   A more-than-additive effect that can occur when two or more environmental influences are operating simultaneously.

systematic error   A source of error that carries through an entire calculation or analytical procedure.

systemic   Refers to a physiological effect that occurs throughout an organism.

tableland   A relatively flat expanse of terrain occurring at high altitude.

tactical weapon   A weapon that is used locally in a battlefield, over a range of less than about 100 km.

tailings   The waste material from the industrial process called milling. See also mill.

tainting   The occurrence of a small residue of hydrocarbons in an edible product, which results in a discernible and unpleasant taste.

tallgrass prairie   A relatively mesic prairie dominated by graminoids and forbs that reach a height of up to about 3 m. In the absence of periodic wildfire, a tallgrass prairie can be replaced by an open forest.

tar balls   Dense, semisolid, asphaltic residuum of the weathering of crude oil spilled at sea.

taxon (pl. taxa) Any identifiable group of taxonomically related organisms.

taxonomy The dicipline of biology that is concerned with the classification of organisms based on their phylogenetic relationships.

temperate Refers to a climate that is intermediate between that of the tropics and the polar regions, characterized by alternating long, warm summers and short, cold winters.

temporal Refers to variation with time.

teratogenic An environmental influence that causes a deformity of a developing fetus.

terrestrial Refers to the land, as opposed to the aquatic or marine environment.

tephra Solid material ejected into the atmosphere during a volcanic eruption.

threatened Refers to a species that is rare, and that could become endangered if conservation activity was not initiated.

thermal energy Kinetic energy of molecular vibration, sometimes called heat energy.

thermocline A vertical zone of rapid temperature change in a thermally stratified water body. See also stratified.

thin A silvicultural treatment in which an overstocked stand has some trees removed, in order to enhance the growth of the remaining trees.

threshold An intensity of an environmental factor at which a measureable biological effect begins to occur.

throughfall Water of precipitation that penetrates the canopy of an ecosystem and arrives at the forest floor without running down the trunks of trees. See also stem flow.

till See glacial till.

titration A chemical manipulation in which a measured volume of one solution is added to a known volume of another, until the reaction between the two is complete, or it has been taken to some desired end point.

tolerance (1) Refers to a genetically based physiological tolerance of an environmental stress or combination of stresses. (2) Refers specifically to tolerance of the stressful environmental conditions of the understory of a closed forest. See also shade-tolerant.

total concentration The amount of the nutrient or toxic element content of a sample that is solubilized by a heat-assisted, strong acid digest. This represents virtually all of the quantity present in the sample. See also available concentration.

toxic element Toxic metals such as Cu, Hg, Ni, etc., and toxic nonmetals such As and Se.

toxic threshold See threshold.

toxicity Refers to the quality of being poisonous. See also acute toxicity and chronic toxicity.

toxicology The study of the biological effects of poisons.

trajectory analysis The prediction of the future location of a moving object or air or water mass, based on measurements of the location in the recent past.

transect A straight sampling line that runs in a particular direction from a central place.

transformation (1) A change in physical state or in chemical structure. (2) In statistics, a calculated change of a variable, usually used to transform the actual frequency distribution to a normal distribution so that a parametric statistical test can be applied.

transition See ecotone.

transparency Relates to the degree to which visible radiation can pass through a water column.

transpiration The evaporation of water from foliage.

tree ring An annual, concentric growth ring of a tree that is located in a site with a strongly seasonal climate, and that is easily visible in a cross section of the stem or in a horizontal core taken from the stem.

trophic response Refers to the response of an ecosystem to the addition of a nutrient or combination of nutrients. Usually measured in terms of productivity.

trophic status Refers to the rate of nutrient supply and productivity of a system. See also hypertrophic, eutrophic, mesotrophic, oligotrophic, and ultraoligotrophic.

tropical Refers to a low-latitude climate that is characterized by consistently warm and humid conditions.

troposphere The atmospheric layer that extends from the ground surface up to the lower limit of the stratosphere at an altitude of 8–17 km. Within the troposphere, air temperature typically decreases with increasing altitude, and there are many convective air currents. As a result, the troposphere is sometimes called the "weather atmosphere." Compare with stratosphere.

tumor An undifferentiated tissue mass formed by a recent growth of cells. See also cancer and neoplasia.

tundra Treeless vegetation that occurs at high latitude or high altitude.

turbidity Refers to a condition of poor transparency caused by a large concentration of suspended particulates.

turnover time The time required for the replacement of the quantity of material or energy in a compartment, with an equivalent quantity of material or energy.

ultraoligotrophic Refers to an extremely unproductive aquatic system with a very restricted nutrient supply.

ultraviolet radiation Electromagnetic radiation having a wavelength less than about 0.4 μm, but longer than that of X-rays.

unavailable concentration See available concentration.

understory Vegetation that occurs beneath the principal canopy of a vertically complex ecosystem such as a forest.

undescribed Refers to a taxon that has not yet been identified and given a scientific binomial by a taxonomist.

undetectable concentration Refers to a concentration of a substance that is smaller than the detection limit of the available analytical technology. This does not necessarily imply a concentration of zero.

upland game Hunted wildlife that do not occur in wetlands. Upland game usually refers to relatively small species such as rabbits, hares, pheasants, and grouse.

vagile Refers to an organism that is easily capable of undertaking a long-distance movement.

variable A measurement that has a range of possible values. See also parameter.

variance A statistical measurement that is related to the amount of variation within a population of measurements.

variety A subspecific taxonomic category.

vascular plant A plant with specialized tissue for the internal transport of water.

vector A mobile species that transports a pathogen among hosts. For example, mosquitoes are the vector between the malaria-causing *Plasmodium* and humans.

vegetation A generic term to describe plants on the landscape.

vegetative growth Asexual propagation.

vigor The capacity of an organism for healthy growth and survival.

visible radiation Electromagnetic radiation within the wavelength band of about 0.4–0.7 μm, comprising "light" that is visually detected by the human eye.

visual predator A predator that locates its prey by sight.

volatilization The process of evaporation from a solid or liquid state to a vapor.

volcanic Referring to material ejected from an erupting volcano, or some other volcanic influence.

volume growth The increase in volume or weight of biomass of a tree or forest.

volume-weighted Refers to a calculation that involves the summation or averaging of values of chemical concentration of a number of water samples, in which the contribution of each sample is assigned a weighting factor determined by its volume.

vulnerable Apt to suffer damage from some stressful environmental agent.

washout The removal of particulate or gaseous air pollutants by their impaction with falling droplets of precipitation. See also rainout.

water body Any depression of the landscape that is filled with water, such as a lake, pond, river, or stream.

water column A unit area of water that extends from the surface to the bottom of a water body.

water table The height of the water-saturated level of the soil.

watershed The total expanse of terrain from which water flows into a water body or stream.

water-use efficiency The amount of water transpired by a plant or crop, per unit of biomass that is produced. See also nutrient-use efficiency.

water yield The amount or volume of water that flows in a given period of time from a watershed.

weather The relatively short-term, day-to-day meteorological conditions at a place. Compare with climate.

weathering (1) Refers to chemical and biological processes by which inorganic nutrients are transformed from a water-insoluble to a water-soluble form. (2) Refers to the progressive volatilization of the relatively light hydrocarbon fractions of spilled petroleum, so that the residues become progressively heavier and more viscous, and eventually transform into a tarry substance.

weed Any organism that occurs in a situation where people regard it as a pest. Usually refers to plants.

wet deposition The flux of chemicals to the surface of the earth during an event of precipitation. Compare with dry deposition.

wet meadow A hydric, graminoid-dominated ecosystem.

wet scrubber A device used to remove pollutants from the waste flue gases of a smelter or fossil-fueled power plant. A wet scrubber traps particulates by impacting them with small water droplets, and removes sulfur dioxide by causing it to form calcium sulfate by reaction with calcium carbonate in solution.

whole-lake Refers to a situation where an entire lake is manipulated for some experimental purpose, as in a whole-lake fertilization experiment.

whole-tree clear-cut   A clear-cut in which all of the above-ground biomass is removed during the harvest. See also clear-cut and complete-tree clear-cut.

windbreak   A line of trees or a fence that protects from the prevailing wind.

wood supply   The rate at which forest biomass can be provided for an industrial or other anthropic purpose.

xeric   Refers to dry site conditions.

yard   See deer yard.

yarding   The practice of dragging or skidding logs to a central place (the landing) during a forest harvest.

yield   (1) The amount of biomass that is harvested. (2) The amount of energy liberated during an explosion.

yield decrement   A decrease in yield caused by some environmental stress.

zooplankton   Small animals that occur in the water column.

BIBLIOGRAPHY

Aber, J. D., Botkin, D. B., and Melillo, J. M. (1978). Predicting the effects of different harvesting regimes on forest floor dynamics in northern hardwoods. *Can. J. For. Res.* **8,** 306–315.

Aber, J. D., Botkin, D.B., and Melillo, J. M. (1979). Predicting the effects of different harvesting regimes on productivity and yield in northern hardwoods. *Can. J. For. Res.* **9,** 10–14.

Abercrombie, M., Hickman, C. J., and Johnson, M. L. (1978). "The Penguin Dictionary of Biology." Penguin Books, New York.

Abernethy, S., Bobra, A. M., Shiu, W. Y., Wells, P. G., and Mackay, D. (1986). Acute lethal toxicity of hydrocarbons and chlorinated hydrocarbons to two planktonic crustaceans: The key role of organism-water partitioning. *Aquat. Toxicol.* **8,** 163–174.

Abrahamsen, G. (1980). Acid precipitation, plant nutrients, and forest growth. *In* "Ecological Impact of Acid Precipitation" (C. D. Drablos and A. Tollan, eds.), pp. 58–63. SNSF Project, Oslo, Norway.

Abrahamsen, G., and Stuanes, A. O. (1980). Effects of simulated rain on the effluent from lysimeters with acid, shallow soil, rich in organic matter. *In* "Ecological Impact of Acid Precipitation" (D. Drablos and A. Tollan, eds.), pp. 152–153. SNSF Project, Oslo, Norway.

Abrahamsen, G., Bjor, K., Horntvedt, R., and Tveite, B. (1976). "Impact of Acid Precipitation on Forest and Freshwater Ecosystems in Norway," Res. Rep. FR6/76. SNSF Project, Oslo, Norway.

Abrahamsen, G., Horntvedt, R., and Tveite, B. (1977). Impacts of acid precipitation on coniferous forest ecosystems. *Water, Air, Soil Pollut.* **8,** 57–73.

Adams, H. S., Stephenson, S. L., Blasing, T. J., and Divick, D. N. (1985). Growth-trend declines of spruce and fir in mid-Appalachian subalpine forests. *Environ. Exp. Bot.* **25,** 315–325.

Advisory Committee on Oil Pollution of the Sea (ACOPS) (1980). "Annual Report 1979." ACOPS, London.

Agosti, J. M., and Agosti, T. E. (1973). The oxidation of certain Prudhoe Bay hydrocarbons by microorganisms indigenous to a natural oil seep at Umiat, Alaska. *In* "Proceedings of the Symposium on the Impact of Oil Resource Development on Northern Plant Communities," pp. 80–85. Institute of Arctic Biology, Fairbanks, Alaska.

Agrios, G. N. (1969). "Plant Pathology." Academic Press, New York.

Ahlgren, I. (1978). Response of Lake Norrviken to reduced nutrient loading. *Verh.—Int. Ver. Theor. Angew. Limnol.* **20,** 846–850.

Alabaster, J. S. (1967). Survival of salmon (*Salmo salar*) and sea trout (*S. trutta*) in fresh and saline waters at high temperatures. *Nat. Resour.* **1,** 717–730.

Alban, D. H. (1982). Effects of nutrient accumulation by aspen, spruce, and pine on soil properties. *Soil Sci. Soc. Am. J.* **46,** 853–861.

Alban, D. H., Perala, D. A., and Schlaegel, B. E. (1978). Biomass and nutrient distribution in aspen, pine, and spruce stands on the same soil type in Minnesota. *Can. J. For. Res.* **8,** 290–299.

Alcock, B., and Allen, L. H. (1985). Crop responses to elevated carbon dioxide concentrations. *In* "Direct Effects of Increasing Carbon Dioxide on Vegetation," DOE/ER-0238, pp. 53–97. U.S. Department of Energy, Washington, D.C.

Alexander, V., and Van Cleve, K. (1983). The Alaska pipeline: A success story. *Annu. Rev. Ecol. Syst.* **14,** 443–463.

Allan, A. A., Schlueter, R. S., and Mikolaj, P. J. (1970). Natural oil seepage at Coal Oil Point, Santa Barbara, California. *Science* **170,** 974–977.

Allan, R. J. (1971). Lake sediments: A medium for regional geochemical exploration of the Canadian Shield. *CIM Bull.* **64**(715), 43–59.

Allen, S. E., Grimshaw, H. M., Parkinson, J. A., and Quarmby, C. (1974). "Chemical Analysis of Ecological Materials." Blackwell, London.

Almer, B., Dickson, W., Ekstrom, C., Hornstrom, E., and Miller, U. (1974). Effects of acidification of Swedish lakes. *Ambio* **3,** 330–336.

Almer, B., Dickson, W., Ekstrom, C., and Hornstrom, E. (1978). Sulfur pollution and the aquatic ecosystem. *In* "Sulfur in the Environment" (J. O. Nriagu, ed.), Part II, pp. 271–311. Wiley, New York.

Altshuller, A. P. (1983). Review: Natural volatile organic substances and their effect on air quality in the United States. *Atmos. Environ.* **17,** 2131–2165.

Alvarez, L. W., Alvarez, W., Asaro, F., and Michel, H. V. (1980). Extraterrestrial causes for the Cretaceous-Tertiary extinction. *Science* **208,** 1095–1108.

Alvarez, W., Kauffman, E. G., Surlyk, F., Alvarez, L. W., Asaro, F., and Michel, H. V. (1984). Impact theory of mass extinctions and the invertebrate fossil record. *Science* **223,** 1135–1141.

Ambler, D. C. (1983). Surface water hydrology. *In* "Kejimkujik Calibrated Catchments Program" (J. Kerekes, ed.), pp. 15–23. Environmental Protection Service, Dartmouth, Nova Scotia.

Amdur, M. O. (1980). Air pollutants. *In* "Toxicology: The Basic Science of Poisons" (J. Doull, C. D. Klaassen, and M. O. Amdur, eds.), pp. 608–631. Macmillan, New York.

American Conference of Governmental and Industrial Hygienists (ACGIH) (1982). "Threshold Limit Values for Chemical Substances in Work Air Adopted by ACGIH for 1982." ACGIH, Washington, D.C.

American National Standards Institute (ANSI) (1976). Evaluation of total sulfation in atmosphere by the lead peroxide candle. *In* "Annual Book of ASTM Standards," pp. 533–535. ANSI, Washington, D.C.

American Ornithologists' Union (AOU) (1983). "Checklist of North American Birds." AOU, Washington, D.C.

Ames, P. L. (1966). DDT residues in the eggs of osprey in the north-eastern United States and their relation to nesting success. *J. Appl. Ecol.* **3,** Suppl., 87–97.

Amiro, B. D., and Courtin, G. M. (1981). Patterns of vegetation in the vicinity of an industrially disturbed ecosystem, Sudbury, Ontario. *Can. J. Bot.* **59,** 1623–1639.

Amthor, J. S. (1984). Does acid rain directly influence plant growth? Some comments and observations. *Environ. Pollut., Ser. A* **36,** 1–6.

Amundson, R. S., and MacLean, D. C. (1982). Influence of oxides of nitrogen on crop growth and yield: An overview. *In* "Air Pollution by Nitrogen Oxides" (T. Schneider and L. Grant, eds.), pp. 501–510. Elsevier, Amsterdam.

Anderson, D. W., and Hickey, J. J. (1970). Eggshell changes in certain North American birds. *Proc. Int. Ornithol. Congr.* **15,** 514–540.

Anderson, H. W., Hoover, M. D., and Reinhart, K. G. (1976). "Forests and water: Effects of forest management on floods, sedimentation, and water supply." USDA For. Serv. Gen. Tech. Rep. Pacific Southwest Forest and Range Experiment Station, Berkeley, California.

Andersson, F. (1986). Acidic deposition and its effects on the forests of Nordic Europe. *Water, Air, Soil Pollut.* **30,** 17–29.

Andresen, A. M., Johnson, A. H., and Siccama, T. G. (1980). Levels of lead, copper, and zinc in the forest floor in the northeastern United States. *J. Environ. Qual.* **9,** 293–296.

Aneja, V. P., Adams, D. F., and Pratt, C. D. (1984). Environmental impact of natural emissions. *J. Air Pollut. Control Assoc.* **34,** 799–803.

Anlauf, K. G., Fellin, P., Wiebe, H. A., and Melo, O. T. (1982). The Nanticoke shoreline diffusion experiments, June 1978. IV. A. Oxidation of sulfur dioxide in a power plant plume. B. Ambient concentrations and transport of sulfur dioxide, particulate sulfate and nitrate, and ozone. *Atmos. Environ.* **16,** 455–466.

Anonymous (1973). "National Inventory of Sources and Emissions of Lead (1970)," Publ. No. APCD 73-7. Air Pollution Control Directorate, Environment Canada, Ottawa, Ontario.

Anonymous (1976a). "A Nationwide Inventory of Emissions of Air Contaminants (1974),"Rep. EPS 3-AP-76-1. Air Pollution Control Directorate, Environment Canada, Ottawa, Ontario.

Anonymous (1976b). "Report to the Task-force for Evaluation of Budworm Control Alternatives." Prepared for the Cabinet Committee on Economic Development, Province of New Brunswick, Fredericton.

Anonymous (1981). "National Inventory of Natural Sources and Emissions of Primary Pollutants." Air Pollution Control Directorate, Environment Canada, Ottawa, Ontario.

Anonymous (1982). "Ecological Study of the Amoco Cadiz Oil Spill." U.S. Department of Commerce, National Oceanic and Atmospheric Administration. Washington, D.C.

Anonymous (1987). "National *anatum* Peregrine Falcon Recovery Plan." Federal-Provincial Wildlife Conference of Government Wildlife Directors, Ottawa, Ontario.

Anonymous (1988). Untitled cartoons. *Bull. Br. Ecol. Soc.* **2,** 78.

Antonovics, J., Bradshaw, A. D., and Turner, R. G. (1971). Heavy metal tolerance in plants. *Adv. Ecol. Res.* **7,** 1–85.

Armstrong, F. A. J. (1979). Mercury in the aquatic environment. *In* "Effects of Mercury in the Canadian Environment," NRCC No. 16739, pp. 84–100. Associate Committee on

Scientific Criteria for Environmental Quality, National Research Council of Canada, Ottawa, Ontario.

Armstrong, J. A. (1985a). Tactics and strategies for larval suppression and prevention of damage using chemical insecticides. *In* "Recent Advances in Spruce Budworms Research," pp. 301–319. Canadian Forestry Service, Ottawa, Ontario.

Armstrong, J. A. (1985b). Spruce budworm control program in eastern Canada. *In* "Recent Advances in Spruce Budworms Research," pp. 384–385. Canadian Forestry Service, Ottawa, Ontario.

Ashton, P. S. (1984). Long-term changes in dense and open inland forests following herbicide attack. *In* "Herbicides in War: The Long-term Ecological and Human Consequences" (A. H. Westing, ed.), pp. 33–37. Taylor & Francis, London.

Atkins, R. P. (1969). Lead in a suburban environment. *J. Air Pollut. Control Assoc.* **19,** 591–595.

Atlas, R. M. (1982). Microbial hydrocarbon degradation within sediment impacted by the Amoco Cadiz oil spill. *In* "Ecological Study of the Amoco Cadiz Oil Spill," pp. 1–25. U.S. Department of Commerce, National Oceanic and Atmospheric Administration, Washington, D.C.

Atlas, R. M. (1985). Effects of hydrocarbons on microorganisms and petroleum biodegradation in Arctic ecosystems. *In* "Petroleum Effects in the Arctic Environment" (F. R. Engelhardt, ed.), pp. 63–99. Elsevier, Am. New York.

Aubertin, G. M., and Patric, J. H. (1974). Water quality after clear-cutting a small watershed in West Virginia. *J. Environ. Qual.* **3,** 243–249.

Auclair, A. N. D. (1987). The distribution of forest declines in eastern Canada. *In* "Proceedings of the Workshop on Forest Decline and Reproduction: Regional and Global Consequences" (L. Kairiukstis, S. Nilsson, and A. Stroszak, eds.), pp. 307–319. IIASA, Laxenburg, Austria.

Auerbach, S. I. (1981). Ecosystem response to stress: A review of concepts and approaches. *In* "Stress Effects on Natural Ecosystems" (G. W. Barrett and R. Rosenberg, eds.), pp. 29–41. Wiley, New York.

Aumento, F. (1970). "Serpentine Mineralogy of Ultrabasic Intrusions in Canada and on the Mid-Atlantic Ridge," Pap. 69-53. Geological Survey of Canada, Department of Energy, Mines and Resources, Ottawa, Ontario.

Ayensu, E. S. (1978). Calling the roll of the world's vanishing plants. *Smithson. Mag.* **9**(8), 122–129.

Ayer, W. C., Brun, G. L., Eidt, D. C., Ernst, W., Mallet, V. N., Matheson, R. A., and Silk, P. J. (1984). "Persistence of Aerially Applied Fenitrothion in Water, Soil, Sediment, and Balsam Fir Foliage," Inf. Rep. M-X-153. Maritimes Forest Research Centre, Fredericton, New Brunswick.

Bache, B. W. (1980). The acidification of soil. *In* "Effects of Acid Precipitation on Terrestrial Ecosystems" (T. C. Hutchinson and M. Havas, eds.), pp. 183–202. Plenum, New York.

Baes, C. F., Bjorkstrom, A., and Mulholland, P. J. (1985). Uptake of carbon dioxide by the oceans. *In* "Atmospheric Carbon Dioxide and the Global Carbon Cycle," DOE/ER-0239, pp. 81–111. U.S. Department of Energy, Washington, D.C.

Bailey, R. E. (1982). "The Current Status of the Softwood Resource on Cape Breton Island," For. Res. Note 3. Nova Scotia Department of Lands and Forests, Truro.

Baker, J., and Schofield, C. L. (1982). Aluminum toxicity to fish in acidic waters. *Water, Air, Soil Pollut.* **18,** 289–301.

Baker, J. M. (1971a). The effects of a single oil spillage. *In* "The Ecological Effects of Oil Pollution in Littoral Communities," pp. 16–20. Institute of Petroleum, London.

Baker, J. M. (1971b). Successive spillages. *In* "The Ecological Effects of Oil Pollution in Littoral Communities," pp. 21–32. Institute of Petroleum, London.

Baker, J. M. (1971c). Seasonal effects. *In* "The Ecological Effects of Oil Pollution in Littoral Communities," pp. 44–51. Institute of Petroleum, London.

Baker, J. M. (1971d). Effects of cleaning. *In* "The Ecological Effects of Oil Pollution in Littoral Communities," pp. 52–57. Institute of Petroleum, London.

Baker, J. M. (1971e). Growth stimulation following oil pollution. *In* "The Ecological Effects of Oil Pollution in Littoral Communities," pp. 71–77. Institute of Petroleum, London.

Baker, J. M. (1971f). The effects of oils on plant physiology. *In* "The Ecological Effects of Oil Pollution in Littoral Communities," pp. 88–98. Institute of Petroleum, London.

Baker, J. M. (1971g). Seasonal effects of oil pollution on salt marsh vegetation. *Oikos* **22,** 106–110.

Baker, J. M. (1973). Recovery of salt marsh vegetation from successive oil spillages. *Environ. Pollut.* **4,** 223–230.

Baker, J. M. (1976). Ecological changes in Milford Haven during its history as an oil port. *In* "Marine Ecology and Oil Pollution" (J. M. Baker, ed.), pp. 55–66. Wiley, New York.

Baker, J. M. (1983). "Impact of Oil Pollution on Living Resources," Comm. Ecol. Pap. No. 4. International Union for Conservation of Nature and Natural Resources, Gland, Switzerland.

Baker, R. H. (1946). Some effects of the war on the wildlife of Micronesia. *Trans. North Am. Wildl. Conf.* **11,** 205–213.

Balda, R. P. (1975). Vegetation structure and breeding bird diversity. *In* "Management of Forest and Range Habitats for Non-game Birds," U.S.D.A. For. Serv. Gen. Tech. Rep. WO-1, pp. 59–80. USDA, Washington, D.C.

Banfield, A. W. F. (1974). "The Mammals of Canada." National Museum of Natural Sciences, National Museums of Canada, Univ. of Toronto Press, Toronto.

Barica, J. (1980). Why hypertrophic ecosystems? *In* "Hypertrophic Ecosystems" (J. Barica and L. R. Mur, eds.), pp. IX–XI. Junk, The Hague, The Netherlands.

Barnaby, F. (1982). The effects of a global nuclear war: The arsenals. *Ambio* **11,** 76–83.

Barnaby, F., and Rotblat, J. (1982). The effects of nuclear weapons. *Ambio* **11,** 84–93.

Barnard, W. R. (1986). Alkaline materials flux from unpaved

roads: Source strength, chemistry, and potential for acid rain neutralization. *Water, Air, Soil Pollut.* **30**, 285–293.

Barney, G. O., dir. (1980). "The Global 2000 Report to the President." U.S. Gov. Printing Office, Washington, D.C.

Barsdate, R. J., Alexander, V., and Benoit, R. E. (1973). Natural oil seeps at Cape Simpson, Alaska: Aquatic effects. *In* "Proceedings of the Symposium on the Impact of Oil Resource Development on Northern Plant Communities," pp. 91–95. Institute of Arctic Biology, Fairbanks, Alaska.

Baskerville, G. L. (1975a). Spruce budworm: Super silviculturalist. *For. Chron.* **51**, 138–140.

Baskerville, G. L. (1975b). Spruce budworm: The answer is forest management: or is it? *For. Chron.* **51**, 157–160.

Batcheler, C. L. (1983). The possum and rata-kamaki dieback in New Zealand: A review. *Pac. Sci.* **37**, 415–426.

Batzer, H. O. (1976). Silvicultural control techniques for the spruce budworm. *Misc. Publ.—U.S. Dep. Agric.* **1327**, 110–116.

Baumhover, A. H., Graham, A. J., Bitter, B. A., Hopkins, D. E., New, W. D., Dudley, F. H., and Bushland, R. C. (1955). Screw-worm control through release of sterilized flies. *J. Econ. Entomol.* **48**, 462–466.

Bazzaz, F. A., Garbutt, K., and Williams, W. E. (1985). Effect of increasing carbon dioxide concentration on plant communities. *In* "Direct Effects of Increasing Carbon Dioxide on Vegetation," DOE/ER-0238, pp. 155–170. U.S. Department of Energy, Washington, D.C.

Beall, R. C. (1976). Elk habitat selection in relation to thermal radiation. *In* "Proceedings of the Elk—Logging—Roads Symposium," pp. 97–100. University of Idaho, Moscow.

Beamish, R. J. (1974). Growth and survival of white suckers (*Catostomus commersoni*) in an acidified lake. *J. Fish. Res. Board Can.* **31**, 49–54.

Beamish, R. J., and Harvey, H. H. (1972). Acidification of the La Cloche Mountain Lakes, Ontario, and resulting fish mortalities. *J. Fish. Res. Board Can.* **29**, 1131–1143.

Beamish, R. J., Lockhart, W. L., Van Loon, J. C., and Harvey, H. H. (1975). Long-term acidification of a lake and resulting effects on fishes. *Ambio* **4**, 98–102.

Beanlands, G. E., and Duinker, P. N. (1983). "An Ecological Framework for Environmental Impact Assessment in Canada." Institute for Resource and Environmental Studies, Dalhousie University, Halifax, Nova Scotia.

Beaton, J. D., Tisdale, S. L., and Platov, J. (1971). "Crop Response to Sulfur in North America," Tech. Bull. No. 18. Sulfur Institute, Washington, D.C.

Beaton, J. D., Bixby, D. W., Tisdale, S. L., and Matov, J. S. (1974). "Fertilizer Sulfur: Status and Potential in the U.S.," Tech. Bull. No. 21. Sulfur Institute, Washington, D.C.

Beaver, D. L. (1976). Avian populations in herbicide treated brush fields. *Auk* **93**, 543–553.

Beeton, A. M., and Edmondson, W. T. (1972). The eutrophication problem. *J. Fish. Res. Board Can.* **29**, 673–682.

Beggs, G. L., and Gunn, J. M. (1986). Response of lake trout

(*Salvelinus namaycush*) and brook trout (*S. fontinalis*) to surface water acidification in Ontario. *Water, Air, Soil Pollut.* **30**, 711–717.

Beilke, S., and Gravenhorst, G. (1978). Heterogenous SO_2-oxidation in the droplet phase. *Atmos. Environ.* **12**, 231–239.

Beljan, J. R., Irey, N. S., Kilgore, W. W., Kimura, K., Suskind, R. R., Vostal, J. J., and Wheater, R. H. (1981). "The Health Effects of 'Agent Orange' and Polychlorinated Dioxin Contaminants." Council on Scientific Affairs, American Medical Association, Chicago, Illinois.

Bellrose, F. C. (1959). Lead poisoning as a mortality factor in waterfowl populations. *Bull.—Ill. Nat. Hist. Surv.* **27**, 235–288.

Bellrose, F. C. (1976). "Ducks, Geese, and Swans of North America." Stackpole Books, Harrisburg, Pennsylvania.

Bengston, G., Nordstrom, S., and Rundgren, S. (1983). Population density and tissue metal concentration of lumbricids in forest soils near a brass mill. *Environ. Pollut. Ser. A* **30**, 87–108.

Bengston, J. L., and Laws, R. M. (1985). Trend in crabeater seal age at maturity: An insight into Antarctic marine invertebrates? *In* "Antarctic Nutrient Cycles and Food Webs" (W. R. Siegfield, P. R. Candy, and R. M. Laws, eds.), pp. 661–665. Springer-Verlag, Berlin and New York.

Bengtsson, B., Dickson, D., and Nyberg, P. (1980). Liming lakes in Sweden. *Ambio* **9**, 34–36.

Benkovitz, C. M. (1982). Compilation of an inventory of anthropogenic emissions in the United States and Canada. *Atmos. Environ.* **16**, 1551–1563.

Benoit, L. F., Skelly, J. M., Moore, L. D., and Dochinger, L. S. (1982). Radial growth reductions of *Pinus strobus* L. correlated with foliar ozone sensitivity as an indicator of ozone-induced losses in eastern forests. *Can. J. For. Res.* **12**, 673–678.

Benson, D. W., and Dodds, G. D. (1977). "Deer of Nova Scotia." Nova Scotia Department of Lands and Forests, Halifax.

Berger, D. D., Anderson, R. W., Weaver, J. D., and Risebrough, R. W. (1970). Shell thinning in eggs of Ungava peregrines. *Can. Field-Nat.* **84**, 265–267.

Bergerud, A. T. (1988). Caribou, wolves, and man. *TREE* **3**, 68–72.

Berggren, B., and Oden, S. (1972). "1. Analysresultat Rorande Fungmetaller Och Klorerade Kolvaten I. Rotslam Fran Svenska Reningsverk 1968–1971." Institutionen fur Markvetenskap Lantbrukshogskolan, Uppsala, Sweden.

Berne, S., and Bodennec, G. (1984). Evolution of hydrocarbons after the Tanio oil spill—a comparison with the Amoco Cadiz accident. *Ambio* **13**, 109–114.

Berne, S., Marchand, M., and D'Ozouville, L. (1980). Pollution of sea water and marine sediments in control areas. *Ambio* **9**, 287–293.

Berrow, M. L., and Webber, J. (1972). Trace elements in sewage sludges. *J. Sci. Food Agric.* **23**, 93–100.

Berry, C. R. (1973). The differential susceptibility of eastern white pine to three types of air pollution. *Can. J. For. Res.* **3**, 543–547.

Beschta, R. L., and Jackson, W. L. (1979). The intrusion of fine sediments into a stable gravel bed. *J. Fish. Res. Board Can.* **36**, 204–210.

Beyer, W. N., Pattee, O. H., Sileo, L., Hoffman, D. J., and Mulhern, B. M. (1985). Metal contamination in wildlife living near two zinc smelters. *Environ. Pollut., Ser. A* **38**, 63–86.

Bidleman, T. F., and Leonard, R. (1982). Aerial transport of pesticides over the northern Indian Ocean and adjacent seas. *Atmos. Environ.* **16**, 1099–1107.

Bierhuizen, J. F. H., and Prepas, E. E. (1985). Relationship between nutrients, dominant ions, and phytoplankton standing crop in prairie saline lakes. *Can. J. Fish. Aquat. Sci.* **42**, 1588–1594.

Birge, W. J. (1983). Structure—activity relationships in environmental toxicology. *Fundam. Appl. Toxicol.* **3**, 341–342.

Bjor, K., and Tiegen, O. (1980). Effects of acid precipitation on soil and forest. 6. Lysimeter experiment in greenhouse. *In* "Ecological Impact of Acid Precipitation" (D. Drablos and A. Tollan, eds.), pp. 200–201. SNSF Project, Oslo, Norway.

Bjork, S. (1977). Recovery and restoration of damaged lakes in Sweden. *In* "Recovery and Restoration of Damaged Ecosystems" (J. Cairns, K. L. Dickson, and E. E. Herricks, eds.), pp. 110–133. Univ. Press of Virginia, Charlottesville.

Black, C. A., Evans, D. D., White, J. L., Ensminger, L. E., and Clark, F. E., eds. (1965). "Methods of Soil Analysis," Monogr. No. 9. Am. Soc. Agron., Madison, Wisconsin.

Blais, J. R. (1965). Spruce budworm outbreaks in the past three centuries in the Laurentide Park, Quebec. *For. Sci.* **11**, 130–138.

Blais, J. R. (1981). Mortality of balsam fir and white spruce following a spruce budworm outbreak in the Ottawa River watershed in Quebec. *Can. J. For. Res.* **11**, 620–629.

Blais, J. R. (1983). Trends in the frequency, extent, and severity of spruce budworm outbreaks in eastern Canada. *Can. J. For. Res.* **13**, 539–547.

Blais, J. R. (1985). The ecology of the eastern spruce budworm: A review and discussion. *In* "Recent Advances in Spruce Budworms Research," pp. 49–59. Canadian Forestry Service, Ottawa, Ontario.

Blake, D. R., and Rowland, F. S. (1988). Continuing worldwide increase in tropospheric methane, 1978 to 1987. *Science* **239**, 1129–1131.

Blakeslee, P. A. (1973). "Monitoring Considerations for Municipal Wastewater Effluent and Sludge Application to the Land." U.S. Environmental Protection Agency, U.S. Department of Agriculture, Universities Workshop, July 9–13, 1973, Urbana, Illinois.

Blank, L. W. (1985). A new type of forest decline in Germany. *Nature (London)* **314**, 311–314.

Blank, L. W., Roberts, T. M., and Skeffington, R. A. (1988). New perspectives on forest decline. *Nature (London)* **336**, 27–30.

Blankenship, D. R., and King, K. A. (1984). A probable sighting of 23 eskimo curlews in Texas. *Am. Birds* **38**, 1066–1067.

Blasing, T. J. (1985). Background: Carbon cycle, climate, and vegetation responses. *In* "Characterization of Information Requirements for Studies of CO_2 Effects: Water Resources, Agriculture, Fisheries, Forests, and Human Health," DOE/ER-0236, pp. 9–22. U. S. Department of Energy, Washington, D.C.

Bliss, L. C. (1971). Arctic and alpine plant life cycles. *Annu. Rev. Ecol. Syst.* **2**, 405–438.

Bliss, L. C., ed. (1977). "Truelove Lowland, Devon Island, Canada: A High Arctic Ecosystem." Univ. of Alberta Press, Edmonton, Alta.

Bliss, L. C., Courtin, G. M., Pattie, D. L., Riewe, R. R., Whitfield, D. W. A., and Whidden, P. (1973). Arctic tundra ecosystems. *Annu. Rev. Ecol. Syst.* **4**, 359–399.

Blockstein, D. E., and Tordoff, H. B. (1985). Gone forever. A contemporary look at the extinction of the passenger pigeon. *Am. Birds* **39**, 845–851.

Blouin, A. C. (1985). Comparative patterns of plankton communities under different regimes of pH in Nova Scotia, Canada. Unpublished Ph.D. Thesis, Department of Biology, Dalhousie University, Halifax, Nova Scotia.

Blouin, A. C., Collins, T. M., and Kerekes, J. (1984). Plankton of an acid-stressed lake (Kejimkujik National Park, Nova Scotia, Canada). Part 2. Population dynamics of an exclosure experiment. *Verh. Internat. Verein. Limnol.* **22**, 22–27.

Blum, B. M., and MacLean, D. A. (1985). Potential silviculture, harvesting, and salvage practices in eastern North America. *In* "Recent Advances in Spruce Budworms Research," pp. 264–280. Canadian Forestry Service, Ottawa, Ontario.

Bobra, A., Shiu, W. Y., and Mackay, D. (1985). Quantitative structure-activity relationships for the acute toxicity of chlorobenzenes to *Daphnia magna. Environ. Toxicol. Chem.* **4**, 297–305.

Boehm, P. (1982). The Amoco Cadiz analytical chemistry program. *In* "Ecological Study of the Amoco Cadiz Oil Spill," pp. 35–99. U.S. Department of Commerce, National Oceanic and Atmospheric Administration, Washington, D.C.

Boer, A. H. (1978). Management of deer wintering areas in New Brunswick. *Wildl. Soc. Bull.* **6**, 200–205.

Boffey, P. M. (1971). Herbicides in Vietnam: AAAS study finds widespread devastation. *Science* **171**, 43–47.

Bolandrin, M. F., Klocke, J. A., Wurtele, E. S., and Bollinger, W. H. (1985). Natural plant chemicals: Sources of industrial and medicinal materials. *Science* **228**, 1154–1160.

Bolin, B., Doos, B. R., Jager, J., and Warrick, R. A. (1986). "The Greenhouse Effect, Climatic Change, and Ecosystems," SCOPE Rep. 29. Wiley, Chichester, U.K.

Bonnor, G. M. (1978). "A Guide to Canadian Forest Inventory Terminology and Usage." Forest Management Institute, Canadian Forestry Service, Ottawa, Ontario.

Bordeleau, C. (1986). "Evaluation of Sugar Maple Dieback in Quebec." Presented at the meeting of Northeastern Forest Pest Control Council, Albany, N.Y., March 11–12, 1986.

Bormann, F. H. (1982). The effects of air pollution on the New England landscape. *Ambio* **11**, 338–346.

Bormann, F. H., and Likens, G. E. (1979). "Pattern and Process in a Forested Ecosystem." Springer-Verlag, New York.

Bormann, F. H., Likens, G. E., Siccama, T. G., Pierce, R. S., and Eaton, J. S. (1974). The export of nutrients and recovery of stable conditions following deforestation at Hubbard Brook. *Ecol. Monogr.* **44**, 255–277.

Borrecco, J. E., Black, H. C., and Hooven, E. F. (1979). Response of small mammals to herbicide-induced habitat changes. *Northwest Sci.* **53**, 97–106.

Borror, D. J., DeLong, D. M., and Triplehorn, C. A. (1976). "An Introduction to the Study of Insects." Holt, Reinhart, & Winston, New York.

Bottrell, D. G., and Smith, R. F. (1982). Integrated pest management. *Environ. Sci. Technol.* **16**, 282A–288A.

Bourne, W. R. P. (1976). Seabirds and pollution. *In* "Marine Pollution" (R. Johnston, ed.), pp. 403–502. Academic Press, London.

Bowen, H. J. M. (1979). "Environmental Chemistry of the Elements." Academic Press, New York.

Bowman, K. P. (1988). Global trends in total ozone. *Science* **239**, 48–50.

Box, E. O. (1982). Biosphere processes and natural emissions to the atmosphere: A quantitative geographic modelling basis. *J. Air Pollut. Control Assoc.* **32**, 954–957.

Boyle, J. R., and Ek, A. R. (1972). An evaluation of some effects of bole and branch pulpwood harvesting on site macronutrients. *Can. J. For. Res.* **2**, 407–412.

Boyle, J. R., Phillips, J. J., and Ek, A. R. (1973). Whole tree harvesting: Nutrient budget evaluation. *J. For.* **71**, 760–762.

Boyle, R. W. (1971). Environmental control. The land environment. *CIM Bull.* **64**(712), 46–50.

Boyles, D. T. (1980). Toxicity of hydrocarbons and their halogenated derivatives in an aqueous environment. *In* "Hydrocarbons and Halogenated Hydrocarbons in the Aquatic Environment" (B. K. Afghan and D. Mackay, eds.), pp. 545–557. Plenum, New York.

Brakke, F.H., ed. (1976). "Impact of Acid Precipitation on Forest and Freshwater Ecosystems in Norway," Res. Rep. FR6/76. SNSF Project, Oslo, Norway.

Bradshaw, A. D. (1977). The evolution of metal tolerance and its significance for vegetation establishment on metal contaminated sites. *Int. Conf. Heavy Met. Environ. [Symp. Proc.], 1st, 1975* Vol. 2, pp. 599–622.

Bradshaw, A. D., and Chadwick, M. J. (1981). "The Restoration of Land." Blackwell, Oxford.

Bradshaw, A. D., McNeilly, T. S., and Gregory, R. P. G. (1965). Industrialization, evolution, and the development of heavy metal tolerance in plants. *Symp. Br. Ecol. Soc.* **5**, 327–343.

Brand, D. G., Kehoe, P., and Connors, M. (1986). Coniferous afforestation leads to soil acidification in central Ontario. *Can. J. For. Res.* **16**, 1289–1391.

Breen, P. A., and Mann, K. H. (1976). Destructive grazing of kelp beds by sea urchins off eastern Canada. *J. Fish. Res. Board Can.* **33**, 1278–1283.

Brooks, J. L. (1969). Eutrophication and changes in the composition of the zooplankton. *In* "Eutrophication: Causes, Consequences, Correctives," pp. 236–255. National Academy of Sciences, Washington, D.C.

Brooks, M. N., Colbert, J. J., Mitchell, R. G., and Stark, R. W. Tech. Co-ord. (1985). "Managing Trees and Stands Susceptible to Western Spruce Budworm," Tech. Bull. 1695. U.S.D.A. For. Serv., Washington, D.C.

Brooks, R. R. (1987). "Serpentine and Its Vegetation." Dioscorides Press, Portland, Oregon.

Brooks, R. R., Wither, E. D., and Zepernick, B. (1977). Cobalt and nickel in *Rinorea* species. *Plant Soil* **47**, 707–712.

Brown, C. P. (1946). Food of Maine ruffed grouse by seasons and cover types. *J. Wildl. Manage.* **10**, 17–28.

Brown, E. R., Sinclair, T., Keith, L., Beamer, P., Hazdra, J. J., Nair, V., and Callaghan, O. (1977). Chemical pollutants in relation to diseases in fish. *Ann. N.Y. Acad. Sci.* **298**, 535–546.

Brown, G. W. (1969). Predicting temperatures of small streams. *Water Resour. Res.* **5**, 68–75.

Brown, G. W. (1970). Predicting the effect of clear-cutting on stream temperature. *J. Soil Water Conserv.* **25**, 11–13.

Brown, G. W., and Krygier, J. T. (1970). Effects of clear-cutting on stream temperature. *Water Resour. Res.* **6**, 1133–1140.

Brown, G. W., Gahler, A. R., and Marston, R. B. (1973). Nutrient losses after clear-cut logging and slash burning in the Oregon Coast Range. *Water Resour. Res.* **9**, 1450–1453.

Brown, L. R., and Jacobson, J. (1987). Assessing the future of urbanization. *In* "State of the World 1987," pp. 38–56. Worldwatch Institute, Norton, New York.

Brown, L. R., and Postel, S. (1987). Thresholds of change. *In* "State of the World 1987," pp. 3–19. Worldwatch Institute, Norton, New York.

Brown, R. G. B., Gillespie, D. I., Lock, A. R., Pearce, P. A., and Watson, G. H. (1973). Bird mortality from oil slicks off eastern Canada, February–April 1970. *Can. Field-Nat.* **87**, 225–234.

Brown, R. S., Wolke, R. E., Saila, S. B., and Brown, C. W. (1977). Prevalence of neoplasia in 10 New England populations of the soft-shell clam (*Mya arenaria*). *Ann. N.Y. Acad. Sci.* **298**, 522–534.

Bruck, R. I. (1985). Boreal montane ecosystem decline in the Southern Appalachian Mountains: Potential role of anthropogenic pollution. *In* "Air Pollutant Effects on Forest Ecosystems," pp. 137–155. Acid Rain Foundation, St. Paul, Minnesota.

Bruns, H. (1960). The economic importance of birds in forests. *Bird Study* **7**, 193–208.

Bryan, G. W. (1976). Some aspects of heavy metal tolerance in aquatic organisms. *In* "Effects of Pollutants on Aquatic Organisms" (A. P. M. Lockwood, ed.), pp. 7–34. Cambridge, Univ. Press, London and New York.

Buchauer, M. J. (1973). Contamination of soil and vegetation near a zinc smelter by zinc, cadmium, and lead. *Environ. Sci. Technol.* **7**, 131–135.

Buckner, C. H., and McLeod, B. B. (1975). The impact of insecticides on small forest mammals. *In* "Aerial Control of Forest Insects in Canada" (M. L. Prebble, ed.), pp. 314–318. Department of the Environment, Ottawa, Ontario.

Buckner, C. H., and McLeod, B. B. (1977). "Ecological Impact Studies of Experimental and Operational Spruce Budworm (*Choristoneura fumiferana* Clemens) Control Programs on Selected Non-target Organisms in Quebec, 1976," Rep. CC-X-137. Forest Pest Management Institute, Environment Canada, Sault Ste. Marie, Ontario.

Buech, R. R. (1980). Vegetation of a Kirtland's warbler breeding area and 10 nest sites. *Jack Pine Warbler* **58**, 59–72.

Buhler, D. R., Claeys, R. R., and Mate, B. R. (1975). Heavy metal and chlorinated hydrocarbon residues in California sea-lions *Zalophus californianus*. *J. Fish. Res. Board Can.* **32**, 2391–2397.

Bull, E. L., Twombly, A. D., and Quigley, T. M. (1980). Perpetuating snags in managed mixed conifer forests of the Blue Mountains, Oregon. *In* "Management of Western Forests and Grasslands for Non-game Birds," U.S.D.A. For. Serv. Gen. Tech. Rep. INT-86, pp. 325–336. Intermountain Forest and Range Experiment Station, Ogden, Utah.

Bull, E. L., Peterson, S. R., and Thomas, J. W. (1986). Resource partitioning among woodpeckers in northeastern Oregon. *USDA For. Serv. Res. Note* PNW-444. Pacific Northwest Forest and Range Experiment Station, Portland, Oregon.

Bunce, H. F. (1979). Fluoride emissions and forest growth. *J. Air Pollut. Control Assoc.* **29**, 642–643.

Bunce, H. F. (1984). Fluoride emissions and forest survival, growth, and regeneration. *Environ. Poll.* (series A) **35**, 169–188.

Burger, J. A., and Pritchett, W. L. (1979). Clearcut harvesting and site preparation can dramatically decrease nutrient reserves of a forest site. *In* "Impact of Intensive Harvesting on Forest Nutrient Cycling," p. 393. College of Environmental Science and Forestry, State University of New York, Syracuse.

Burnaby, F. (1988). Refrigerants and the ozone layer. *Ambio* **17**, 354.

Burns, J. W. (1972). Some effects of logging and associated road construction on northern California streams. *Trans. Am. Fish. Soc.* **101**, 1–17.

Busby, D. G., Pearce, P. A., and Garrity, N. R. (1981). Brain cholinesterase response in songbirds exposed to experimental fenitrothion spraying in New Brunswick, Canada. *Bull. Environ. Contam. Toxicol.* **26**, 401–406.

Busby, D. G., Pearce, P. A., and Garrity, N. R. (1982). Brain cholinesterase inhibition in forest passerines exposed to experimental aminocarb spraying. *Bull. Environ. Contam. Toxicol.* **28**, 225–229.

Busby, D. G., Pearce, P. A., Garrity, N. R., and Reynolds, L. M. (1983a). Effect of an organophosphorus insecticide on brain cholinesterase activity in white-throated sparrows exposed to aerial forest spraying. *J. Appl. Ecol.* **20**, 255–264.

Busby, D. G., Pearce, P. A., and Garrity, N. R. (1983b). Brain ChE response in forest songbirds exposed to aerial spraying of aminocarb and possible influence of application methodology and insecticide formulation. *Bull. Environ. Contam. Toxicol.* **31**, 125–131.

Butler, J. N., Morris, B. F., and Sass, J. (1973). "Pelagic Tar from Bermuda and the Sargasso Sea," Spec. Publ. No. 10. Bermuda Biological Station for Research, St. George's West, Bermuda.

Buttrick, P. L. (1917). A forester at the fighting front. *Am. For.* **23**, 710–716.

Cabioch, L. (1980). Pollution of subtidal sediments and disturbance of benthic animal communities. *Ambio* **9**, 294–296.

Cade, T. J. (1988). Using science and technology to re-establish species lost in nature. *In* "Biodiversity" (E. O. Wilson, ed.), pp. 279–288. National Academy Press, Washington, D.C.

Cade, T. J., and Fyfe, R. (1970). The North American peregrine survey, 1970. *Can. Field-Nat.* **84**, 231–240.

Cadle, S.H., Dasch, J. M., and Grossnickle, N. E. (1984). Retention and release of chemical species by a northern Michigan snowpack. *Water, Air, Soil Pollut.* **22**, 303–319.

Cain, M. D., and Yaussy, D. A. (1984). Can hardwoods be eradicated from pine sites? *South. J. Appl. For.* **8**, 7–13.

Cairns, D. K., and Elliot, R. D. (1987). Oil spill impact assessment for seabirds: The role of refugia and growth centers. *Biol. Conserv.* **40**, 1–9.

Campbell, B., and Lack, E. (1985). "A Dictionary of Birds." British Ornithologists' Union, London.

Campbell, P. G. C., Bisson, M., Bengie, R., Tessier, A., and Villeneuve, J. P. (1983). Speciation of aluminum in acidic waters. *Anal. Chem.* **55**, 2246–2252.

Campbell, W. B., Harris, R. W., and Benoit, R. E. (1973). Response of Alaskan tundra microflora to crude oil spill. *In* "Proceedings of the Symposium on the Impact of Oil Resource Development on Northern Plant Communities," pp. 53–62. Institute of Arctic Biology, Fairbanks, Alaska.

Canadian Environmental Assessment Research Council/National Research Council (CEARC/NRC) (1986). "Cumulative Environmental Effects." CEARC/NRC, Board on Basic Ecology, Ottawa, Ontario.

Canfield, D. E., and Bachman, R. W. (1981). Prediction of total phosphorus concentrations, chlorophyll a, and secchi depths in natural and artificial lakes. *Can. J. Fish. Aquat. Sci.* **38,** 414–423.

Cann, D. B., MacDougall, J. I., and Hilchey, J. D. (1965). "Soil Survey of Kings County, Nova Scotia," Rep. No. 15. Nova Scotia Soil Survey, Nova Scotia Department of Agriculture and Marketing, Truro.

Cannon, H. L. (1960). Botanical prospecting for ore deposits. *Science* **132,** 591–598.

CANSAP (1983). "CANSAP Data Summary 1982." Environment Canada, Ottawa, Ontario.

CANSAP (1984). "CANSAP Data Summary 1983." Environment Canada, Ottawa, Ontario.

Canter, L. (1977). "Environmental Impact Assessment." McGraw-Hill, New York.

Cantwell, E. N., Jacobs, E. S., Gunz, W. G., and Liberi, V. E. (1972). Control of particulate lead emissions from automobiles. *In* "Cycling and Control of Metals" (M. G. Curry and G. M. Gigliotti, eds.), pp. 95–107. U.S. Environmental Protection Agency, Cincinnati, Ohio.

Carey, A. C., Miller, E. A., Geballe, G. T., Wargo, P. M., Smith, W. H., and Siccama, T. G. (1984). *Armillaria mellea* and decline of red spruce. *Plant Dis.* **68,** 794–795.

Carrier, L. (1986). "Decline in Quebec's Forests: Assessment of the Situation." Service de la Recherche Applique, Ministère de l'Energie et des Ressources, Quebec.

Carson, R. (1962). "Silent Spring." Houghton-Mifflin, Boston, Massachusetts.

Carter, N. E., and Lavigne, D. R. (1986). "Protection Spraying against Spruce Budworm in New Brunswick 1985." Timber Management Branch, New Brunswick Department of Forests, Mines, and Energy, Fredericton.

Case, A. B., and Donnelly, R. G. (1979). "Type and Extent of Ground Disturbance Following Skidder Logging in Newfoundland and Labrador," Inf. Rep. N-X-176. Newfoundland Forest Research Centre, St. John's.

Castell, C. P. (1944). The ponds and their vegetation. *London Nat.* **24,** 15–22.

Castell, C. P. (1954). The bomb-crater ponds of Brookham Common. *London Nat.* **34,** 16–21.

Catling, P. M., Freedman, B., Stewart, C., Kerekes, J. J., and Lefkovitch, L. P. (1986). Aquatic plants of acid lakes in Kejimkujik National Park; floristic composition and relation to water chemistry. *Can. J. Bot.* **64,** 724–729.

Cess, R. D. (1985). Nuclear war: Illustrative effects of atmospheric smoke and dust upon solar radiation. *Clim. Change* **7,** 237–251.

Chambers, L. A. (1976). Classification and extent of air pollution problems. *In* "Air Pollution" (A. C. Stern, ed.), 3rd ed., Vol. 1, pp. 3–22. Academic Press, New York.

Chan, W. H., and Lusis, M. A. (1985). Post-superstack Sudbury smelter emissions and their fate in the atmosphere: An overview of the Sudbury environmental study. *Water, Air, Soil Pollut.* **20,** 211–220.

Chan, W. H., Ro, C. U., Lusis, M. A., and Vet, R.J. (1982).

Impact of the INCO nickel smelter emissions on precipitation quality in the Sudbury area. *Atmos. Environ.* **16,** 801–814.

Chan, W. H., Vet, R. J., Lusis, M. A., and Skelton, G. B. (1983). Airborne particulate size distribution measurements in nickel smelter plumes. *Atmos. Environ.* **17,** 1173–1181.

Chan, W. H., Vet, R. J., Ro, C.-U., Tang, A. J. S., and Lusis, M. A. (1984a). Impact of INCO smelter emissions on wet and dry depositions in the Sudbury area. *Atmos. Environ.* **18,** 1001–1008.

Chan, W. H., Vet, R. J., Ro, C.-U., Tang, A. J. S., and Lusis, M. A. (1984b). Long-term precipitation quality and wet deposition fields in the Sudbury basin. *Atmos. Environ.* **18,** 1175–1188.

Chapin, F. S., III, and Shaver, G. R. (1985). Individualistic growth response of tundra plant species to environmental manipulation in the field. *Ecology* **66,** 564–576.

Chapin, G., and Wasserstrom, R. (1981). Agricultural production and malaria resurgence in Central America and India. *Nature (London)* **293,** 181–185.

Chapman, A. R. O. (1981). Stability of sea urchin dominated barren grounds following destructive grazing of kelp in St. Margaret's Bay, eastern Canada. *Mar. Biol.* **62,** 307–311.

Charlson, R.J., Lovelock, J. E., Andreae, M. O., and Warren, S. G. (1987). Oceanic phytoplankton, atmospheric sulfur, cloud albedo, and climate. *Nature (London)* **326,** 655–661.

Charlton, M. N. (1979). "Hypolimnetic Oxygen Depletion in Central Lake Erie: Has There Been Any Change? Sci. Ser. No. 110. Inland Waters Directorate, National Water Research Institute, Environment Canada, Burlington, Ontario.

Charlton, M. N. (1980). Oxygen depletion in Lake Erie: Has there been any change? *Can. J. Fish. Aquat. Sci.* **37,** 72–81.

Chung, Y. S., and Dann, T. (1985). Observations of stratospheric ozone at the ground level in Regina, Canada. *Atmos. Environ.* **19,** 157–162.

Cicerone, R. J. (1987). Changes in stratospheric ozone. *Science* **237,** 35–42.

Clair, T. A., and Komadina, V. (1984). "Aluminum Speciation in Waters of Nova Scotia and their Impact on WQB Analytical and Field Method," IWD-AR-WQB-84-69. Inland Waters Directorate, Atlantic Region, Moncton, New Brunswick.

Clark, D. R. (1979). Lead concentrations: Bats vs. terrestrial small mammals collected near a major highway. *Environ. Sci. Technol.* **13,** 338–340.

Clark, R. B. (1984). Impact of oil pollution on seabirds. *Environ. Pollut., Ser. A* **33,** 1–22.

Clark, R. C., and Brown, D. W. (1977). Petroleum: Properties and analyses in biotic and abiotic systems. *In* "Effects of Petroleum on Arctic and Subarctic Marine Environments and Organisms" (D. C. Malins, ed.), Vol. 1, pp. 1–90. Academic Press, New York.

Clark, R. C., and MacLeod, W. D. (1977). Inputs, transport

mechanisms, and observed concentrations of petroleum in the marine environment. *In* "Effects of Petroleum on Arctic and Subarctic Marine Environments and Organisms" (D. C. Malins, ed.), Vol. 1, pp. 91–224. Academic Press, New York.

Clark, R. S., Chair. (1947). Scientific Meeting on the Effect of the War on the Stocks of Commercial Food Fishes. Copenhagen, Oct. 24, 1947.

Clark, T. H., and Stern, C. W. (1968). "Geological Evolution of North America." Ronald Press, New York.

Clowater, W. G., and Andrews, P. W. (1981). "An Assessment of Damage Caused by the Spruce Budworm on Spruce and Balsam Fir Trees in New Brunswick." New Brunswick Department of Natural Resources, Forest Management Branch, Fredericton.

Clymo, R. S. (1963). Ion exchange in *Sphagnum* and its relation to bog ecology. *Ann. Bot. (London)* [N.S.] **27,** 309–324.

Clymo, R. S. (1964). The origin of acidity in *Sphagnum* bogs. *Bryologist* **67,** 427–431.

Clymo, R. S. (1984). *Sphagnum*-dominated peat bogs: A naturally acidic ecosystem. *Philos. Trans. R. Soc. London* **305,** 487–499.

Coffin, D. L., and Stokinger, H. E. (1978). Biological effects of air pollutants. *In* "Air Pollution" (A. C. Stern, ed.), 3rd ed., Vol. 2, pp. 232–360. Academic Press, New York.

Cogbill, C. V. (1976). The history and character of acid precipitation in eastern North America. *Water, Air, Soil Pollut.* **6,** 407–413.

Cogbill, C. V. (1977). The effects of acid precipitation on tree growth in eastern North America. *Water, Air, Soil Pollut.* **8,** 89–93.

Cogbill, C. V., and Likens, G. E. (1974). Acid precipitation in the northeastern United States. *Water Resour. Res.* **10,** 1133–1137.

Cohen, C. J., Grothaus, L. C., and Perrigan, S. C. (1981). "Effects of Simulated Acid Rain on Crop Plants" Spec. Rep. 619. Agricultural Experiment Station, Oregon State University, Corvallis.

Cohen, D. B., Webber, M. D., and Bryant, D. N. (1978). Land application of chemical sewage sludge—lysimeter studies. *Conf. Proc.—Res. Program Abatement Munic. Pollut. Prov. Can.-Ont. Agreement Great Lakes Water Qual.* **6,** 108–137.

Colbeck, I., and Harrison, R. M. (1985). The photochemical pollution episode of 5–16 July 1983 in north-west England. *Atmos. Environ.* **19,** 1921–1929.

Cole, D. W., and Gessel, S. P. (1965). Movement of elements through forest soils as influenced by tree removal and fertilizer additions. *In* "Forest Soil Relationships in North America" (C. T. Youngberg, ed.), pp. 95–104. Oregon State Univ. Press, Corvallis.

Cole, D. W., Crane, W. J. B., and Grier, C. C. (1975). The effect of forest management practices on water chemistry in a second-growth Douglas-fir ecosystem. *In* "Forest Soils and

Forest Land Management" (B. Bernier and C. H. Winget, eds.), pp. 195–207. Laval Univ. Press, Quebec.

Coleman, R. (1966). The importance of sulfur as a plant nutrient in world crop production. *Soil Sci.* **101,** 230–239.

Collins, S. L., James, F. C., and Risser, P. G. (1982). Habitat relationships of wood warblers (Parulidae) in northern central Wisconsin. *Oikos* **39,** 50–58.

Conner, R. N. (1978). Snag management for cavity-nesting birds. *In* "Proceedings of the Workshop on Management of Southern Forests for Non-game Birds," U.S.D.A. For. Serv. Gen. Tech. Rep. SE-14, pp. 120–128. Southeastern Forest Experiment Station, Asheville, North Carolina.

Conner, R. N., Hooper, R. G., Crawford, H. S., and Mosby, H. S. (1975). Woodpecker nesting habitat in cut and uncut woodlands in Virginia. *J. Wildl. Manage.* **39,** 144–150.

Conner, R. N., Kroll, J. C., and Kulhavy, D. L. (1983). The potential of girdled and 2,4-D injected southern red oaks as woodpecker nesting and foraging sites. *South. J. Appl. For.* **7,** 125–128.

Conroy, N., Hawley, K., Keller, W., and LaFrance, C. (1975). Influence of the atmosphere on lakes in the Sudbury area. *In* "Proceedings of the First Specialty Symposium on Atmospheric Contributions to the Chemistry of Lake Waters, Sept. 28–Oct. 1, 1975, pp. 146–165. Toronto, Ontario.

Conroy, N., Hawley, K., Keller, W., and LaFrance, C. (1976). Influences of the atmosphere on lakes in the Sudbury area. *J. Great Lakes Res.* **2,** Suppl. 1, 146–165.

Cook, R. B., and Schindler, D. W. (1983). The biogeochemistry of sulfur in an experimentally acidified lake. *Ecol. Bull.* **35,** 115–127.

Cook, R. B., Kelly, C. A., Schindler, D. W., and Turner, M. A. (1986). Mechanisms of hydrogen ion neutralization in an experimentally acidified lake. *Limnol. Oceanogr.* **31,** 134–148.

Cooke, A. S. (1979). Changes in egg shell characteristics of the sparrowhawk (*Accipiter nisus*) and peregrine (*Falco peregrinus*) associated with exposure to environmental pollutants during recent decades. *J. Zool.* **187,** 245–263.

Cooke, A. S., and Frazer, J. F. D. (1976). Characteristics of newt breeding sites. *J. Zool.* **178,** 223–236.

Cooper, C. F. (1983). Carbon storage in managed forests. *Can. J. For. Res.* **13,** 155–166.

Corbett, E. S., Lynch, J. A., and Sopper, W. E. (1978). Timber harvesting practices and water quality in the eastern United States. *J. For.* **76,** 484–488.

Cordone, A. J., and Kelley, D. W. (1961). The influence of inorganic sediment on the aquatic life of streams. *Calif. Fish Game* **47,** 189–228.

Cormack, D. (1983). "Response to Oil and Chemical Marine Pollution." Appl. Sci. Pub., London.

Corner, E. D. S. (1978). Pollution studies with marine plankton. Part 1. Petroleum hydrocarbons and related compounds. *Adv. Mar. Biol.* **15,** 289–380.

Costescu, L. W. (1974). The ecological consequences of airborne metallic contaminants from the Sudbury smelters. Ph.D. Thesis, Department of Botany, University of Toronto. Toronto, Ontario.

Council on Environmental Quality (CEQ) (1975). "Environmental Quality." U.S. Govt. Printing Office, Washington, D.C.

Council on Environmental Quality (CEQ) (1979). "Environmental Quality." Executive Office of the President, Washington, D.C.

Covey, C. (1985). Climatic effects of nuclear war. *BioScience* **35**, 563–569.

Cowling, E. B. (1985). Comparison of regional declines of forests in Europe and North America: The possible role of airborne chemicals. *In* "Air Pollutant Effects on Forest Ecosystems," pp. 217–234. Acid Rain Foundation, St. Paul, Minnesota.

Cox, R. M., and Hutchinson, T. C. (1979). Metal co-tolerances in the grass *Deschampsia caespitosa*. *Nature (London)* **279**, 231–233.

Cox, R. M., and Hutchinson, T. C. (1980). Multiple metal tolerances in the grass *Deschampsia caespitosa* (L.) Beauv. from the Sudbury smelting area. *New Phytol.* **84**, 631–647.

Cox, R. M., and Hutchinson, T. C. (1981). Environmental factors influencing the rate of spread of the grass *Deschampsia caespitosa* invading areas around the Sudbury nickel-copper smelters. *Water, Air, Soil Pollut.* **16**, 83–106.

Crawford, H. S., and Titterington, R. W. (1979). Effects of silvicultural practices on bird communities in upland spruce-fir stands. *In* "Management of North Central and Northeastern Forests for Non-game Birds," U.S.D.A. For. Serv. Gen. Tech. Rep. NC-51, pp. 110–119. North Central Forest Experiment Station, St. Paul, Minnesota.

Crawford, H. S., Titterington, R. W., and Jennings, D. T. (1983). Bird predation and spruce budworm populations. *J. For.* **81**, 433–435.

Crete, M., and Bédard, J. (1975). Daily browse consumption by moose in the Gaspe Peninsula, Quebec. *J. Wildl. Manage.* **39**, 368–373.

Crocker, R. L., and Major, J. (1955). Soil development in relation to vegetation and surface age at Glacier Bay, Alaska. *J. Ecol.* **43**, 427–448.

Cromartie, E., Reichel, W. L., Locke, N. L., Belisle, A. A., Kaiser, T. E., Lamont, T. G., Mulhern, B. M., Prouty, R. M., and Swineford, D. M. (1975). Residues of organochlorine pesticide and polychlorinated biphenyls and autopsy data for bald eagles, 1971–1972. *Pestic. Monit. J.* **9**, 11–14.

Cronan, C. S., and Reiners, W. A. (1983). Canopy processing of acid precipitation by coniferous and hardwood forests in New England. *Oecologia* **59**, 216–223.

Cronan, C. S., and Schofield, C. L. (1979). Aluminum leaching response to acid precipitation: Effects on high-elevation watersheds in the northeast. *Science* **204**, 304–305.

Crouch, G. L. (1985). "Effects of Clearcutting a Subalpine Forest in Central Colorado on Wildlife Habitat," Res. Rep. RM-258. Rocky Mountain Forest and Range Experiment Station, Fort Collins, Colorado.

Crutzen, P. J., and Arnold, F. (1986). Nitric acid cloud formation in the cold Antarctic stratosphere: A major cause for the springtime "ozone hole." *Nature (London)* **324**, 651–655.

Crutzen, P. J., and Birks, J. W. (1982). The atmosphere after a nuclear war: Twilight at noon. *Ambio* **11**, 114–125.

Cullis, C. F., and Hirschler, M. M. (1980). Atmospheric sulfur: Natural and man-made sources. *Atmos. Environ.* **14**, 1263–1278.

Cunningham, J. C. (1985). Biorationals for control of spruce budworms. *In* "Recent Advances in Spruce Budworms Research," pp. 320–349. Canadian Forestry Service, Ottawa, Ontario.

Cunningham, J. D., Keeney, D. R., and Ryan, J. A. (1975a). Yield and metal composition of corn and rye grown on sewage sludge-amended soil. *J. Environ. Qual.* **4**, 448–454.

Cunningham, J. D., Ryan, J. A., and Keeney, D. R. (1975b). Phytotoxicity in and metal uptake from soil treated with metal-amended sewage sludge. *J. Environ. Qual.* **4**, 455–460.

Cure, J. D. (1985). Carbon dioxide doubling responses: A crop survey. *In* "Direct Effects of Increasing Carbon Dioxide on Vegetation," DOE/ER-0238, pp. 99–116. U.S. Department of Energy, Washington, D.C.

Currie, D. J., and Kalff, J. (1984). The relative importance of bacterioplankton and phytoplankton in phosphorus uptake in freshwater. *Limnol. Oceanogr.* **29**, 311–321.

Currie, D.J., Bentzen, E., and Kalff, J. (1986). Does algal-bacterial phosphorus partitioning vary among lakes? A comparative study of orthophosphate uptake and alkaline phosphatase activity in freshwater. *Can. J. Fish. Aquat. Sci.* **43**, 311–318.

Currier, H. B., and Peoples, S. A. (1954). Phytotoxicity of hydrocarbons. *Hilgardia* **23**, 155–173.

Curry-Lindahl, K. (1971). War and the white rhinos. *Oryx* **11**, 263–267.

Curry-Lindahl, K. (1972). "Let Them Live: A Worldwide Survey of Animals Threatened With Extinction." William Morrow, New York.

Czapowskyj, M. M. (1977). Status of fertilization and nutrition research in northern forest types. *In* "Proceedings of the Symposium on Intensive Culture of Northern Forest Types," U.S.D.A. For. Serv. Gen. Tech. Rep. NE-29, pp. 185–203. Northeastern Forest Experiment Station, Broomall, Pennsylvania.

Dale, J. M., and Freedman, B. (1982). Lead and zinc contamination of roadside soil and vegetation in Halifax, Nova Scotia. *Proc. N.S. Inst. Sci.* **32**, 327–336.

Dale, J. M., Freedman, B., and Kerekes, J. (1985). Acidity and associated water chemistry of amphibian habitats in Nova Scotia. *Can. J. Zool.* **63**, 97–105.

Dale, J. M., Freedman, B., and Kerekes, J. (1986). Experimental studies of the effects of acidity and associated water chemistry on amphibians. *Proc. N.S. Inst. Sci.* **35**, 35–54.

Dana, S. T. (1914). French forests in the war zone. *Am. For.* **20**, 769–785.

Daniel, T. W., Helms, J. A., and Baker, F. S. (1979). "Principles of Silviculture," 2nd ed. McGraw-Hill, Toronto, Ontario.

Davidson, C. I. (1979). Air pollution in Pittsburgh: A historical perspective. *J. Air Pollut. Control Assoc.* **29,** 1035–1042.

Davis, A. M. (1972). Selenium accumulation in *Astragalus* species. *Agron. J.* **64,** 751–754.

Davis, J. W., Gregory, G. A., and Ockenfels, R. A., eds. (1983). "Snag Habitat Management," Tech. Rep. RM-99. U.S.D.A. For. Serv., Rocky Mountain Forest and Range Experiment Station, Fort Collins, Colorado.

Davis, M. B. (1978). Climatic interpretation of pollen in quaternary sediments. *In* "Biology and Quaternary Environments" (D. Walker and J. C. Guppy, eds.), pp. 35–51. Australian Academy of Science, Canberra.

Davis, M. B. (1981). Outbreaks of forest pathogens in quaternary history. *Proc. Int. Conf. Palynol. 4th,* pp. 216–228.

Dawson, W. R., Ligon, T. D., Murphy, J. R., Myers, J. P., Simberloff, D., and Vernon, J. (1987). Report of the scientific advisory panel on the spotted owl. *Condor* **89,** 205–229.

Day, R. T. (1983). A survey and census of the endangered Furbish lousewort, *Pedicularis furbishae,* in New Brunswick. *Can. Field-Nat.* **93,** 325–327.

Daye, P. G., and Garside, E. T. (1979). Development and survival of embryos and alevins of the Atlantic salmon, *Salmo salar* L., continuously exposed to acidic levels of pH, from fertilization. *Can. J. Zool.* **57,** 1713–1718.

Dean, L. C., Havens, R., Harper, K. T., and Rosenbaum, J. B. (1973). Vegetative stabilization of mill mineral wastes. *In* "Ecology and Reclamation of Devastated Land" (R. J. Hutnik and G. Davis, eds.), Vol. 2, pp. 119–136. Gordon & Breach, New York.

deBauer, M., Tejeda, T. H., and Manning, W. J. (1985). Ozone causes needle injury and tree decline in *Pinus hartwegii* at high altitudes in the mountains around Mexico City. *J. Air Pollut. Control Assoc.* **35,** 838.

DeByle, N. V., and Packer, P. E. (1972). Plant nutrient and soil losses in overland flow from burned forest clear-cuts. *In* "Watersheds in Transition," pp. 296–307. American Water Resources Association, Urbana, Illinois.

DeCosta, J., Janicki, A., Shellito, G., and Wilcox, G. (1983). The effect of phosphorus additions in enclosures on the phytoplankton and zooplankton of an acid lake. *Oikos* **40,** 283–294.

DeGraaf, R. M., and Payne, B. R. (1975). Economic values of non-game birds and some urban wildlife research needs. *Trans. North Am. Wildl. Nat. Resour. Conf.* **40,** 281–287.

de Gusmao-Camara, T. (1983). Tropical moist forest conservation in Brazil. *In* "Tropical Rain Forest: Ecology and Management" (S. L. Sutton, T. C. Whitmore, and A. C. Chadwick, eds.), pp. 413–421. Blackwell, Boston, Massachusetts.

Dekker, W. L., Jones, V., and Achutuni, R. (1985). The impact of CO_2-induced climate change on U.S. agriculture. *In* "Characterization of Information Requirements for Studies

of CO_2 Effects: Water Resources, Agriculture, Fisheries, Forests, and Human Health," DOE/ER-0236, pp. 69–94. U. S. Department of Energy, Washington, D.C.

DeLaune, R. D., Patrick, W. H., and Buresh, R. J. (1979). Effect of crude oil on a Louisiana *Spartina alterniflora* salt marsh. *Environ. Pollut., Ser. A* **20,** 21–31.

Del Moral, R. (1983). Initial recovery of vegetation on Mount St. Helens, Washington. *Am. Midl. Nat.* **109,** 72–80.

Delorme, L. D. (1982). Lake Erie oxygen: The prehistoric record. *Can. J. Fish. Aquat. Sci.* **39,** 1021–1029.

Deneke, F. J., McCown, B. H., Coyne, P. I., Rickard, W., and Brown, J. (1975). "Biological Aspects of Terrestrial Oil Spills," Res. Rep. 346. U.S. Army Corps of Engineers, Cold Regions Research and Engineering Laboratory, Hanover, New Hampshire.

Detwiler, R. P., and Hall, C. A. S. (1988). Tropical forests and the global carbon cycle. *Science* **239,** 42–47.

de Zafra, R. L., Jaramillo, M., Parrish, A., Solomon, P., Connor, B., and Barrett, J. (1987). High concentrations of chlorine monoxide at low altitudes in the Antarctic spring stratosphere: Diurnal variation. *Nature (London)* **328,** 408–411.

Diamond, J. M. (1972). Biogeographic kinetics: Estimation of relaxation times for avifaunas of southwest Pacific islands. *Proc. Natl. Acad. Sci. U.S.A.* **69,** 3199–3203.

Diamond, J. M. (1982). Man the exterminator. *Nature (London)* **298,** 787–789.

Diamond, J. M. (1984). "Natural" extinctions of isolated populations. *In* "Extinctions" (M. H. Nitecki, ed.), pp. 191–246. Univ. of Chicago Press, Chicago, Illinois.

Diamond, J. M., and May, R. M. (1976). Island biogeography and the design of nature reserves. *In* "Theoretical Ecology: Principles and Applications" (R. M. May, ed.), pp. 163–186. Saunders, Philadelphia, Pennsylvania.

Dicks, B. (1977). Changes in the vegetation of an oiled Southampton Water salt marsh. *In* "Recovery and Restoration of Damaged Ecosystems" (J. Cairns, K. L. Dickson, and E. E. Herricks, eds.), pp. 208–240. Univ. Press of Virginia, Charlottesville.

Dillon, P. J., and Rigler, F. H. (1974). The phosphorus-chlorophyll relationship in lakes. *Limnol. Oceanogr.* **19,** 767–773.

Dillon, P. J., Jefferies, D. S., Snyder, W., Reid, R., Yan, N. D., Evans, D., Moss, J., and Scheider, W. A. (1977). "Acid Rain in South-central Ontario: Present Conditions and Future Consequences." Ontario Ministry of the Environment, Rexdale.

Dillon, P. J., Yan, N. D., Scheider, W. A., and Conroy, N. (1979). Acidic lakes in Ontario, Canada: Characterization, extent, and responses to base and nutrient additions. *Arch. Hydrobiol., Suppl.* **13,** 317–336.

Dilworth, T. G., Pearce, P. A., and Dobell, J. V. (1974). DDT in New Brunswick woodcocks. *J. Wildl. Manage.* **38,** 331–337.

Dimitriades, B. (1981). The role of natural organics in photochemical air pollution. *J. Air Pollut. Control Assoc.* **31,** 229–235.

Doane, C. C., and McManus, M. L., eds. (1981). "The Gypsy Moth: Research Toward Integrated Pest Management," Tech. Bull. 1584. U.S.D.A. For. Serv., Washington, D.C.

Dobson, A. P., and Hudson, P. G. (1986). Parasites, disease, and the structure of ecological communities. *TREE* **1,** 11–15.

Dobzhansky, T., Ayala, F. J., Stebbins, G. L., and Valentine, J. W. (1977). "Evolution." Freeman, San Francisco, California.

Dodds, D. G. (1974). Distribution, habitat, and status of moose in the Atlantic Provinces of Canada and northeastern United States. *Nat. Can.* **101,** 51–65.

Donaghho, W. R. (1950). Observations of some birds of Guadalcanal and Tulagi. *Condor* **52,** 127–132.

Donaldson, D., ed. (1978). "Forestry: Sector Policy Paper." World Bank, Washington, D.C.

dos Santos, J. M. (1987). Climate, natural vegetation, and soils in Amazonia: An overview. *In* "The Geophysiology of Amazonia" (R. E. Dickinson, ed.), pp. 25–36. Wiley, New York.

Dost, H., ed. (1973). "Acid Sulphate Soils." Wageningen, F.R.G.

Douglas, R. J. W. (1970). "Geology and Economic Minerals of Canada." Department of Energy, Mines and Resources, Ottawa, Ontario.

Douglass, J. E., and Swank, W. T. (1972). Streamflow modification through management of eastern forests. USDA For. Serv. Res. Pap. SE-94. Southeast Forest Experiment Station, Asheville, North Carolina.

Douglass, J. E., and Swank, W. T. (1975). Effects of management practices on water quality and quantity: Coweeta Laboratory, North Carolina. *In* "Municipal Watershed Management," U.S.D.A. For. Serv. Gen. Tech. Rep. NE-13, pp. 1–13. Northeastern Forest Experiment Station, Broomall, Pennsylvania.

Dreisinger, B. R. (1967). "The Impact of Sulfur Pollution of Crops and Forests," Conf. Background Pap. A4-2-1. Pollution in Our Environment. Canadian Council of Resource Ministers, Montreal.

Dreisinger, B. R., and McGovern, P. C. (1971). "Sulfur Dioxide Levels and Vegetation Injury in the Sudbury Area during the 1970 Season, Annual Report." Air Management Branch, Ontario Department of Energy and Resources Management, Sudbury.

Driscoll, C. T., Baker, J. P., Bisogni, J. J., and Schofield, C. L. (1980). Effect of aluminum speciation on fish in dilute acidified waters. *Nature (London)* **284,** 161–164.

Duce, R. A. (1978). Speculations on the budget of particulate and vapour phase non-methane organic carbon in the global troposphere. *Pageophysics* **116,** 244–273.

Dugle, J. R., and Mayoh, K. R. (1984). Responses of 56 naturally-growing shrub taxa to chronic gamma irradiation. *Environ. Exp. Bot.* **24,** 267–276.

Duthie, J. R. (1972a). Detergents: Nutient considerations and total assessment. *In* "Nutrients and Eutrophication" (G. E. Likens, ed.), pp. 205–216. Am. Soc. Limnol. Oceanogr., Lawrence, Kansas.

Duthie, J. R. (1972b). Detergent developments and their impact on water quality. *In* "Nutrients in Natural Waters" (H. E. Allen and J. R. Kramer, eds.), pp. 333–352. Wiley, New York.

Dvorak, A. J., and Lewis, B. G. (1978). "Impacts of Coal-fired Power Plants on Fish, Wildlife, and their Habitats," FWS/OB7-78. Fish and Wildlife Service, U.S. Department of the Interior, Washington, D.C.

Dyrness, C. T. (1967). "Mass Soil Movements in the H. J. Andrews Experimental Forest," Res. Pap. PNW-42. Pacific Northwest Forest and Range Experiment Station, Portland, Oregon.

Eaton, J. S., Likens, G. E., and Bormann, F. H. (1973). Throughfall and stemflow chemistry in a northern hardwoods forest. *J. Ecol.* **61,** 495–508.

Eaton, J. S., Likens, G. E., and Bormann, F. H. (1980). Wet and dry deposition of sulfur at Hubbard Brook. *In* "Effects of Acid Precipitation on Terrestrial Ecosystems" (T. C. Hutchinson and M. Havas, eds.), pp. 69–75. Plenum, New York.

Ebregt, A., and Boldewijn, J. M. A. M. (1977). Influence of heavy metals in spruce forest soil on amylase activity, CO_2 evolution from starch, and soil respiration. *Plant Soil* **47,** 137–148.

Edminster, F. C. (1947). "The Ruffed Grouse: Its Life History, Ecology, and Management." Macmillan, New York.

Edmonds, J. A., and Reilly, J. M. (1985). Future global energy and carbon dioxide emissions. *In* "Atmospheric Carbon Dioxide and the Global Carbon Cycle." pp. 215–245. DOE/ER-0239. United States Department of Energy, Washington, D.C.

Edmondson, W. T. (1972). Nutrients and phytoplankton in Lake Washington. *In* "Nutrients and Eutrophication" (G. E. Likens, ed.), pp. 172–188. Am. Soc. Limnol. Oceanogr., Lawrence, Kansas.

Edmondson, W. T. (1977). Recovery of Lake Washington from eutrophication. *In* "Recovery and Restoration of Damaged Ecosystems" (J. Cairns, K. L. Dickson, and E. E. Herricks, eds.), pp. 102–110. Univ. Press of Virginia, Charlottesville.

Edmondson, W. T., and Lehman, J. T. (1981). The effect of changes in the nutrient income on the condition of Lake Washington. *Limnol. Oceanogr.* **26,** 1–29.

Edmondson, W. T., and Litt, A. H. (1982). *Daphnia* in Lake Washington. *Limnol. Oceanogr.* **27,** 272–293.

Edwards, C. A. (1975). "Persistent Pesticides in the Environment." CRC Press, Cleveland, Ohio.

Ehrlich, A. H. (1985). The human population: Size and dynamics. *Am. Zool.* **25,** 395–406.

Ehrlich, P. R. (1983). Genetics and the extinction of butterfly populations. *In* "Genetics and Conservation" (C. M. Schonewald-Cox, ed.), pp. 152–163. Benjamin/Cummings, Menlo Park, California.

Ehrlich, P. R., and Ehrlich, A. (1981). "Extinction: The Causes and Consequences of the Disappearance of Species." Ballantine Books, New York.

Ehrlich, P. R., Breedlove, D. E., Brussard, P. F., and Sharpe, M. A. (1972). Weather and the "regulation" of subalpine populations. *Ecology* **53**, 243–247.

Ehrlich, P. R., Ehrlich, A. H., and Holdren, J. P. (1977). "Ecoscience: Population, Resources, Environment." Freeman, San Francisco, California.

Eidt, D. C. (1975). The effect of fenitrothion from large-scale forest spraying on benthos in New Brunswick headwater streams. *Can. Entomol.* **107**, 743–760.

Eidt, D. C. (1977). "Effects of Fenitrothion on Benthos in the Nashwaak Project Study Streams in 1976," Inf. Rep. M-X-70. Maritimes Forest Research Centre, Fredericton, New Brunswick.

Eidt, D. C. (1985). Toxicity of *Bacillus thuringiensis* var. *kurstaki* to aquatic insects. *Can. Entomol.* **117**, 829–837.

Eidt, D. C., and Mallet, V. N. (1986). Accumulation and persistence of fenitrothion in needles of balsam fir and possible effects on abundance of *Neodiprion abietis* (Harris) (Hymenoptera: Diprionidae). *Can. Entomol.* **118**, 69–77.

Eidt, D. C., and Pearce, P. A. (1986). The biological consequences of lingering fenitrothion residues in conifer foliage—a synthesis. *For. Chron.* **62**, 246–249.

Eidt, D. C., and Weaver, C. A. A. (1986). "Frequency of Forest Respraying and Use of *B.t.* in New Brunswick, 1975–1986," Inf. Rep. M-X-158. Maritimes Forest Research Centre, Fredericton, New Brunswick.

Eisenreich, S. J., Thornton, J. D., Munger, J. W., and Gorham, E. (1980). Impact of land-use on the chemical composition of rain and snow in northern Minnesota. *In* "Ecological Impact of Acid Precipitation" (D. Drablos and A. Tollan, eds.), pp. 110–111. SNSF Project, Oslo, Norway.

Eldredge, N., and Gould, S. J. (1972). Punctuated equilibria: An alternative to phyletic gradualism. *In* "Models in Paleobiology" (T. J. M. Schopf, ed.), pp. 82–115. Freeman, Cooper, San Francisco, California.

Ellenberg, H. (1986). The effects of environmental factors and use alternatives upon the species diversity and regeneration of tropical rain forests. *Appl. Geogr. Dev.* **28**, 20–36.

Endler, J. A. (1982). Pleistocene forest refuges: Fact or fancy. *In* "Biological Diversification in the Tropics" (G. T. Prance, ed.), pp. 641–657. Columbia Univ. Press, New York.

Engelhardt, F. R., ed. (1985). "Petroleum Effects in the Arctic Environment." Am. Elsevier, New York.

Enger, L., and Hogstrom, U. (1979). Dispersion and wet deposition of sulfur from a power plant. *Atmos. Environ.* **13**, 797–810.

Entomological Society of America (ESA) (1980). "Pesticide Handbook." ESA, College Park, Maryland.

Environmental and Social Systems Analysts Ltd. (ESSA) (1982). "Review and Evaluation of Adaptive Environmental Assessment and Management." Environment Canada, Ottawa, Ontario.

Erickson, J. D., Mulinare, J., McClain, P. W., Fitch, T. G., James, L. M., McClearn, A. B., and Adams, M. J. (1984). "Vietnam Veterans' Risks for Fathering Babies with Birth Defects." U.S. Department of Health and Human Services, Centers for Disease Control, Atlanta, Georgia.

Eriksson, M. O. G. (1984). Acidification of lakes: Effects on waterbirds in Sweden. *Ambio* **13**, 260–262.

Eriksson, O. (1976). Silvicultural practices and reindeer grazing in northern Sweden. *Ecol. Bull.* **21**, 107–120.

Erwin, T. L. (1982). Tropical forests: Their richness in Coleoptera and other arthropod species. *Coleopterists Bull.* **36**, 74–75.

Erwin, T. L. (1983a). Beetles and other insects of tropical forest canopies at Manaus, Brazil, sampled by insecticidal fogging. *In* "Tropical Rain Forest: Ecology and Management" (S. L. Sutton, T. C. Whitmore, and A. C. Chadwick, eds.), pp. 59–75. Blackwell, Boston, Massachusetts.

Erwin, T. L. (1983b). Tropical forest canopies: The last biotic frontier. *Bull. Entomol. Soc. Am.* **29**(1), 14–19.

Erwin, T. L. (1988). The tropical forest canopy. The heart of biotic diversity. *In* "Biodiversity" (E. O. Wilson, ed.), pp. 123–129. National Academy Press, Washington, D.C.

Eto, M. (1974). "Organophosphorus Pesticides: Organic and Biological Chemistry." CRC Press, Cleveland, Ohio.

Euler, D. (1978). "Vegetation Management for Wildlife in Ontario." Ministry of Natural Resources, Toronto, Ontario.

Evans, C. S., Asher, C. J., and Johnson, C. M. (1968). Isolation of dimethyl diselenide and other volatile selenium compounds from *Astragalus racemosus* (Pursh.). *Aust. J. Biol. Sci.* **21**, 13–20.

Evans, K. E. (1978). Forest management opportunities for songbirds. *Trans. North Am. Wildl. Nat. Resour. Conf.* **43**, 69–77.

Evans, K. E., and Conner, R. N. (1979). Snag management. *In* "Management of North Central and Northeastern Forests for Non-game Birds," U.S.D.A. For. Serv. Gen. Tech. Rep. NC-51, pp. 214–225. North Central Forest Experiment Station, St. Paul, Minnesota.

Evans, L. S. (1982). Effects of acidity in precipitation on terrestrial vegetation. *Water, Air, Soil Pollut.* **18**, 395–403.

Evans, L. S. (1986). Proposed mechanisms of initial injury-causing apical dieback in red spruce at high elevation in eastern North America. *Can. J. For. Res.* **16**, 1113–1116.

Evans, P. G. H., and Nettleship, D. M. (1986). Conservation of the Atlantic Alcidae. *In* "The Atlantic Alcidae" (D. N. Nettleship and T. R. Birkhead, eds.), pp. 427–488. Academic Press, New York.

Fabacher, D. L., Schmitt, C. J., Besser, J. M., Baumann, P. C., and Mac, M. J. (1986). Great Lakes fish—neoplasia investigations. *Tox. Chem. Aquat. Life: Res. Manage. 1986* p. 87.

Fairfax, J. A. W., and Lepp, N. W. (1975). Effect of simulated "acid rain" on cation loss from leaves. *Nature (London)* **255**, 324–325.

Farman, J. C., Gradiner, B. G., and Shanklin, J. D. (1985).

Large losses of total ozone in Antarctica reveals seasonal ClO_x/NO_x interaction. *Nature (London)* **315,** 207–210.

Farmer, C. B., Toon, G. C., Schaper, P. W., Blavien, J. F., and Lowes, L. L. (1987). Stratospheric trace gases in the spring 1986 Antarctic atmosphere. *Nature (London)* **329,** 126–130.

Farnsworth, N. R. (1988). Screening plants for new medicines. *In* "Biodiversity" (E. O. Wilson, ed.), pp. 83–97. National Academy Press, Washington, D.C.

Farrell, E. P., Nilsson, I., Tamm, C. O., and Wiklander, G. (1980). Effects of artificial acidification with sulfuric acid on soil chemistry in a Scots pine forest. *In* "Ecological Impact of Acid Precipitation" (D. Drablos and A. Tollan, eds.), pp. 186–187. SNSF Project, Oslo, Norway.

Fay, J. (1978). Hawaii: Extinction unmerciful. *Garden* **103**(7), 22–27.

Ferman, M. A., Wolff, G. T., and Kelly, N. A. (1981). The nature and sources of haze in the Shenandoah Valley/Blue Ridge Mountains area. *J. Air. Pollut. Control Assoc.* **31,** 1074–1082.

Fimreite, N. (1970). Mercury uses in Canada and their possible hazards as sources of mercury contamination. *Environ. Pollut.* **1,** 119–131.

Fimreite, N., Feif, R. W., and Keith, J. A. (1970). Mercury contamination of Canadian prairie seed eaters and their avian predators. *Can. Field-Nat.* **84,** 269–276.

Findlay, D. L., and Kling, H. J. (1975). "Seasonal Successions of Phytoplankton in Seven Lake-basins in the Experimental Lakes Area, Northwestern Ontario, Following Artificial Eutrophication," Fish. Mar. Serv. Tech. Rep. No. 513. Environment Canada, Freshwater Institute, Winnipeg, Manitoba.

Findlay, D. L., and Saesura, G. (1980). Effects on phytoplankton biomass, succession, and composition in Lake 223 as a result of lowering pH levels from 7.0 to 5.6. Data from 1974 to 1979. *Fish. Mar. Serv. Manuscr. Rep.* 1585. Department of Fisheries and Oceans, Winnipeg, Manitoba.

Finkelstein, K., and Gundlach, E. R. (1981). Method of estimating spilled oil quantity on the shoreline. *Environ. Sci. Technol.* **15,** 545–549.

Fischhoff, D. A., Bowdish, K. S., Perlak, F. J., Morrone, P. G., McCormick, S. M., Niedermeyer, J. G., Dean, D. A., Kusmo-Kretzmen, K., Mayer, E. J., Rochester, D. E., Rogers, S. G., and Fraley, R. T. (1987). Insect tolerant transgenic tomato plants. *Biotechnology* **5,** 807–814.

Fisher, H. I., and Baldwin, P. H. (1946). War and the birds of Midway Atoll. *Condor* **48,** 3–15.

Fisher, K. C., and Elson, P. F. (1950). The selected temperature of the Atlantic salmon and speckled trout and the effect of temperature on the response to electrical stimulus. *Physiol. Zool.* **1,** 27–34.

Fisher, J., Simon, N., and Vincent, J. (1969). "Wildlife in Danger." Viking Press, New York.

Fitter, R. (1968). "Vanishing Wild Animals of the World." Kaye & Ward Ltd., London.

Fitter, R. (1974). Most endangered mammals: An action program. *Oryx* **12,** 436–449.

Fletcher, J. L., and Busnel, R. G., eds. (1978). "Effects of Noise on Wildlife." Academic Press, New York.

Flick, W. A., Schofield, C. L., and Webster, D. A. (1982). Remedial actions for interim maintenance of fish stocks in acidified water. *Acid. Rain/Fish., Proc. Int. Symp., 1981* pp. 287–306.

Flint, M. L., and van den Bosch, R. (1983). "Introduction to Integrated Pest Management." Plenum, New York.

Floc'h, J.-Y., and Diouris, M. (1980). Initial effects of Amoco Cadiz oil on intertidal algae. *Ambio* **9,** 284–286.

Folkeson, L. (1984). Deterioration of the moss and lichen vegetation in a forest polluted by heavy metals. *Ambio* **13,** 37–39.

Food and Agriculture Organization (FAO) (1982). "World Forest Products Demand and Supply 1990 and 2000." FAO, Rome, Italy.

Food and Agriculture Organization (FAO) (1986). "Yearbook of Forest Products, 1973–1984," FAO For. Ser. 19. FAO, Rome, Italy.

Forgeron, F. D. (1971). Soil geochemistry in the Canadian Shield. *CIM Bull.* **64**(715), 37–42.

Formann, R. T. T., Galli, A. E., and Leck, C. F. (1976). Forest size and avian diversity in New Jersey woodlots with some land use implications. *Oecologia* **26,** 1–8.

Fortescue, J. A. C. (1980). "Environmental Geochemistry: A Holistic Approach." Springer-Verlag, New York.

Fosberg, F. R. (1959). Plants and fall-out. *Nature (London)* **183,** 1448.

Foster, M. S., and Holmes, R. W. (1977). The Santa Barbara oil spill: An ecological disaster? *In* "Recovery and Restoration of Damaged Ecosystems" (J. Cairns, K. L. Dickson, and E. E. Herricks, eds.), pp. 166–190. Univ. Press of Virginia, Charlottesville.

Foster, N. W. (1983). Acid precipitation and soil solution chemistry within a maple-birch forest in Canada. *For. Ecol. Manage.* **12,** 215–231.

Foster, N. W., and Morrison, I. K. (1983). "Soil Fertility, Fertilization, and Growth of Canadian Forests," Inf. Rep. O-X-353. Great Lakes Forest Research Centre, Sault Ste. Marie, Ontario.

Fowells, H. A., comp. (1965). "Silvics of Forest Trees in the United States," Agric. Handb. No. 271. U.S. Department of Agriculture, Forest Service, Washington, D.C.

Fowler, D. (1980a). Wet and dry deposition of sulfur and nitrogen compounds from the atmosphere. *In* "Effects of Acid Precipitation on Terrestrial Ecosystems" (T. C. Hutchinson and M. Havas, eds.), pp. 9–28. Plenum, New York.

Fowler, D. (1980b). Removal of sulfur and nitrogen compounds from the atmosphere in rain and by dry deposition. *In* "Ecological Impact of Acid Precipitation" (D. Drablos and A. Tollan, eds.), pp. 22–32. SNSF Project, Oslo, Norway.

Fox, D. L. (1986). The transformation of pollutants. *In* "Air Pollution" (A. C. Stern, ed.), 3rd ed., Vol. 6, pp. 61–94. Academic Press, New York.

Foy, C. D., Chaney, R. L., and White, M. C. (1978). The physiology of metal toxicity in plants. *Annu. Rev. Plant Physiol.* **29,** 511–566.

France, R. L. (1987). Reproductive impairment of the crayfish *Orconectes virilis* in response to acidification of Lake 223. *Can. J. Fish. Aquat. Sci.* **44**, Suppl. 1, 97–106.

France, R. L., and Graham, L. (1985). Increased microsporidian parasitism of the crayfish *Orconectes virilis* in an experimentally acidified lake. *Water, Air, Soil Pollut.* **26**, 129–136.

Frank, R., Ishida, K., and Suda, P. 1976a). Metals in agricultural soils in Ontario. *Can. J. Soil Sci.* **56**, 181–196.

Frank, R., Braun, H. E., Ishida, K., and Suda, P. (1976b). Persistent organic and inorganic pesticide residues in orchard soils and vineyards of southern Ontario. *Can. J. Soil Sci.* **56**, 463–484.

Frank, R., Braun, H. E., Van Hove Holdrinet, M., Sirans, S. J., and Ripley, B. D. (1982). Agriculture and water quality in the Canadian Great Lakes basin. V. Pesticide use in 11 agricultural watersheds and presence in streamwater, 1975–1977. *J. Environ. Qual.* **11**, 497–505.

Freda, J. (1986). The influence of acidic pond water on amphibians: A review. *Water, Air, Soil Pollut.* **30**, 439–450.

Fredricksen, R. L., Moore, D. G., and Norris, L. A. (1975). The impact of timber harvest, fertilization, and herbicide treatment on streamwater quality in western Oregon and Washington. *In* "Forest Soils and Forest Land Management" (B. Bernier and C. H. Winget, eds.), pp. 283–313. Laval Univ. Press, Quebec.

Freed, L. A., Conant, S., and Fleischer, R. C. (1987). Evolutionary ecology and radiation of Hawaiian passerine birds. *TREE* **2**, 196–203.

Freedman, B. (1981). "Intensive Forest Harvest—A Review of Nutrient Budget Considerations," Inf. Rep. M-X-121. Maritimes Forest Research Centre, Fredericton, New Brunswick.

Freedman, B. (1981). Trace elements and organics associated with coal combustion in power plants: Emissions and environmental impacts. *Coal: Phoenix '80s [Eighties], Proc. CIC Coal Symp., 64th, 1981* pp. 616–631.

Freedman, B. (1982). "An Overview of the Environmental Impacts of Forestry, with Particular Reference to the Atlantic Provinces." Institute for Resource and Environmental Studies, Dalhousie University, Halifax, Nova Scotia.

Freedman, B., and Clair, T. A. (1987). Ion mass balances and seasonal fluxes from four acidic brownwater streams in Nova Scotia. *Can. J. Fish. Aquat. Sci.* **44**, 538–548.

Freedman, B., and Hutchinson, T. C. (1976). Physical and biological effects of experimental crude oil spills on low arctic tundra in the vicinity of Tuktoyaktuk, N.W.T., Canada. *Can. J. Bot.* **54**, 2219–2230.

Freedman, B., and Hutchinson, T. C. (1980a). Pollutant inputs from the atmosphere and accumulations in soils and vegetation near a nickel-copper smelter at Sudbury, Ontario, Canada. *Can. J. Bot.* **58**, 108–132.

Freedman, B., and Hutchinson, T. C. (1980b). Effects of smelter pollutants on forest leaf litter decomposition near a nickel-copper smelter at Sudbury, Ontario. *Can. J. Bot.* **58**, 1722–1736.

Freedman, B., and Hutchinson, T. C. (1980c). Long-term effects of smelter pollution at Sudbury, Ontario, on forest community composition. *Can. J. Bot.* **58**, 2123–2140.

Freedman, B., and Hutchinson, T. C. (1981). Sources of metal and elemental contamination of terrestrial environments. *In* "Effects of Heavy Metal Pollution on Plants," Vol. 2, pp. 35–94. Appl. Sci. Publ., London.

Freedman, B., and Hutchinson, T. C. (1986). Aluminum in terrestrial ecosystems. *In* "Aluminum in the Canadian Environment," pp. 129–152. National Research Council of Canada, Ottawa, Ontario.

Freedman, B., and Prager, U. (1986). Ambient bulk deposition, throughfall, and stemflow in a variety of forest stands in Nova Scotia. *Can. J. For. Res.* **16**, 854–860.

Freedman, B., and Svoboda, J. (1982). Populations of breeding birds at Alexandra Fiord, Ellesmere Island, Northwest Territories, compared with other arctic localities. *Can. Field-Nat.* **96**, 56–60.

Freedman, B., Beauchamp, C., McLaren, I. A., and Tingley, S. I. (1981a). Forestry management practices and populations of breeding birds in Nova Scotia. *Can. Field-Nat.* **95**, 307–311.

Freedman, B., Morash, R., and Hanson, A. J. (1981b). Biomass and nutrient removals by conventional and whole-tree clearcutting of a red spruce—balsam fir stand in central Nova Scotia. *Can. J. For. Res.* **11**, 249–257.

Freedman, B., Duinker, P. N., Morash, R., and Prager, U. (1982). "Standing Crops of Biomass and Nutrients in a Variety of Forest Stands in Central Nova Scotia," Inf. Rep. M-X-134. Maritimes Forest Research Centre, Fredericton, New Brunswick.

Freedman, B., Svoboda, J., Labine, C., Muc, M., Henry, G., Nams, M., Stewart, J., and Woodley, E. (1983). Physical and ecological characteristics of Alexandra Fiord, a high arctic oasis on Ellesmere Island. *Int. Conf. Permafrost, 4th* pp. 301–306.

Freedman, B., Stewart, C., and Prager, U. (1985). "Patterns of Water Chemistry of Four Drainage Basins in Central Nova Scotia," Tech. Rep. IWD-AR-WQB-85-93. Water Quality Branch, Inland Waters Directorate, Environment Canada, Moncton, New Brunswick.

Freedman, B., Duinker, P. N., and Morash, R. (1986). Biomass and nutrients in Nova Scotia forests, and implications of intensive harvesting for future site productivity. *For. Ecol. Manage.* **15**, 103–127.

Freedman, B., Poirier, A. M., and Scott, F. (1988a). Effects of 2,4,5-T on breeding birds, small mammals, and their habitat in conifer clearcuts in Nova Scotia. *Can. Field-Nat.* **102**, 6–11.

Freedman, B., Zobens, V., and Hutchinson, T. C. (1988b). Effects of naturally occurring pollution on vegetation at the Smoking Hills, Northwest Territories, Canada. submitted for publication.

Freeman, M. M. R. (1985). Effects of petroleum activities on the ecology of Arctic man. *In* "Petroleum Effects in the Arctic Environment" (F. R. Engelhardt, ed.), pp. 245–273. Elsevier, Am. New York.

Freeth, S. J., and Kay, R. L. F. (1987). The Lake Nyos gas disaster. *Nature (London)* **325,** 104–105.

Freidel, F. (1964). "Over There: The Story of America's First Great Overseas Crusade." Little, Brown, Boston, Massachusetts.

Friedland, A. J., and Battles, J. J. (1987). Red spruce (*Picea rubens*) decline in the northeastern United States: Review and recent data from Whiteface Mountain. *In* "Proceedings of the Workshop on Forest Decline and Reproduction: Regional and Global Consequences" (L. Kairiukstis, S. Nilsson, and A. Stroszak, eds.), pp. 287–296. IIASA, Laxenburg, Austria.

Friedland, A. J., and Johnson, A. H. (1985). Lead distribution and fluxes in a high-elevation forest in northern Vermont. *J. Environ. Qual.* **14,** 332–336.

Friedland, A. J., Johnson, A. H., Siccama, T. G., and Mader, D. L. (1984). Trace metal profiles in the forest floor of New England. *Soil Sci. Soc. Am. J.* **48,** 422–425.

Friedland, A. J., Hawley, G. J., and Gregory, R. A. (1985). Investigations of nitrogen as a possible contributor to red spruce (*Picea rubens* Sarg.) decline. *In* "Air Pollutant Effects on Forest Ecosystems," pp. 95–106. Acid Rain Foundation, St. Paul, Minnesota.

Frink, C. R., and Voigt, G. K. (1977). Potential effects of acid precipitation on soils of the humid temperate zone. *Water, Air, Soil Pollut.* **7,** 371–388.

Frost, D. V., and Lish, P. M. (1975). Selenium in biology. *Annu. Rev. Pharmacol.* **15,** 259–284.

Fry, F. E., Hart, J. S., and Walder, K. F. (1946). Lethal temperature for a sample of young speckled trout. *Ont. Fish. Res. Lab.* **66,** 5–35.

Fyfe, R. W. (1976). Rationale and success of the Canadian Wildlife Service peregrine breeding project. *Can. Field-Nat.* **90,** 308–319.

Fyfe, R. W., Temple, S. A., and Cade, T. J. (1976). The 1975 North American peregrine falcon survey. *Can. Field-Nat.* **90,** 228–273.

Gage, S. H., and Miller, C. A. (1978). "A Long-term Bird Census in Spruce Budworm—Prone Balsam Fir Habitats in Northwestern New Brunswick," Inf. Rep. M-X-84. Maritimes Forest Research Centre, Fredericton, New Brunswick.

Galli, A. E., Leck, C. F., and Forman, R. T. T. (1976). Avian distribution patterns in forest islands of different sizes in central New Jersey. *Auk* **93,** 356–364.

Galloway, J. N., Thornton, J. D., Norton, S. A., Volchok, H. L., and McLean, R. A. N. (1982). Trace metals in atmospheric deposition: A review and assessment. *Atmos. Environ.* **16,** 1677–1700.

Galloway, J. N., Hendry, G. R., Schofield, C. L., Peters, N. E., and Johannes, A. H. (1987a). Processes and causes of lake acidification during spring snowmelt in the west-central Adirondack Mountains, New York. *Can. J. Fish. Aquat. Sci.* **44,** 1595–1602.

Galloway, J. N., Dianwu, Z., Jiling, X., and Likens, G. E.

(1987b). Acid rain: China, United States, and a remote area. *Science* **236,** 1559–1562.

Galston, A. W., and Richards, P. W. (1984). Terrestrial plant ecology and forestry: An overview. *In* "Herbicides in War: The Long-term Ecological and Human Consequences" (A. H. Westing, ed.), pp. 39–42. Taylor & Francis, London.

Gammon, R. H., Sundquist, E. T., and Fraser, P. J. (1985). History of carbon dioxide in the atmosphere. *In* "Atmospheric Carbon Dioxide and the Global Carbon Cycle," DOE/ER-0239, pp. 25–62. U.S. Department of Energy, Washington, D.C.

Ganning, B., Reish, D. J., and Straughan, D. (1984). Recovery and restoration of rocky shores, sandy beaches, tidal flats, and shallow subtidal bottoms impacted by oil spills. *In* "Restoration of Habitats Impacted by Oil Spills" (J. L. Cairns and A. L. Buikema, Jr., eds.), pp. 7–36. Butterworth, Boston, Massachusetts.

Ganther, H. E., and Sunde, M. L. (1974). Effect of tuna fish and selenium on the toxicity of methylmercury: A progress report. *J. Food Sci.* **39,** 1–5.

Ganther, H. E., Goudie, C., Sundi, M. L., Kopecky, M. J., Wagner, P., Oh, S., and Hoekstra, W. G. (1972). Selenium: Relation to decreased toxicity of methylmercury added to diets containing tuna. *Science* **175,** 1122–1124.

Gaskin, D. E., Ishida, K., and Frank, R. (1972). Mercury in harbour porpoises (*Phoecena phoecena*) in the Bay of Fundy region. *J. Fish Res. Board Can.* **29,** 1644–1646.

Gates, D. M. (1962). "Energy Exchange in the Biosphere." Harper & Row, New York.

Gates, D. M. (1985). "Energy and Ecology." Sinauer Assoc., New York.

Gatz, D. F., Barnard, W. R., and Stensland, G. J. (1986). The role of alkaline materials in precipitation chemistry: A brief review of the issues. *Water, Air, Soil Pollut.* **30,** 245–251.

Gaynor, J. D., and Halstead, R. L. (1976). Chemical and plant extractability of metals and plant growth on soils amended with sludge. *Can. J. Soil Sci.* **11,** 1194–1201.

General Accounting Office (GAO) (1981). Better data needed to determine the extent to which herbicides should be used on forest lands. GAO, Washington, D.C.

Gentry, A. H. (1986). Endemism in tropical vs. temperate plant communities. *In* "Conservation Biology" (M. E. Soule, ed.), pp. 153–181. Sinauer Assoc., Sunderland, Massachusetts.

Gentry, A. H. (1988). Tree species of upper Amazonian forests. *Proc. Natl. Acad. Sci.* **85,** 156–159.

George, J. L., Snyder, A. P., and Hanley, G. (1981). The value of the wild-bird products industry. *Trans. North Am. Wildl. Nat. Resour. Conf.* **46,** 463–471.

Gerardi, M. H., and Grimm, J. K. (1979). "The History, Biology, Damage, and Control of the Gypsy Moth." Associated University Press, London.

Gerrish, G., Mueller-Dombois, D., and Bridges, K. W. (1988). Nutrient limitation and *Metrosideros* forest dieback in Hawaii. *Ecology* **69,** 723–727.

Getter, C. D., Cintron, G., Dicks, B., Lewis, R. R., and Seneca, E. D. (1984). The recovery and restoration of salt marshes and mangroves following an oil spill. *In* "Restoration of Habitats Impacted by Oil Spills" (J. L. Cairns and A. L. Buikema, Jr., eds.), pp. 65–113. Butterworth, Boston, Massachusetts.

Gillis, G. F. (1980). Assessment of the effects of insecticide contamination in streams on the behaviour and growth of fish. *In* "Environmental Surveillance in New Brunswick" (I. W. Varty, ed.), pp. 49–50. Department of Forest Resources, University of New Brunswick, Fredericton.

Gilroy, N. T. (1983). More than just pipes in the ground. *Ambio* **12**, 245–251.

Gizyn, W. I. (1980). The chemistry and environmental impact of the bituminous shale fires at the Smoking Hills, N.W.T. M.Sc. Thesis, Department of Botany, University of Toronto, Toronto, Ontario.

Gladstone, H. S. (1919). "Birds and the War." Skeffington & Son, London.

Glass, G. E., Brydges, T. G., and Loucks, O. L., eds. (1981). "Impact Assessment of Airborne Acidic Deposition on the Aquatic Environment of the United States and Canada," EPA-600/10-81-000. U.S. Environmental Protection Agency, Duluth, Minnesota.

Glemarec, M., and Hussenot, E. (1982). Reponses des peuplements subtidaux à la perturbation creee par L'Amoco Cadiz dans les Abers Benoit et Wrac'h. *In* "Ecological Study of the Amoco Cadiz Oil Spill," pp. 191–203. U.S. Department of Commerce, National Oceanic and Atmospheric Administration, Washington, D.C.

Gochfeld, M. (1980). Mercury levels in some seabirds of the Humboldt Current, Peru. *Environ. Pollut.* **22**, 197–205.

Golding, D. L., and Swanson, R. H. (1986). Snow distribution patterns in clearings and adjacent forest. *Water Resour. Res.* **22**, 1931–1940.

Goldman, C. R. (1961). The contribution of alder trees (*Alnus tenuifolia*) to the primary productivity of Castle Lake, California. *Ecology* **42**, 282–288.

Goldman, C. R. (1965). Micronutrient limiting factors and their detection in natural phytoplankton populations. *In* "Primary Productivity in Aquatic Environments," pp. 123–135. Univ. of California Press, Berkeley.

Goldman, C. R. (1972). The role of minor nutrients in limiting the productivity of aquatic ecosystems. *In* "Nutrients and Eutrophication" (G. E. Likens, ed.), pp. 21–33. American Society of Limnology and Oceanography, Lawrence, Kansas.

Goldsmith, J. R. (1986). Effects on human health. *In* "Air Pollution" (A. C. Stern, ed.), 3rd ed., Vol. 6, pp. 391–463. Academic Press, New York.

Goldsmith, J. R., and Friberg, L. T. (1978). Effects of air pollution on human health. *In* "Air Pollution" (A. C. Stern, ed.), 3rd ed., Vol. 2, pp. 458–611. Academic Press, New York.

Goldwater, L. J., and Clarkson, T. W. (1972). Mercury. *In* "Metallic Contaminants and Human Health" (D. H. K. Lee, ed.), pp. 17–56. Academic Press, New York.

Gollop, J. B., Barry, T. W., and Iversen, E. H. (1986). "Eskimo Curlew. A Vanishing Species?" Spec. Publ. No. 17. Saskatchewan Natural History Society, Regina.

Goodison, B. E., Louie, P. Y. T., and Metcalf, J. R. (1986). Snowmelt acidic shock study in south central Ontario. *Water, Air, Soil Pollut.* **31**, 131–138.

Goodman, G. T., Pitcairn, C. E. R., and Gemmell, R. P. (1973). Ecological factors affecting growth on sites contaminated with heavy metals. *In* "Ecology and Reclamation of Devastated Land" (R. J. Hutnik and G. Davis, eds.), Vol. 2, pp. 149–171. Gordon & Breach, New York.

Gordon, A. G. (1981). Impacts of harvesting on nutrient cycling in the boreal mixedwood forest. *In* "Boreal Mixedwood Symposium" COJFRC Symp. O-P-9, pp. 121–140. Great Lakes Forest Research Centre, Sault Ste. Marie, Ontario.

Gordon, A. G. (1985). Budworm! What about the forest. *In* "Spruce-Fir Management and Spruce Budworm," Gen. Techn. Rep. NE-99, pp. 3–29. Northeastern Forest Experiment Station, Broomall, Pennsylvania.

Gordon, A. G., and Gorham, E. (1963). Ecological aspects of air pollution from an iron-sintering plant at Wawa, Ontario. *Can. J. Bot.* **41**, 1063–1078.

Gorham, E. (1976). Acid precipitation and its influence upon aquatic ecosystems—an overview. *USDA For. Serv. Gen. Tech. Rep.* NE-23, 425–458. Northeastern Forest Experiment Station, Broomall, Pennsylvania.

Gorham, E., and Gordon, A. G. (1960a). Some effects of smelter pollution northeast of Falconbridge, Ontario. *Can. J. Bot.* **38**, 307–312.

Gorham, E., and Gordon, A. G. (1960b). The influence of smelter fumes upon the chemical composition of lake waters near Sudbury, Ontario and upon the surrounding vegetation. *Can. J. Bot.* **38**, 477–487.

Gorham, E., and Gordon, A. G. (1963). Some effects of smelter pollution upon aquatic vegetation near Sudbury, Ontario. *Can. J. Bot.* **41**, 371–378.

Gorham, E., Bayley, S. E., and Schindler, D. W. (1984). Ecological effects of acid deposition upon peatlands—a neglected field in "acid rain" research. *Can. J. Fish. Aquat. Sci.* **41**, 1256–1268.

Gosner, K. L., and Black, I. H. (1957). The effects of acidity on the development and hatching of New Jersey frogs. *Ecology* **38**, 256–262.

Gould, S. J., and Eldredge, N. (1980). Punctuated equilibria: The tempo and mode of evolution reconsidered. *Paleobiology* **3**, 115–151.

Grahn, O. (1977). Macrophyte succession in Swedish lakes caused by deposition of airborne acid substances. *Water, Air, Soil Pollut.* **7**, 295–305.

Grahn, O., Hultberg, H., and Landner, L. (1974). Oligotrophication—a self-accelerating process in lakes subjected to excessive supply of acid substances. *Ambio* **3**, 93–94.

Grainger, A. (1983). Improving the monitoring of deforestation in the humid tropics. *In* "Tropical Rain Forest: Ecology and Management" (S. L. Sutton, T. C. Whitmore, and A. C.

Chadwick, eds.), pp. 387–395. Blackwell, Boston, Massachusetts.

Granat, L., and Rodhe, H. (1973). A study of fallout by precipitation around an oil-fired power plant. *Atmos. Environ.* **7**, 781–792.

Graves, H. S. (1918). Effect of the war on forests of France. *Am. For.* **24**, 707–717.

Gregor, D. J., and Johnson, M. G. (1980). Nonpoint source phosphorus inputs to the Great Lakes. *In* "Phosphorus Management Strategies for Lakes" (R. C. Loehr, C. S. Martin, and W. Rast, eds.), pp. 37–59. Ann Arbor Sci. Publ., Ann Arbor, Michigan.

Gregory, R. P. G., and Bradshaw, A. O. (1965). Heavy metal tolerance in populations of *Agrostis tenuis* and other grasses. *New Phytol.* **69**, 131–143.

Grennfelt, P., and Hultberg, H. (1986). Effects of nitrogen deposition on the acidification of terrestrial and aquatic ecosystems. *Water, Air, Soil Pollut.* **30**, 945–965.

Grennfelt, P., and Schjoldager, J. (1984). Photochemical oxidants in the troposphere: A mounting menace. *Ambio* **13**, 61–67.

Grennfelt, P., Bengston, C., and Skarby, L. (1980). An estimation of the atmospheric input of acidifying substances to a forest ecosystem. *In* "Effects of Acid Precipitation on Terrestrial Ecosystems" (T. C. Hutchinson and M. Havas, eds.), pp. 29–40. Plenum, New York.

Grier, C. C. (1975). Wildfire effects on nutrient distribution and leaching in a coniferous ecosystem. *Can. J. For. Res.* **5**, 599–607.

Grier, J. W. (1982). Ban of DDT and subsequent recovery of reproduction in bald eagles. *Science* **218**, 1232–1235.

Grime, J. P. (1979). "Plant Strategies and Vegetation Processes." Wiley, Toronto, Ontario.

Grime, J. P. (1986). Predictions of terrestrial vegetation responses to nuclear winter conditions. *Int. J. Environ. Stud.* **28**, 11–19.

Groombridge, B. (1982). "The IUCN Amphibia-Reptilia Red Data Book." International Union for Conservation of Nature and Natural Resources, Gland, Switzerland.

Grose, P. L., and Matton, J. S. (1977). "The Argo Merchant Oil Spill: A Preliminary Scientific Report." U.S. Department of Commerce, National Oceanic and Atmospheric Administration, Washington, D.C.

Grover, H. D., and Harwell, M. A. (1985). Biological effects of nuclear war. II. Impact on the biosphere. *BioScience* **35**, 576–583.

Grover, H. D., and White, G. F. (1985). Toward understanding the effects of nuclear war. *Bioscience* **35**, 552–556.

Grue, C. E., Fleming, W. J., Busby, D. G., and Hill, E. F. (1983). Assessing hazards of organophosphate pesticides to wildlife. *Trans. Wild. Nat. Resour. Conf.* **48**, 200–220.

Grzimek, B. (1972). "Grzimek's Animal Life Encyclopedia." Van Nostrand-Reinhold, Toronto, Ontario.

Guicherit, R., and van den Hout, D. (1982). The global NO_x cycle. *In* "Air Pollution by Nitrogen Oxides" (T. Schneider and L. Gront, eds.), pp. 15–29. Elsevier, Amsterdam.

Gullion, G. W. (1967). Selection and use of drumming sites by male ruffed grouse. *Auk* **84**, 87–112.

Gullion, G. W. (1969). "Aspen-Ruffed Grouse Relationships." Presented at 31st Midwest Wildlife Conference, Dec. 8, 1969, St. Paul, Minnesota (cited in Johnsgard, 1983)

Gullion, G. W. (1977). Forest manipulation for ruffed grouse. *Trans. North Am. Wildl. Nat. Resour. Conf.* **42**, 449–458.

Gullion, G. W. (1986). "Northern Forest Management for Wildlife," For. Ind. Lect. No. 17. Faculty of Forestry, University of Alberta, Edmonton.

Gutierrez, R. J., and Carey, A. B., eds. (1985). "Ecology and Management of the Spotted Owl in the Pacific Northwest," U.S.D.A. For. Serv., Gen. Tech. Rep. PNW-185. Pacific Northwest Forest and Range Experiment Station, Portland, Oregon.

Haagen-Smit, A. J., and Wayne, L. G. (1976). Atmospheric reactions and scavenging processes. *In* "Air Pollution" (A. C. Stern, ed.), 3rd ed., Vol. 1, pp. 235–288. Academic Press, New York.

Haapala, H., Seepuren, E., and Meskus, E. (1975). Effects of spring floods on water acidity in the Kiiminkijoki area, Finland. *Oikos* **26**, 26–31.

Haefner, J. D., and Wallace, J. B. (1981). Shifts in aquatic insect populations in a first-order southern Appalachian stream following a decade of old field succession. *Can. J. Fish. Aquat. Sci.* **38**, 353–359.

Haertl, L. (1976). Nutrient limitation of algal standing crops in shallow prairie lakes. *Ecology* **57**, 664–678.

Haffer, J. (1969). Speciation in Amazonian forest birds. *Science* **165**, 131–137.

Haffer, J. (1982). General aspects of the refuge theory. *In* "Biological Diversification in the Tropics" (G. T. Prance, ed.), pp. 6–24. Columbia Univ. Press, New York.

Haines, T. A., and Baker, J. P. (1986). Evidence of fish population responses to acidification in the eastern United States. *Water, Air, Soil Pollut.* **31**, 605–629.

Hall, H. A., Eidt, D. C., Symons, P. E. K., and Banks, D. (1975). Biological consequences in streams of aerial spraying with fenitrothion against spruce budworm in New Brunswick. *Water Pollut. Res. Can.* pp. 84–88.

Hall, J. D., and Lantz, R. L. (1969). Effects of logging on the habitat of coho salmon and cutthroat trout in coastal streams. *In* "Symposium on Salmon and Trout in Streams" (T. G. Northcotte, ed.), pp. 355–375. University of British Columbia, Vancouver.

Hall, R. J., and Ide, F. P. (1987). Evidence of acidification on stream insect communities in central Ontario between 1937 and 1985. *Can. J. Fish. Aquat. Sci.* **44**, 1652–1657.

Hall, R. J., and Likens, G. E. (1980). Ecological effects of experimental acidification on a stream ecosystem. *In* "Ecological Impact of Acid Precipitation" (D. Drablos and A. Tollan, eds.), pp. 375–376. SNSF Project, Oslo, Norway.

Hall, R. J., and Likens, G. E. (1984). Effect of discharge rate on biotic and abiotic chemical flux in an acidifed stream. *Can. J. Fish. Aquat. Sci.* **41,** 1132–1138.

Hallbacken, L., and Tamm, C. O. (1985). Changes in soil acidity from 1927 to 1982–84 in a forest area of south-west Sweden. *Soil Sci. Soc. Am. J.* **49,** 1280–1282.

Halls, L. K. (1978). Effect of timber harvesting on wildlife, wildlife habitat, and recreation values. *In* "Complete Tree Utilization of Southern Pine," Proc. Symp., pp. 108–114. Forest Products Research Laboratory, Madison, Wisconsin.

Halls, L. K., and Epps, E. A. (1969). Browse quality influenced by tree overstory in the south. *J. Wildl. Manage.* **32,** 1028–1031.

Hamburg, S. P., and Cogbill, C. V. (1988). Historical decline of red spruce populations and climatic change. *Nature (London)* **331,** 428–431.

Hammer, U. T. (1969). Blue-green algal blooms in Saskatchewan lakes. *Verh.—Int. Ver. Theor. Angerv. Limnol.* **17,** 116–125.

Hanley, P. T., Hemming, J. E., Morsell, J. W., Morehouse, T. A., Leask, L. E., and Harrison, G. S. (1981). "Natural Resource Protection and Petroleum Development in Alaska," Publ. FWS/OBS-80/22. Fish and Wildlife Service, U.S. Department of the Interior, Washington, D.C.

Hansen, J., and Lebedeff, S. (1987). Global trends of measured surface air temperature. *J. Geophys. Res.* **82,** 13,345–13,372.

Hansen, J., and Lebedeff, S. (1988). Global surface air temperatures: update through 1987. *Geophys. Res. Lett.* **15,** 323–326.

Hargreaves, J. W., and Whitten, B. A. (1976). Effect of pH on growth of acid stream algae. *Br. Phycol. J.* **11,** 215–223.

Hargreaves, J. W., Lloyd, E. J. H., and Whitten, B. A. (1975). Chemistry and vegetation of highly acidic streams. *Freshwater Biol.* **5,** 563–576.

Harlow, R. F., and Palmer, Z. F. (1967). Clearcutting in coordinated deer-timber management in the southern Appalachians. *Wildl. N. C.* **31,** 14–15.

Harrington, J. B. (1987). Climatic change: A review of causes. *Can. J. For. Res.* **17,** 1313–1339.

Harris, G. P., and Vollenweider, R. A. (1982). Palaeolimnological evidence of early eutrophication in Lake Erie. *Can. J. Fish. Aquat. Sci.* **39,** 618–626.

Harris, L. D. (1984). "The Fragmented Forest." Univ. of Chicago Press, Chicago, Illinois.

Harris, M. M., and Jurgensen, M. F. (1977). Development of *Salix* and *Populus* mycorrhizae in metallic mine tailings. *Plant Soil* **47,** 509–517.

Harrison, A. D. (1958). The effects of sulfuric acid pollution on the biology of streams in the Transvaal, South Africa. *Verh.—Int. Ver. Theor. Angerv. Limnol.* **13,** 603–610.

Hartman, W. L. (1972). Lake Erie: effects of exploitation, environmental changes and new species on the fishery resources. *J. Fish. Res. Board Can.* **29,** 899–912.

Hartman, W. L. (1973). "Effects of Exploitation, Environmental Changes, and New Species on the Fish Habitats and Resources of Lake Erie," Tech. Rep. No. 22. Great Lakes Fisheries Commission, Ann Arbor, Michigan.

Hartshorn, G. S. (1983). Wildlands conservation in Central America. *In* "Tropical Rain Forest: Ecology and Management" (S. L. Sutton, T. C. Whitmore, and A. C. Chadwick, eds.), pp. 423–444. Blackwell, Boston, Massachusetts.

Harvey, A. E., Jurgensen, M. F., and Larsen, M. J. (1976). Intensive fibre utilization and prescribed fire: Effects on the microbial ecology of forests. *USDA For. Serv. Gen. Techn. Rep.* INT-28. Intermountain Forest and Range Experiment Station, Ogden, Utah.

Harvey, A. E., Jurgensen, M. F., and Larsen, M. J. (1980). Biological implications of increasing harvest intensity on the maintenance of productivity of forest soils. *In* "Environmental Consequences of Timber Harvesting in Rocky Mountain Coniferous Forests," U.S.D.A. For. Serv., Gen. Tech. Rep. INT-90, pp. 211–220. Intermountain Forest and Range Experiment Station. Ogden, Utah.

Harvey, H. H., and Lee, C. (1982). Historical fisheries changes related to surface water pH changes in Canada. *Acid. Rain/Fish., Proc. Int. Symp., 1981* pp. 45–55.

Harvey, H. H., Pierce, R. C., Dillon, P. J., Kramer, J. P., and Whelpdale, D. M. (1981). "Acidification of the Canadian Environment: Scientific Criteria for Assessment of the Effects of Acidic Deposition on Aquatic Ecosystems," Rep. No. 18475. National Research Council of Canada, Ottawa, Ontario.

Harwell, M. A., and Grover, H. D. (1985). Biological effects of nuclear war. I. Impact on humans. *BioScience* **35,** 570–575.

Harwell, M. A., and Hutchinson, T. C. (1985). "Environmental Consequences of Nuclear War," Vol. 2, Scope Rep. 28. Wiley, Toronto, Ontario.

Hasler, A. D., Brynildson, O. M., and Helm, W. T. (1951). Improving conditions for fish in brown-water bog lakes by alkalization. *J. Wildl. Manage.* **15,** 347–352.

Hatcher, J. D., and White, F. M. M. (1984). "Task Force on Chemicals in the Environment and Human Reproductive Problems in New Brunswick," Report to Department of Health, Province of New Brunswick. Faculty of Medicine, Dalhousie University, Halifax, Nova Scotia.

Hauhs, M., and Wright, R. F. (1986). Regional pattern of acid deposition and forest decline along a cross section through Europe. *Water, Air, Soil Pollut.* **31,** 463–474.

Haupt, H. F., and Kidd, W. J. (1965). Good logging practices reduce sedimentation in central Ohio. *J. For.* **63,** 664–670.

Hausenbuiller, R. L. (1978). "Soil Science." Wm. C. Brown, Dubuque, Iowa.

Havas, M. (1986). Aluminum in the aquatic environment. *In* "Aluminum in the Canadian Environment," pp. 79–126. National Research Council of Canada, Associate Committee on Scientific Criteria for Environmental Quality, Ottawa, Ontario.

Havas, M., and Hutchinson, T. C. (1982). Aquatic invertebrates from the Smoking Hills, N.W.T.: Effect of pH and metals on mortality. *Can. J. Fish. Aquat. Sci.* **39,** 890–893.

Havas, M., and Hutchinson, T. C. (1983a). The Smoking Hills: Natural acidification of an aquatic ecosystem. *Nature (London)* **301,** 23–27.

Havas, M., and Hutchinson, T. C. (1983b). Effect of low pH on the chemical composition of aquatic invertebrates from tundra ponds at the Smoking Hills, N.W.T., Canada. *Can. J. Zool.* **61,** 241–249.

Havas, M., Hutchinson, T. C., and Likens, G. E. (1984a). Effect of low pH on sodium regulation in two species of *Daphnia. Can. J. Zool.* **62,** 1965–1970.

Havas, M., Hutchinson, T. C., and Likens, G. E. (1984b). Red herrings in acid rain research. *Environ. Sci. Technol.* **18,** 176A–186A.

Hawksworth, D. L., and Rose, F. (1976). "Lichens as Air Pollution Monitors," Stud. Biol. No. 66. Institute of Biology, Edward Arnold, London.

Hay, A. (1982). "The Chemical Scythe: Lessons of 2,4,5-T and Dioxin." Plenum, New York.

Heath, R. T. (1986). Dissolved organic phosphorus compounds: Do they satisfy planktonic phosphate demands in summer? *Can. J. Fish. Aquat. Sci.* **43,** 343–350.

Heck, W. W., and Brandt, C. S. (1978). Effects on vegetation: Native, crops, forest. *In* "Air Pollution" (A. C. Stern, ed.), 3rd ed., Vol. 2, pp. 158–231. Academic Press, New York.

Heck, W. W., Taylor, O. C., Adams, R., Bingham, G., Preston, E., and Weinstein, L. (1982). Assessment of crop loss from ozone. *J. Air Pollut. Control Assoc.* **32,** 353–361.

Heck, W. W., Adams, R. M., Cure, W. W., Heagle, A. S., Heggestad, H. E., Kohut, R. J., Kress, L. W., Rawlings, J. O., and Taylor, O. C. (1983). A reassessment of crop loss from ozone. *Environ. Sci. Technol.* **17,** 573A–581A.

Heck, W. W., Heagle, A. S., and Shriner, D. S. (1986). Effects on vegetation: Native, crops, and forests. *In* "Air Pollution" (A. C. Stern, ed.), 3rd ed., Vol. 6, Academic Press, New York.

Heggestad, H. E. (1980). "Field Assessment of Air Pollution Impacts on the Growth and Productivity of Crop Species." Presented at Air Pollution Control Association annual meeting, Montreal (cited in Roberts, 1984).

Heinrichs, H., and Mayer, R. (1977). Distribution and cycling of major and trace elements in two central European forest ecosystems. *J. Environ. Qual.* **6,** 402–406.

Hellquist, C. B., and Crow, G. E. (1980). "Aquatic Vascular Plants of New England," Part 1, Stn. Bull. 515. New Hampshire Agricultural Experiment Station, Durham.

Henderson, J. W. (1978). "The Effects of Forest Operations on the Water Resources of the Shubenacadie-Stewiacke River Basin," Tech. Rep. No. 11, Shubenacadie-Stewiacke River Basin Board, Truro, Nova Scotia.

Henderson-Sellers, B. (1984). "Pollution of Our Atmosphere." Adam Hilger, Techno House, Bristol, U.K.

Hendry, G. R., and Vertucci, F. (1980). Benthic plant communities in acidic Lake Colden, New York: *Sphagnum* and the algal mat. *In* "Ecological Impact of Acid Precipitation" (D. Drablos and A. Tollan, eds.), pp. 314–315. SNSF Project, Oslo, Norway.

Hendry, G. R., and Wright, R. F. (1976). Acid precipitation in Norway: Effects on aquatic fauna. *J. Great Lakes Res.* **2,** Suppl., 192–207.

Hendry, G. R., Yan, N. D., and Baumgartner, B. J. (1980a). Responses of freshwater plants and invertebrates to acidification. *In* "Restoration of Lakes and Inland Rivers," EPA 440/5-81/010, pp. 457–466. U.S. Environmental Protection Agency, Washington, D.C.

Hendry, G. R., Galloway, J. N., Norton, S. A., Schofield, C. L., Schoffer, P. W., and Burns, D. A. (1980b). "Geological and Hydrochemical Sensitivity of the Eastern United States to Acid Precipitation," EPA-600/3-80-024. U.S. Environmental Protection Agency, Environmental Research Laboratory, Corvallis, Oregon.

Henriksen, A. (1980). Acidification of freshwaters—a large scale titration. *In* "Ecological Impacts of Acid Precipitation" (D. Drablos and A. Tollan, eds.), pp. 68–74. SNSF Project, Oslo, Norway.

Henriksen, A. (1982). Susceptibility of surface waters to acidification. *Acid. Rain/Fish., Proc. Int. Symp., 1981* pp. 103–121.

Henry, G. H. R., Freedman, B., and Svoboda, J. (1986a). Effects of fertilization on three tundra plant communities of a polar desert oasis. *Can. J. Bot.* **64,** 2502–2507.

Henry, G. H. R., Freedman, B., and Svoboda, J. (1986b). Survey of vegetated areas and muskox populations in east-central Ellesmere Island. *Arctic* **39,** 78–81.

Hepting, G. H. (1971). "Diseases of Forest and Shade Trees of the United States," Agric. Handb. No. 386. U.S. Department of Agriculture, Forest Service, Washington, D.C.

Hermens, J., Konemann, H., Leeuwangh, P., and Mursch, A. (1985). Quantitative structure-activity relationships in aquatic toxicity studies and complex mixtures of chemicals. *Environ. Toxicol. Chem.* **4,** 273–279.

Heron, J. (1961). The seasonal variation of phosphate, silicate, and nitrate in waters of the English Lake District. *Limnol. Oceanogr.* **6,** 338–346.

Hesser, R., Hooper, R., Weirich, C. B., Selcher, J., Hollender, B., and Snyder, R. (1975). The aquatic biota. *In* "Clearcutting in Pennsylvania," pp. 9–20. Pennsylvania State School of Forest Resources, University Park.

Hetherington, E. D. (1976). "Dennis Creek: A Look at Water Quality Following Logging in the Okanagan Basin," Rep. BC-X-147. Pacific Forest Research Centre, Victoria, British Columbia.

Hewlett, J. F., and Helvey, J. D. (1970). Effects of forest clear-felling on the storm hydrograph. *Water Resour. Res.* **6,** 768–782.

Heyerdahl, T. (1971). Atlantic Ocean pollution and biota observed by the "Ra" expeditions. *Biol. Conserv.* **3,** 164–167.

Hibber, C. R. (1964). Identity and significance of certain organisms associated with sugar maple decline in New York woodlands. *Phytopathology* **74,** 1389–1392.

Hibbert, A. R. (1967). Forest treatment effects on water yield. *In* "International Symposium on Forest Hydrology Proceedings" (W. E. Sopper and H. W. Hull, eds.), pp. 527–543. Pergamon, New York.

Hickey, J. J., ed. (1969). "Peregrine Falcon Populations: Their Biology and Decline." Univ. of Wisconsin Press, Madison.

Hickey, J. J., and Anderson, D. W. (1968). Chlorinated hydrocarbons and eggshell changes in raptorial and fish-eating birds. *Science* **162,** 271–273.

Hiep, D. (1984). Long-term changes in the mangrove habitat following herbicidal attack. *In* "Herbicides in War: The Long-term Ecological and Human Consequences" (A. H. Westing, ed.), pp. 89–90. Taylor & Francis, London.

Hileman, B. (1983). Acid fog. *Environ. Sci. Technol.* **17,** 117A–120A.

Hinrichsen, D. (1986). Multiple pollutants and forest decline. *Ambio* **15,** 258–265.

Hinrichsen, D. (1987). The forest decline enigma. *BioScience* **37,** 542–546.

Hobaek, A., and Raddum, G. G. (1980). "Zooplankton Communities in Acidified Lakes in South Norway," IR 75/80. SNSF Project, Oslo, Norway.

Hodges, C. S., Adee, K. T., Stein, J. D., Wood, H. B., and Doty, R. D. (1986). Decline of Ohia (*Metrosideros polymorpha*) in Hawaii: A review. *USDA For. Serv. Gen. Tech. Rep. PSW-86.* Pacific Southwest Forest and Range Experiment Station, Berkeley, California.

Hoffman, D. J., Harder, J. W., Rolf, R. S., and Rosen, J. M. (1987). Balloon-borne observations of the development and vertical structure of the Antarctic ozone hole in 1986. *Nature (London)* **326,** 59–62.

Hogan, G. D., Courtin, G. M., and Rauser, W. E. (1977a). The effects of soil factors on the distribution of *Agrostis gigantea* on a mine waste site. *Can. J. Bot.* **55,** 1038–1042.

Hogan, G. D., Courtin, G. M., and Rauser, W. E. (1977b). Copper tolerance in clones of *Agrostis gigantea* from a mine waste site. *Can. J. Bot.* **55,** 1043–1050.

Hoggan, M. C., Davidson, A., Brunelle, M. F., Neuitt, J. S., and Gins, J. D. (1978). Motor vehicle emissions and atmospheric lead concentrations in the Los Angeles area. *J. Air Pollut. Control Assoc.* **28,** 1200–1206.

Hokenson, K. E. F., McCormick, J. H., Jones, B. R., and Tucker, J. H. (1973). Thermal requirements for maturation, spawning, and embryo survival of the brook trout (*Salvelinus fontinalis*). *J. Fish. Res. Board Can.* **30,** 975–984.

Holling, C. S., and Walters, C. J. (1977). Fenitrothion or not fenitrothion: That is not the question. *Natl. Res. Counc. Can., NRC Assoc. Comm. Sci. Criter. Environ. Qual. [Rep.] NRCC, NRCC/CNRC* **16073,** 279–297.

Holloway, G. T. (1917). "Report of the Ontario Nickel Commission." Legislative Assembly of Ontario, Toronto.

Holmes, S. B. (1979). "Aquatic Impact Studies of a Spruce Budworm Control Program in the Lower St. Lawrence Region of Quebec in 1978," Rep. FPM-X-26. Forest Pest Management Institute, Canadian Forestry Service, Sault Ste. Marie, Ontario.

Holmes, W. N. (1984). Petroleum pollutants in the marine environment and their possible effects on seabirds. *Rev. Environ. Toxicol.* **1,** 251–317.

Holmes, W. N., and Cronshaw, J. (1977). Biological effects of petroleum on marine birds. *In* "Effects of Petroleum on Arctic and Subarctic Marine Environments" (D. C. Malins, ed.), Vol. 2, pp. 359–398. Academic Press, New York.

Holtby, L. B. (1988). Effects of logging on stream temperatures in Carnation Creek, British Columbia, and associated impacts on the coho salmon (*Onchorhynchus kisutch*). *Can. J. Fish. Aquat. Sci.* **45,** 502–515.

Honer, T. G., and Bickerstaff, A. (1985). "Canada's Forest Area and Wood Volume Balance 1977–1981," BC-X-272. Canadian Forestry Service, Pacific Forestry Centre, Victoria, British Columbia.

Hong, P. N. (1984a). Characteristics of mangroves in the region of Mekong River mouths. *In* "The First Vietnam National Symposium on Mangrove Ecosystems, 27–28 Dec., 1984," pp. 55–69. Hanoi, Vietnam.

Hong, P. N. (1984b). Effects of chemical warfare on mangrove forests on tip of the Canau Peninsula, Minhhai Province. *In* "The First Vietnam National Symposium on Mangrove Ecosystems, 27–28 Dec., 1984," pp. 163–174. Hanoi, Vietnam.

Hong, P. N. (1987). "Mangrove Ecology in Viet Nam." Presented at the International Conference on Ecology in Viet Nam, May 28–30, 1987, New Paltz, New York.

Horn, M. H., Teal, J. M., and Backus, R. H. (1970). Petroleum lumps on the surface of the sea. *Science* **168,** 245–246.

Hornbeck, J. W. (1973). Storm flow from hardwood-forested and cleared watersheds in New Hampshire. *Water Resour. Res.* **9,** 346–354.

Hornbeck, J. W. (1977). Nutrients: A major consideration for intensive forest management. *USDA For. Serv. Gen. Tech. Rep.* NE-29, 241–250. Northeastern Forest Experiment Station, Broomall, Pennsyvlania.

Hornbeck, J. W., and Ursic, S. J. (1979). Intensive harvest and forest streams: Are they compatible? *In* "Impacts of Intensive Harvesting on Forest Nutrient Cycling," pp. 249–262. College of Environmental Science and Forestry, State University of New York, Syracuse, New York.

Hornbeck, J. W., Likens, G. E., Pierce, R. S., and Bormann, F. H. (1975). Strip-cutting as a means of protecting site and streamflow quality when clear-cutting northern hardwoods. *In* "Forest Soils and Forest Land Management" (B. Bernier and C. W. Winget, eds.), pp. 209–225. Laval Univ. Press, Quebec.

Hornbeck, J. W., Smith, R. B., and Federer, C. A. (1986). Growth decline in red spruce and balsam fir relative to natural processes. *Water, Air, Soil Pollut.* **31,** 425–430.

Hornbeck, J. W., Martin, C. W., Pierce, R. S., Bormann, F. H., Likens, G. E., and Eaton, J. S. (1987a). "The Northern Hardwood Forest Ecosystem: Ten Years of Recovery from Clearcutting," Publ. NE-RP-596. U.S.D.A. For. Serv., Northeastern Forest Experiment Station, Broomall, Pennsylvania.

Hornbeck, J. W., Smith, R. B., and Federer, C. A. (1987b). Extended growth decreases in New England are limited to red spruce and balsam fir. *In* "Proceedings of the International Symposium on Ecological Aspects of Tree-ring Analysis," CONF-8608144, pp. 38–44. U.S. Dept. of Commerce, Springfield, Virginia.

Hosker, R. P., and Lindberg, S. E. (1982). Review: Atmospheric deposition and plant assimilation of gases and particles. *Atmos. Environ.* **16,** 889–910.

Houghton, H. G. (1955). On the chemical composition of fog and cloud water. *J. Meteorol.* **12,** 355–357.

Houghton, R. A., Hobbie, J. E., Melillo, J. M., Moore, B., Peterson, B. J., Shaver, G. R., and Woodwell, G. M. (1983). Changes in the carbon content of terrestrial biota and soils between 1860 and 1980: A net release of CO_2 to the atmosphere. *Ecol. Monogr.* **53,** 235–262.

Houghton, R. A., Schlesinger, W. H., Brown, S., and Richards, J. F. (1985). Carbon dioxide exchange between the atmosphere and terrestrial ecosystems. *In* "Atmospheric Carbon Dioxide and the Global Carbon Cycle," DOE/ER-0239, pp. 113–140. U.S. Department of Energy, Washington, D.C.

Houston, D. B. (1974). Response of selected *Pinus strobus* L. clones to fumigations with sulfur dioxide and ozone. *Can. J. For. Res.* **4,** 65–68.

Houston, D. B., and Stairs, G. R. (1973). Genetic control of sulfur dioxide and ozone tolerance in eastern white pine. *For. Sci.* **19,** 267–271.

Hov, O. (1984). Ozone in the troposphere: High level pollution. *Ambio* **13,** 73–79.

Howell, R. K., Koch, E. J., and Rose, L. (1979). Field assessment of air pollution induced soybean yield losses. *Agron. J.* **71,** 285–288.

Howitt, R. E., Gossard, T. W., and Adams, R. M. (1984). Effects of alternative ozone concentrations and response data on economic assessments: the case of California crops. *J. Air Pollut. Control Assoc.* **34,** 1122–1127.

Hubbell, S. P., and Foster, R. B. (1983). Diversity of canopy trees in a neotropical forest and implications for conservation. *In* "Tropical Rain Forest: Ecology and Management" (S. L. Sutton, T. C. Whitmore, and A. C. Chadwick, eds.), pp. 25–41. Blackwell, Boston, Massachusetts.

Hubendick, B. (1987). Tropical diseases and human ecology. *Ambio.* **16,** 218–219.

Huettl, R. F. (1986a). "New type" of forest decline and diagnostic fertilization. *Int. Conf. Environ. Contam. 2nd, 1986.*

Huettl, R. F. (1986b). "Forest Decline and Nutritional Disturbances." IUFRO World Congress, Ljubjana, Yugoslavia.

Huettl, R. F., and Wisniewski, J. (1987). Fertilization as a tool to mitigate forest decline. (Manuscript).

Huey, N. A. (1968). The lead dioxide estimation of sulfur dioxide pollution. *J. Air Pollut. Control Assoc.* **18,** 610–611.

Huhn, F. J. (1974). Lake sediment records of industrialization in the Sudbury area of Ontario. M.Sc. Thesis, Department of Botany, University of Toronto, Toronto, Ontario.

Hulme, M. A., Ennis, T. J., and Lavallee, A. (1983). Current status of *Bacillus thuringiensis* for spruce budworm control. *For. Chron.* **59,** 58–61.

Hultberg, H., and Grahn, O. (1975a). Effects of acidic precipitation on macrophytes in oligotrophic Swedish lakes. *In* "Atmospheric Contribution to the Chemistry of Lake Waters," pp. 208–221. Int. Assoc. Great Lakes Res.

Hultberg, H., and Grahn, O. (1975b). "Some Effects of Adding Lime to Lakes in Western Sweden," Transl. Ser. No. 3607. Department of the Environment, Fisheries and Marine Services, Ottawa, Ontario.

Hultberg, H., and Grennfelt, P. (1986). Gardsjon project: Lake acidification, chemistry in catchment runoff, lake liming and microcatchment manipulations. *Water, Air, Soil Pollut.* **30,** 31–46.

Hunt, E. G., and Bischoff, A. I. (1960). Inimical effects on wildlife of periodic DDD applications to Clear Lake. *Calif. Fish Game* **46,** 91–106.

Hunter, J. G., and Vergnano, O. (1952). Nickel toxicity in plants. *Ann. Appl. Biol.* **39,** 279–284.

Hunter, M. L., Jones, J. J., Gibbs, K. E., Moring, J. R., and Brett, M. (1985). "Interactions Among Waterfowl, Fishes, Invertebrates, and Macrophytes in Four Maine Lakes of Different Acidity," Biol. Rep. No. 80. Eastern Energy and Land Use Team, U.S. Fish and Wildlife Service, Washington, D.C.

Huntsman, A. G. (1942). Death of salmon and trout with high temperature. *J. Fish. Res. Board Can.* **5,** 485–501.

Hutcheson, M. R., and Hall, F. P. (1974). Sulphate washout from a coal-fired power plant plume. *Atmos. Environ.* **8,** 23–28.

Hutchinson, G. E. (1957). "A Treatise on Limnology," Vol. 1. Wiley, New York.

Hutchinson, G. E. (1969). Eutrophication, past and present. *In* "Eutrophication: Causes, Consequences, Correctives," pp. 17–26. National Academy of Sciences, Washington, D.C.

Hutchinson, G. E. (1975). "A Treatise on Limnology," Vol. 3. Wiley, New York.

Hutchinson, T. C. (1967). Comparative studies of the ability of species to withstand prolonged periods of darkness. *J. Ecol.* **55,** 291–299.

Hutchinson, T. C. (1972). "The Occurrence of Lead, Cadmium, Nickel, Vanadium, and Chloride in Soils and Vegetation of Toronto in Relation to Traffic Density," Publ. EH-2. Institute for Environmental Studies, University of Toronto, Toronto, Ontario.

Hutchinson, T. C., and Freedman, B. (1978). Effects of experimental crude oil spills on subarctic boreal forest vegetation near Norman Wells, N.W.T., Canada. *Can. J. Bot.* **56,** 2424–2433.

Hutchinson, T. C., and Havas, M. (1985). Recovery of previously acidified lakes near Coniston, Canada following reductions in atmospheric sulfur and metal emissions. *Water, Air, Soil Pollut.* **20,** 20–32.

Hutchinson, T. C., and Whitby, L. M. (1974). Heavy metal pollution in the Sudbury mining and smelting region of

Canada. 1. Soil and vegetation contamination by nickel, copper, and other metals. *Environ. Conserv.* **1,** 123–132.

Hutchinson, T. C., Gizyn, W., Havas, M., and Zobens, V. (1978). Effect of long-term lignite burns on arctic ecosystems at the Smoking Hills, N.W.T. *Trace Subst. Environ. Health* **12,** 317–332.

Hutchinson, T. C., Hellebust, J. A., Mackay, D., Tam, D., and Kauss, P. (1979). Relationships of hydrocarbon solubility to toxicity in algae and cellular membrane effects. *API Publ.* **4308,** 541–547.

Hutchinson, T. C., Nakatsu, C., and Tam, D. (1981). Multiple metal tolerances and co-tolerances in algae. *Heavy Met. Environ., Int. Conf. 3rd, 1981,* pp. 300–304.

Hutchinson, T. C., Bozic, L., and Munoz-Vega, G. (1986). Responses of five species of conifer seedlings to aluminum stress. *Water, Air, Soil Pollut.* **31,** 283–294.

Hutton, M. (1980). Metal contamination of feral pigeons *Columba livia* from the London area. Biological effects of lead exposure. *Environ Pollut. Ser. A* **22,** 281–293.

Hutton, M. (1984). Impact of airborne metal contamination on a deciduous woodland system. *In* "Effects of Pollutants at the Ecosystem Level" (P. J. Sheehan, D. R. Miller, G. C. Butler, and P. Bourdeau, eds.), Scope Rep. 22, pp. 365–375. Wiley, New York.

Hutton, M., and Goodman, G. T. (1980). Metal contamination of feral pigeons *Columba livia* from the London area. Part 1. Tissue accumulation of lead, cadmium, and zinc. *Environ. Pollut.* **22,** 207–217.

Huynh, D. H., Can, D. N., Anh, Q., and Thang, N. V. (1984). Long-term changes in the mammalian fauna following herbicidal attack. *In* "Herbicides in War: The Long-term Ecological Consequences" (A. H. Westing, ed.), pp. 49–52. Taylor & Francis, London.

Hynes, H. B. M. (1971). "The Biology of Polluted Waters." Univ. of Toronto Press, Toronto, Ontario.

Illmavirta, V. (1980). Phytoplankton in 35 Finnish brown-water lakes of different trophic status. *Dev. Hydrobiol.* **3,** 121–130.

Imai, M., Yoshida, K., Kotchmar, D. J., and Lee, S. D. (1985). A survey of health effects studies of photochemical air pollution in Japan. *J. Air Pollut. Control Assoc.* **35,** 103–108.

International Electric Research Exchange (IERE) (1981). "Effects of SO_2 and Its Derivatives on Health and Ecology," Vol. 1. Canadian Electrical Association, Montreal.

International Union for the Conservation of Nature and Natural Resources (IUCN) (1980). "World Conservation Strategy: Living Resource Conservation for Sustainable Development." IUCN, Gland, Switzerland.

Inving, P. M. (1983). Acidic precipitation effects on crops: A review and analysis of research. *J. Environ. Qual.* **12,** 442–453.

Isidorov, V. A., Zenkevich, I. G., and Ioffe, B. V. (1985). Volatile organic compounds in the atmosphere of forests. *Atmos. Environ.* **19,** 1–8.

Jacobson, J. (1980). The influence of rainfall composition on the yield and quality of agricultural crops. *In* "Ecological Impacts of Acid Precipitation" (D. Drablos and A. Tollan, eds.), pp. 41–46. SNSF Project, Oslo, Norway.

Jacobson, J. (1982a). Ozone and the growth and productivity of agricultural crops. *In* "Effects of Gaseous Air Pollution in Agriculture and Horticulture" (M. H. Unsworth and D. P. Ormrod, eds.), pp. 293–304. Butterworth, London.

Jacobson, J. (1982b). Economics of biological assessment. *J. Air Pollut. Control Assoc.* **32,** 145–146.

Jacobson, J., and Hill, A. C., eds. (1970). "Recognition of Air Pollution Injury to Plants: A Pictorial Atlas." Air Pollution Control Association, Pittsburgh, Pennsylvania.

Jacobson, J., and Showman, R. E. (1984). Field surveys of vegetation during a period of rising electric power generation in the Ohio Valley. *J. Air Pollut. Control Assoc.* **34,** 48–51.

Jacoby, J. M., Lynch, D. D., Welch, E. B., and Perkins, M. A. (1982). Internal phosphorus loading in a shallow eutrophic lake. *Water Res.* **16,** 911–919.

Jaffre, T., Brooks, R. R., Lee, J., and Reeves, R. D. (1976). *Sebertia acuminata:* A hyperaccumulator of nickel from New Caledonia. *Science* **193,** 579–580.

Jain, S. K., and Bradshaw, A. D. (1966). Evolutionary divergence among adjacent plant populations. 1. The evidence and its theoretical analysis. *Heredity* **21,** 407–441.

James, F. C. (1971). Ordinations of habitat relationships among breeding birds. *Wilson Bull.* **83,** 215–236.

James, G. I., and Courtin, G. M. (1985). Stand structure and growth form of the birch transition community in an industrially damaged ecosystem, Sudbury, Ontario. *Can. J. For. Res.* **15,** 809–817.

Janzen, D. H. (1987). Insect diversity in a Costa Rican dry forest: why keep it, and how. *Biol. J. Linn. Soc.* **30,** 343–356.

Jeffries, D. S., Cox, C. M., and Dillon, P. J. (1976). Depression of pH in lakes and streams in central Ontario during snowmelt. *J. Fish. Res. Board Can.* **36,** 640–646.

Jenny, H. (1980). "The Soil Resource." Springer-Verlag, New York.

Jensen, K. W., and Snekvik, E. (1972). Low pH levels wipe out salmon and trout populations in southernmost Norway. *Ambio* **1,** 223–225.

Johannessen, M., and Henriksen, A. (1978). Chemistry of snow meltwater: Changes in concentration during melting. *Water Resour. Res.* **14,** 615–619.

Johnels, A., Tyler, G., and Westermark, T. (1979). A history of mercury levels in Swedish fauna. *Ambio* **8,** 160–168.

Johnsen, I., and Sochting, V. (1980). Distribution of cryptogamic epiphytes in a Danish city in relation to air pollution and bark properties. *Bryologist* **79,** 86–92.

Johnsgard, P. A. (1983). "The Grouse of the World." Univ. of Nebraska Press, Lincoln.

Johnson, A. H. (1979). Evidence of acidification of headwater streams in the New Jersey pinelands. *Science* **206,** 834–836.

Johnson, A. H. (1983). Red spruce decline in the northeastern

U.S.: Hypotheses regarding the role of acid rain. *J. Air Pollut. Control Assoc.* **33,** 1049–1054.

Johnson, A. H., and Siccama, T. G. (1984). Decline of red spruce in the northern Appalachians: Assessing the possible role of acid deposition. *Tappi J.* **67,** 68–72.

Johnson, A. H., Siccama, T. G., Wong, D., Turner, R. S., and Barringer, T. H. (1981). Recent changes in pattern of tree growth rate in the New Jersey pinelands: A possible effect of acid rain. *J. Environ. Qual.* **10,** 427–430.

Johnson, A. H., Siccama, T. G., and Friedland, A. J. (1982). Spatial and temporal patterns of lead accumulation in the forest floor in the northeastern United States. *J. Environ. Qual.* **11,** 577–580.

Johnson, A. H., Friedland, A. J., and Dushoff, J. G. (1986). Recent and historical red spruce mortality: Evidence of climatic influence. *Water, Air, Soil Pollut.* **30,** 319–330.

Johnson, N. E., and Lawrence, W. H. (1977). Role of pesticides in the management of American forests. *In* "Pesticides in the Environment" (R. White-Stevens, ed.), Vol. 3, pp. 135–255. Dekker, New York.

Johnson, N. M. (1979). Acid rain: Neutralization within the Hubbard Brook ecosystem and regional implications. *Science* **204,** 497–499.

Johnsson, B., and Sundberg, R. (1972). "Has the Acidification by Atmospheric Pollution Caused a Growth Reduction in Swedish Forests?" Res. Note 20. Department of Forest Yield Research, Royal College of Forestry, Stockholm, Sweden.

Johnston, D. W., and Odum, E. P. (1956). Breeding bird populations in relation to plant succession in the Piedmont of Georgia. *Ecology* **37,** 50–62.

Jones, A. W. (1957). The flora of the city of London bombed sites. *London Nat.* **37,** 189–210.

Jones, D., Ronald, K., Levigne, D. M., Frank, R., Holdrinet, M., and Uthe, J. F. (1975). Organochlorine and mercury residues in the harp seal (*Pagophilus groenlandicus*). *Sci. Total Environ.* **5,** 181–195.

Jones, P., Allen, L. H., Jones, J. W., and Valle, R. V. (1985). Photosynthesis and transpiration responses of soybean canopies to short- and long-term CO_2 treatments. *Agron. J.* **77,** 119–126.

Jones, P. D., Wigley, T. M. L., and Wright, P. B. (1986). Global temperature variations between 1861 and 1984. *Nature (London)* **322,** 430–434.

Jordan, M. J. (1975). Effects of zinc smelter emissions and fire on a chestnut-oak woodland. *Ecology* **56,** 78–91.

Jordan, M. J., and Lechevalier, M. P. (1975). Effects of zinc-smelter emissions on forest soil microflora. *Can. J. Microbiol.* **21,** 1855–1865.

Jowett, D. (1964). Population studies on lead tolerant *Agrostis tenuis*. *Evolution (Lawrence, Kansas.)* **18,** 70–80.

Joyal, R. (1976). Winter foods of moose in La Verendre Park, Quebec: An evaluation of two browse survey methods. *Can. J. Zool.* **54,** 1765–1770.

Jurgensen, M. F., Larsen, M. J., and Harvey, A. E. (1979). Forest soil biology—timber harvesting relationships. *USDA*

For. Serv. Gen. Tech. Rep. INT-69. Intermountain Forest and Range Experiment Station, Ogden, Utah.

Kadono, Y. (1982). Occurrence of aquatic macrophytes in replation to pH, alkalinity, Ca^{++}, Cl^-, and conductivity. *Jpn. J. Ecol.* **32,** 39–44.

Kalff, J., and Welch, H. E. (1974). Phytoplankton production in Char Lake, a natural polar lake, and in Meretta Lake, a polluted polar lake, Cornwallis Island, Northwest Territories. *J. Fish. Res. Board Can.* **31,** 621–636.

Kalff, J., Kling, H. J., Holmgren, S. H., and Welch, H. E. (1975). Phytoplankton, phytoplankton growth, and biomass cycles in an unpolluted and in a polluted polar lake. *Verh.— Int. Ver. Theor. Angerv. Limnol.* **19,** 487–495.

Karr, J. R. (1968). Habitat and avian diversity on strip-mined land in east-central Illinois. *Condor* **70,** 348–367.

Karr, J. R. (1982). Avian extinction on Barro Colorado Island, Panama: A reassessment. *Am. Nat.* **119,** 220–239.

Karr, J. R., and Roth, R. R. (1971). Vegetation structure and avian diversity in several New World areas. *Am. Nat.* **105,** 423–435.

Katz, M., ed. (1939). "Effects of Sulfur Dioxide on Vegetation." National Research Council of Canada, Ottawa, Ontario.

Keeney, D. R. (1975). Toxic elements in agriculture. Unpublished manuscript, Department of Soil Science, University of Wisconsin, Madison.

Keeves, A. (1966). Some evidence of loss of productivity with successive rotations of *Pinus radiata* in the south-east of south Australia. *Aust. For.* **30,** 51–63.

Kellogg, W. W., Cadle, R. D., Allen, E. R., Lazrus, A. L., and Martell, E. A. (1972). The sulfur cycle. *Science* **175,** 587–596.

Kelly, C. A., Rudd, J. W. M., Furutani, A., and Schnidler, D. W. (1984). Effects of lake acidification on rates of organic matter decomposition in sediments. *Limnol. Oceanogr.* **29,** 687–694.

Kelly, J. M. (1984). Power plant influences on bulk precipitation, throughfall, and stemflow nutrient inputs. *J. Environ. Qual.* **13,** 405–409.

Kenaga, E. E. (1975). The evaluation of the safety of 2,4,5-T to birds in areas treated for vegetation control. *Residue Rev.* **59,** 1–19.

Kendall, R. J. (1982). Wildlife toxicology. *Environ. Sci. Technol.* **16,** 448a–453a.

Kendall, R. J., and Scanlon, P. F. (1982). Tissue lead concentrations and blood characteristics of mourning doves from southwestern Virginia. *Arch. Environ. Contam. Toxicol.* **11,** 269–272.

Kendeigh, S. C. (1948). Bird populations and biotic communities in northern lower Michigan. *Ecology* **29,** 101–114.

Kerekes, J., and Freedman, B. (1988). Physical, chemical, and biological characteristics of three watersheds in Kejimkujik National Park, Nova Scotia. *Arch. Environ. Contam. Toxicol.* **18,** 183–200.

Kerekes, J., Howell, G., Beauchamp, S., and Pollock, T.

(1982). Characterization of three lake basins sensitive to acid precipitation in central Nova Scotia. *Int. Rev. Gesamten Hydrobiol.* **67,** 679–694.

Kerekes, J., Freedman, B., Howell, G., and Clifford, P. (1984). Comparison of the characteristics of an acidic eutrophic and an acidic oligotrophic lake near Halifax, Nova Scotia. *Water Pollut. Res. J. Can.* **19,** 1–10.

Kerr, R. A. (1988a). Is the greenhouse here? *Science* **239,** 559–561.

Kerr, R. A. (1988b). Stratospheric ozone is decreasing. *Science* **239,** 1489–1491.

Kettela, E. (1983). "A Cartographic History of Spruce Budworm Defoliation from 1967 to 1981 in Eastern North America," Inf. Rep. DPC-X-14. Maritimes Forest Research Centre, Canadian Forestry Service, Fredericton, New Brunswick.

Kevan, P. G. (1975). Forest application of the insecticide fenitrothion and its effect on wild bee pollinators (Hymenoptera: Apoidea) of lowbush blueberries (*Vaccinium* spp.) in southern New Brunswick, Canada. *Biol. Conserv.* **7,** 301–309.

Kilpatrick, R., chair. (1979). "Review of the Safety for Use of the Herbicide 2,4,5-T." Advisory Committee on Pesticides, U.K. Ministry of Agriculture, Fisheries, and Food, London.

Kilpatrick, R., chair. (1980). "Further Review of the Safety for Use in the U.K. of the Herbicide 2,4,5-T." Advisory Committee on Pesticides, U.K. Ministry of Agriculture, Fisheries, and Food, London.

Kimmins, J. P. (1977). Evaluation of the consequences for future tree productivity of the loss of nutrients in whole-tree harvesting. *For. Ecol. Manage.* **1,** 169–183.

Kimmins, J. P., Scoular, K. A., and Feller, M. C. (1981). "FORCYTE—A Computer Approach to Evaluating the Effects of Whole Tree Harvesting on Nutrient Budgets and Future Tree Productivity," Report to Canadian Forestry Service, ENFOR Programme. Faculty of Forestry, University of British Columbia, Vancouver.

King, D. L., Simmler, J. J., Decker, D. S., and Ogg, C. W. (1974). Acid strip mine lake recovery. *J.—Water Pollut. Control Fed.* **10,** 2301–2316.

King, W. B. (1981). "Endangered Birds of the World: The ICBP Bird Red Data Book." Smithsonian Institution Press, Washington, D.C.

Kingsbury, P. D. (1975). Effects of aerial forest spraying on aquatic fauna. *In* "Aerial Control of Forest Insects in Canada" (M. L. Prebble, ed.), pp. 280–283. Department of the Environment, Ottawa, Ontario.

Kingsbury, P. D. (1977). "Fenitrothion in a Lake Ecosystem," Rep. CC-X-146. Chemical Control Research Institute, Environment Canada, Ottawa, Ontario.

Kingsbury, P. D., and McLeod, B. B. (1980). "The Impact of Spruce Budworm Control Operations Involving Sequential Applications upon Forest Avifauna in the Lower St. Lawrence Region of Quebec," Rep. FPM-X-34. Forest Pest Management Institute, Canadian Forestry Service, Sault Ste. Marie, Ontario.

Kingsbury, P. D., and McLeod, B. B. (1981). "Fenitrothion and Forest Avifauna Studies on the Effects of High Dosage Applications," Rep. FPM-X-43. Forest Pest Management Institute, Canadian Forestry Service, Sault Ste. Marie, Ontario.

Kingsbury, P. D., McLeod, B. B., and Millikin, R. L. (1981). "The Environmental Impact of Nonyl Phenol and the Matacil Formulation," Part 2, Rep. FPM-X-36. Forest Pest Management Institute, Canadian Forestry Service, Sault Ste. Marie, Ontario.

Kirkby, E. A. (1969). Ion uptake and ionic balance in plants in relation to the form of nitrogen nutrition. *In* "Ecological Aspects of the Mineral Nutrition of Plants" (I. H. Rorison, ed.), pp. 215–235. Blackwell, Oxford.

Kirkham, I. R., and Montevecchi, W. A. (1982). The breeding birds of Funk Island: An historical perspective. *Am. Birds* **36,** 111–118.

Klein, D. R. (1974). Reaction of reindeer to obstructions and disturbances. *Science* **173,** 393–398.

Klein, R. M., and Perkins, T. D. (1987). Cascades of causes and effects of forest decline. *Ambio* **16,** 86–93.

Kling, G. W., Clark, M. A., Compton, H. R., Devine, J. D., Evans, W. C., Humphry, A. M., Koenigsberg, E. J., Lockewood, J. P., Tuttle, M. L., and Wagner, G. N. (1987). The 1986 Lake Nyos gas disaster in Cameroon, West Africa. *Science* **236,** 169–175.

Knoll, A. H. (1984). Patterns of extinction in the fossil record of vascular plants. *In* "Extinctions" (M. H. Nitecki, ed.), pp. 21–68. Univ. of Chicago Press, Chicago, Illinois.

Kochenderfer, J. N. (1970). Erosion control on logging roads in the Appalachians. *USDA For. Serv. Res. Pap.* NE-158. Northeastern Forest Experiment Station, Broomall, Pennsylvania.

Kochenderfer, J. N., and Aubertin, G. M. (1975). Effects of management practices on water quality and quantity: Fernow Experimental Forest, West Virginia. *In* "Municipal Watershed Management," U.S.D.A. For. Serv., Gen. Tech. Rep. NE-13, pp. 14–24. Northeastern Forest Experiment Station, Broomall, Pennsylvania.

Kochenderfer, J. N., and Wendel, G. W. (1983). Plant succession and hydrological recovery on a deforested and herbicided watershed. *For. Sci.* **29,** 545–558.

Koeman, J. H., Peeters, W. H. M., Koudstaal-Hol, C. H. M., Tjioe, P. S., and Degoeij, J. J. M. (1973). Mercury-selenium correlations in marine mammals. *Nature (London)* **245,** 385–386.

Koivusaari, J., Nuuja, I., Polokangas, R., and Finnlund, M. (1980). Relationships between productivity, eggshell thickness, and pollutant contents of addled eggs in the population of white-tailed eagle *Haliaeetus labicilla* L. in Finland during 1969–78. *Environ. Pollut., Ser. A* **23,** 41–52.

Koons, B. B. (1984). Input of petroleum to the marine environment. *Mar. Technol. Soc. J.* **18,** 97–112.

Kornberg, H., chair. (1981). "Royal Commission on Environmental Pollution," 8th Rep. H. M. Stationery Office, London.

Kramer, J., and Tessier, A. (1982). Acidification of aquatic ecosystems: A critique of chemical approaches. *Environ. Sci. Technol.* **16**, 606A–615A.

Krause, G. H. M., Arndt, U., Brandt, C. J., Bucher, J., Kenk, G., and Matzner, E. (1986). Forest decline in Europe: Development and possible causes. *Water, Air, Soil Pollut.* **31**, 647–668.

Krause, H. H. (1982). Nitrate formation and movement before and after clear-cutting of a monitored watershed in central New Brunswick, Canada. *Can. J. For. Res.* **12**, 922–930.

Krefting, L. W. (1962). Use of silvicultural techniques for improving deer habitat in the United States. *J. For.* **16**, 40–42.

Krefting, L. W. (1974). Moose distribution and habitat selection in north central North America. *Nat. Can.* **101**, 81–100.

Krefting, L. W., and Hansen, H. L. (1969). Increasing browse for deer by aerial applications of 2,4-D. *J. Wildl. Manage.* **33**, 784–790.

Kruckeberg, A. R. (1954). The ecology of serpentine soils. III. Plant species in relation to serpentine soils. *Ecology* **35**, 267–274.

Kruckeberg, A. R. (1984). "California Serpentine: Flora, Vegetation, Geology, Soils, and Management Problems." Univ. of California Press, Los Angeles.

Krug, E. C., Isaacson, P. J., and Frink, C. R. (1985). Appraisal of some current hypotheses describing acidification of watersheds. *J. Air Pollut. Control Assoc.* **35**, 109–114.

Kuja, A. L. (1980). Revegetation of mine tailings using native species from disturbed sites in northern Canada. M.Sc. Thesis, Department of Botany, University of Toronto, Toronto, Ontario.

Kunke, D. H., and Brace, L. G. (1986). "Silvicultural Statistics for Canada, 1975–76 to 1982–83," Inf. Rep. NOR-X-275. Northern Forestry Centre, Canadian Forestry Service, Edmonton, Alberta.

Kwain, W. H. (1975). Effects of temperature on development and survival of rainbow trout, *Salmo gairdneri*, in acid waters. *J. Fish. Res. Board Can.* **32**, 493–497.

Lacroix, G. L. (1985). Survival of eggs and alevins of Atlantic salmon (*Salmo salar*) in relation to the chemistry of interstitial water in redds in some acidic streams of Atlantic Canada. *Can. J. Fish. Aquat. Sci.* **42**, 292–299.

Lacroix, G. L., and Townsend, D. R. (1987). Responses of juvenile Atlantic salmon (*Salmo salar*) to episodic increases in acidity of Nova Scotia rivers. *Can. J. Fish. Aquat. Sci.* **44**, 1475–1484.

Lagerwerff, J. V., and Specht, A. W. (1970). Contamination of roadside soils and vegetation with cadmium, nickel, lead, and zinc. *Environ. Sci. Technol.* **4**, 583–586.

Lagerwerff, J. V., Biersdorf, G. T., and Brower, D. L. (1976). Retention of metals in sewage sludge. I. Constituent heavy metals. *J. Environ. Qual.* **5**, 19–22.

Land Use Regulatory Commission (LURC) (1979). "A Survey of Erosion and Sedimentation Problems Associated with Logging in Maine." LURC, Maine Department of Conservation, Augusta.

Larcher, W., and Bauer, H. (1981). Ecological significance of resistance to low temperature. *In* "Physiological Plant Ecology." (O. L. Lange, P. S. Nobel, C. B. Osmond, and H. Ziegler, eds.), pp. 403–437. Springer-Verlag, New York.

Last, F. J. (1987). The nature, and elucidation of causes of forest decline. *In* "Proceedings of the Workshop on Forest Decline and Reproduction: Regional and Global Consequences" (L. Kairiukstis, S. Nilsson, and A. Stroszak, ed.), pp. 63–78. IIASA, Laxenburg, Austria.

Lathrop, G. D., Wolfe, W. H., Albanese, R. A., and Moynahan, P. M. (1984). "An Epidemiologic Investigation of Health Effects in Air Force Personnel Following Exposure to Herbicides: Baseline Morbidity Study Results." The Surgeon General, U.S. Air Force, Washington, D.C.

Lavelle, P. (1987). Biological processes and productivity of soils in the humid tropics. *In* "The Geophysiology of Amazonia" (R. E. Dickensen, ed.), pp. 175–223. Wiley, New York.

Laws, R. M. (1977). The significance of vertebrates in the Antarctic marine ecosystems. *In* "Adaptations Within Antarctic Systems" (G. A. Llano, ed.), pp. 411–438. Smithsonian Institution, Washington, D.C.

Lezerte, B. D. (1984). Forms of aqueous aluminum in acidified catchments of central Ontario: A methodological analysis. *Can. J. Fish. Aquat. Sci.* **41**, 766–776.

Leaf, C. F., and Brink, G. E. (1972). Simulating effects of harvest cutting on snowmelt in Colorado subalpine forests. *In* "Watersheds in Transition," pp. 191–196. American Water Resources Association, Urbana, Illinois.

LeBlanc, F., and De Sloover, J. (1970). Relation between industrialization and the distribution and growth of epiphytic lichens and mosses in Montreal. *Can. J. Bot.* **48**, 1485–1496.

LeBlanc, F., and Rao, D. N. (1966). Reaction of several lichens and epiphytic mosses to sulfur dioxide in Sudbury, Ontario. *Bryologist* **69**, 338–346.

LeBlanc, F., Rao, D. M., and Comeau, G. (1972). The epiphytic vegetation of *Populus tremuloides* and its significance as an air pollution indicator in Sudbury, Ontario. *Can. J. Bot.* **50**, 519–528.

Lee, P. L. (1985). History and current status of spotted owl (*Strix occidentalis*) habitat management in the Pacific Northwest region, U.S.D.A., Forest Service. *In* "Ecology and Management of the Spotted Owl in the Pacific Northwest," U.S.D.A. For. Serv., Gen. Tech. Rep. PNW-185, pp. 5–10. Pacific Northwest Forest and Range Experiment Station, Portland, Oregon.

Lee, R. F. (1977). Accumulation and turnover of petroleum hydrocarbons in marine organisms. *In* "Fate and Effects of Petroleum Hydrocarbons in Marine Ecosystems and Organisms" (D. A. Wolfe, ed.), pp. 60–70. Pergamon, New York.

Leigh, E. G. (1982). Why are there so many kinds of tropical trees? *In* "The Ecology of a Tropical Forest" (E. G. Leigh, A. S. Rand, and D. M. Windsor, eds.), pp. 63–66. Smithsonian Institution Press, Washington, D.C.

Leighton, F. A., Butler, R. G., and Peakall, D. B. (1985). Oil and Arctic marine birds: An assessment of risk. *In* "Petroleum Effects in the Arctic Environment" (F. R. Engelhardt, ed.), pp. 183–215. Am. Elsevier, New York.

Leivestad, H., and Muniz, I. P. (1976). Fish kills at low pH in a Norwegian river. *Nature (London)* **259**, 391–392.

Leivestad, H., Hendry, G., Muniz, I. P., and Snekvik, E. (1976). Effects of acid precipitation on freshwater organisms. *In* "Impact of Acid Precipitation on Forest and Freshwater Ecosystems in Norway" (F. H. Brakke, ed.), Res. Rep. FR 6/76, pp. 87–111. SNSF Project, Oslo, Norway.

Lepp, N. W., ed. (1981). "Effects of Heavy Metal Pollution on Plants," Vol. 1. Appl. Sci. Publ., London.

Lepp, N. W., and Fairfax, J. A. W. (1976). The role of acid rain as a regulator of foliar nutrient uptake and loss. *In* "Microbiology of Aerial Plant Surfaces" (C. H. Dickinson and T. F. Preece, eds.), pp. 107–118. Academic Press, London.

LeRoy, J. C., and Keller, H. (1972). How to reclaim mined areas, tailings ponds, and dumps into valuable land. *World Min.* **25**(1), 34–41.

Lessmark, O., and Thornelof, E. (1986). Liming in Sweden. *Water, Air, Soil Pollut.* **31**, 809–815.

Levasseur, J. E., and Jory, M. L. (1982). "Rétablissement naturel d'une végétation de marais maritimes altérée par les hydrocarbures de l'Amoco Cadiz: Modalités et tendances. *In* "Ecological Study of the Amoco Cadiz Oil Spill," pp. 329–362. U.S. Department of Commerce, National Oceanic and Atmospheric Administration, Washington, D.C.

Levine, S. N., Stainton, M. P., and Schindler, D. W. (1986). A radiotracer study of phosphorus cycling in a eutrophic Canadian Shield lake, Lake 227, northwestern Ontario. *Can. J. Fish. Aquat. Sci.* **43**, 366–378.

Lewis, W. M., and Grant, M. C. (1979). Relationships between stream discharges and yield of dissolved substances from a Colorado mountain watershed. *Soil Sci.* **128**, 353–363.

Li, T. A., and Landsberg, H. E. (1975). Rainwater pH close to a major power plant. *Atmos. Environ.* **9**, 81–88.

Liebsch, E. J., and de Pena, R. G. (1982). Sulfate aerosol production in coal-fired power plant plumes. *Atmos. Environ.* **16**, 1323–1331.

Likens, G. E. (1985). An experimental approach for the study of ecosystems. *J. Ecol.* **73**, 381–396.

Likens, G. E., and Butler, T. J. (1981). Recent acidification of precipitation in North America. *Atmos. Environ.* **15**, 1103–1110.

Likens, G. E., Bormann, F. H., Johnson, N. M., Fisher, D. W., and Pierce, R. S. (1970). Effects of forest cutting and herbicide treatment on nutrient budgets in the Hubbard Brook Watershed ecosystem. *Ecol. Monogr.* **40**, 23–47.

Likens, G. E., Bormann, F. H., Pierce, R. J., Eaton, J. S., and Johnson, N. M. (1977). "Biogeochemistry of a Forested Ecosystem." Springer-Verlag, New York.

Likens, G. E., Bormann, F. H., Pierce, R. S., and Reiners, W. A. (1978). Recovery of a deforested ecosystem. *Science* **199**, 492–496.

Likens, G. E., Bormann, F. H., Pierce, R. S., Eaton, J. S., and Munn, R. E. (1984). Long-term trends in precipitation chemistry at Hubbard Brook, New Hampshire. *Atmos. Environ.* **18**, 2641–2647.

Liljestrand, H. M. (1985). Average rainwater pH, concepts of atmospheric acidity, and buffering in open systems. *Atmos. Environ.* **19**, 487–499.

Lincer, J. L., Cade, T. J., and Devine, J. M. (1970). Organochlorine residues in Alaskan peregrine falcons, rough-legged hawks, and their prey. *Can. Field-Nat.* **84**, 255–263.

Lincoln, R. J., Boxshall, G. A., and Clark, P. F. (1982). "A Dictionary of Ecology, Evolution and Systematics." Cambridge University Press. Cambridge, U.K.

Lind, C. T., and Cottam, G. (1969). The submerged aquatics of University Bay: A study of eutrophication. *Am. Midl. Nat.* **81**, 353–369.

Lindberg, S. E., Lovett, G. M., Richter, D. D., and Johnson, D. W. (1986). Atmospheric deposition and canopy interactions of major ions in a forest. *Science* **231**, 141–145.

Linkins, A. E., Johnson, L. A., Everett, K. R., and Atlas, R. M. (1984). Oil spills: Damage and recovery in tundra and taiga. *In* "Restoration of Habitats Impacted by Oil Spills" (J. L. Cairns and A. L. Buikema, Jr., eds.), pp. 135–155. Butterworth, Boston, Massachusetts.

Linzon, S. N. (1971). Economic effects of SO_2 on forest growth. *J. Air Pollut. Control Assoc.* **21**, 81–86.

Linzon, S. N., and Temple, P. J. (1980). Soil resampling and pH measurements after an 18-year period in Ontario. *In* "Ecological Impact of Acid Precipitation" (D. Drablos and A. Tollan, eds.), pp. 176–177. SNSF Project, Oslo, Norway.

Lioy, P. J., and Samson, P. J. (1979). Ozone concentration patterns observed during the 1976–1977 long range transport study. *Environ. Int.* **2**, 77–83.

Little, P., and Martin, H. (1972). A survey of zinc, lead, and cadmium in soil and natural vegetation around a smelting complex. *Environ. Pollut.* **3**, 241–254.

Lockie, J. D., Ratcliffe, D. A., and Balharry, R. (1969). Breeding success and dieldrin contamination of golden eagles in west Scotland. *J. Appl. Ecol.* **6**, 381–389.

Loehr, R. C., Martin, C. S., and Rast, W. (1980). "Phosphorus Management Strategies for Lakes." Ann Arbor Sci. Publ., Ann Arbor, Michigan.

Logan, J. A. (1983). Nitrogen oxides in the atmosphere: Global and regional budgets. *JGR, J. Geophys. Res.* **88**, 785–807.

Logie, R. R. (1975). Effects of aerial spraying of DDT on salmon populations of the Miramichi River. *In* "Aerial Control of Forest Insects in Canada" (M. L. Prebble, ed.), pp. 293–300. Department of the Environment, Ottawa, Ontario.

Loucks, O. L. (1985). Looking for surprise in managed stressed systems. *BioScience* **35**, 428–432.

Lovejoy, T. E. (1985). Amazonia, people and today. *In* "Amazonia" (G. T. Prance and T. E. Lovejoy, eds.), pp. 328–338. Pergamon, New York.

Lovejoy, T. E., Bierregaard, R. O., Jr., Rylands, A. B., Malcolm, J. R., Quintela, C. E., Harper, L. H., Brown, K. S., Jr., Powell, A. H., Schubart, H. O. R., and Hays, M. B. (1986). Edge and other effects of isolation on Amazon forest fragments. *In* "Conservation Biology" (M. E. Soule, ed.), pp. 257–285. Sinauer, Sunderland, Massachusetts.

Lovejoy, T. E., Ramkin, J. M., Bierregaard, R. O., Brown, K. S., Emmons, L. H., and Van der Voort, M. E. (1984). Ecosystem decay of Amazon forest remnants. *In* "Extinctions" (M. H. Nitecki, ed.), pp. 297–325. Univ. of Chicago Press, Chicago, Illinois.

Lovett, G. M., Reiners, W. A., and Olson, R. K. (1982). Cloud droplet deposition in subalpine balsam fir forests: Hydrological and chemical budgets. *Science* **218**, 1303–1304.

Lozano, F. C., and Morrison, I. K. (1981). Disruption of hardwood nutrition by sulfur dioxide, nickel, and copper air pollution near Sudbury, Ontario. *J. Environ. Qual.* **10**, 198–204.

Lucas, G., and Synge, H. (1978). "The IUCN Plant Red Data Book." International Union for Conservation of Nature and Natural Resources, Morges, Switzerland.

Lugo, A. E. (1988). Estimating reductions in the diversity of tropical forest species. *In* "Biodiversity" (E. O. Wilson, ed.), pp. 58–70. National Academy Press, Washington, D.C.

Lund, J. W. G. (1950). Studies on *Asterionella formosa* Hass. II. Nutrient depletion and the spring maximum. *J. Ecol.* **38**, 1–35.

Lund, J. W. G., Mackereth, F. J. H., and Mortimer, C. H. (1963). Changes in depth and time of certain chemical and physical conditions and of the standing crop of *Asterionella formosa* Hass. in the North Basin of Windermere in 1947. *Philos. Trans. R. Soc. London, Ser. B* **246**, 255–290.

Lusis, M. A., Chan, W. H., Tang, A. J. S., and Johnson, N. D. (1983). Scavenging rates of sulfur and trace metals from a smelter plume. *In* "Precipitation Scavenging, Dry Deposition, and Resuspension," pp. 369–382. Elsevier, London.

Luther, F. M., and Ellingson, R. G. (1985). Carbon dioxide and the radiation budget. *In* "Projecting the Climatic Effects of Increasing Carbon Dioxide," DOE/ER-0237, pp. 25–56. U.S. Department of Energy, Washington, D.C.

Lynch, M., and Shapiro, J. (1981). Predation, enrichment, and phytoplankton community structure. *Limnol. Oceanogr.* **26**, 86–102.

Lyon, J. J. (1976). Elk use as related to characteristics of clearcuts in western Montana. *In* "Proceedings of the Elk—Logging—Roads Symposium," pp. 69–72. University of Idaho, Moscow.

Lyon, L. J., and Basile, J. V. (1980). Influences of timber harvesting and residue management on big game. *In* "Environmental Consequences of Timber Harvesting in Rocky Mountain Coniferous Forests," U.S.D.A. For. Serv., Gen. Tech. Rep. INT-90, pp. 441–453. Intermountain Forest and Range Experiment Station, Ogden, Utah.

Lyon, L. J., and Jensen, C. E. (1980). Management implications of elk and deer use of clear-cuts in Montana. *J. Wildl. Manage.* **44**, 352–362.

Lyon, L. J., and Mueggler, W. F. (1966). Herbicide treatment of north Idaho browse evaluated six years later. *J. Wildl. Manage.* **31**, 538–541.

MacArthur, R. H. (1964). Environmental factors affecting bird species diversity. *Am. Nat.* **98**, 387–397.

MacArthur, R. H., and MacArthur, J. W. (1961). On bird species diversity. *Ecology* **42**, 594–598.

MacArthur, R. H., and Wilson, E. O. (1967). "The Theory of Island Biogeography." Princeton Univ. Press, Princeton, New Jersey.

MacArthur, R. H., Recher, H., and Cody, M. (1966). On the relation between habitat selection and species diversity. *Am. Nat.* **100**, 319–322.

McBride, J. R., Miller, P. R., and Laver, R. D. (1985). Effects of oxidant air pollutants on forest succession in the mixed conifer forest type of southern California. *In* "Air Pollutant Effects on Forest Ecosystems," pp. 157–167. Acid Rain Foundation, St. Paul, Minnesota.

McClenahan, J. R. (1978). Community changes in a deciduous forest exposed to air pollution. *Can. J. For. Res.* **8**, 432–438.

McClung, R. M. (1969). "Lost Wild America. The Story of Our Extinct and Vanishing Wildlife." William Morrow, New York.

McColl, J. G. (1978). Ionic composition of forest soil solutions and effects of clear-cutting. *Soil. Sci. Soc. Am. J.* **42**, 358–363.

McColl, J. G., and Grigall, D. F. (1979). Nutrient losses by leaching and erosion by intensive forest harvesting. *In* "Impact of Intensive Harvesting on Forest Nutrient Cycling," pp. 231–248. College of Environmental Science and Forestry, State University of New York, Syracuse, New York.

McCormick, J. F., and Platt, R. B. (1962). Effects of ionizing radiation on a natural plant community. *Radiat. Bot.* **3**, 161–188.

McCormick, L. H., and Steiner, K. C. (1978). Variation in aluminum tolerance among six genera of trees. *For. Sci.* **24**, 565–568.

McCown, B. H., Deneke, F. J., Rickard, W., and Tieszen, L. L. (1973a). The response of Alaskan terrestrial plant communities to the presence of petroleum. *In* "Proceedings of the Symposium on the Impact of Oil Resource Development on Northern Plant Communities," pp. 34–43. Institute of Arctic Biology, Fairbanks, Alaska.

McCown, B. H., Brown, J., and Barsdate, R. J. (1973b). Natural oil seeps at Cape Simpson, Alaska: Localized influences on terrestrial habitat. *In* "Proceedings of the Symposium on the Impact of Oil Resource Development on Northern Plant Communities," pp. 86–90. Institute of Arctic Biology, Fairbanks, Alaska.

McCracken, F. I. (1985a). Observations on the decline of black willow. *J. Miss. Acad. Sci.* **30**, 1–5.

McCracken, F. I. (1985b). Oak decline and mortality in the

south. *Proc. Symp. Southeast. Hardwoods, 3rd, 1985* pp. 77–81.

MacCracken, M. C., and Luther, F. M. eds. (1985). "Projecting the Climatic Effects of Increasing Carbon Dioxide," DOE/ER-0237. U.S. Department of Energy, Washington, D.C.

MacCrimmon, H. R., Wren, C. D., and Gots, B. L. (1983). Mercury uptake by lake trout, *Salvelinus namaycush,* relative to age, growth, and diet in Tadenac Lake with comparative data from other Canadian Shield lakes. *Can. J. Fish. Aquat. Sci.* **40,** 114–120.

McEwen, F. L., and Stephenson, G. R. (1979). "The Use and Significance of Pesticides in the Environment." Wiley, New York.

McFee, W. W., Kelley, J. M., and Beck, R. H. (1977). Acid precipitation effects on soil pH and base saturation of exchange sites. *Water, Air, Soil Pollut.* **7,** 401–408.

McGovern, P. C., and Balsillie, D. (1972). "SO$_2$ Levels and Environmental Studies in the Sudbury Area during 1971." Ontario Ministry of the Environment, Air Quality Branch, Sudbury.

McIlveen, W. D., Rutherford, S. T., and Linzon, S. N. (1986). "A Historical Perspective of Sugar Maple Decline Within Ontario and Outside of Ontario," ARB-141-86-Phyto. Ontario Ministry of the Environment, Toronto.

McKay, C. (1985). "Freshwater Fish Contamination in Canadian Waters," Unpublished report. Chemical Hazards Division, Fish Habitat Management Branch, Department of Fisheries and Oceans, Ottawa, Ontario.

Mackay, D., and Wells, P. G. (1981). Factors influencing the aquatic toxicity of chemically dispersed oils. *Proc. Arct. Mar. Oilspill Program Tech. Semin., 4th* pp. 445–467.

Mackinnon, D., and Freedman, B. (1988). Effects of silvicultural glyphosate spraying on breeding birds and their habitat in central Nova Scotia. Research Report to World Wildlife Fund Canada. Department of Biology, Dalhousie University, Halifax, Nova Scotia.

McLaughlin, S. B. (1985). Effects of air pollution on forests. A critical review. *J. Air Pollut. Control Assoc.* **35,** 512–534.

MacLean, A. J., and Dekker, A. J. (1978). Availability of zinc, copper, and nickel to plants grown in sewage-treated soils. *Can. J. Soil Sci.* **58,** 381–389.

MacLean, A. J., Store, B., and Cordukes, W. B. (1973). Amounts of mercury in soil of some golf course sites. *Can. J. Soil Sci.* **53,** 130–132.

MacLean, D. A. (1980). Vulnerability of fir-spruce stands during uncontrolled spruce budworm outbreaks: A review and discussion. *For. Chron.* **56,** 213–221.

MacLean, D. A. (1984). Effects of spruce budworm outbreaks on the productivity and stability of balsam fir forests. *For. Chron.* **60,** 273–279.

MacLean, D. A. (1985). Effects of spruce budworm outbreaks on forest growth and yield. *In* "Recent Advances in Spruce Budworms Research," pp. 148–175. Canadian Forestry Service, Ottawa, Ontario.

MacLean, D. A., and Erdle, T. A. (1984). A method to determine effects of spruce budworm on stand yield and wood supply projections for New Brunswick. *For. Chron.* **60,** 167–173.

MacLean, D. A., and Morgan, M. G. (1983). Long-term growth and yield responses of young fir to manual and chemical release from shrub competition. *For. Chron.* **59,** 177–183.

MacLean, D. A., Kline, A. W., and Levine, D. R. (1984). Effectiveness of spruce budworm spraying in New Brunswick in protecting the spruce component of spruce-fir stands. *Can. J. For. Res.* **14,** 163–176.

MacLean, D. C., and Schneider, R. E. (1976). Photochemical oxidants in Yonkers, New York. Effects on yield of bean and tomato. *J. Environ. Qual.* **5,** 75–78.

McLeod, J. M. (1967). The effect of phosphamidon on bird populations in jack pine stands in Quebec. *Can. Field-Nat.* **81,** 102–106.

McLeod, B., and Millikin, R. (1982). "Environmental Impact Assessment of Experimental Spruce Budworm Adulticide Trials," Part 1, Rep. FPM-X-54. Forest Pest Management Institute, Canadian Forestry Service, Sault Ste. Marie, Ontario.

McLuhan, M. (1967). "The Medium is the Massage: An Inventory of Effects." Bantam Books, New York.

McNeil, D. K., Bendell, B. E., and Ross, R. K. (1987). "Studies of the Effects of Acidification on Aquatic Wildlife in Canada: Waterfowl and Trophic Relationships in Small Lakes in Northern Ontario," Occas. Pap. No. 62. Canadian Wildlife Service, Ottawa, Ontario.

McNicol, J. G., and Timmermann, H. R. (1981). Effects of forestry practices on ungulate populations in the boreal mixedwood forest. *In* "Boreal Mixedwood Symposium," COJFRC Symp. Proc. O-P-9, pp. 141–154. Great Lakes Forest Research Centre, Sault Ste. Marie, Ontario.

McQueen, E. G., Veale, A. M. O., Alexander, W. S., and Bates, M. N. (1977). "2,4,5-T and Human Birth Defects." Department of Health, New Zealand, Auckland.

Maessen, O., Freedman, B., and Nams, M. L. N. (1983). Resource allocation in high-arctic vascular plants of differing growth form. *Can. J. Bot.* **61,** 1680–1691.

Mahendrappa, M. K. (1983). Chemical characteristics of precipitation and hydrogen input in throughfall and stemflow under eastern Canadian forest stands. *Can. J. For. Res.* **13,** 948–955.

Malik, N., and Vanden Born, W. H. (1986). "Use of Herbicides in Forest Management," Inf. Rep. NOR-X-282. Northern Forest Research Centre, Edmonton, Alberta.

Malingreau, J. P., Stephens, G., and Fellows, L. (1985). Remote sensing of forest fires: Kalimantan and North Borneo in 1982–83. *Ambio* **14,** 314–321.

Malins, D. C., ed. (1977). "Effects of Petroleum on Arctic and Subarctic Marine Environments," Vol. 2. Academic Press, New York.

Malley, D. F., Findlay, D. L., and Chang, P. S. S. (1982). Ecological effects of acid precipitation on zooplankton. *In*

"Acid Precipitation Effects on Ecological Systems" (F. M. D'Itri, ed.), pp. 297–327. Ann Arbor Science Publ., Ann Arbor, Michigan.

Mallins, D. C., McCain, B. B., Brown, D. W., Chan, S.-L., Myers, M. S., Landahl, J. J., Prohaska, P. G., Friedman, A. J., Rhodes, L. D., Burrows, D. G., Gronlund, W. D., and Hodgins, H. O. (1983). Chemical pollutants in sediments and diseases of bottom-dwelling fish in Puget Sound, Washington. *Environ. Sci. Technol.* **18**, 705–713.

Malyuga, D. P. (1964). "Biogeochemical Methods of Prospecting." Academic Science Press, Moscow (transl.). Consultants Bureau, New York.

Manion, P. D. (1985). Factors contributing to the decline of forests, a conceptual overview. *In* "Air Pollutant Effects on Forest Ecosystems," pp. 63–73. Acid Rain Foundation, St. Paul, Minnesota.

Mann, K. H. (1977). Destruction of kelp-beds by sea urchins: A cyclical phenomenon or irreversible degradation. *Helgol. Wiss. Meeresunters.* **30**, 455–467.

Marks, P. L. (1974). The role of pin cherry (*Prunus pensylvanica*) in the maintenance of stability in northern hardwood ecosystems. *Ecol. Monogr.* **44**, 73–88.

Marks, P. L., and Bormann, F. H. (1972). Revegetation following forest cutting: mechanisms for return to steady-state nutrient cycling. *Science* **176**, 914–915.

Marland, G., and Rotty, R. M. (1985). Greenhouse gases in the atmosphere: What do we know. *J. Air Pollut. Control Assoc.* **35**, 1033–1038.

Martin, A. C., Zim, H. S., and Nelson, A. L. (1951). "American Wildlife and Plants: A Guide to Wildlife Food Habits." Dover, New York.

Martin, C., Pierce, R. S., Likens, G. E., and Bormann, F. H. (1986). "Clearcutting Affects Stream Chemistry in the White Mountains of New Hampshire," Res. Pap. NE-579. Northeastern Forest Experiment Station, Broomall, Pennsylvania.

Martin, E. S., and Hiebert, M. (1985). Explosive remnants of the Second Indochina War in Viet Nam. *In* "Explosive Remnants of War: Mitigating the Environmental Impacts" (A. H. Westing, ed.), pp. 39–50. Taylor & Francis, Philadelphia, Pennsylvania.

Martin, N. D. (1960). An analysis of bird populations in relation to forest succession in Algonquin Provincial Park, Ontario. *Ecology* **41**, 126–140.

Martin, P. S. (1967). Prehistoric overkill. *In* "Pleistocene Extinctions: The Search for a Cause" (P. S. Martin and H. E. Wright, eds.), pp. 75–120. Yale Univ. Press, New Haven, Connecticut.

Martin, P. S. (1984). Catastrophic extinctions and late Pleistocene blitzkrieg: Two radiocarbon tests. *In* "Extinctions" (M. H. Nitecki, ed.), pp. 153–189. Univ. of Chicago Press, Chicago, Illinois.

Maser, C., Anderson, R. G., Cromack, K., Williams, J. T., and Martin, R. E. (1979). Dead and down woody material. *U.S., Dep. Agric., Agric. Handb.* **553**, 78–95. Washington, D.C.

Mathews, W. H., and Bustin, R. M. (1984). Why do the Smoking Hills burn? *Can. J. Earth Sci.* **21**, 737–742.

Matida, Y., Furuta, Y., Kumada, H., Tanaka, H., Yokote, M., and Kimura, S. (1975). Effects of some herbicides applied in the forest to freshwater fishes and other aquatic organisms. I. Survey of the effects of aerially applied sodium chlorate and a mixture of 2,4-D and 2,4,5-T on the stream community. *Bull. Freshwater Fish. Res. Lab.* **25**, 41–52.

Matzner, E., Murach, D., and Fortmann, H. (1986). Soil acidity and its relationship to root growth in declining forest stands in Germany. *Water, Air, Soil Pollut.* **31**, 273–282.

Mayer, R., and Ulrich, B. (1974). Conclusions on the filtering action of forests from ecosystem analysis. *Oecol. Plant.* **9**, 157–168.

Mayer, R., and Ulrich, B. (1977). Acidity of precipitation as influenced by the filtering of atmospheric sulfur and nitrogen compounds—its role in the element balance and effect on soil. *Water, Air, Soil Pollut.* **7**, 409–416.

Mayfield, H. F. (1960). "The Kirtland's Warbler," Bull. 40. Cranbrook Institute of Science, Bloomfield Hills, Michigan.

Mayfield, H. F. (1961). Cowbird parasitism and the population of the Kirtland's warbler. *Evolution (Lawrence, Kans.)* **15**, 174–179.

Mayfield, H. F. (1977). Brood parasitism: Reducing interactions between Kirtland's warblers and brown-headed cowbirds. *In* "Endangered Birds, Management Techniques for Preserving Threatened Species" (S. A. Temple, ed.), pp. 85–91. Univ. of Wisconsin Press, Madison.

Means, J. E., and Winjum, J. K. (1983). Road to recovery after eruption of Mt. St. Helens. *In* "Using Our Natural Resources: 1983 Yearbook of Agriculture," pp. 204–215. U.S. Department of Agriculture, Washington, D.C.

Mearns, A. J., and Sherwood, M. J. (1977). Distributions of neoplasms and other diseases in marine fishes relative to the discharge of waste water. *Ann. N.Y. Acad. Sci.* **298**, 210–223.

Megahan, W. F., and Kidd, W. J. (1972). Effects of logging and logging roads on erosion and sediment deposition from steep terrain. *J. For.* **70**, 136–141.

Melillo, J. M., Palm, C. A., Houghton, R. A., Woodwell, G. M., and Myers, N. (1985). A comparison of two recent estimates of disturbance in tropical forests. *Environ. Conserv.* **12**, 37–40.

Meszaros, E. (1981). "Atmospheric Chemistry. Fundamental Aspects. Studies in Environmental Science." Am. Elsevier, New York.

Metcalf, R. L. (1971). The chemistry and biology of pesticides. *In* "Pesticides in the Environment" (R. White-Stevens, ed.), Vol. 1, Part 1, pp. 1–144. Dekker, New York.

Michael, A. D. (1977). The effects of petroleum hydrocarbons on marine populations and communities. *In* "Fate and Effect of Petroleum Hydrocarbons in Marine Ecosystems and Organisms" (D. A. Wolfe, ed.), pp. 129–137. Pergamon, New York.

Millan, M., Barton, S. C., Johnson, N. D., Weisman, B., Lusis, M., Chan, W., and Vet, R. (1982). Rain scavenging from tall stack plumes: A new experimental approach. *Atmos. Environ.* **16,** 2709–2714.

Miller, C. A. (1975). The spruce budworm: How it lives and what it does. *For. Chron.* **51,** 136–138.

Miller, D. R. (1984). Chemicals in the environment. *In* "Effects of Pollutants at the Ecosystem Level" (P. J. Sheehan, D. R. Miller, G. C. Butler, and P. Boudreau, eds.), pp. 7–14. Wiley, New York.

Miller, E., and Miller, D. R. (1980). Snag use by birds. *In* "Management of Western Forests and Grasslands for Non-game Birds," U.S.D.A. For. Serv., Gen. Tech. Rep. INT-86, pp. 337–356. Intermountain Forest and Range Experiment Station, Ogden, Utah.

Miller, E. L., Beasley, R. S., and Lawson, E. R. (1985). Stormflow, sedimentation, and water quality responses following silvicultural treatments in the Ouachita Mountains. *In* "Proceedings, Forestry and Water Quality: A Mid-South Symposium" (B. G. Blackman, ed.), pp. 117–129. University of Arkansas, Little Rock.

Miller, G. E., Grant, P. M., Kishore, R., Steinbruger, F. J., Rowland, F. S., and Guinn, V. P. (1972). Mercury concentrations in museum specimens of tuna and swordfish. *Science* **175,** 1121–1122.

Miller, G. T. (1985). "Living in the Environment." Wadsworth, Belmont, California.

Miller, J. H., and Sirois, D. L. (1986). Soil disturbance by skyline yarding vs. skidding in a loamy hill forest. *Soil Sci. Soc. Am. J.* **50,** 1579–1583.

Miller, L. K., Lingg, A. J., and Bulla, L. A. (1983). Bacterial, viral, and fungal insecticides. *Science* **219,** 715–721.

Miller, M., Wasik, S. P., Huang, G. L., Shiu, W. Y., and McKay, D. (1985). Relationships between octanol-water partition coefficients and aqueous solubility. *Environ. Sci. Technol.* **19,** 522–529.

Miller, P. R. (1973). Oxidant-induced community change in a mixed conifer forest. *Adv. Chem. Ser.* **122,** 101–117.

Miller, P. R., McCutchan, M. H., and Milligan, H. P. (1972). Oxidant air pollution in the Central Valley, Sierra Nevada foothills, and Mineral King Valley of California. *Atmos. Environ.* **6,** 623–633.

Mills, K. H. (1984). Fish population responses to experimental acidification of a small Ontario lake. *In* "Early Biotic Responses to Advancing Lake Acidification" (G. R. Hendry, ed.), pp. 117–131. Butterworth, Toronto, Ontario.

Mills, K. H. (1985). Responses of lake whitefish (*Coregonus clupeaformis*) to fertilization of Lake 223, the Experimental Lakes Area. *Can. J. Fish. Aquat. Sci.* **42,** 129–138.

Mills, K. H., and Chalanchuk, S. M. (1987). Population dynamics of lake whitefish (*Coregonus clupeaformis*) during and after the fertilization of Lake 226, the Experimental Lakes Area. *Can. J. Fish. Aquat. Sci.* **44,** Suppl. 1, 55–63.

Mills, K. H., Chalanchuk, S. M., Mohr, L. C., and Davies, I. J.

(1987). Responses of fish populations in Lake 223 to 8 years of experimental acidification. *Can. J. Fish. Aquat. Sci.* **44,** Suppl. 1, 114–125.

Milne, A. R., and Herlinveaux, R. H. (1977). "Crude Oil in Cold Water." Department of Fisheries and Department of the Environment, Sidney, British Columbia.

Ministry of Agriculture and Environment '82 Committee (MAEC) (1983). MAEC, Royal Ministry of Agriculture, Stockholm, Sweden.

Mishra, D., and Kar, M. (1974). Nickel in plant growth and metabolism. *Bot. Rev.* **40,** 395–452.

Mitchell, G. C. (1986). Vampire bat control in Latin America. *In* "Ecological Knowledge and Environmental Problem-Solving," pp. 151–163. National Academy Press, Washington, D.C.

Mitchell, M. F., and Roberts, J. R. (1984). A case study of the use of fenitrothion in New Brunswick: The evolution of an advanced approach to ecological monitoring. *In* "Effects of Pollutants at the Ecosystem Level" (P. J. Sheehan, D. R. Miller, G. C. Butler, and P. Boudreau, eds.), pp. 377–402. Wiley, New York.

Mittermeier, R. A. (1988). Primate diversity and the tropical forest. Case studies from Brazil and Madagascar and the importance of megadiversity countries. *In* "Biodiversity" (E. O. Wilson, ed.), pp. 145–154. National Academy Press, Washington, D.C.

Mix, M. C. (1986). Cancerous diseases in aquatic animals and their association with environmental pollutants: A critical literature review. *Mar. Environ. Res.* **20,** 1–141.

Moller, D. (1984). Estimation of the global man-made sulfur emission. *Atmos. Environ.* **18,** 19–27.

Mollitor, A. V., and Raynal, D. J. (1983). Atmospheric deposition and ionic input in Adirondack forests. *J. Air Pollut. Control Assoc.* **33,** 1032–1036.

Monk, C. D. (1966). Effects of short-term gamma irradiation on an old field. *Radiat. Bot.* **6,** 329–335.

Moore, G. C., and Boer, A. H. (1979). "New Brunswick Deer Yard Management Projects: A Summary." New Brunswick Department of Natural Resources, Fredericton.

Moore, N. W., and Hooper, M. D. (1975). On the number of bird species in British woods. *Biol. Conserv.* **8,** 239–250.

Moore, T. R., and Zimmermann, R. C. (1972). Establishment of vegetation on serpentine asbestos mine wastes, southeastern Quebec, Canada. *J. Appl. Ecol.* **14,** 589–599.

Morash, R., and Freedman, B. (1987). "Initial Impacts of Silvicultural Herbicide Spraying on the Vegetation of Regenerating Clearcuts in Central Nova Scotia," Research Report to Maritimes Forest Research Centre. School for Resource and Environmental Studies, Dalhousie University, Halifax, Nova Scotia.

Morash, R., Freedman, B., and McCurdy, R. (1987). Persistence of glyphosate in foliage and litter. *In* "Measurement of the Environmental Effects Associated with Forestry Use of Roundup." pp. 15–19. Environment Canada, Environmental Protection Service, Dartmouth, Nova Scotia.

Morgan, K., and Freedman, B. (1986). Breeding bird communities in a hardwood forest succession in Nova Scotia. *Can. Field-Nat.* **100**, 506–519.

Morris, O. N. (1982). Bacteria as pesticides: Forest applications. *In* "Microbial and Viral Pesticides" (E. Kurstak, ed.), pp. 239–287. Dekker, New York.

Morris, R. F., Cheshire, W. F., Miller, C. A., and Mott, D. G. (1958). The numerical response of avian and mammalian predators during a gradation of the spruce budworm. *Ecology* **39**, 487–494.

Morrison, I. K. (1981). Effect of simulated acid precipitation on composition of percolate from reconstructed profiles of two northern Ontario forest soils. *Can. For. Serv., Res. Notes* **1**, 6–8.

Morrison, I. K. (1983). Composition of percolate from reconstructed profiles of jack pine forest soils as influenced by acid input. *In* "Effects of Accumulations of Air Pollutants in Forest Ecosystems" (B. Ulrich and J. Pankrath, eds.), pp. 195–206. Reidel, Berlin.

Morrison, I. K. (1984). A review of literature on acid deposition effects on forest ecosystems. *Commonw. For. Bur., For. Abstr.* **45**, 483–505.

Morrison, M. L., and Meslow, E. C. (1983). Impacts of forest herbicides on wildlife: Toxicity and habitat alteration. *Trans. North Am. Wildl. Nat. Resour. Conf.* **48**, 175–185.

Morrison, M. L., and Meslow, E. C. (1984a). Response of avian communities to herbicide-induced habitat changes. *J. Wildl. Manage.* **48**, 14–22.

Morrison, M. L., and Meslow, E. C. (1984b). Effects of the herbicide glyphosate on bird community structure, western Oregon. *For. Sci.* **30**, 95–106.

Morrison, R. G., and Yarranton, G. A. (1973). Diversity, richness, and evenness during a primary sand dune succession at Grand Bend, Ontario. *Can. J. Bot.* **51**, 2401–2411.

Mueller-Dombois, D. (1980). The 'Ohia'a dieback phenomenon in the Hawaiian rain forest. *In* "The Recovery Process in Damaged Ecosystems" (J. Cairns, ed.), pp. 153–161. Ann Arbor Science Publ., Ann Arbor, Michigan.

Mueller-Dombois, D. (1986). Perspectives for an etiology of stand-level dieback. *Annu. Rev. Ecol. Syst.* **17**, 221–243.

Mueller-Dombois, D. (1987a). Natural dieback in forests. *Bio Science* **37**, 575–583.

Mueller-Dombois, D. (1987b). Forest dynamics in Hawaii. *TREE* **2**, 216–220.

Mueller-Dombois, D., Canfield, J. E., Holt, R. A., and Buelow, G. P. (1983). Tree-group death in North America and Hawaiian forests: A pathological problem or a new problem for vegetation ecology? *Phytocoenologia* **11**, 117–137.

Muir, P. S., Kimberley, K. A., Carter, B. H., Armentano, T. V., and Pribush, R. A. (1986). Fog chemistry at an urban midwestern site. *J. Air Pollut. Control Assoc.* **36**, 1359–1361.

Muller, E. F., and Kramer, J. R. (1977). Precipitation scavenging in central and northern Ontario. *ERDA Symp. Ser.* **41**, 590–601.

Muller, P. (1980). Effects of artificial acidification on the growth of periphyton. *Can. J. Fish. Aquat. Sci.* **37**, 355–363.

Mullinson, W. R. (1981). "Public Concerns about the Herbicide 2,4-D." Dow Chemical, Midland, Michigan.

Munger, J. W. (1982). Chemistry of atmospheric precipitation in the north-central United States: Influence of sulfate, nitrate, ammonium, and calcareous soil particulates. *Atmos. Environ.* **16**, 1633–1645.

Munger, J. W., and Eisenreich, S. J. (1983). Continental-scale variations in precipitation chemistry. *Environ. Sci. Technol.* **17**, 32A–42A.

Munn, R. E., Likens, G. E., Weisman, B., Hornbeck, J. W., Martin, C. W., and Bormann, F. H. (1984). A meteorological analysis of precipitation chemistry event samples at Hubbard Brook, N.H. *Atmos. Environ.* **18**, 2775–2779.

Murphy, J. E. (1987). "The 1985–86 Canadian Peregrine Falcon Survey." Canadian Wildlife Service, Ottawa, Ontario.

Murphy, M. L., and Hall, J. D. (1983). Varied effects of clearcut logging on predators and their habitat in small streams in the Cascade Mountains, Oregon. *Can. J. Fish. Aquat. Sci.* **38**, 137–145.

Murphy, M. L., Hawkins, C. P., and Anderson, N. H. (1981). Effects of canopy modification and accumulated sediment on stream communities. *Trans. Am. Fish. Soc.* **110**, 469–478.

Murphy, S. D. (1980). Pesticides. *In* "Toxicology: The Basic Science of Poisons" (J. Doull, C. D. Klaassen, and M. O. Amdur, eds.), pp. 357–408. Macmillan, New York.

Murray, F. (1981). Effects of fluorides on plant communities around an aluminum smelter. *Environ. Pollut., Ser. A* **24**, 45–56.

Myers, N. (1979). "The Sinking Ark: A New Look at the Problem of Disappearing Species." Pergamon, Oxford.

Myers, N. (1983). "A Wealth of Wild Species." Westview Press, Boulder, Colorado.

Myers, N. (1988). Tropical forests and their species. Going . . . Going . . . *In* "Biodiversity" (E. O. Wilson, ed.), pp. 28–35. National Academy Press, Washington, D.C.

Nakatsu, C. H. (1983). The algal flora and ecology of three brown water systems. M.Sc. Thesis, Department of Botany, University of Toronto, Toronto, Ontario.

Nams, M. L. N., and Freedman, B. (1987a). Ecology of heath communities dominated by *Cassiope tetragona* at Alexandra Fiord, Ellesmere Island, Canada. *Holarctic Ecol.* **10**, 22–32.

Nams, M. L. N., and Freedman, B. (1987b). Phenology and resource allocation in a high arctic evergreen dwarf shrub, *Cassiope tetragona. Holarctic Ecol.* **10**, 128–136.

Narver, D. W. (1970). Effects of logging debris on fish production. *In* "Forest Land Uses and Stream Environment," pp. 100–111. Oregon State University, Corvallis.

National Academy of Sciences (NAS) (1971). "Fluorides." NAS, Washington, D.C.

National Academy of Sciences (NAS) (1972). "Lead—Airborne Lead in Perspective." Committee on Biological Effects of Atmospheric Pollutants, NAS, Washington, D.C.

National Academy of Sciences (NAS) (1974). "The Effects of Herbicides in South Vietnam," Part A. Committee on the Effects of Herbicides in Vietnam, NAS, Washington, D.C.

National Academy of Sciences (NAS) (1975a). "Forest Pest Control." NAS, Washington, D.C.

National Academy of Sciences (NAS) (1975b). "Underexploited Tropical Plants." NAS, Washington, D.C.

National Academy of Sciences (NAS) (1975c). "Contemporary Pest Control Practices and Prospects." NAS, Washington, D.C.

National Academy of Sciences (NAS) (1975d). "Petroleum in the Marine Environment." NAS, Washington, D.C.

National Academy of Sciences (NAS) (1976). "Halocarbons: Effects on Stratospheric Ozone." NAS, Washington, D.C.

National Academy of Sciences (NAS) (1979a). "Stratospheric Ozone Depletion by Halocarbons: Chemistry and Transport." NAS, Washington, D.C.

National Academy of Sciences (NAS) (1979b). "Tropical Legumes: Resources for the Future." NAS, Washington, D.C.

National Academy of Sciences (NAS) (1985). "Oil in the Sea." NAS, Washington, D.C.

National Acid Precipitation Assessment Program (NAPAP) (1983). "Annual Report 1983." NAPAP, Interagency Task Force on Acid Precipitation, Washington, D.C.

National Oceanic and Atmospheric Administration (NOAA-CNEXO) (1982). "Ecological Study of the Amoco Cadiz Oil Spill," Report of the NOAA-CNEXO Joint Scientific Commission. U.S. Department of Commerce, Washington, D.C.

National Research Council (NRC) (1976). "Vapour-Phase Organic Pollutants." NRC, Committee on Medical and Biologic Effects of Environmental Pollutants, Washington, D.C.

National Research Council (NRC) (1982). "Carbon Dioxide and Climate: A Second Assessment." National Research Council, National Academy Press, Washington, D.C.

National Research Council (NRC) (1983). "Changing Climate." National Research Council, National Academy Press, Washington, D.C.

National Research Council (NRC) (1985a). "The Effects on the Atmosphere of a Major Nuclear Exchange." National Research Council, National Academy Press, Washington, D.C.

National Research Council (NRC) (1985b). "Oil in the Sea. Inputs, Fates, and Effects." National Research Council, National Academy Press, Washington, D.C.

National Research Council (NRC) (1986). "Pesticide Resistance." National Research Council, National Academy Press, Washington, D.C.

Nauwerck, A. (1963). Die neziehungen zwischen zooplankton and phytoplankton in See Erken. *Symb. Bot. Ups.* **17,** 1–163 (cited in Ostrofsky and Duthie, 1975).

Neary, D. G., and Currier, J. B. (1982). Impact of wildfire and watershed management on water quality in South Carolina's Blue Ridge Mountains. *South. J. Appl. For.* **6,** 81–90.

Neff, J. M., and Anderson, J. W. (1981). "Responses of Marine Mammals to Petroleum and Specific Petroleum Hydrocarbons." Appl. Sci. Publ., London.

Neff, J. M., and Haensly, W. E. (1982). Long-term impact of the Amoco Cadiz crude oil spill on oysters *Crassostrea gigas* and plaice *Pleuronectes platessa* from Aber Benoit and Aber Wrac'h, Brittany, France. *In* "Ecological Study of the Amoco Cadiz Oil Spill." U.S. Department of Commerce, National Oceanic and Atmospheric Administration, Washington, D.C.

Nelson, P. F., and Quigley, S. M. (1984). The hydrocarbon composition of exhaust emitted from gasoline fueled vehicles. *Atmos. Environ.* **18,** 79–87.

Nelson, R. W. (1976). Behavioural aspects of egg breakage in peregrine falcons. *Can. Field-Nat.* **90,** 320–329.

Nelson-Smith, A. (1977). Recovery of some British rocky seashores from oil spills and cleanup operations. *In* "Recovery and Restoration of Damaged Ecosystems" (J. Cairns, K. L. Dickson, and E. E. Herricks, eds.), pp. 191–207. Univ. Press of Virginia, Charlottesville.

Nero, R. W., and Schindler, D. W. (1983). Decline of *Mysis relicta* during the acidification of Lake 223. *Can. J. Fish. Aquat. Sci.* **40,** 1905–1911.

Nettleship, D. N., and Evans, P. G. H. (1986). Distribution and status of the Atlantic Alcidae. *In* "The Atlantic Alcidae" (D. N. Nettleship and T. R. Birkhead, eds.), pp. 54–154. Academic Press, New York.

Newman, H. C., and Schmidt, W. C. (1980). Silviculture and residue treatment affect water used by a larch/fir forest. *In* "Environmental Consequences of Timber Harvesting in Rocky Mountain Coniferous Forests," U.S.D.A. For. Serv., Gen. Tech. Rep. INT-90, pp. 75–100. Intermountain Forest and Range Experiment Station, Ogden, Utah.

Newman, L. (1981). Atmospheric oxidation of sulfur dioxide: A review as viewed from power plant and smelter plume studies. *Atmos. Environ.* **15,** 2231–2239.

Newton, M. (1975). Constructive use of herbicides in forest resource management. *J. For.* **73,** 329–336.

Newton, M., and Knight, F. B. (1981). "Handbook of Weed and Insect Control Chemicals for Forest Resource Managers." Timber Press, Beaverton, Oregon.

Nigam, P. C. (1975). Chemical insecticides. *In* "Aerial Control of Forest Insects in Canada" (M. L. Prebble, ed.), pp. 8–24. Department of the Environment, Ottawa, Ontario.

Nihlgard, B. (1985). The ammonium hypothesis—an additional explanation to the forest dieback in Europe. *Ambio* **14,** 2–8.

Nilssen, J. P. (1980). Acidification of a small watershed in southern Norway and some characteristics of acidic aquatic environments. *Int. Rev. Gesamten Hydrobiol.* **65,** 177–207.

Nilsson, S., and Duinker, P. (1987). The extent of forest decline in Europe. *Environment* **29**(9), 4–31.

Nilsson, S. I., Miller, H. G., and Miller, J. D. (1982). Forest growth as a possible cause of soil and water acidification: an examination of the concepts. *Oikos* **39,** 40–49.

Noon, B. R., Bingham, V. P., and Noon, J. P. (1979). The effects of changes in habitat on northern hardwood forest bird communities. *In* "Management of North Central and Northeastern Forests for Non-game Birds," U.S.D.A. For. Serv., Gen. Tech. Rep. NC-51, pp. 33–48. North Central Forest Experiment Station, St. Paul, Minnesota.

Norheim, G., Somme, L., and Holt, G. (1982). Mercury and persistent chlorinated hydrocarbons in Antarctic birds from Bouvetoya and Dronning Maud Land. *Environ. Pollut., Ser. A* **28**, 233–240.

Norris, L. A. (1981). The movement, persistence, and fate of the phenoxy herbicides and TCDD in the forest floor. *Residue Rev.* **80**, 65–135.

Norris, L. A., Montgomery, M. L., and Johnson, E. R. (1977). The persistence of 2,4,5-T in a Pacific Northwest forest. *Weed Sci.* **25**, 417–422.

Norris, L. A., Montgomery, M. L., Warren, L. E., and Mosher, W. D. (1982). Brush control with herbicides on hill pasture sites in southern Oregon. *J. Range Manage.* **35**, 75–80.

Norris, L. A., Lorz, H. W., and Gregory, S. V. (1983). Influence of forest and rangeland management on anadromous fish habitat in North America. *USDA For. Gen. Tech. Rep.* PNW-149. Pacific Northwest Forest and Range Experiment Station, Portland, Oregon.

Northridge, S. P. (1984). "World Review of Interactions Between Marine Mammals and Fisheries," FAO Fish. Tech. Pap. 251. Food and Agricultural Organization of the United Nations, Rome, Italy.

Norton, S. A., and Young, H. E. (1976). Forest biomass utilization and nutrient budgets. *In* "Oslo Biomass Studies," pp. 56–73. University of Maine, Orono.

Nounou, P. (1980). The oil spill age. *Ambio* **9**, 297–302.

Nova Scotia Department of Lands and Forests (NSLF) (1975). "Green Belts for Wildlife in Nova Scotia." Wildlife Division, NSLF, Truro.

Nunn, M. D. (1983). "Decision in the Supreme Court of Nova Scotia, Trial Division, between Palmer, Googoo, Mullendore, Francis, MacGillivary, Sampson, Shaw, MacIntyre, MacInnes, Mustard, Dauphney, Haldeman, Grose, Epifano, and Calvert; and Nova Scotia Forest Industries." Supreme Court of Nova Scotia, Trial Division, Halifax.

O'Brien, P. Y., and Dixon, P. S. (1976). The effects of oils and oil components on algae: A review. *Br. Phycol. J.* **11**, 115–142.

Oden, S. (1976). The acidity problem—an outline of concepts. *USDA For. Serv. Gen. Tech. Rep.* NE-23, 1–36. Northeastern Forest Experiment Station, Upper Darby, Pennsylvania.

Oden, S., and Andersson, R. (1971). "The Longterm Changes in the Chemistry of Soils in Scandinavia due to Acid Precipitation," Sect. 5.1. Sweden's Case Study for United Nations Conference on the Human Environment, Stockholm.

Odum, E. P. (1981). The effects of stress on the trajectory of ecological succession. *In* "Stress Effects on Natural Eco-systems" (G. W. Barrett and R. Rosenberg, ed.), pp. 43–47. Wiley, New York.

Odum, E. P. (1983). "Basic Ecology." Saunders College Publishing, New York.

Odum, E. P. (1985). Trends expected in stressed ecosystems. *BioScience* **35**, 419–422.

Oechel, W. C., and Strain, B. R. (1985). Native species responses to increased atmospheric carbon dioxide concentration. *In* "Direct Effects of Increasing Carbon Dioxide on Vegetation," DOE/ER-0238, pp. 117–155. U.S. Department of Energy, Washington, D.C.

Ogner, G., and Tiegen, O. (1980). Effects of acid irrigation and liming on two clones of Norway spruce. *Plant Soil* **57**, 305–321.

Ohlendorf, H. M., Risebrough, R. W., and Vermeer, K. (1978). "Exposure of Marine Birds to Environmental Pollutants," Wildl. Res. Rep. 9. U.S. Department of the Interior, Fish and Wildlife Service, Washington, D.C.

Oliver, B. G., Thurman, E. M., and Malcolm, R. L. (1983). The contribution of humic substances to the acidity of coloured natural waters. *Geochim. Cosmochim. Acta* **47**, 2031–2035.

Olson, J. S. (1958). Rates of succession and soil changes on southern Lake Michigan sand dunes. *Bot. Gaz. (Chicago)* **119**, 125–170.

Olson, S. L., Swift, C. C., and Mokhiber, C. (1979). An attempt to determine the prey of the great auk, *Pinguinis impennis*. *Auk* **96**, 790–792.

Organization for Economic Cooperation and Development (OECD) (1985). "The State of the Environment 1985." OECD, Paris, France.

Orians, G. H., and Pfeiffer, E. W. (1970). Ecological effects of the war in Vietnam. *Science* **168**, 544–554.

Ostaff, D. P. (1985). Quantifying effects of spruce budworm damage in eastern Canada. *In* "Recent Advances in Spruce Budworms Research," pp. 247–248. Canadian Forestry Service, Ottawa, Ontario.

Ostrovsky, M. L., and Duthie, H. C. (1975). Primary productivity and phytoplankton of lakes on the eastern Canadian Shield. *Verh.—Int. Ver. Theor. Angew. Limnol.* **19**, 732–738.

Otvos, I. S., and Raske, A. G. (1980). "Effects of Aerial Application of Matacil on Larval and Pupal Parasites of the Eastern Spruce Budworm," Inf. Rep. N-X-189. Newfoundland Forest Research Centre, St. John's.

Overrein, L. N. (1972). Sulphur pollution patterns observed; leaching of calcium in forest soil determined. *Ambio* **1**, 145–147.

Ovington, J. D. (1959). Mineral content of plantations of *Pinus sylvestris*. *Ann. Bot. (London)* [N.S.] **23**, 75–88.

Ovington, J. D. (1962). Quantitative ecology and the woodland ecosystem concept. *Adv. Ecol. Res.* **1**, 103–192.

Owadally, A. W. (1979). The dodo and the tambalaque tree. *Science* **203**, 1363–1364.

Packer, P. E. (1967a). Criteria for designing and locating logging roads to control sediment. *For. Sci.* **13**, 2–18.

Packer, P. E. (1967b). Forest treatment effects on water quality. *Int. Symp. For. Hydrol.* pp. 687–699.

Packer, P. E., and Williams, B. D. (1976). Logging and prescribed burning effects on the hydrologic and soil stability behaviour of larch/Douglas-fir forests in the northern Rocky Mountains. *Proc. Tall Timbers Fire Conf.* **14**, 465–479.

Page, A. L. (1974). "Fate and Effects of Trace Elements in Sewage Sludge Applied to Agricultural Lands," Publ. EPA-670/2-74-005. Office of Research and Development, U.S. Environmental Protection Agency, Cincinnati, Ohio.

Page, A. L., Ganje, T. J., and Joshi, M. S. (1971). Lead quantities in plants, soil, and air near some major highways in southern California. *Hilgardia* **41**, 1–31.

Paijmans, J. (1970). An analysis of four tropical rain forest sites in New Guinea. *J. Ecol.* **58**, 77–101.

Palumbo, R. F. (1962). Recovery of the land plants at Eniwetok Atoll following a nuclear explosion. *Radiat. Bot.* **1**, 182–189.

Parker, C. A., Freegarde, M., and Hatchard, C. G. (1971). The effect of some chemical and biological factors on the degradation of crude oil at sea. *In* "Water Pollution by Oil," pp. 237–244. Institute of Petroleum, London.

Parker, G. R., and Morton, L. D. (1978). The estimation of winter forage and its use by moose on clearcuts in north-central Newfoundland. *J. Range Manage.* **31**, 300–304.

Pastor, J., Gardner, R. H., Dale, V. H., and Post, W. M. (1987). Seasonal changes in nitrogen availability as a potential factor contributing to spruce declines in boreal North America. *Can. J. For. Res.* **17**, 1394–1400.

Patton, A. (1986). "1985 Deer Harvest Report." Nova Scotia Department of Lands and Forests, Truro.

Patton, D. R. (1974). Patch cutting increased deer and elk use of a pine forest in Arizona. *J. For.* **72**, 764–766.

Payne, B. R., and DeGraaf, R. M. (1975). Economic values and recreational trends associated with human enjoyment of non-game birds. *In* "Management of Forest and Range Habitats for Non-game Birds," U.S.D.A. For. Serv., Gen. Tech. Rep. WO-1, pp. 6–10. USDA, Washington, D.C.

Payne, J. R., and McNabb, G. D. (1984). Weathering of petroleum in the marine environment. *Mar. Technol. Soc. J.* **18**(3), 24–42.

Peakall, D. B. (1976). The peregrine falcon (*Falco peregrinus*) and pesticides. *Can. Field-Nat.* **90**, 301–307.

Pearce, P. A. (1975). Effects on birds. *In* "Aerial Control of Forest Insects in Canada" (M. L. Prebble, ed.), pp. 306–313. Department of the Environment, Ottawa, Ontario.

Pearce, P. A., and Busby, D. G. (1980). Research on the effects of fenitrothion on the white-throated sparrow. *In* "Environmental Surveillance in New Brunswick," pp. 24–28. EMOFICO, University of New Brunswick, Fredericton.

Pearce, P. A., and Peakall, D. B. (1977). The impact of fenitrothion on bird populations in New Brunswick. *Natl. Res. Counc. Can., NRC Assoc. Comm. Sci. Criter. Environ. Qual. [Rep.] NRCC, NRCC/CNRC* **16073**, 299–305.

Pearce, P. A., Peakall, D. B., and Erskine, A. J. (1979). "Impact on Forest Birds of the 1976 Spruce Budworm Spray Operation in New Brunswick," Prog. Note No. 97. Canadian Wildlife Service, Environment Canada, Ottawa, Ontario.

Peek, J. M. (1974). A review of moose food habits in North America. *Nat. Can.* **101**, 195–215.

Pengelly, W. L. (1972). Clearcutting: Detrimental aspects for wildlife resources. *J. Soil Water Conserv.* **27**, 255–258.

Penner, J. E., Haselman, L. C., and Edwards, L. L. (1986). Smoke-plume distribution above large-scale fires: Implications for simulations of "nuclear winter." *J. Clim. Appl. Meteorol.* **25**, 1434–1444.

Percy, J. A., and Wells, P. G. (1984). Effects of petroleum in polar marine environments. *Mar. Technol. Soc. J.* **18**(3), 51–61.

Percy, K. E. (1983). Sensitivity of eastern Canadian forest tree species to simulated acid precipitation. *Aquilo, Ser. Bot.* **19**, 41–49.

Percy, K. E. (1986). The effects of simulated acid rain on germinative capacity, growth, and morphology of forest tree seedlings. *New Phytol.* **104**, 473–484.

Perkins, H. C. (1974). "Air Pollution." McGraw-Hill, New York.

Perry, I., and Moore, P. D. (1987). Dutch elm disease as an analogue of Neolithic elm decline. *Nature (London)* **326**, 72–73.

Persson, G., chair. (1982). "Acidification Today and Tomorrow." Ministry of Agriculture, Environment '82 Committee, Stockholm, Sweden.

Persson, H. (1948). On the discovery of *Merceya ligulata* in the Azores with a discussion of the so-called copper mosses. *Rev. Bryol. Lichenol.* **17**, 75–78.

Persson, R. (1974). "Review of the World's Forest Resources in the Early 1970s." Skogshogsholm, Stockholm, Sweden (cited in Donaldson, 1978).

Peters, R. L., and Darling, J. D. S. (1985). The greenhouse effect and nature reserves. *BioScience* **35**, 707–717.

Peters, T. H. (1984). Rehabilitation of mine tailings: A case of complete ecosystem reconstruction and revegetation of industrially stressed lands in the Sudbury area, Ontario, Canada. *In* "Effects of Pollutants at the Ecosystem Level" (P. J. Sheehan, D. R. Miller, and P. Bourdeau, eds.), pp. 403–421. Wiley, New York.

Peterson, H. B., and Nielson, R. F. (1973). Toxicities and deficiencies in mine tailings. *In* "Ecology and Reclamation of Devastated Land" (R. J. Hutnik and G. Davis, eds.), Vol. 1, pp. 15–24. Gordon & Breach, New York.

Peterson, J. (1984). Global population projections through the 21st century. *Ambio* **13**, 134–141.

Peterson, P. J. (1975). Element accumulation by plants and their tolerance of toxic mineral soils. In "Symposium Proceeding, Vol. II, International Conference on Heavy Metals in the Environment." (T. C. Hutchinson, ed.), pp. 39–59. Toronto, Canada.

Peterson, P. J. (1978). Land and vegetation. *In* "The Biogeochemistry of Lead in the Environment" (J. O. Nriagu, ed.), Part B. pp. 355–384. Elsevier/North-Holland Biomedical Press, Amsterdam, The Netherlands.

Peterson, P. J., and Butler, G. W. (1962). The uptake and assimilation of selenite by higher plants. *Aust. J. Bot.* **15**, 126–146.

Peterson, P. J., and Butler, G. W. (1967). Significance of selenocystathionine in an Australian selenium-accumulating plant, *Neptunia amplexicaulis. Nature (London)* **213**, 599–600.

Peterson, R. H., Daye, P. G., and Metcalfe, J. L. (1980). Inhibition of Atlantic salmon (*Salmo salar*) hatching at low pH. *Can. J. Fish. Aquat. Sci.* **37**, 770–774.

Peterson, R. J. (1969). Population trends of ospreys in the northeastern U.S. *In* "Peregrine Falcon Populations: Their Biology and Decline" (J. J. Hickey, ed.), pp. 341–350. Univ. of Wisconsin Press, Madison.

Phillips, R. W. (1970). Effects of sediment on the gravel environment and fish production. *In* "Proceedings of the Conference on Forest Land Uses and Stream Environments," pp. 64–74. Oregon State University, Corvallis.

Piene, H. (1974). "Factors Influencing Organic Matter Decomposition and Nutrient Turnover in Cleared and Spaced, Young Conifer Stands on the Cape Breton Highlands, Nova Scotia," Inf. Rep. M-X-41. Maritimes Forest Research Centre, Fredericton, New Brunswick.

Pierce, R. C., Martin, C. W., Reeves, C. C., Likens, G. E., and Bormann, F. H. (1972). Nutrient loss from clearcuttings in New Hampshire. *In* "Watersheds in Transition," pp. 285–295. American Water Resources Association, Urbana, Illinois.

Pimlott, D., Brown, D., and Sam, K. (1976). "Oil Under the Ice." Canadian Arctic Resources Committee, Ottawa, Ontario.

Pimm, S. L. (1987). The snake that ate Guam. *TREE* **2**, 293–294.

Pitelka, F. A. (1941). Distribution of birds in relation to major biotic communities. *Am. Midl. Nat.* **25**, 113–137.

Pittock, A. B., Ackerman, T. B., Crutzen, P. J., MacCracken, M. C., Shapiro, C. S., and Turco, R. P. (1985). "Environmental Consequences of Nuclear War," Vol. 1, Scope 28. Wiley, Toronto, Ontario.

Planas, D., and Heckey, R. E. (1984). Comparison of phosphorus turnover times in northern Manitoba reservoirs with lakes of the Experimental Lakes Area. *Can. J. Fish. Aquat. Sci.* **41**, 605–612.

Platt, R. B. (1965). Ionizing radiation and homeostasis of ecosystems. *In* "Ecological Effects of Nuclear War" (G. M. Woodwell, ed.), Brookhaven Natl. Lab., AEC Rep. BNC-917(C-43), pp. 39–60. U.S. Atomic Energy Commission, Washington, D.C.

Plotkin, M. J. (1988). The outlook for new agricultural and industrial products from the tropics. *In* "Biodiversity" (E. O. Wilson, ed.), pp. 106–116. National Academy Press, Washington, D.C.

Plowright, R. C., Pendrel, B. A., and McLaren, I. A. (1978). The impact of aerial fenitrothion spraying upon the population biology of bumble bees (*Bombus* LATR: Hym.) in south-western New Brunswick. *Can. Entomol.* **110**, 1145–1156.

Porcella, D. B. (1978). Eutrophication. *J. Water Pollut. Control Fed.* **50**, 1313–1319.

Postel, S. (1987). Stabilizing chemical cycles. *In* "State of the World 1987," pp. 157–176. Worldwatch Institute, Norton, New York.

Prakash, A. (1976). "NTA (Nitrilotriacetic Acid)—An Ecological Appraisal," Rep. EPA 3-WP-76-8. Environmental Protection Service, Environment Canada, Ottawa, Ontario.

Prance, G. T. (1982). "Biological Diversification in the Tropics." Columbia Univ. Press, New York.

Prinz, B. (1984). "Recent Forest Decline in the Federal Republic of Germany and Contribution of the LIS for its Explanation." German/American Information Exchange on Forest Dieback, Excursion Guide, Essen, F.R.G.

Proctor, J. (1971a). The plant ecology of serpentine. II. Plant response to serpentine soils. *J. Ecol.* **59**, 397–410.

Proctor, J. (1971b). The plant ecology of serpentine. III. The influence of a high magnesium/calcium ratio and high nickel and chromium levels in some British and Swedish serpentine soils. *J. Ecol.* **59**, 827–842.

Proctor, J., and Woodell, S. R. J. (1975). The ecology of serpentine soils. *Adv. Ecol. Res.* **9**, 255–265.

Quy, V. (1984). Birds in the Mekong Delta. *In* "The First Vietnam National Symposium on Mangrove Ecosystems, 27–28 Dec., 1984," pp. 175–181. Hanoi, Vietnam.

Quy, V., Huynh, D. H., Landa, V., Leighton, M., Medvedev, L. W., Mototani, I., Pfeiffer, E. W., Sokolov, V. E., and Thang, T. T. (1984). Terrestrial animal ecology: Symposium summary. *In* "Herbicides in War: The Long-term Ecological and Human Consequences" (A. H. Westing, ed.), pp. 45–47. Taylor & Francis, London.

Raddum, G. (1978). Invertebrates: Quality and quantity as fish food. *In* "Limnological Aspects of Acid Precipitation" (G. R. Hendry, ed.), Rep. No. 51074. Brookhaven National Laboratory, Brookhaven, New York.

Raddum, G. G., and Fjellheim, A. (1984). Acidification and early warning organisms in freshwater in western Norway. *Int. Assoc. Theor. Appl. Limnol., Proc.* **2**, 1973–1980.

Rahel, F. J., and Magnuson, J. J. (1983). Low pH and the absence of fish species in naturally acidic Wisconsin lakes: Inferences for cultural acidification. *Can. J. Fish. Aquat. Sci.* **40**, 3–9.

Ramanathan, V. (1988). The greenhouse theory of climate change: a test by an inadvertent global experiment. *Science* **240**, 293–299.

Rampino, M. R., and Stothers, R. B. (1984). Geological rhythms and cometary impacts. *Science* **226**, 1427–1431.

Rapport, D. J. (1983). The stress-response environmental statistical system and its applicability to the Laurentian Lower Great Lakes. *Stat. J. U.N.* **ECE1**, 377–405.

Rapport, D. J., Regier, H. A., and Hutchinson, T. C. (1985). Ecosystem behaviour under stress. *Am. Nat.* **125**, 617–640.

Ratcliffe, D. A. (1967). Decrease in eggshell weight in certain birds of prey. *Nature (London)* **215**, 208–210.

Ratcliffe, D. A. (1970). Changes attributable to pesticides in egg breakage frequency and eggshell thickness in some British birds. *J. Appl. Ecol.* **7**, 67–115.

Raup, D. M. (1984). Death of species. *In* "Extinctions" (M. H. Nitecki, ed.), pp. 1–19. Univ. of Chicago Press, Chicago, Illinois.

Raup, D. M. (1986a). Biological extinctions in Earth history. *Science* **231**, 1528–1533.

Raup, D. M. (1986b). "The Nemesis Affair." Norton, New York.

Raup, D. M. (1988). Diversity crises in the geological past. *In* "Biodiversity" (E. O. Wilson, ed.), pp. 51–57. National Academy Press, Washington, D.C.

Reay, R. F. (1972). The accumulation of arsenic from arsenic-rich natural waters by aquatic plants. *J. Appl. Ecol.* **9**, 557–565.

Reed, F. L. C., and Associates, Ltd. (1980). Recent reductions in the Canadian timber base. *In* "The Forest Imperative: Proceedings of the Canadian Forest Congress," pp. 167–181. Toronto, Ontario.

Reeves, R. D., Brooks, R. R., and MacFarlane, R. M. (1981). Nickel uptake by *Streptanthus* and *Caulanthus* with particular reference to the hyperaccumulator *S. polygaloides* Gray (Brassicaceae). *Am. J. Bot.* **68**, 708–712.

Regier, H. A., and Hartman, W. L. (1973). Lake Erie's fish community: 150 years of cultural stresses. *Science* **180**, 1248–1255.

Regier, H. A., and Loftus, K. H. (1972). Effects of fisheries exploitation on salmonid communities in oligotrophic lakes. *J. Fish. Res. Board Can.* **29**, 959–968.

Reich, P. B., and Amundson, R. G. (1985). Ambient levels of ozone reduce net photosynthesis in tree and crop species. *Science* **230**, 566–570.

Reilly, A., and Reilly, C. (1973). Copper-induced chlorosis in *Becium homblei*. *Plant Soil* **38**, 671–674.

Reilly, C. (1967). Accumulation of copper by some Zambian plants. *Nature (London)* **215**, 666–669.

Reinhart, K. G. (1973). Timber-harvest, clearcutting, and nutrients in the northeastern United States. *USDA For. Serv., Res. Rep.* NE-170. Northeastern Forest Experiment Station, Broomall, Pennsylvania.

Reinhart, K. G., Eschner, A. R., and Trimble, G. R. (1963). Effect on streamflow of four forest practices in the mountains of West Virginia. *USDA For. Serv., Res. Pap.* NE-1. Northeastern Forest Experiment Station, Broomall, Pennsylvania.

Reuss, J. R. (1975). "Chemical/Biological Relationships Relevant to Ecological Effects of Acid Rainfall," Publ. EPA-660/3-75-032. U.S. Environmental Protection Agency, Corvallis, Oregon.

Reuss, J. R. (1976). Chemical and biological relationships relevant to the effect of acid rainfall on the soil-plant system. *USDA For. Serv. Gen. Tech. Rep.* NE-23, 791–813. Northeastern Forest Experiment Station, Broomall, Pennsylvania.

Reuss, J. R., and Johnson, D. W. (1985). Effect of soil processes on the acidification of water by acid precipitation. *J. Environ. Qual.* **14**, 26–31.

Reuss, J. R., Cosby, B. J., and Wright, R. F. (1987). Chemical processes governing soil and water acidification. *Nature (London)* **329**, 27–32.

Reynolds, H. G. (1962). Use of natural openings in a ponderosa pine forest of Arizona by deer, elk, and cattle. *USDA For. Serv., Res. Note* RM-78. Rocky Mountain Forest and Range Experiment Station, Fort Collins, Colorado.

Reynolds, H. G. (1966). Use of openings in spruce-fir forests of Arizona by elk, deer, and cattle. *USDA For. Serv., Res. Note* RM-66. Rocky Mountain Forest and Range Experiment Station, Fort Collins, Colorado.

Reynolds, H. G. (1969). Improvement of deer habitat on southwestern forest lands. *J. For.* **67**, 803–805.

Reynolds, H. W., and Hawley, A. W. L. (1987). "Bison Ecology in Relation to Agricultural Development in the Slave River Lowlands, N.W.T.," Occas. Pap. 63. Canadian Wildlife Service, Ottawa, Ontario.

Rice, R. M., Rothacher, J. S., and Megahan, W. F. (1972). Erosional consequences of timber harvesting: An appraisal. *In* "Watersheds in Transition," pp. 321–329. American Water Resources Association, Urbana, Illinois.

Richards, P. W. (1952). "The Tropical Rain Forest." Cambridge Univ. Press, London and New York.

Richardson, C. J., and Lund, J. A. (1975). Effects of clearcutting on nutrient losses in aspen forests on three soil types in Michigan. *In* "Mineral Cycling in Southeastern Ecosystems" (F. G. Howell, J. B. Gentry, and M. H. Smith, eds.), ERDA Symp. Ser., CONF-740513, pp. 673–686. ERDA, Washington, D.C.

Richman, A. D., Case, T. J., and Schwaner, T. D. (1988). Natural and unnatural extinction rates of reptiles on islands. *Am. Nat.* **131**, 611–630.

Richmond, M. L., Henny, C. J., Floyd, R. L., Mannan, R. W., Finch, D. M., and DeWeese, L. R. (1979). Effects of Sevin-4-oil, dimilin, and orthene on forest birds in northeastern Oregon. *USDA For. Serv., Res. Pap.* PSW-148. Pacific Southwest Forest and Range Experiment Station, Berkeley, California.

Riley, C. E. (1960). The ecology of water areas associated with coal strip-mined lands in Ohio. *Ohio J. Sci.* **60**, 106–121.

Risdale, P. S. (1916). Shot, shell and soldiers devastate forests. *Am. For.* **22**, 333–340.

Risdale, P. S. (1919a). Forest casualties of our allies. *Am. For.* **25**, 899–906.

Risdale, P. S. (1919b). French forests for our army. *Am. For.* **25**, 963–972.

Risdale, P. S. (1919c). War's destruction of British forests. *Am. For.* **25**, 1027–1040.

Risdale, P. S. (1919d). Belgium's forests blighted by the Hun. *Am. For.* **25**, 1251–1258.

Risebrough, R. W., Florant, G. L., and Berger, D. D. (1970). Organochlorine pollutants in peregrines and merlins migrating through Wisconsin. *Can. Field-Nat.* **84**, 247–253.

Rivers, J. B., Pearson, J. E., and Schultz, C. D. (1972). Total and organic mercury in marine fish. *Bull. Environ. Contam. Toxicol.* **8,** 257–266.

Robbins, C. S. (1979). Effects of forest fragmentation on bird populations. *In* "Management of North Central and Northeastern Forests for Non-game Birds," U.S.D.A. For. Serv., Gen. Tech. Rep. NC-51, pp. 198–212. North Central Forest Experiment Station, St. Paul, Minnesota.

Roberts, T. M. (1984). Effects of air pollutants in agriculture and forestry. *Atmos. Environ.* **18,** 629–652.

Robinson, E., and Robbins, R. C. (1972). Emissions, concentrations, and fate of gaseous atmospheric pollutants. *In* "Air Pollution Control" (W. Strauss, ed.), Part II, pp. 1–94. Wiley (Interscience), New York.

Rodhe, H., and Granat, L. (1984). An evaluation of sulfate in European precipitation 1955–1982. *Atmos. Environ.* **18,** 2627–2639.

Roff, D. A., and Bowen, W. D. (1983). Population dynamics and management of the northwest Atlantic harp seal (*Phoca groenlandica*). *Can. J. Fish. Aquat. Sci.* **40,** 919–932.

Roff, J. C., and Kwiatkowski, R. E. (1977). Zooplankton and zoobenthos communities of selected northern Ontario lakes of differing acidities. *Can. J. Zool.* **55,** 899–911.

Rogers, L. L., Mooty, J. J., and Dawson, D. (1981). Foods of the white-tailed deer in the upper Great Lakes region—a review. *USDA For. Serv. Gen. Tech. Rep.* NC-65. North Central Forest Experiment Station, St. Paul, Minnesota.

Rohlich, G. A., and Uttormark, P. D. (1972). Wastewater treatment and eutrophication. *In* "Nutrients and Eutrophication" (G. E. Likens, ed.), pp. 231–243. American Society of Limnology and Oceanography, Lawrence, Kansas.

Roose, M., Roberts, T. M., and Bradshaw, A. D. (1982). Evolution of resistance to gaseous air pollutants. *In* "Effects of Gaseous Pollutants in Agriculture and Horticulture" (M. H. Unsworth and D. P. Ormrod, eds.), pp. 379–410. Butterworth, London.

Rorison, I. H. (1980). The effects of soil acidity on nutrient availability and plant response. *In* "Effects of Acid Precipitation on Terrestrial Ecosystems" (T. C. Hutchinson and M. Havas, eds.), pp. 283–304. Plenum, New York.

Rosenberg, C. R., Hutnik, R. J., and Davis, D. D. (1979). Forest composition at varying distances from a coal-fired power plant. *Environ. Pollut.* **19,** 307–317.

Rosenberg, R., Lindahl, O., and Blank, H. (1988). Silent spring in the sea. *Ambio* **17,** 289–290.

Rosenfeld, C. L. (1980). Observations on the Mount St. Helen's eruption. *Am. Sci.* **68,** 494–509.

Rosenfeld, I., and Beath, O. A. (1964). "Selenium: Geobotany, Biochemistry, Toxicity, and Nutrition." Academic Press, New York.

Rosenqvist, I. T. (1978a). Alternative sources for acidification of river water in Norway. *Sci. Total Environ.* **10,** 39–49.

Rosenqvist, I. T. (1978b). Acid precipitation and other possible sources for acidification of rivers and lakes. *Sci. Total Environ.* **10,** 271–272.

Rosenqvist, I. T. (1980). Influence of forest vegetation and agriculture on the acidity of freshwater. *In* "Advances in Environmental Science and Engineering" (J. R. Pfallin and E. W. Ziegler, eds.), pp. 56–79. Gordon & Breach, New York.

Rosenqvist, I. T., Jorgensen, P., and Rueslatter, H. (1980). The importance of natural H^+ production for acidity in soil and water. *In* "Ecological Impact of Acid Precipitation" (D. Drablos and A. Tollan, eds.), pp. 240–242. SNSF Project, Oslo, Norway.

Ross, S. L., Logan, W. J., and Rowland, W. (1977). "Oil Spill Countermeasures." Department of Fisheries and the Environment, Sidney, British Columbia.

Rosseland, B. O., Skogheim, O. K., and Sevaldrud, I. H. (1986). Acid deposition and effects in northern Europe. Damage to fish populations in Scandinavia continue to apace. *Water, Air, Soil Pollut.* **30,** 65–74.

Roth, R. R. (1976). Spatial heterogeneity and bird species diversity. *Ecology* **57,** 773–782.

Rothacher, J. (1970). Increases in water yield following clearcut logging in the Pacific northwest. *Water Resour. Res.* **6,** 653–657.

Rothewell, R. L. (1971). "Watershed Management Guidelines for Logging and Road Construction," Inf. Rep. A-X-42. Northern Forest Research Centre, Edmonton, Alta.

Rotty, R. M., and Masters, C. D. (1985). Carbon dioxide from fossil fuel combustion: trends, resources, and technological implications. *In* "Atmospheric Carbon Dioxide and the Global Carbon Dioxide Cycle," pp. 63–80. U.S. Dept. Energy, Washington, D.C.

Rowe, R. D., and Chestnut, L. G. (1985). Economic assessment of the effects of air pollution on agricultural crops in the San Joaquin Valley. *J. Air Pollut. Control Assoc.* **35,** 728–734.

Royal Ministry of Foreign Affairs/Royal Ministry of Agriculture (RMFA/RMA) (1971). "Air Pollution Across National Boundaries: The Impact on the Environment of Sulfur in Air and Precipitation." RMFA/RMA, Stockholm, Sweden.

Royal Society of New Zealand (RSNZ) (1980). "Assessment of Toxic Hazards of the Herbicide 2,4,5-T in New Zealand," Misc. Ser. No. 4. RSNZ, Auckland.

Rowland, F. S. (1988). Chlorofluorocarbons, stratospheric ozone, and the Antarctic "ozone hole." *Environ. Conserv.* **15,** 101–116.

Royama, T. (1984). Population dynamics of the spruce budworm, *Choristoneura fumiferana*. *Ecol. Monogr.* **54,** 429–462.

Rozencranz, A. (1988). Bhopal, transnational corporations, and hazardous technologies. *Ambio* **17,** 336–341.

Rudd, J. W. M., Kelly, C. A., Schindler, D. W., and Turner, M. A. (1988). Disruption of the nitrogen cycle in acidified lakes. *Science* **240,** 1515–1517.

Ruhling, A., Baath, E., Nordgren, A., and Soderstrom, B. (1984). Fungi in metal-contaminated soil near the Gusum brass mill, Sweden. *Ambio* **13,** 34–36.

Russell, E. W. (1973). "Soil Conditions and Plant Growth," 10th ed. Longman, London.

Ryan, D. F., and Bormann, F. H. (1982). Nutrient resorption in northern hardwood forests. *BioScience* **32**, 29–32.

Ryding, S. O. (1985). Chemical and microbiological processes as regulators of the exchange of substances between sediments and water in shallow eutrophic lakes. *Int. Rev. Gesemten Hydrobiol.* **70**, 657–702.

Saber, P. A., and Dunson, W. A. (1978). Toxicity of bog water to embryonic and larval anuran amphibians. *J. Exp. Zool.* **204**, 33–42.

Sakamoto, M. (1966). Primary production by phytoplankton community in some Japanese lakes and its dependence on lake depth. *Arch. Hydrobiol.* **63**, 1–28.

Salisbury, E. J. (1943). The flora of bombed areas. *Nature (London)* **151**, 462–466.

Salop, J., Levy, G. F., Wakelyn, N. T., Middleton, E. M., and Gervin, J. C. (1983). The application of forest classification from Landsat data as a basis for natural hydrocarbon emission estimation and photochemical oxidant model simulations in southern Virginia. *J. Air Pollut. Control Assoc.* **33**, 17–22.

Sanders, C. J., Stark, R. W., Mullins, E. J., and Murphy, J., eds. (1985). "Recent Advances in Spruce Budworms Research." Canadian Forestry Service, Ottawa, Ontario.

Savidge, J. A. (1978). Wildlife in a herbicide-treated Jeffrey pine plantation in eastern California. *J. For.* **76**, 476–478.

Savidge, J. A. (1984). Guam: Paradise lost for wildlife. *Biol. Conserv.* **30**, 305–317.

Savidge, J. A. (1987). Extinction of an island forest avifauna by an introduced snake. *Ecology* **68**, 660–668.

Savile, D. B. O. (1972). "Arctic Adaptations in Plants," Monogr. No. 6. Research Branch, Canada Department of Agriculture, Ottawa, Ontario.

Scale, P. R. (1982). The effects of emissions from an iron-sintering plant in Wawa, Canada on forest communities. MES Thesis, Institute for Environmental Studies, University of Toronto, Toronto, Ontario.

Scarborough, A. M., and Flanagan, P. W. (1973). Observations on the effects of mechanical disturbance and oil on soil microbial populations. *In* "Proceedings of the Symposium on the Impact of Oil Resource Development on Northern Plant Communities," pp. 63–71. Institute of Arctic Biology, Fairbanks, Alaska.

Scavia, D., Fahnenstiel, G. L., Evans, M. S., Jude, D. J., and Lehman, J. T. (1986). Influence of salmonine predation and weather on long-term water quality in Lake Michigan. *Can. J. Fish. Aquat. Sci.* **43**, 435–443.

Scheibling, R. E. (1984). Echinoids, epizootics, and ecological stability in the rocky subtidal off Nova Scotia, Canada. *Helgol. Meeresunters.* **37**, 233–242.

Scheibling, R. E., and Stephenson, R. L. (1984). Mass mortality of *Strongylocentrotus droebachiensis* (Echinodermata: Echinoidea) off Nova Scotia, Canada. *Mar. Biol.* **78**, 153–164.

Scheider, W. A., Adamski, J., and Paylor, M. (1975). "Reclamation of Acidified Lakes Near Sudbury, Ontario." Ontario Ministry of the Environment, Rexdale.

Scheider, W. A., Cave, B., and Jones, J. (1976). "Reclamation of Acidified Lakes Near Sudbury, Ontario by Neutralization and Fertilization." Ontario Ministry of the Environment, Rexdale.

Scheider, W. A., Jeffries, D. S., and Dillon, P. J. (1980). "Bulk Deposition in the Sudbury and Muskoka-Haliburton Areas of Ontario During the Shutdown of INCO Ltd. in Sudbury." Ontario Ministry of the Environment, Rexdale.

Schelske, C. L., and Stoermer, E. F. (1972). Phosphorus, silica, and eutrophication of Lake Michigan. *In* "Nutrients and Eutrophication" (G. E. Likens, ed.), pp. 157–170. American Society of Limnology and Oceanography, Lawrence, Kansas.

Schelske, C. L., Stoermer, E. F., Fahnenstiel, G. L., and Haibach, M. (1986). Phosphorus enrichment, silica utilization, and biogeochemical silica depletion in the Great Lakes. *Can. J. Fish. Aquat. Sci.* **43**, 407–415.

Schindler, D. W. (1977). Evolution of phosphorus limitation in lakes: Natural mechanisms compensate for deficiencies of nitrogen and carbon in eutrophied lakes. *Science* **195**, 260–262.

Schindler, D. W. (1978). Factors regulating phytoplankton production and standing crop in the world's freshwaters. *Limnol. Oceanogr.* **23**, 478–486.

Schindler, D. W. (1985). The coupling of elemental cycles by organisms: Evidence from whole-lake chemical perturbations. *In* "Chemical Processes in Lakes" (W. Stumm, ed.), pp. 225–250. Wiley, New York.

Schindler, D. W. (1988). Effects of acid rain on freshwater ecosystems. *Science* **239**, 149–157.

Schindler, D. W., and Fee, E. J. (1974). Experimental Lakes Area: Whole-lake experiments in eutrophication. *J. Fish. Res. Board Can.* **31**, 937–953.

Schindler, D. W., and Holmgren, S. K. (1971). Primary production and phytoplankton of the Experimental Lakes Area, northwestern Ontario, and other low-carbonate waters, and a liquid scintillation method for determining ^{14}C activity in photosynthesis. *J. Fish. Res. Board Can.* **28**, 189–201.

Schindler, D. W., and Noven, B. (1971). Vertical distribution and seasonal abundance of zooplankton in two shallow lakes of the Experimental Lakes Area. *J. Fish. Res. Board Can.* **28**, 245–256.

Schindler, D. W., and Turner, M. A. (1982). Biological, chemical, and physical responses of lakes to experimental acidification. *Water, Air, Soil Pollut.* **18**, 259–271.

Schindler, D. W., Brunskill, G. J., Emerson, S., Broecker, W. S., and Perg, T. H. (1972). Atmospheric carbon dioxide: Its role in maintaining phytoplankton standing crop. *Science* **177**, 1192–1194.

Schindler, D. W., Kling, H., Schmidt, R. V., Prokopowich, J., Frost, V. E., Reid, R. A., and Capel, M. (1973). Eutrophication of Lake 227 by addition of phosphate and nitrate: The second, third, and fourth years of enrichment, 1970, 1971, 1972. *J. Fish. Res. Board Can.* **30**, 1415–1440.

Schindler, D. W., Kalff, J., Welch, H. E., Brunskill, G. J., Kling, H., and Kritsch, N. (1974). Eutrophication in the high arctic-Meretta Lake, Cornwallis Island (75°N Lat.). *J. Fish. Res. Board Can.* **31,** 647–662.

Schindler, D. W., Newbury, R. W., Beaty, K. G., and Campbell, P. (1976). Natural water and chemical budgets for a small Precambrian lake basin in eastern Canada. *J. Fish. Res. Board Can.* **33,** 2526–2543.

Schindler, D. W., Newbury, R. W., Beaty, K. G., Prokopowich, J., Ruszczynski, T., and Dulton, J. A. (1980). Effects of a windstorm and forest fire on chemical losses from forested watersheds and on the quality of receiving streams. *Can. J. Fish. Aquat. Sci.* **37,** 328–334.

Schindler, D. W., Mills, K. H., Malley, D. F., Findlay, D. L., Shearer, J. A., Davies, I. J., Turner, M. A., Linsey, G. A., and Cruikshank, D. A. (1985). Long-term ecosystem stress: The effects of years of experimental acidification on a small lake. *Science* **228,** 1395–1401.

Schindler, D. W., Turner, M. A., Stainton, M. P., and Linsey, G. A. (1986). Natural sources of acid neutralizing capacity in low alkalinity lakes of the Precambrian Shield. *Science* **232,** 843–847.

Schmidt, W. C., Felton, D. G., and Carlson, C. E. (1983). Alternatives to chemical insecticides in budworm-susceptible forests. *West. Wildl.* **9,** 13–19.

Schneider, S. H., and Thompson, S. L. (1988). Simulating the climatic effects of nuclear war. *Nature (London)* **333,** 221–227.

Schnitzer, M. (1978). Humic substances: Chemistry and reactions. *In* "Soil Organic Matter" (M. Schnitzer and S. U. Khan, eds.), pp. 1–64. Am. Elsevier, New York.

Schofield, C. L. (1976a). Lake acidification in the Adirondack Mountains of New York: Causes and consequences. *USDA For. Serv. Gen. Tech. Rep.* NE-23, 477. Northeastern Forest Experiment Station, Broomall, Pennsylvania.

Schofield, C. L. (1976b). Acid precipitation: Effects on fish. *Ambio* **5,** 228–230.

Schofield, C. L. (1982). Historical fisheries changes in the United States related to decreases in surface water pH. *Acid. Rain/Fish., Proc. Int. Symp., 1981* pp. 57–67. American Fisheries Society. Bethesda, MD.

Scoggan, H. J. (1950). "The Flora of Bic and the Gaspe Peninsula, Quebec," Bull. No. 115, Biol. Ser. No. 39. National Museum of Canada, Ottawa, Ontario.

Scott, J. T., Siccama, T. G., Johnson, A. H., and Breish, A. R. (1984). Decline of red spruce in the Adirondacks. *Bull. Torrey Bot. Club* **111,** 438–444.

Scott, V. E., Whelan, J. A., and Svoboda, P. L. (1980). Cavity-nesting birds and forest management. *In* "Management of Western Forests and Grasslands for Non-game Birds," U.S.D.A. For. Serv., Gen. Tech. Rep. INT-86, pp. 311–324. Intermountain Forest and Range Experiment Station. Ogden, Utah.

Scotter, G. W. (1967). The winter diet of barren-ground caribou in northern Canada. *Can. Field-Nat.* **81,** 33–39.

Seddon, B. (1972). Aquatic macrophytes as limnological indicators. *Freshwater Biol.* **2,** 107–130.

Seip, H. M., and Freedman, B. (1980). Effects of acid precipitation on soils. *In* "Effects of Acid Precipitation on Terrestrial Ecosystems." (T. C. Hutchinson and M. Havas, ed.), pp. 591–596. Plenum Press, New York.

Seip, H. M., and Tollan, A. (1978). Acid precipitation and other possible sources for acidification of rivers and lakes. *Sci. Total Environ.* **10,** 253–270.

Seneca, E. D., and Broome, S. W. (1982). Restoration of marsh vegetation impacted by the Amoco Cadiz oil spill and subsequent cleanup operations at Ile Grande, France. *In* "Ecological Study of the Amoco Cadiz Oil Spill," pp. 363–419. U.S. Department of Commerce, National Oceanic and Atmospheric Administration, Washington, D.C.

Shacklette, H. T. (1965). Bryophytes associated with mineral deposits and solutions in Alaska. *Geol. Surv. Bull. (U.S.)* **1198c.**

Shapiro, J. (1980). The importance of trophic-level interactions to the abundance and species composition of algae in lakes. *In* "Hypertrophic Ecosystems" (J. Barica and L. R. Mur, eds.), pp. 105–116. Junk, The Hague, The Netherlands.

Shapiro, J., and Wright, D. I. (1984). Lake restoration by biomanipulation: Round Lake, Minnesota, the first two years. *Freshwater Biol.* **14,** 371–383.

Shearer, J. A., Fee, E. J., DeBruyn, E. R., and DeClercq, D. R. (1987). Phytoplankton primary production and light attenuation responses to the experimental acidification of a small Canadian Shield lake. *Can. J. Fish. Aquat. Sci.* **44,** Suppl. 1, 83–90.

Sheath, R. G., Havas, M., Hellebust, J. A., and Hutchinson, T. C. (1982). Effects of long-term natural acidification on the algal communities of tundra ponds at the Smoking Hills, N.W.T., Canada. *Can. J. Bot.* **60,** 58–72.

Sheridan, D. (1981). "Desertification of the United States." Council on Environmental Quality, Superintendent of Documents, U.S. Govt. Printing Office, Washington, D.C.

Sherwood, M. J., and Mearns, A. J. (1977). Environmental significance of fin erosion in southern California demersal fishes. *Ann. N.Y. Acad. Sci.* **298,** 177–189.

Shields, L. M., and Wells, P. V. (1962). Effects of nuclear testing on desert vegetation. *Science* **135,** 38–40.

Shields, L. M., Wells, P. V., and Rickard, W. H. (1963). Vegetational recovery on atomic target areas in Nevada. *Ecology* **44,** 697–705.

Short, H. L., Blair, R. M., and Epps, E. A. (1975). "Composition and Digestibility of Deer Browse in Southern Forests," U.S.D.A. For. Serv., Res. Pap. SO-111. Southern Forest Experiment Station, New Orleans, Louisiana.

Shortle, W. C., and Smith, K. T. (1988). Aluminum-induced calcium deficiency syndrome in declining red spruce. *Science* **240,** 1017–1018.

Showman, R. E. (1975). Lichens as indicators of air quality around a coal-fired power generating station. *Bryologist* **78,** 1–6.

Showman, R. E. (1981). Lichen recolonization following air quality improvement. *Bryologist* **84,** 492–497.

Shrift, A. (1969). Aspects of selenium metabolism in higher plants. *Annu. Rev. Plant Physiol.* **20,** 475–494.

Shugart, H. H., Smith, T. M., Kitchings, J. T., and Kroodsma,

R. L. (1978). The relationship of nongame birds to southern forest types and successional stages. *In* "Proceedings of the Workshop on Management of Southern Forests for Nongame Birds," U.S.D.A. For. Serv., Gen. Tech. Rep. SE-14, pp. 5–16. Southeastern Forest Experiment Station, Asheville, North Carolina.

Shuman, T. (1984). The aging of the world's population. *Ambio* **13**, 175–181.

Siccama, T. G. (1974). Vegetation, soil, and climate on the Green Mountains of Vermont. *Ecol. Monogr.* **44**, 325–345.

Siccama, T. G., and Smith, W. H. (1978). Lead accumulation in a northern hardwood forest. *Environ. Sci. Technol.* **12**, 593–594.

Siccama, T. G., Bliss, M., and Vogelmann, H. W. (1982). Decline of red spruce in the Green Mountains of Vermont. *Bull. Torrey Bot. Club* **109**, 162–168.

Sidle, R. C., Hook, J. E., and Kardos, L. T. (1976). Heavy metals application and plant uptake in a land disposal system for waste water. *J. Environ. Qual.* **5**, 97–102.

Sigal, L. L., and Nash, T. H., III (1983). Lichen communities on conifers in southern California mountains: An ecological survey relative to oxidant air pollution. *Ecology* **64**, 1343–1354.

Simmons, F. C. (1979). "Handbook for Eastern Timber Harvesting." Superintendent of Documents, U.S. Gov. Printing Office, Washington, D.C.

Simon, C. (1987). Hawaiian evolutionary biology: An introduction. *TREE* **2**, 175–178.

Singer, R., Roberts, D. A., and Boylen, C. W. (1983). The macrophytic community of an acidic lake in Adirondack (New York: USA): A new depth record for aquatic angiosperms. *Aquat. Bot.* **16**, 49–57.

Singh, H. B., Viezee, W., Johnson, W. B., and Ludwig, F. L. (1980). The impact of stratospheric ozone on tropospheric air quality. *J. Air Pollut. Control Assoc.* **30**, 1009–1017.

Sivard, R. L. (1982). "World Military and Social Expenditures 1982." World Priorities, Leesburg, Virginia.

Sivard, R. L. (1986). "World Military and Social Expenditures 1986," 11th ed. World Priorities, Washington, D.C.

Sivard, R. L. (1987). "World Military and Social Expenditures 1987–1988." World Priorities, Washington, D.C.

Skarby, L., and Sellden, G. (1984). The effects of ozone on crops and forests. *Ambio* **13**, 68–72.

Skene, M. (1915). The acidity of *Sphagnum* and its relation to chalk and mineral salts. *Ann. Bot. (London)* **29**, 65–90.

Skinner, W. R., and Telfer, E. S. (1974). Spring, summer, and fall foods of deer in New Brunswick. *J. Wildl. Manage.* **38**, 210–214.

Slagsvold, T. (1977). Bird population changes after clearance of deciduous scrub. *Biol. Conserv.* **12**, 229–244.

Slaney, P. A., Smith, T. A., and Halsey, T. G. (1977a). "Physical Alterations to Small Stream Channels Associated with Streamside Logging Practices in the Central Interior of British Columbia," Fish. Manage. Rep. No. 31. Fish and Wildlife Branch, Fisheries Research and Technical Services Division, Environment Canada, Vancouver, British Columbia.

Slaney, P. A., Smith, T. A., and Halsey, T. G. (1977b). "Some Effects of Forest Harvesting on Salmonid Rearing Habitat in Two Streams in the Central Interior of British Columbia," Fish. Manage. Rep. No. 71. Fish and Wildlife Branch, Fisheries Research and Technical Services Division, Environment Canada, Vancouver, British Columbia.

Small, R. D., and Bush, B. W. (1985). Smoke production from multiple nuclear explosions in nonurban areas. *Science* **229**, 465–469.

Smith, A. E. (1982). Herbicides and the soil environment in Canada. *Can. J. Soil Sci.* **62**, 433–460.

Smith, A. G., and Pilcher, J. R. (1973). Radiocarbon dates and vegetational history of the British Isles. *New Phytol.* **72**, 903–914.

Smith, C. T. (1985). "Literature Review and Approaches to Studying the Impacts of Forest Harvesting and Residue Management on Forest Nutrient Cycles," Inf. Rep. 13. College of Forest Resources, University of Maine, Orono.

Smith, D. W. (1970). Concentration of soil nutrients before and after fire. *Can. J. Soil Sci.* **50**, 17–29.

Smith, J. E. (1968). "Torrey Canyon Pollution and Marine Life." Cambridge Univ. Press, London and New York.

Smith, S. H. (1972a). Factors of ecologic succession in oligotrophic fish communities of the Laurentian Great Lakes. *J. Fish. Res. Board Can.* **29**, 717–730.

Smith, S. H. (1972b). The future of Salmonid communities in the Laurentian Great Lakes. *J. Fish. Res. Board Can.* **29**, 951–957.

Smith, T. G., and Armstrong, F. A. G. (1978). Mercury and selenium in ringed and bearded seal tissues. *Arctic* **31**, 75–84.

Smith, V. H. (1980). Nutrient dependence of primary productivity in lakes. *Limnol. Oceanogr.* **24**, 1051–1064.

Smith, V. H. (1983). Low nitrogen to phosphorus ratios favour dominance by bluegreen algae in lake phytoplankton. *Science* **221**, 669–671.

Smith, W. H. (1981). "Air Pollution and Forests." Springer-Verlag, New York.

Smith, W. H. (1984). Ecosystem pathology: A new perspective for phytopathology. *For. Ecol. Manage.* **9**, 193–219.

Smith, W. H., and Siccama, T. G. (1981). The Hubbard Brook ecosystem study: Biogeochemistry of lead in the northern hardwood forest. *J. Environ. Qual.* **10**, 323–333.

Snedaker, S. C. (1984). Coastal, marine, and aquatic ecology: An overview. *In* "Herbicides in War: The Long-term Ecological and Human Consequences" (A. H. Westing, ed.), pp. 95–107. Taylor & Francis, London.

Soane, B. O., and Saunder, D. H. (1959). Nickel and chromium toxicity of serpentine soils in southern Rhodesia. *Soil Sci.* **88**, 322–330.

Soikkeli, S., and Karenlampi, L. (1984). The effects of nitrogen fertilization on the ultrastructure of mesophyll cells of conifer needles in northern Finland. *Eur. J. For. Pathol.* **14**, 129–136.

Solomon, A. M., and West, D. C. (1985). Potential responses of forests to CO_2-induced climate change. *In* "Characterization of Information Requirements for Studies of CO_2 Effects: Water Resources, Agriculture, Fisheries, Forests, and Human Health," DOE/ER-0236, pp. 145–169. U.S. Department of Energy, Washington, D.C.

Solomon, A. M., Trabolka, J. R., Reichle, D. E., and Voorhees, L. D. (1985). The global cycle of carbon. *In* "Atmospheric Carbon Dioxide and the Global Carbon Cycle," DOE/ER-0239, pp. 1–13. U.S. Department of Energy, Washington, D.C.

Solomon, P. M., Connor, B., de Zafra, R. L., Parrish, A., Barrett, J., and Jaramillo, M. (1987). High concentrations of chlorine monoxide at low altitudes in the Antarctic spring stratosphere: Secular variation. *Nature (London)* **328**, 411–413.

Solomon, S. (1987). More news from Antarctica. *Nature (London)* **326**, 20.

Sonstegard, R. A. (1977). Environmental carcinogenesis studies in fishes of the Great Lakes of North America. *Ann. N.Y. Acad. Sci.* **298**, 261–269.

Sonzogni, W. C., Robertson, A., and Beeton, A. M. (1983). Great Lakes management: Ecological factors. *Environ. Manage.* **7**, 531–542.

Sopper, W. E. (1975). Effects of timber harvesting and related management practices on water quality in forested watersheds. *J. Environ. Qual.* **4**, 24–29.

Soule, M. E. (1983). What do we really know about extinction? *In* "Genetics and Conservation" (C. M. Schonewald-Cox, ed.), pp. 111–124. Benjamin/Cummings, Menlo Park, California.

Soule, M. E., ed. (1986). "Conservation Biology: The Science of Scarcity and Diversity." Sinauer Assoc., Sunderland, Massachusetts.

Southward, A. J., and Southward, E. C. (1978). Recolonization of rocky shores in Cornwall after use of toxic dispersants to clean up the Torrey Canyon spill. *J. Fish. Res. Board Can.* **35**, 682–706.

Sparrow, A. H., Schwemmer, S. S., Klug, E. E., and Pugliellio, L. (1970). Woody plants: Changes in survival in response to long-term (8 years) chronic gamma irradiation. *Science* **169**, 1082–1084.

Spencer, C. N., and King, D. L. (1984). Role of fish in regulation of plant and animal communities in eutrophic ponds. *Can. J. Fish. Aquat. Sci.* **41**, 1851–1855.

Spitzer, P. R., and Poole, A. (1980). Coastal ospreys between New York City and Boston: A decade of reproductive recovery 1969–1979. *Am. Birds* **34**, 234–241.

Spitzer, P. R., Risebrough, R. W., Walker, W., Hernandez, R., Poole, A., Puleston, D., and Nisbet, I. C. T. (1978). Productivity of ospreys in Connecticut—Long Island increases as DDE residues decline. *Science* **202**, 333–335.

Spray, C. J., Crick, H. Q. P., and Harty, A. D. M. (1987). Effects of aerial applications of fenitrothion on bird populations of a Scottish pine plantation. *J. Appl. Ecol.* **24**, 29–47.

Sprugel, D. G. (1976). Dynamic structure of wave-regenerated *Abies balsamea* forests in the northeastern United States. *J. Ecol.* **64**, 889–911.

Sprules, W. G. (1975a). Midsummer crustacean zooplankton communities in acid-stressed lakes. *J. Fish. Res. Board Can.* **32**, 389–395.

Sprules, W. G. (1975b). Factors affecting the structure of limnetic crustacean zooplankton communities in central Ontario lakes. *Verh.—Int. Ver. Theor. Angew. Limnol.* **19**, 635–643.

Sprules, W. G. (1977). Crustacean zooplankton communities as indicators of limnological conditions: An approach using principal components analysis. *J. Fish. Res. Board Can.* **34**, 962–975.

Spurr, S. H., and Barnes, B. V. (1980). "Forest Ecology," 3rd ed. Wiley, Toronto, Ontario.

Stadtman, T. C. (1974). Selenium biochemistry. *Science* **183**, 915–922.

Staley, J. M. (1965). Decline and mortality of red and scarlet oaks. *For. Sci.* **11**, 2–17.

Stanley, S. M. (1984). Marine mass extinctions. *In* "Extinctions" (M. H. Nitecki, ed.), pp. 69–118. Univ. of Chicago Press, Chicago, Illinois.

Stark, N. (1980). Changes in soil water quality resulting from three timber cutting methods and three levels of fibre utilization. *J. Soil Water Conserv.* **35**, 183–187.

Steiner, K. C., McCormick, L. H., and Canavera, D. J. (1980). Differential response of paper birch provenances to aluminum in solution culture. *Can. J. For. Res.* **10**, 25–29.

Steinhart, C. E., and Steinhart, J. S. (1972). "Blowout: A Case Study of the Santa Barbara Oil Spill." Duxbury Press, Belmont, California.

Stephens, S. L., and Birks, J. W. (1985). After nuclear war: Perturbations in atmospheric chemistry. *BioScience* **35**, 557–562.

Stephenson, G. R. (1983). "Expert Report in the Case of Victoria Palmer *et al.* v. Nova Scotia Forest Industries." Supreme Court of Nova Scotia, Trial Division, Halifax.

Stephenson, M., and Mackie, G. L. (1986). Lake acidification as a limiting factor in the distribution of the freshwater amphipod *Hyalella azteca. Can. J. Fish. Aquat. Sci.* **43**, 288–292.

Stewart, G. H., and Veblen, T. T. (1983). Forest instability and canopy tree mortality in Westland, New Zealand. *Pac. Sci.* **37**, 427–431.

Stewart, R. E., and Aldrich, J. W. (1951). Removal and repopulation of breeding birds in a spruce-fir forest community. *Auk* **68**, 471–482.

Stewart, R. E., Gross, L. L., and Hankala, B. H. (1984). "Effects of Competing Vegetation on Forest Trees: A Bibliography with Abstracts," U.S.D.A. For. Serv., Gen. Tech. Rep. WO-43. USDA, Washington, D.C.

Stocks, B. J. (1985). Forest fire behaviour in spruce budworm-killed balsam fir. *In* "Recent Advances in Spruce Budworms Research," pp. 188–199. Canadian Forestry Service, Ottawa, Ontario.

Stocks, B. J. (1987). Fire potential in the spruce budworm-killed forests of Ontario. *For. Chron.* **63,** 8–14.

Stokes, P. M. (1980). Benthic algal communities in acidic lakes. *In* "Effects of Acidic Precipitation on Benthos" (R. Singer, ed.), pp. 119–133. North American Benthological Society, Springfield, Illinois.

Stone, E. L., and Timmer, V. R. (1975). On the copper content of some northern conifers. *Can. J. Bot.* **53,** 1453–1456.

Stothers, R. B. (1984). The great Tambora eruption in 1815 and its aftermath. *Science* **224,** 1191–1198.

Strain, B., and Cure, J. D., eds. (1985). "Direct Effects of Increasing Carbon Dioxide on Vegetation," DOE/ER-0238. U.S. Department of Energy, Washington, D.C.

Strand, L. (1983). Acid precipitation and forest growth in Norway. *Aquilo, Ser. Bot.* **19,** 32–39.

Strijbosch, H. (1979). Habitat selection of amphibians during their aquatic phase. *Oikos* **33,** 363–372.

Straughan, D., and Abbott, B. C. (1971). The Santa Barbara oil spills: Ecological changes and natural oil leaks. *In* "Water Pollution by Oil," pp. 257–262. Institute of Petroleum, London.

Strojan, C. L. (1978a). The impact of zinc smelter emissions on forest litter arthropods. *Oikos* **31,** 41–46.

Strojan, C. L. (1978b). Forest leaf litter decomposition in the vicinity of a zinc smelter. *Oecologia* **32,** 203–212.

Stross, R. G., and Hasler, A. D. (1960). Some lime-induced changes in lake metabolism. *Limnol. Oceanogr.* **5,** 265–272.

Stross, R. G., Neess, J. C., and Hasler, A. D. (1961). Turnover time and production of planktonic crustacea in limed and reference portion of a bog lake. *Ecology* **42,** 237–245.

Stroud, R. H. (1967). "Water Quality Criteria to Protect Aquatic Life—A Summary," Spec. Publ. No. 4. American Fisheries Society, Washington, D.C.

Stuanes, A. O. (1980). Effects of acid precipitation on soil and forest. 5. Release and loss of nutrients from a Norwegian forest soil due to artificial rain of varying acidity. *In* "Ecological Impact of Acid Precipitation" (D. Drablos and A. Tollan, eds.), pp. 198–199. SNSF Project, Oslo, Norway.

Sundaram, A., Sundaram, K. M. S., and Cadogan, B. L. (1985). Influence of formulation properties on droplet spectra and soil residues of aminocarb aerial sprays in conifer forests. *J. Environ. Sci. Health* **2,** 167–186.

Sundaram, K. M. S., and Nott, R. (1984). "Fenitrothion Residues in Selected Components of a Conifer Forest Following Aerial Application of Tank Mixes Containing Triton X-100," Inf. Rep. FPM-X-65. Forest Pest Management Institute, Canadian Forestry Service, Ottawa, Ontario.

Sundaram, K. M. S., and Nott, R. (1985). "Distribution and Persistence of Aminocarb in Terrestrial Components of the Forest Environment after Semi-operational Application of Two Mixtures of Matacil 180F," Inf. Report FPM-X-67. Forest Pest Management Institute, Canadian Forestry Service, Sault Ste. Marie, Ontario.

Sundaram, K. M. S., and Nott, R. (1986). "Persistence of Fenitrothion Residues in a Conifer Forest Environment," Inf. Rep. FPM-X-75. Forest Pest Management Institute, Sault Ste. Marie, Ontario.

Sussman, V. H., and Mulhern, J. J. (1964). Air pollution from coal refuse disposal areas. *J. Air Pollut. Control Assoc.* **14,** 279–284.

Sutcliffe, D. W., and Carrick, T. R. (1973). Studies on mountain streams in the English Lake District. I. pH, calcium, and the distribution of invertebrates in the River Duddon. *Freshwater Biol.* **3,** 437–462.

Swan, L. A., and Papp, C. S. (1972). "The Common Insects of North America." Harper & Row, New York.

Swanson, R. H., Golding, D. L., Rothwell, R. L., and Bernier, P. Y. (1986). "Hydrologic Effects of Clear-cutting at Marmot Creek and Streeter Watersheds, Alberta," Inf. Rep. NOR-X-278. Northern Forestry Research Centre, Edmonton, Alta.

Swift, L. W., and Messer, J. B. (1971). Forest cutting raises water temperatures of small streams in the southern Appalachians. *J. Soil Water Conserv.* **26,** 111–115.

Swift, M. J., ed. (1984). "Soil Biological Processes and Tropical Soil Fertility," News Mag., Spec. Issue. International Union of Biological Sciences.

Tabatabai, M. A. (1987). Physicochemical fate of sulfate in soils. *J. Air Pollut. Control Assoc.* **37,** 34–38.

Takano, B. (1987). Correlation of volcanic activity with sulfur oxyanion speciation in a crater lake. *Science* **235,** 1633–1635.

Tamm, C. O., and Cowling, E. B. (1976). Acidic precipitation and forest vegetation. *USDA For. Serv. Gen. Tech. Rep.* NE-23, 845–855. Northeastern Forest Experiment Station, Broomall, Pennsylvania.

Tamm, C. O., and Hallbacken, L. (1986). Changes in soil pH over a 50-year period under different canopies in SW Sweden. *Water, Air, Soil Pollut.* **31,** 337–341.

Tamm, C. O., and Wiklander, G. (1980). Effects of artificial acidification with sulphuric acid on tree growth in Scots pine forest. *In* "Ecological Impacts of Acid Precipitation" (D. Drablos and A. Tollan, eds.), pp. 188–189. SNSF Project, Oslo, Norway.

Tamm, C. O., Holmen, H., Popovic, B., and Wiklander, G. (1974). Leaching of plant nutrients from soils as a consequence of forestry operations. *Ambio* **3,** 211–221.

Tamm, C. O., and Hallbacken, L. (1988). Changes in soil acidity in two forest areas with different acid deposition: 1920s to 1980s. *Ambio* **17,** 56–61.

Tang, A. J. S., Yap, D., Kurtz, J., Chan, W. H., and Lusis, M. A. (1987). An analysis of the impact of the Sudbury smelters on wet and dry deposition in Ontario. *Atmos. Environ.* **21,** 813–824.

Tannock, J., Howells, W. W., and Phelps, R. J. (1983). Chlorinated hydrocarbon pesticide residues in eggs of some birds in Zimbabwe. *Environ. Pollut., Ser. B* **5,** 147–155.

Taylor, A. W. (1978). Post-application volatilization of

pesticides under field conditions. *J. Air Pollut. Control Assoc.* **28**, 922–927.

Taylor, R. J., and Basake, F. A. (1984). Patterns of fluoride accumulation and growth reduction exhibited by Douglas-fir in the vicinity of an aluminum-reduction plant. *Environ. Pollut., Ser. A* **33**, 221–235.

Tejning, S. (1967). Mercury in pheasant (*Phasianus colchicus* L.) deriving from seed grain dressed with methyl and ethyl mercury compounds. *Oikos* **18**, 334–344.

Telfer, E. S. (1967). Comparison of a deer yard and a moose yard in Nova Scotia. *Can. J. Zool.* **45**, 485–490.

Telfer, E. S. (1972). Browse selection by deer and hares. *J. Wildl. Manage.* **36**, 1344–1349.

Telfer, E. S. (1978). Cervid distribution, browse, and snow cover in Alberta. *J. Wildl. Manage.* **42**, 352–361.

Temple, S. A. (1977). Plant-animal mutualism: Coevolution with dodo leads to near extinction of plant. *Science* **197**, 885–886.

Temple, S. A. (1979). Reply to: The dodo and the tambaloque tree, by Owadally (1979). Science **203**, 1363–1364.

Temple, S. A., Mossman, M. J., and Ambuel, B. (1979). The ecology and management of avian communities in mixed hardwood-coniferous forest. *In* "Management of North Central and Northeastern Forests for Non-game Birds," U.S.D.A. For. Serv., Gen. Tech. Rep. NC-51, pp. 132–151. North Central Forest Experiment Station, St. Paul, Minnesota.

Terborgh, J. (1974). Preservation of natural diversity: The problem of extinction prone species. *BioScience* **24**, 715–722.

Terborgh, J., and Winter, B. (1980). Some causes of extinction. *In* "Conservation Biology" (M. E. Soule and B. A. Wilcox, eds.), pp. 119–133. Sinauer, Sunderland, Massachusetts.

Terman, G. L. (1978). "Atmospheric Sulfur—The Agronomic Aspects," Tech. Bull. No. 23. Sulfur Institute, Washington, D.C.

Terraglio, F. P., and Manganelli, R. M. (1967). The absorption of atmospheric SO_2 by water solutions. *J. Air Pollut. Control Assoc.* **17**, 403–406.

Thaler, G. R., and Plowright, R. C. (1980). The effect of aerial insecticide spraying for spruce budworm control on the fecundity of entomophilous plants in New Brunswick. *Can. J. Bot.* **58**, 2022–2027.

Thalken, C. E., and Young, A. L. (1983). Long-term field studies of a rodent population continuously exposed to TCDD. *In* "Human and Environmental Risks of Chlorinated Dioxins and Related Compounds" (R. E. Tucker, A. L. Young, and A. P. Gray, eds.), pp. 357–372. Plenum, New York.

The Institute of Ecology (TIE) (1973). "An Ecological Glossary for Engineers and Resource Managers." TIE, Madison, Wisconsin.

The Times (1918). "The Times History of the War," Vol. XVI. The Times, London.

The Times (1919). "The Times History of the War," Vol. XIX. The Times, London.

Thomas, J. W., Anderson, R. G., Maser, C., and Bull, E. L. (1979). Snags. *U.S., Dep. Agric., Agric. Handb.* **553**, 60–77. USDA, Washington, D.C.

Thomas, M. L. H. (1977). Long term biological effects of bunker C oil in the intertidal zone. *In* "Fate and Effect of Petroleum Hydrocarbons in Marine Ecosystems and Organisms" (D. A. Wolfe, ed.), pp. 238–245. Pergamon, New York.

Thomas, N. A., Robertson, A., and Sonzogni, W. C. (1980). Review of control objectives: New target loads and input controls. *In* "Phosphorus Management Strategies for Lakes" (R. C. Loehr, C. S. Martin, and W. Rast, eds.), pp. 61–90. Ann Arbor Sci. Publ., Ann Arbor, Michigan.

Thompson, C. R., and Taylor, O. C. (1969). Effects of air pollutants on growth, leaf drop, fruit drop, and yield of citrus trees. *Environ. Sci. Technol.* **3**, 934–940.

Thompson, L. K., Sidhu, S. S., and Roberts, B. A. (1979). Fluoride accumulations in soil and vegetation in the vicinity of a phosphorus plant. *Environ. Pollut., Ser. A* **18**, 221–234.

Thompson, M. E., Elder, F. C., Davis, A. R., and Whitlow, S. (1980). Evidence of acidification of rivers in eastern Canada. *In* "Ecological Impact of Acid Precipitation" (D. Drablos and A. Tollan, eds.), pp. 244–245. SNSF Project, Oslo, Norway.

Thompson, S. L., Aleksandrov, V. V., Stenchikov, G. L., Schneider, S. H., Covey, C., and Chervin, R. M. (1984). Global climatic consequences of nuclear war: Simulations with three dimensional models. *Ambio* **13**, 236–243.

Thorington, R. W., Tannenbaum, B., Tarak, A., and Rudran, R. (1982). Distribution of trees on Barro Colorado Island: A five hectare sample. *In* "The Ecology of a Tropical Forest" (E. G. Leigh, A. S. Rand, and D. M. Windsor, eds.), pp. 83–94. Smithsonian Institution Press, Washington, D.C.

Thornback, J., and Jenkins, M. (1982). "The IUCN Mammal Red Data Book." International Union for Conservation of Nature and Natural Resources," Gland, Switzerland.

Thornton, F. C., Schaedle, M., and Raynal, D. J. (1986). Effect of aluminum on the growth of sugar maple in solution culture. *Can. J. For. Res.* **16**, 892–896.

Thornton, I. W. B. (1984). Krakatau—the development and repair of a tropical ecosystem. *Ambio* **13**, 216–225.

Tiedemann, A. R., Helvey, J. D., and Anderson, T. D. (1978). Stream chemistry and watershed nutrient economy following wildlife and fertilization in eastern Washington. *J. Environ. Qual.* **7**, 580–588.

Tome, M. A., and Pough, F. H. (1982). Responses of amphibians to acid precipitation. *Acid. Rain/Fish., Proc. Int. Symp., 1981* pp. 245–253.

Trelease, S. F., and Martin, A. L. (1936). Plants made poisonous by selenium absorbed from the soil. *Bot. Rev.* **2**, 373–396.

Trelease, S. F., and Trelease, H. M. (1938). Selenium as a stimulating and possible essential element for indicator plants. *Am. J. Bot.* **25**, 373–380.

Trelease, S. F., and Trelease, H. M. (1939). Physiological differentiation in *Astragalus* with reference to selenium. *Am. J. Bot.* **26**, 530–535.

Trelease, S. F., di Somma, A. A., and Jacobs, A. L. (1960). Seleno-amino acid found in *Astragalus bisulcatus. Science* **132**, 168.

Trimble, G. R., and Sartz, J. R. (1957). How far from a stream should a logging road be located. *J. For.* **55**, 339–341.

Troedsson, T. (1980). Ten year acidification of Swedish forest soils. *In* "Ecological Impact of Acid Precipitation" (D. Drablos and A. Tollan, eds.), p. 184. SNSF Project, Oslo, Norway.

Troendle, C. A. (1987). "The Potential Effect of Partial Cutting and Thinning on Streamflow from the Subalpine Forest," Res. Pap. RM-274. U.S.D.A. For. Serv., Rocky Mountain Forest and Range Experiment Station, Fort Collins, Colorado.

Tukey, H. B. (1980). Some effects of rain and mist on plants, with implications for acid precipitation. *In* "Effects of Acid Precipitation on Terrestrial Ecosystems" (T. C. Hutchinson and M. Havas, eds.), pp. 140–150. Plenum, New York.

Turner, D. J. (1977). "The Safety of the Herbicides 2,4-D and 2,4,5-T," For. Comm. Bull. 57. H. M. Stationery Office, London.

Turner, M. A., Jackson, M. B., Findlay, D. L., Graham, R. W., DeBruyn, E. R., and Vandermeer, E. M. (1987). Early responses of periphyton to experimental lake acidification. *Can. J. Fish. Aquat. Sci.* **44**, Suppl. 1, 135–149.

Tveite, B. (1980a). Effects of acid precipitation on soil and forest. 8. Foliar nutrient concentrations in field experiments. *In* "Ecological Impact of Acid Precipitation" (D. Drablos and A. Tollan, eds.), pp. 204–205. SNSF Project, Oslo, Norway.

Tveite, B. (1980b). Effects of acid precipitation on soil and forest. 9. Tree growth in field experiments. *In* "Ecological Impact of Acid Precipitation" (D. Drablos and A. Tollan, eds.), pp. 206–207. SNSF Project, Oslo, Norway.

Tver, D. F. (1981). "Dictionary of Dangerous Pollutants, Ecology, and Environment." Industrial Press, New York.

Tyler, G. (1974). Heavy metal pollution and soil enzymatic activity. *Plant Soil* **41**, 303–311.

Tyler, G. (1975). Heavy metal pollution and mineralization of nitrogen in forest soils. *Nature (London)* **255**, 701–702.

Tyler, G. (1976). Heavy metal pollution, phosphatase activity, and mineralization of organic phosphorus in forest soils. *Soil Biol. Biochem.* **8**, 327–332.

Tyler, G. (1981). Leaching of metals from the A-horizon of a spruce forest soil. *Water, Air, Soil Pollut.* **15**, 353–369.

Tyler, G. (1984). The impact of heavy metal pollution on forests: a case study of Gusum, Sweden. *Ambio* **13**, 18–24.

Ueno, M. (1958). The disharmonious lakes of Japan. *Verh.— Int. Ver. Theor. Limnol. Angew. Limnol.* **13**, 217–226.

Uhl, C. (1983). You can keep a good forest down. *Nat. Hist.* **4**, 71–78.

Uhl, C. (1987). Factors controlling succession following slash-and-burn agriculture in Amazonia. *J. Ecol.* **75**, 377–407.

Uhlmann, D. (1980). Stability and multiple steady states of hypereutrophic ecosystems. *In* "Hypertrophic Ecosystems" (J. Barica and L. R. Mur, eds.), pp. 235–247. Junk, The Hague, The Netherlands.

Ulrich, B. (1983). Soil acidity and its relation to acid deposition. In "Effect of Accumulation of Air Pollutants in Forest Ecosystems." (B. Ulrich and J. Pankrath, eds.), pp. 127–146. D. Reidel Publ. Co. Boston, Massachusetts.

Ulrich, B., Mayer, R., and Khanna, P. K. (1980). Chemical changes due to acid precipitation in a loess-derived soil in central Europe. *Soil Sci.* **130**, 193–199.

United Nations Environment Program (UNEP) (1982). "English-Russian Glossary of Selected Terms in Preventive Toxicology." UNEP, Moscow.

Unsworth, M. H. (1979). Dry deposition of gases and particulate inputs onto vegetation: A review. *In* "Methods Involved in Studies of Acid Precipitation to Forest Ecosystems," pp. 9–15. Edinburgh, U.K.

Urone, P. (1976). The primary air pollutants—gaseous. Their occurrence, sources, and effects. *In* "Air Pollution" (A. C. Stern, ed.), 3rd ed., Vol. 1, pp. 23–75. Academic Press, New York.

Urone, P. (1986). The pollutants. *In* "Air Pollution" (A. C. Stern, ed.), 3rd ed., Vol. 6, pp. 1–60. Academic Press, New York.

U.S. Department of Agriculture (USDA) (1984). "Pesticide Background Statements," Vol. 1, Agric. Handb. No. 633. USDA, Washington, D.C.

Vallentyne, J. R. (1974). "The Algal Bowl: Lakes and Man," Misc. Spec. Publ. No. 22. Department of the Environment, Fisheries and Marine Service, Ottawa, Ontario.

Van Dach, H. (1943). The effect of pH on pure cultures of *Euglena mutabilis. Ohio J. Sci.* **43**, 47–48.

Vandermeulen, J. A. (1987). Toxicity and sublethal effects of petroleum hydrocarbons in freshwater biota. *In* "Oil in Freshwater: Chemistry, Biology, Countermeasure Technology," pp. 267–303. Pergamon, New York.

Van Groenwoud, H. (1977). "Interim Recommendations for the Use of Buffer Strips for the Protection of Small Streams in the Maritimes," Inf. Rep. M-X-74. Canadian Forestry Service, Maritimes Forest Research Centre, Fredericton, New Brunswick.

Van Loon, J. C. (1974). "Analysis of Heavy Metals in Sewage Sludge and Liquids Associated with Sludges." Presented at Canada/Ontario Sludge Handling and Disposal Seminar, Toronto, Ontario, Sept. 18–19.

Varty, I. W. (1975). Side effects of pest control projects on terrestrial arthropods other than the target species. *In* "Aerial Control of Forest Insects in Canada" (M. L. Prebble, ed.), pp. 266–275. Department of the Environment, Ottawa, Ontario.

Varty, I. W. (1977). Long-term effects of fenitrothion spray programs on non-target terrestrial arthropods. *Natl. Res. Counc. Can., NRC Assoc. Comm. Sci. Criter. Environ. Qual. [Rep.] NRCC, NRCC/CNRC* **16073,** 343–375.

Veith, G. D., DeFoe, D. L., and Bergstedt, B. V. (1979). Measuring and estimating the bioconcentration factor of chemicals in fish. *J. Fish. Res. Board Can.* **36,** 1040–1048.

Veneman, P. L. M., Murray, J. M., and Baker, J. H. (1983). Spatial distribution of pesticide residues in a former apple orchard. *J. Environ. Qual.* **12,** 101–104.

Vergnano, O., and Hunter, J. G. (1952). Nickel and cobalt toxicities in oat plants. *Ann. Bot. (London) [N.S.]* **17,** 317–328.

Vermeij, G. J. (1986). The biology of human-caused extinctions. *In* "The Preservation of Species" (B. G. Norton, ed.), pp. 29–49. Princeton Univ. Press, Princeton, New Jersey.

Verner, J. (1975). Avian behaviour and habitat management. *In* "Symposium on Management of Forest and Range Habitats for Nongame Birds," U.S.D.A. For. Serv., Gen. Tech. Rep. WO-1, pp. 39–58. Southwestern Forest and Range Experiment Station, Tuscon, Arizona.

Verry, E. S. (1972). Effect of aspen clearcutting on water yield and quality in northern Minnesota. *In* "Watersheds in Transition," pp. 276–284. American Water Resources Association, Urbana, Illinois.

Virupaksha, T. K., and Shrift, A. (1965). Biochemical differences between selenium accumulator and non-accumulator *Astragalus* species. *Biochim. Biophys. Acta* **107,** 69–80.

Vitousek, P. M. (1988). Diversity and biological invasions of oceanic islands. *In* "Biodiversity" (E. O. Wilson, ed.), pp. 181–189. National Academy Press, Washington, D.C.

Vitousek, P. M., Gosz, J. R., Grier, C. C., Melillo, J. M., Reiners, W. A., and Todd, R. C. (1979). Nitrate losses from disturbed ecosystems. *Science* **204,** 469–474.

Vitousek, P. M., Ehrlich, P. R., Ehrlich, A. H., and Matson, P. A. (1986). Human appropriation of the products of photosynthesis. *BioScience* **36,** 368–373.

Vogelmann, H. W., Siccama, T., Leedy, D., and Ovitt, D. C. (1968). Precipitation from fog moisture in the Green Mountains of Vermont. *Ecology* **49,** 1205–1207.

Vollenweider, R. A., and Kerekes, J. J. (1981). Background and summary results of the OECD cooperative program on eutrophication. *In* "Restoration of Inland Lakes and Waters," pp. 25–36. U.S. Environmental Protection Agency, Washington, D.C.

Vollenweider, R. A., and Kerekes, J. J. (1982). "Eutrophication of Waters: Monitoring, Assessment and Control." Organization for Economic Co-operation and Development, Paris, France.

Wagle, R. F., and Kitchen, J. H. (1972). Influence of fire on soil nutrients in a ponderosa pine type. *Ecology* **53,** 118–125.

Walker, A. (1984). Extinction in hominid evolution. *In* "Extinctions" (M. H. Nitecki, ed.), pp. 119–152. Univ. of Chicago Press, Chicago, Illinois.

Walker, E. P., Warnick, F., Hamlet, S. E., Lange, K. I., Davis,
M. A., Vible, H. E., and Wright, P. F. (1968). "Mammals of the World," 2nd ed. Johns Hopkins Press, Baltimore, Maryland.

Walker, H. M. (1985). Ten-year ozone trends in California and Texas. *J. Air Pollut. Control Assoc.* **35,** 903–912.

Wallace, E. S., and Freedman, B. (1986). Forest floor dynamics in a chronosequence of hardwood stands in Nova Scotia. *Can. J. For. Res.* **16,** 293–302.

Wallin, K. (1984). Decrease and recovery patterns of some raptors in relation to the introduction and ban of alkyl-mercury and DDT in Sweden. *Ambio* **13,** 263–265.

Walsh, R. J., Donovan, J. W., Adena, M. A., Rose, G., and Battistutta, D. (1983). "Case-control Studies of Congenital Anomalies and Vietnam Service (Birth Defects Study)," Report to Minister for Veteran's Affairs. Australian Government Publishing Service, Canberra.

Walstad, J. D., and Dost, F. N. (1984). "The Health Risks of Herbicides in Forestry: A Review of the Scientific Literature," Spec. Publ. No. 10. Forest Research Laboratory, University of Oregon, Corvallis.

Ward, D. M., Atlas, R. M., Boehm, P. D., and Calder, J. A. (1980). Microbial degradation and chemical evolution of oil from the Amoco spill. *Ambio* **9,** 277–283.

Ward, N. I., Brooks, R. R., Roberts, E., and Boswell, C. R. (1977). Heavy metal pollution from automotive emissions and its effect on roadside soils and pasture species in New Zealand. *Environ. Sci. Technol.* **11,** 917–923.

Wardle, J. A., and Allen, R. B. (1983). Dieback in New Zealand *Nothofagus* forests. *Pac. Sci.* **37,** 397–404.

Wargo, P. M. (1985). Interactions of stress and secondary organisms in decline of forest trees. *In* "Air Pollutant Effects on Forest Ecosystems," pp. 75–86. Acid Rain Foundation, St. Paul, Minnesota.

Wargo, P. M., and Houston, D. R. (1974). Infection of defoliated sugar maple trees by *Armillaria mellea. Phytopathology* **64,** 817–882.

Warnock, J. W., and Lewis, J. (1978). "The Other Face of 2,4-D: A Citizens Report." South Okanagan Environmental Coalition, Penticton, British Columbia.

Warren, H. V., Delevault, R. E., and Barakso, J. (1966). Some observations on the geochemistry of mercury as applied to prospecting. *Econ. Geol. Ser. Can.* **61,** 1010–1028.

Waters, T. F. (1956). The effects of lime application to acid bog lakes in northern Michigan. *Trans. Am. Fish. Soc.* **86,** 329–344.

Waters, T. F., and Ball, R. C. (1957). Lime application to a softwater, unproductive lake in northern Michigan. *J. Wildl. Manage.* **21,** 385–391.

Watson, W. Y., and Richardson, D. H. (1972). Appreciating the potential of a devastated land. *For. Chron.* **48,** 312–315.

Watt, W. D., Scott, D., and Ray, S. (1979). Acidification and other chemical changes in Halifax County lakes after 21 years. *Limnol. Oceanogr.* **24,** 1154–1161.

Watt, W. D., Scott, C. D., and White, W. J. (1983). Evidence of acidification of some Nova Scotia rivers and its impact on Atlantic salmon, *Salmo salar. Can. J. Fish. Aquat. Sci.* **40,** 462–473.

Webb, J. M., Alexander, S. K., and Winters, J. K. (1985). Effects of autumn application of oil on *Spartina alterniflora* in a Texas salt marsh. *Environ. Pollut., Ser. A* **24,** 321–337.

Webber, L. R., and Beauchamp, E. G. (1977). Heavy metals in corn grown on waste amended soils. *Int. Conf. Heavy Met. Environ. [Symp. Proc.], 1st, 1975* Vol. 2, pp. 443–452.

Weetman, G. F., and Algar, D. (1983). Low site-class black spruce and jack pine nutrient removals after full-tree and tree-length logging. *Can. J. For. Res.* **13,** 1030–1036.

Weetman, G. F., and Webber, B. (1972). The influence of wood harvesting on the nutrient status of two spruce stands. *Can. J. For. Res.* **2,** 351–369.

Weetman, G. F., Krause, H. H., Timmer, V. R., and Hoyt, J. S. (1974). Some fertilization response data for the Maritimes and Quebec. *For. Tech. Rep. (Can. For. Serv.)* **5,** 77–82.

Wein, R. W., and MacLean, D. A. (1983). An overview of fire in northern ecosystems. *In* "The Role of Fire in Northern Circumpolar Ecosystems" (R. W. Wein and D. A. MacLean, eds.), pp. 1–18. Wiley, New York.

Weiss, M. J., and Rizzo, D. M. (1987). Forest declines in major forest types of the eastern United States. *In* "Proceedings of the Workshop on Forest Decline and Reproduction: Regional and Global Consequences" (L. Kairiukstis, S. Nilsson, and A. Stroszak, eds.), pp. 297–305. IIASA, Laxenburg, Austria.

Welch, H. E., Symons, P. E. K., and Narver, D. W. (1977). "Some Effects of Potato Farming and Forest Clear-cutting on Small New Brunswick Streams," Tech. Rep. 745. Environment Canada, Fisheries and Marine Service, St. Andrews, New Brunswick.

Wells, P. G. (1984). Marine ecotoxicological tests with zooplankton. *Ecotoxicol. Test. Mar. Environ.* **1,** 215–256.

Wells, P. G., and Percy, J. A. (1985). Effects of oil on Arctic invertebrates. *In* "Petroleum Effects in the Arctic Environment" (F. R. Engelhardt, ed.), pp. 101–156. Am. Elsevier, New York.

Wells, S. M., Pyle, R. M., and Collins, N. M. (1983). "The IUCN Invertebrate Red Data Book." International Union for Conservation of Nature and Natural Resources, Gland, Switzerland.

Welsh, D., and Fillman, D. R. (1980). The impact of forest cutting on boreal bird populations. *Am. Birds* **34,** 84–94.

Wenger, K. F., ed. (1984). "Forestry Handbook," 2nd ed. Wiley, New York.

Westing, A. H. (1966). Sugar maple decline: An evaluation. *Econ. Bot.* **20,** 196–212.

Westing, A. H. (1971). Ecological effects of military defoliation on the forests of South Vietnam. *BioScience* **21,** 893–898.

Westing, A. H. (1976). "Ecological Consequences of the Second Indochina War." Almqvist & Wiksell, Stockholm, Sweden.

Westing, A. H. (1977). "Weapons of Mass Destruction and the Environment." Crane, Russak, New York.

Westing, A. H. (1980). "Warfare in a Fragile World: Military Impact on the Human Environment." Taylor & Francis, London.

Westing, A. H. (1982). The environmental aftermath of warfare in Viet Nam. *In* "SIPRI Yearbook 1982," pp. 363–389. Stockholm International Peace Research Institute, Stockholm, Sweden.

Westing, A. H. (1984a). The remnants of war. *Ambio* **13,** 14–17.

Westing, A. H., ed. (1984b). "Herbicides in War: The Long-term Ecological and Human Consequences." Taylor & Francis, London.

Westing, A. H. (1984c). Herbicides in war: Past and present. *In* "Herbicides in War: The Long-term Ecological and Human Consequences" (A. H. Westing, ed.), pp. 3–24. Taylor & Francis, London.

Westing, A. H. (1985a). Explosive remnants of war: An overview. *In* "Explosive Remnants of War. Mitigating the Environmental Impacts" (A. H. Westing, ed.), pp. 1–16. Taylor & Francis, Philadelphia, Pennsylvania.

Westing, A. H. (1985b). Misspent energy: Munitions expenditures past and future. The world annual arsenal of nuclear weapons. *Bull. Peace Proposals* **16,** 9–10.

Westing, A. H. (1987). "The Ecological Dimension of Nuclear War" (unpublished manuscript). Stockholm International Peace Research Institute, Stockholm, Sweden.

Wetzel, R. G. (1975). "Limnology." Saunders, Toronto, Ontario.

Whelpdale, D. M., and Munn, R. E. (1976). Global sources, sinks, and transport of air pollution. *In* "Air Pollution" (A. C. Stern, ed.), 3rd ed., Vol. 1, pp. 289–324. Academic Press, New York.

Whicker, F. W., and Schultz, V. (1982). "Radioecology: Nuclear Energy and the Environment." CRC Press, Boca Raton, Florida.

Whitby, L. M., and Hutchinson, T. C. (1974). Heavy metal pollution in the Sudbury mining and smelting region of Canada. II. Soil toxicity tests. *Environ. Conserv.* **1,** 191–200.

Whitby, L. M., Stokes, P. M., Hutchinson, T. C., and Myslik, G. (1976). Ecological consequence of acidic and heavy-metal discharges from the Sudbury smelters. *Can. Mineral.* **14,** 47–57.

White, F. M. M., Cohen, F. G., McCurdy, R. F., and Sherman, G. (1984). The chlorophenoxy herbicides and human reproductive problems: A review of the epidemiological evidence. *In* "Chlorophenates in the Wood Industry," pp. 41–50. University of British Columbia, Vancouver.

Whiteside, T. (1979). "The Pendulum and the Toxic Cloud: The Course of Dioxin Contamination." Yale Univ. Press, New Haven, Connecticut.

Whittaker, R. H. (1954). The ecology of serpentine soils. IV.

The vegetational response to serpentine soils. *Ecology* **35**, 275–288.

Whittaker, R. H., Likens, G. E., Bormann, F. H., Eaton, J. S., and Siccama, T. G. (1979). The Hubbard Brook ecosystem study. Nutrient cycling and element behaviour. *Ecology* **60**, 203–220.

Whyte, A. G. D. (1973). Productivity of first and second crops of *Pinus radiata* on the Moutere gravel soils of Nelson. *N. Z. J. For.* **18**, 87–103.

Wiklander, L. (1975). The role of neutral salts in the ion exchange between acid precipitation and soil. *Geoderma* **14**, 93–105.

Wiklander, L. (1980). Interaction between cations and anions influencing adsorption and leaching. *In* "Effects of Acid Precipitation on Terrestrial Ecosystems" (T. C. Hutchinson and M. Havas, eds.), pp. 239–254. Plenum, New York.

Wilander, A., and Ahl, T. (1972). The effects of lime treatment to a small lake in Bergslagen, Sweden. *Vatten* **5**, 431–445.

Wile, I., and Miller, G. (1983). "The Macrophyte Flora of 46 Acidified and Acid-sensitive Soft Water Lakes in Ontario." Ontario Ministry of the Environment, Rexdale.

Wile, I., Miller, G. E., Hitchin, G. G., and Yan, N. D. (1985). Species composition and biomass of the macrophyte vegetation of one acidified and two acid-sensitive lakes in Ontario. *Can. Field-Nat.* **99**, 308–312.

Will, G. (1985). Productivity forcasting—the need to consider factors other than nitrogen and direct nutrient removal in produce. *In* "Proceedings, Workshop on Nutritional Consequences of Intensive Harvesting on Site Quality," pp. 27–30. Forest Research Institute, Rotorua, New Zealand.

Williams, B. L., Cooper, J. M., and Pyatt, D. F. (1979). Effects of afforestation with *Pinus contorta* on nutrient content, acidity, and exchangeable cations in peat. *Forestry* **51**, 29–35.

Williams, W. T. (1980). Air pollution disease in the Californian forests. A base line for smog disease on ponderosa and jeffrey pines in the Sequoia and Los Padres National Forests, California. *Environ. Sci. Technol.* **14**, 179–182.

Williams, W. T. (1983). Tree growth and smog disease in the forests of California: Case history, ponderosa pine in the southern California Nevada. *Environ. Pollut., Ser. A* **30**, 59–75.

Williams, W. T., and Williams, J. A. (1986). Effects of oxidant air pollution on needle health and annual-ring width in a ponderosa pine forest. *Environ. Conserv.* **13**, 229–234.

Williams, W. T., Brady, M., and Willison, S. C. (1977). Air pollution damage to the forests of the Sierra Nevada Mountains of California. *J. Air Pollut. Control Assoc.* **27**, 230–234.

Will-Wolf, S. (1980a). Structure of corticolous lichen communities before and after exposure to emissions from a "clean" coal-fired generating station. *Bryologist* **83**, 281–295.

Will-Wolf, S. (1980b). Effects of a "clean" coal-fired power generating station on four common Wisconsin lichen species. *Bryologist* **83**, 296–300.

Wilson, E. O. (1988). The current state of biological diversity. *In* "Biodiversity" (E. O. Wilson, ed.), pp. 3–18. National Academy Press, Washington, D.C.

Wilson, W. E. (1981). Sulfate formation in point source plumes: a review of recent field studies. *Atmos. Environ.* **15**, 2573–2581.

Windholz, M. (1983). "The Merck Index. Tenth Edition." Merck and Co., Ltd., Rahway, New Jersey.

Winterhalder, E. K. (1978). A historical perspective of mining and reclamation in Sudbury. *Proc. 3rd Annu. Meet., Can. Land Reclam. Assoc.* pp. 1–13.

Wise, W. (1968). "Killer Smog." Ballantyne Books, New York.

Wolff, L. (1958). "In Flanders Fields." Viking Press, New York.

Wolfe, W. J. (1971). Biogeochemical prospecting in glaciated terrain of the Canadian Precambrian Shield. *CIM Bull.* **64**(715), 72–80.

Wood, B. W., Wittwer, R. F., and Carpenter, S. B. (1977). Nutrient element accumulation and distribution in an intensively cultured American sycamore plantation. *Plant Soil* **48**, 417–433.

Wood, C. W., and Nash, T. N. III (1976). Copper smelter effluent effects on Sonoran Desert vegetation. *Ecology* **57**, 1311–1316.

Wood, G. L. (1972). "Animal Facts and Figures." Guinness Superlatives Limited, Enfield, U.K.

Wood, G. W. (1979). Recuperation of native bee populations in blueberry fields exposed to drift of fenitrothion from forest spray operations in New Brunswick. *J. Econ. Entomol.* **72**, 36–39.

Wood, T., and Bormann, F. H. (1975). Increases in foliar leaching caused by acidification of an artificial mist. *Ambio* **4**, 169–171.

Wood, T., and Bormann, F. H. (1976). Short-term effects of a simulated acid rain upon the growth and nutrient relations of *Pinus strobus* L. *USDA For. Serv. Gen. Tech. Rep.* NE-23, 815–824. Northeastern Forest Experiment Station, Broomall, Pennsylvania.

Woodwell, G. M. (1970). Effects of pollution on the structure and physiology of ecosystems. *Science* **168**, 429–433.

Woodwell, G. M. (1982). The biotic effects of ionizing radiation. *Ambio* **11**, 143–148.

Woodwell, G. M., and Whittaker, R. H. (1968). Effects of chronic gamma irradiation on plant communities. *Q. Rev. Biol.* **43**, 42–55.

Woodwell, G. M., Wurster, C. F., and Isaacson, P. A. (1967). DDT residues in an East Coast estuary: A case of biological concentration of a persistent insecticide. *Science* **156**, 821–824.

Woodwell, G. M., Whittaker, R. H., Reiners, W. A., Likens, G. E., Delwiche, C. C., and Botkin, D. B. (1978). The biota and the world carbon budget. *Science* **199**, 141–146.

World Resources Institute (WRI) (1986). "World Resources 1986: An Assessment of the Resource Base That Supports the Global Economy." WRI, New York.

World Resources Institute (WRI) (1987). "World Resources 1987: An Assessment of the Resource Base That Supports the Global Economy." WRI, New York.

World Wildlife Fund (WWF) (1984). "Plantpac Number 4. Threatened Plants." WWF, London.

Wright, R. F. (1976). The impact of forest fire on the nutrient fluxes to small watersheds in northeastern Minnesota. *Ecology* **57**, 649–663.

Wright, R. F., and Gjessing, E. T. (1976). Acid precipitation: Changes in the chemical composition of lakes. *Ambio* **5**, 219–223.

Wright, R. F., and Snekvik, E. (1978). Acid precipitation: Chemistry and fish populations in 700 lakes in southernmost Norway. *Verh.—Int. Ver. Theor. Angew. Limnol.* **20**, 765–775.

Wrighton, F. E. (1947). Plant ecology at Crippengate, 1947. *London Nat.* **27**, 44–48.

Wrighton, F. E. (1948). Plant ecology at Crippengate, 1948. *London Nat.* **28**, 29–44.

Wrighton, F. E. (1949). Plant ecology at Crippengate, 1949. *London Nat* **29**, 85–88.

Wrighton, F. E. (1950). Plant ecology at Crippengate, 1950. *London Nat.* **30**, 73–79.

Wrighton, F. E. (1952). Plant ecology at Crippengate, 1951–2. *London Nat.* **32**, 90–93.

Wu, L., and Bradshaw, A. D. (1972). Aerial pollution and the rapid evolution of copper tolerance. *Nature (London)* **238**, 167–169.

Wurster, D. H., Wurster, C. F., and Strickland, W. N. (1965). Bird mortality following DDT spray for Dutch elm disease. *Ecology* **46**, 488–499.

Yan, N. D. (1979). Phytoplankton community of an acidified, heavy metal-contaminated lake near Sudbury, Ontario: 1973–1977. *Water, Air, Soil Pollut.* **11**, 43–55.

Yan, N. D., and Stokes, P. M. (1978). Phytoplankton of an acidic lake, and its response to experimental alterations of pH. *Environ. Conserv.* **5**, 93–100.

Yan, N. D., and Strus, R. (1980). Crustacean zooplankton communities of acidic metal-contaminated lakes near Sudbury, Ontario. *Can. J. Fish. Aquat. Sci.* **37**, 2282–2293.

Yan, N. D., Girard, R. E., and Lafrance, C. J. (1979). "Survival of Rainbow Trout, *Salmo gairdneri*, in Submerged En-closures in Lakes Treated with Neutralizing Agents Near Sudbury, Ontario." Ontario Ministry of the Environment, Rexdale.

Yevich, P. P., and Barszcz, C. A. (1977). Neoplasia in soft-shell clams (*Mya arenaria*) collected from oil-impacted sites. *Ann. N.Y. Acad. Sci.* **298**, 409–426.

Yokouchi, Y., Okaniwa, M., Ambe, Y., and Fuwa, K. (1983). Seasonal variations of monoterpenes in the atmosphere of a pine forest. *Atmos. Environ.* **17**, 743–750.

Yoshimura, S. (1933). Kata-numa, a very strong acid-water lake on Volcano Katanuma, Miyagi Prefecture, Japan. *Arch. Hydrobiol.* **26**, 197–202.

Yoshimura, S. (1935). Lake Akanuma, a siderotrophic lake at the foot of Volcano Baande, Hikisuma Prefecture, Japan. *Proc. Imp. Acad. (Tokyo)* **11**, 426–428.

Yule, W. N. (1975). Persistence and dispersal of insecticide residues. *In* "Aerial Control of Forest Insects in Canada" (M. L. Prebble, ed.), pp. 263–266. Department of the Environment, Ottawa, Ontario.

Zar, H. (1980). Point source loads of phosphorus to the Great Lakes. *In* "Phosphorus Management Strategies for Lakes" (R. C. Loehr, C. S. Martin, and W. Rast, eds.), pp. 27–36. Ann Arbor Science Publ., Ann Arbor, Michigan.

Zieman, J. C., Orth, R., Phillips, R. C., Thayler, G., and Thorhaug, A. (1984). The effects of oil on seagrass ecosystems. *In* "Restoration of Habitats Impacted by Oil Spills" (J. L. Cairns and A. L. Buikema, Jr., eds.), pp. 37–64. Butterworth, Boston, Massachusetts.

Ziff, S. (1985). "The Toxic Time Bomb." Aurora Press, New York.

Zimmerman, P. R. (1979). "Testing of Hydrocarbon Emissions from Vegetation, Leaf Litter, and Aquatic Surfaces, and Development of a Methodology for Compiling Biogenic Emission Inventories," EPA-450/4-79-004. U.S. Environmental Protection Agency, Washington, D.C. (cited by Yokouchi *et al.*, 1983).

Ziswiler, V. (1967). "Extinct and Vanishing Animals." Springer-Verlag, New York.

Zoettl, H. W., and Huettl, R. H. (1986). Nutrient supply and forest decline in southwest Germany. *Water, Air, Soil Pollut.* **31**, 449–462.

BIOLOGICAL INDEX

CHEMICAL INDEX

GEOGRAPHICAL INDEX

SUBJECT INDEX